SUPERUNIFICATION AND EXTRA DIMENSIONS

1ST TORINO MEETING ON

SUPERUNIFICATION AND EXTRA DIMENSIONS

22—28 September 1985
Torino, Italy

Sponsored by
I.S.I. (Institute for Scientific Interchange)
I.N.F.N. (Istituto Nazionale di Fisica Nucleare)
Università di Torino

Edited by **R D'Auria**
P Fré

World Scientific

Published by

World Scientific Publishing Co Pte Ltd.
P. O. Box 128, Farrer Road, Singapore 9128.
242, Cherry Street, Philadelphia PA 19106-1906, USA

Library of Congress Cataloging-in-Publication Data

Meeting on Superunification and Extra Dimensions
(1st: 1985: Turin, Italy)
1st Torino Meeting on Superunification and Extra
Dimensions, September 1985, Torino, Italy.

1. Supersymmetry – Congresses. 2. Supergravity –
Congresses. I. D'Auria, R. II. Fré, P. III. Title.
IV. Title: First Torino Meeting on Superunification and
Extra Dimensions.
QC174.17.S9M44 1985 530.1'42 86-22413
ISBN 9971-50-101-5

Printed in Singapore by Singapore National Printers (Pte) Ltd.

FOREWORD

Supersymmetry is the most vital theoretical idea in the quest for the unification of all the fundamental interactions of Nature. It provides extremely stringent limitations on the otherwise unbounded set of conceivable theories and it selects, in this way, a class of models sharing very interesting and attractive properties. Among these latter prominence must be given to the general trend, common to supersymmetric theories, of improving the quantum behaviour by "miraculous" cancellation of divergences.

Supersymmetry was originally born in the context of the Dual Resonance Models and led to the discovery of Supergravity Theories which, probably, are the most sophisticated and convincing class of "unified local field theories".

Today we witness the closure of a historical loop by the resurgence of interest in string — and particularly supersymmetric string-theory, which is viewed as the model of which supergravity is the "low energy approximation". String theory is indeed the modern name and the modern interpretation of Dual Resonance Models.

From the very beginning supersymmetry has been inextricably linked with the hypothesis that the actual number of dimensions of the Universe is not four but higher: indeed the number D of dimensions is a parameter which characterizes the various supersymmetric models and is fixed by their internal consistency.

Higher dimensions is an old idea which has attracted for a long time the interest of theorists as a natural explanation of the "internal" symmetries of elementary particles. Supersymmetry provides an *a priori* criterion for the choice of D.

It follows that Kaluza-Klein supergravities, where the internal gauge symmetries are related to the isometries of compactified dimensions, have been at the center of interest for some years. Today a new development has been brought in by the resurgence of string theories. In these latter the extra dimensions can be utilized in an even more efficient way for the explanation of gauge symmetries. Indeed the vertex operator of dual models can reconstruct non-abelian Lie algebras just starting from tori of very small dimension (the dimension of the rank of the Lie algebra in question).

The present book contains the proceedings of a one-week meeting where the variegated landscape of supersymmetric and/or higher dimensional theories has been illustrated by many of the principal characters in scientific "drama".

September 1985 — the time of the meeting — looks already behind us due to the spectacular development of the subject.

We hope the new progresses will be the topics of the "2nd Torino Meeting on Superunification and Higher Dimensions" which is approaching.

Riccardo D'Auria
Pietro Fré
Department of Physics
University of Torino, Italy

CONTENTS

Part Three: Phenomenology and Supersymmetry

Part Four: Superstrings and Extra Dimensions

Part Five: Anomalies

Part One: Supergravity and Extra Dimensions

COMPACTIFICATION OF EXTRA DIMENSIONS: NONLINEAR ASPECTS AND RESIDUAL SUPERSYMMETRY

B. de Wit

Institute for Theoretical Physics
University of Utrecht
Princetonplein 5, P.O. Box 80.006
3508 TA Utrecht, The Netherlands

Present theoretical ideas favour the possibility that we live in a higher-dimensional space-time with the extra (spatial) dimensions spontaneously compactified to a manifold whose size is sufficiently small to evade its experimental discovery. In four dimensions this theory describes an infinite number of states. The mass spectrum of these states follows from analyzing small fluctuations about the field configuration corresponding to the ground state. Gauge symmetries are related to the isometry group of this ground state and, in principle, it is straightforward to classify all states according to gauge group representations, chirality, etc. However, there are certain aspects of the theory that cannot be understood within the context of a linearized approximation. We will discuss some of those aspects in the first part of this talk. A second topic concerns the possible supersymmetry of a compactification. As an example we investigate this question for compactifications of $d = 10$, $N = 1$ Einstein-Yang-Mills supergravity to a four-dimensional maximally symmetric space-time, taking into account the so-called "warp factor".

For practical purposes one often wants to discard the massive states and establish the existence of an effective low-energy theory based on a finite number of massless states. Obviously such a theory should exhibit the invariances of the compactified ground state, but in addition it should correspond to a consistent truncation of the higher-dimensional theory in the sense that the discarded modes do not reappear through the symmetry transformations associated with the ground state (i.e. for all $\phi=0$ one should have $\delta\phi=0$), or through the interactions. By the latter we mean that the higher-dimensional lagrangian should not contain terms that are linearly proportinal to discarded modes; this condition implies that solutions of the truncated field equations are also solutions of the full field equations. Recently the question of consistency became relevant in understanding the S^7 compactification of d=11 supergravity; here it was noted that the supersymmetry transformations were not automatically consistent upon truncation to the massless modes [1], while, on the other hand, it was observed that the d=4 Einstein equation was in fact consistent upon truncation to the graviton-Yang-Mills sector, in contradistinction with generic Kaluza-Klein theories [2].

One way to ensure consistency is to impose the restriction that all the fields are invariant under a subgroup of the isometry group. The most obvious example of this situation is the torus truncation, where one only retains the fields that are independent of the extra coordinates y^m (see for instance [3]). Also the gauge parameters should satisfy this restriction, e.g.

$$\varepsilon(x,y) \rightarrow \varepsilon(x), \qquad \text{(supersymmetry)}$$
$$\xi^m(x,y) \rightarrow \xi^m(x), \qquad \text{((abelian) isometries)} \tag{1}$$

so that the transformation rules are consistent upon the truncation (obviously, both ϕ and $\delta\phi$ are now y-independent), and so are the field equations. This way of achieving consistency has been exploited in [4].

Another approach emphasizes covariance under the full isometry group. One expects that all modes are contained in an infinite set of irreducible representations of the isometry group, so that the truncation can be effected by restricting this infinite set to a finite number of representations. However, representations cannot be discarded arbitrarily. Viewed as a four-dimensional theory we are dealing with a gauge theory coupled to an infinite number of matter multiplets, and while it is often possible in such theories to discard all or some of the matter multiplets, it is essential to retain the massless gauge multiplet in order to have a consistent truncation.

In fact it turns out that the naive expectation that the modes transform as an infinite set of irreducible representations of the isometry group is not immediately realized [1], as inconsistencies may already appear at the linearized level. To see how this is possible, let us recall that the modes are identified by examining small fluctuations of the fields about the background, and that the y-dependence of these modes is determined in the context of certain gauge conditions. Consequently there is an inherent ambiguity in the y-dependence of these modes. When calculating the linearized symmetry variation of a particular mode of a certain mass, one may find contributions from modes with different masses suggesting that the anticipated decomposition in terms of irreducible multiplets has not been realized. However, this interpretation is not correct: the (linearized) transformations between different multiplets are due to a mismatch in the y-dependence of the various modes, which is a reflection of the aforementioned ambiguity in the y-dependence. The remedy, at least in linearized approximation, is clear: the transformation parameters, whose y-dependence has been identified by requiring that they leave the ground state invariant (so that they are characterized in terms of Killing spinors or vectors), should be modified by field-dependent terms to correct for this mismatch. This nonlinear modification contributes at the linearized level, because the gauge transformations always contain inhomogeneous terms.

In princple, it is thus clear how to regain consistency at the linearized level, so that all multiplets transform among themselves (for the massless modes of the S^7 compactification of d=11 supergravity this was demonstrated in [5]). However, to determine all the nonlinear modifications is a much more difficult task, as it turns out that also the fields are in general subject to nonlinear redefinitions. These redefinitions are important for determining the interactions and the nonlinearly realized symmetries of the truncated theory. Contrary to what is sometimes stated in the literature the nonlinear modifications, although originating from modes with Planck-size masses, do contribute to the renormalizable sector of the truncated theory in the dimensionless four-point couplings (for a discussion, see for instance [6]).

It is evident that closure of the transformation rules in the truncated theory is only guaranteed if the transformation rules are consistent upon the truncation. On the other hand, consistency of the transformation rules guarantees not necessarily that the truncation will be consistent with regard to the field equations. However, if we are dealing with supersymmetry transformations, which only close modelo the full higher-dimensional field equations, then the consistency of the truncated transformation rules implies closure of the supersymmetry transformations, which in turn implies the validity of the full higher-dimensional field equations (of course this argument can only be used if the ground state exhibits supersymmetry).

The above observation has been used to prove the consistency of the truncation to the massless sector of the S^7 compactification of d=11 supergravity to all orders [7]. In that case a crucial role was played by local field-dependent SU(8) transformations, as will be discussed in Nicolai's lecture at this workshop. The results of this work exhibit all aspects that we have discussed above, and have led to the complete embedding of gauged N=8 supergravity into d=11 supergravity (preliminary accounts of our work can be found in [8]). As an example we present the complete ansatz for the d=11 metric, truncated to gauged N=8 supergravity:

$$g_{MN}(x,y) =$$

$$\begin{bmatrix} \Delta^{-1}(x,y)g_{\mu\nu}(x) + B^m_\mu(x,y)B^n_\nu(x,y)g_{mn}(x,y) & g_{np}(x,y)B^p_\mu(x,y) \\ g_{mp}(x,y)B^p_\nu(x,y) & g_{mn}(x,y) \end{bmatrix} \tag{2}$$

with

$$\Delta(x,y) = \left[\frac{g(x,y)}{\overset{\circ}{g}(y)} \right]^{\frac{1}{2}} , \quad (g \equiv \det g_{mn}) \tag{3}$$

$$\Delta^{-1}(x,y) g^{mn}(x,y) = \frac{1}{8} K^{mIJ}(y) K^{nKL}(y) \times \left(u_{ij}{}^{IJ}(x) + v_{ijIJ}(x)\right)\left(u^{ij}{}_{KL}(x) + v^{ijKL}(x)\right) , \tag{4}$$

$$B_\mu^m(x,y) = - \frac{\sqrt{2}}{4} A_\mu^{IJ}(x)\, K^{mIJ}(y) \ . \tag{5}$$

Here $\overset{\circ}{g}_{mn}(y)$ is the S^7 metric and $K^{mIJ}(y)$ are the 28 normalized Killing vectors on S^7 labelled by antisymmetric index pairs [IJ] (I,J = 1,....8); $g_{\mu\nu}(x)$, $A_\mu^{IJ}(x)$, $u_{ij}{}^{IJ}(x)$ and $v_{ijIJ}(x)$ are the fields of N=8 supergravity associated with the graviton, the 28 spin-1 fields and the 70 spin-0 fields ($u_{ij}{}^{IJ}$, v_{ijIJ} and their complex conjugates form a 56x56 E_7 matrix, sometimes called the 56-bein (cf.[3,9])). In particular the expression for $g^{mn}(x,y)$ shows that the form of the truncation to N=8 supergravity is quite complicated, and deviates substantially from the result for linearized fluctuations about the S^7 background [10]. To see this more explicitly substitute $u_{ij}{}^{IJ} \simeq \delta^I_{[i}\ \delta^J_{j]}$ and $v_{ijIJ} \simeq \phi_{ijIJ}$, where ϕ_{IJKL} is antisymmetric in IJKL and complex selfdual and describes the 70 scalar and pseudoscalar fields; this yields for (4)

$$\left[\frac{\overset{\circ}{g}(y)}{g(x,y)}\right]^{\frac{1}{2}} g^{mn}(x,y) = \overset{\circ}{g}{}^{mn}(y) + \tfrac{1}{4}\mathrm{Re}\, \phi_{IJKL}(x)\, K^{mIJ}(y)\, K^{nKL}(y) \ , \tag{6}$$

where we have used

$$K^{mIJ}(y)\, K^{nIJ}(y) = 8\, \mathring{g}^{mn}(y) \tag{7}$$

Hence linearized fluctuations of the metric describe only the scalar N=8 modes, corresponding to the real part of ϕ_{IJKL}, in contrast with finite deviations of the metric which contain both scalar and pseudoscalar modes. As emphasized before the nonlinear aspects do not affect the mass spectrum, which is only characteristic for the small fluctuations, but they are crucial for understanding the interactions (and possibly non-linearly realized symmetries) of the truncated theory, even of its renormalizable sector.

Another important feature of the above results pertains the so-called warp factor Δ, shown in the decompostion (2) of the higher-dimensional metric. This warp factor must in general be included when investigating Kaluza-Klein solutions for which the higher dimensional space has the topology of a product space,

$$M^d \to M^4 \otimes M^{d-4} \ , \tag{8}$$

with M^4 a maximally symmetric 4-dimensional space-time (i.e. Minkowski or (anti-)de Sitter space). Here the metric takes the form

$$g_{MN}(x,y) = \begin{pmatrix} \Delta^{-1}(y)\ g_{\mu\nu}(x) & 0 \\ 0 & g_{mn}(y) \end{pmatrix} \tag{9}$$

where $g_{\mu\nu}(x)$ is the metric of the maximally symmetric space-time.

The presence of the warp factor is important in the search for field configurations with residual supersymmetry. The relevance of such field configurations has recently been stressed in the context of superstrings, which give rise to d=10 supergravity in the zero-slope limit [11]. However, the possibility of a nontrivial warp factor has not been considered so far [11, 12, 13]; this has motivated us to study this question in some further detail [14].

Before discussing some of the results let us recall that this analysis is usually based on on-shell formulations of supergravity, so that the supersymmetry transformations close modulo the corresponding classical field equations. This may cause a problem, as field configurations with residual "on-shell" supersymmetry must necessarily satisfy a subset of the classical field configurations. On the other hand, if classical supergravity is only viewed as an approximation (for instance of a string theory), one prefers to impose no (classical) field equations at all. In an off-shell formulation these difficulties can be avoided, but such a formulation is only known for pure d = 10, N = 1 supergravity [15], so that for the moment we must content ourselves with an analysis based on on-shell transformation rules.

As an example consider d = 10, N = 1 supergravity, coupled to supersymmetric Yang-Mills theory [16,17,18], in a background (9) with four-dimensional Riemann curvature

$$R_{\mu\nu}{}^{\alpha\beta} = m_4{}^2 (e_\mu{}^\alpha e_\nu{}^\beta - e_\nu{}^\alpha e_\mu{}^\beta) \tag{10}$$

where $m_4{}^2$ is zero, positive or negative for Minkowski, anti-de Sitter and de Sitter space, respectively. In such a background the supersymmetry transformations of the fermionic fields are

$$\delta\psi_\mu = D_\mu\varepsilon - i\gamma_\mu\gamma_5\,\Delta^{-\frac{1}{2}}\,T\,\varepsilon\ , \tag{11}$$

$$\delta\psi_m = D_m\varepsilon + \frac{\sqrt{2}}{32}\,e^{2\phi}(\Gamma_m H - 12H_m)\varepsilon\ , \tag{12}$$

$$\delta\lambda = \left(\sqrt{2}\,i\,\Gamma^m\partial_m\phi + \frac{i}{8}\,e^{2\phi}H\right)\varepsilon\ , \tag{13}$$

$$\delta\chi^A = -\,\tfrac{1}{4}\,F^A_{mn}\,\Gamma^{mn}\,\varepsilon\ , \tag{14}$$

where

$$H_m = H_{mnp}\,\Gamma^{np}\ ,$$

$$H = H_{mnp}\,\Gamma^{mnp}\ ,$$

$$T = -\,\frac{i}{4}\,\Gamma^m\partial_m \ln\Delta + i\,\frac{\sqrt{2}}{32}\,e^{2\phi}H\ . \tag{15}$$

Here, ψ_μ and ψ_m originate from the d = 10 gravitino field, while λ, ϕ and H_{mnp} are the spinor, scalar and antisymmetric field-strength tensor of d = 10 supergravity. Because the 4-dimensional space-time has maximal symmetry ψ_μ, ψ_m and λ vanish and ϕ and H_{mnp} are x-independent. For the same reason the components of H_{MNP} with indices in the 4-dimensional subspace must vanish. Similar arguments apply to the fields of d = 10 supersymmetric Yang-Mills theory; χ^A denotes the spinor field, which must vanish, while the nonvanishing components of the Yang-Mills field strengths, F^A_{mn}, are x-independent. The covariant derivatives D_μ and D_m in (11) and (12) contain the standard spin connections $\omega_\mu^{\alpha\beta}$ and ω_m^{ab} for the 4- and 6-dimensional subspaces. Furthermore we recall that ψ_μ, ψ_m, λ, χ^A and ε are Majorana-Weyl spinors.

In a background with residual supersymmetry there must be one or several spinors ε for which the right-hand sides of (11)-(14) vanish. From (11) one derives an integrability condition, which leads to

$$\left(T^2 + \tfrac{1}{4}\, m_4{}^2\Delta\right)\varepsilon = 0 \ , \tag{16}$$

where we made use of (10). Combining (16) with (13) it follows that

$$\left\{\tfrac{1}{2}i\Gamma^m\partial_m\left(\phi+\tfrac{1}{2}\ln \Delta\right) \ , \frac{i\sqrt{2}}{32}\, e^{2\phi}H\right\}\varepsilon = \tfrac{1}{4}\Delta\left(m_4{}^2 - \partial_m(\Delta^{-\frac{1}{2}}e^{\phi})\ \partial^m(\Delta^{-\frac{1}{2}}e^{-\phi})\right)\varepsilon . \tag{17}$$

As the matrix multiplying ε on the left-hand side of (17) is antihermitean, it must have imaginary eigenvalues in contrast with the right-hand side of this equation which yields a real eigenvalue. Therefore one concludes that both sides of (17) should vanish. If the extra coordinates parametrize a compact space, this requires

$$m_4{}^2 = 0 \ , \tag{18}$$

so that the 4-dimensional space-time must be Minkowski space, and

$$\Delta^{-\frac{1}{2}} e^{\pm\phi} = \text{constant} . \tag{19}$$

Choosing the minus sign in (19), so that

$$\tfrac{1}{2}\partial_m \ln \Delta = - \partial_m \phi , \tag{20}$$

both sides of (17) vanish identically, while

$$T \varepsilon = \left(\tfrac{1}{2} i\Gamma^m \partial_m \phi + \frac{i\sqrt{2}}{32} e^{2\phi} H \right)\varepsilon = 0 , \tag{21}$$

by virtue of (13). Adopting the plus sign in (19) implies that $T \varepsilon \neq 0$ (unless both Δ and ϕ are constant which is just a special case of (20)); nontrivial solutions of this type lead to supersymmetry parameters ε in Minkowski space, satisfying

$$D_\mu \varepsilon = i\gamma_\mu \gamma_5 \Delta^{-\frac{1}{2}} T \varepsilon . \tag{22}$$

with

$$T^2 \varepsilon = 0 \quad , \quad T \varepsilon \neq 0 . \tag{23}$$

As such spinors give rise to a Poincaré superalgebra that cannot be implemented in a positive-definite Hilbert space [19], we disregard such solutions and assume (20).

The spinors ε can now be decomposed as a product of a 4-dimensional anticommuting spinor and a 6-dimensional commuting (Majorana) spinor η.

Absorbing a factor $\exp(\frac{1}{2}\phi)$ into the definition of η, supersymmetry implies the following conditions on η

$$D_m(\hat{\omega})\,\eta = 0\ , \tag{24}$$

$$\left(\Gamma^m\partial_m\phi + \frac{\sqrt{2}}{16}\,e^{2\phi}H\right)\eta = 0\ , \tag{25}$$

where the d = 6 spin connection $\hat{\omega}_m^{ab}$ appearing in $D_m(\hat{\omega})$ contains torsion, viz.

$$\hat{\omega}_{mab} = \omega_{mab} + K_{mab}\ , \tag{26}$$

with the contortion tensor equal to

$$K_{mab} = 2e_{m[a}\,e^n_{b]}\,\partial_n\phi + \frac{3\sqrt{2}}{2}\,e^{2\phi}\,H_{mab}\ . \tag{27}$$

From (24) one derives

$$R_{mn}^{ab}(\hat{\omega})\Gamma_{ab}\,\eta = 0\ ,$$

$$\left(D_m(\hat{\omega})(\Gamma^n\partial_n\phi) + \frac{\sqrt{2}}{16}\,D_m(\hat{\omega})(e^{2\phi}H)\right)\eta = 0\ , \tag{28}$$

where $R_{mn}^{ab}(\hat{\omega})$ is the curvature tensor associated with the connection $\hat{\omega}_m^{ab}$. Contracting (27) with Γ-matrices leads to a variety of equations, such as

$$(D_m + 8\partial_m\phi)H^m \eta = 0 \; , \tag{29}$$

$$\frac{\sqrt{2}}{32} e^{2\phi} D_m H_{npq} \Gamma^{mnpq} \eta$$

$$= \left[- 2(\partial_m\phi)^2 - \tfrac{1}{2}D^m D_m\phi + \frac{3}{16} e^{4\phi} (H_{mnp})^2\right]\eta \; , \tag{30}$$

$$R = 14D^m D_m\phi - 52(\partial_m\phi)^2 + \frac{3}{2} e^{4\phi}(H_{mnp})^2 \; . \tag{31}$$

For constant ϕ these equations reduce to those found in [11], but for arbitrary ϕ they are clearly less restrictive. However, when combined with (some of) the classical field equations, one is usually led to Ricci-flat spaces with a trivial warp factor and vanishing field - strengths H_{mnp} and F^A_{mn}. For instance, one may impose the Bianchi identity for H_{mnp}, which implies that $D_{[m}H_{npq]}$ is proportional to $F^A_{[mn} F^A_{pq]}$ with a negative proportionality factor[17,18]. Then the left-hand side of (30) can be replaced by a term proportional to $F^A_{mn} F^A_{pq} \Gamma^{mnpq} \eta$, which is equal to $2(F^A_{mn})^2\eta$ due to the fact that also (14) must vanish. The proportionality factors are such that $(F^A_{mn})^2$ combines with the $(H_{mnp})^2$ term on the right-hand side of (30) with equal sign (cf. [11]). On the other hand it can be shown that the Ricci scalar R must be positive. Integrating both (30) and (31) over the 6-dimensional space then leads to the conclusion that $F^A_{mn} = H_{mnp} = \partial_m\phi = R = 0$. As is well-known, in the anomaly-free theory [20] the Bianchi identity must acquire a second term quadratic in the Riemann tensor, which invalidates the above conclusion.

Another notable feature discovered in [11] was the existence of an almost complex structure

$$J_{mn} = - i\eta^{\dagger}\Gamma_{mn}\tilde{\Gamma}\eta \ , \tag{32}$$

where η is a 6-dimensional spinor that satisfies the above restrictions (24) and (25) imposed by supersymmetry, and

$$\tilde{\Gamma} \equiv -i\, \Gamma^1\Gamma^2 \ldots \Gamma^6 \ . \tag{33}$$

It follows from (24) that $\eta^{\dagger}\eta$ is constant, so that η may be normalized to unity: $\eta^{\dagger}\eta = 1$. One then proves by Fierz rearrangement

$$J_m{}^n\, J_n{}^p = - \delta_m{}^p \ , \tag{34}$$

and from (24)

$$D_p J_{mn} = - 2g_{p[m}\, J_{n]}{}^q\, \partial_q\phi + 2\partial_{[m}\phi\, J_{n]p} - 3\sqrt{2}\, e^{2\phi}\, J_{[m}{}^q\, H_{n]pq} \ . \tag{35}$$

The latter is, of course, equivalent to $D_p(\hat{\omega})J_{mn} = 0$. From (35) one derives

$$D^m\, J_m{}^n = - 4\, \partial^m\phi\, J_m{}^n \ , \tag{36}$$

which shows that the six-dimensional space is not a semi-Kähler space. Furthermore, the first equation (28) implies

$$R_{mnp}{}^{q}(\hat{\omega})\; J_q{}^p = 0\ . \tag{37}$$

In order for J to be a complex structure on the 6-dimensional space, it should be integrable. A necessary and sufficient condition for this is that the Nijenhuis tensor

$$N^p_{mn} \equiv J_m{}^q\; D_{[q}J_{n]}{}^p - J_n{}^q\; D_{[q}J_{m]}{}^p \tag{38}$$

vanishes. To show that this is indeed the case, even for a nontrivial warp factor, one first derives from (25)

$$J^{mn}\, \partial_n\phi = \frac{3\sqrt{2}}{16}\, e^{2\phi}\; H^{mnp}\; J_{np}\ , \tag{39}$$

$$J_{[mn}\; \partial_{p]}\phi = \frac{\sqrt{2}}{48}\, e^{2\phi}\left(\sqrt{g}\;\varepsilon_{mnpqrs}H^{qrs} + 18\; H^{q[mn}\; J_q{}^{p]}\right)\ . \tag{40}$$

With the help of these equations, and (35) one proves

$$N^p_{mn} = 6\left(J_{[mn}\; J_{pq]} + \frac{1}{6}\sqrt{g}\;\varepsilon_{mnpqrs}J^{rs}\right)\partial^q\phi\ , \tag{41}$$

which vanishes as can be shown by repeated Fierz rearrangement. Hence the six extra coordinates parametrize a complex space. According to the first equation (28) the holomony group associated with the connection (26) leaves the spinor η invariant, so it must be contained in U(3); from (37) one concludes that U(3) is in fact restricted to SU(3).

From (35) and (40) one can also derive an expression for H_{mnp} in terms of the metric J_{mn} and ϕ, namely

$$H_{mnp} = \frac{\sqrt{2}}{12} e^{-2\phi} \sqrt{g}\, \varepsilon_{mnpqrs} \left(D^q J^{rs} - 6\, \partial^q \phi\, J^{rs}\right) . \qquad (42)$$

Although the presence of a nontrivial warp factor complicates matters and leads to certain modifications, two important results of [11] remain unaffected, namely the zero cosmological constant in the 4-dimensional space-time, and the complex structure with SU(3) holonomy group of the 6-simensional space. A possible explanation for this fact follows from our previous observation that (residual) supersymmetry based on on-shell transformation rules always imposes a subset of the classical field equations, which are known to be very restrictive for this particular theory [21]. Actually, in Minkowski space ($m_4{}^2=0$) , one can show for a large class of theories that the warp factor must be trivial as a result of the (classical) Einstein equations (cf. [14]). This situation does not arise for classical compactifications of d = 11 (or d = 10, N = 2 nonchiral) supergravity, which lead to an anti-de Sitter space-time [10,22,23].

We have also investigated the effect of a nontrivial warp factor for d = 10, N = 2 supergravity, both for the nonchiral (type 2a) and chiral (type 2b) version. Even without the warp factor the requirement of residual supersymmetry is not very restrictive in these theories, unless one assumes the 4-dimensional space-time to be flat and imposes (classical) Bianchi identities on the background. In that case it follows that the space parametrized by the extra dimensions must be Ricci-flat [12,13]. With a nontrivial warp factor it is not always possible to derive such a result.

This lecture is based on results obtained in collaboration with H. Nicolai and with N.D. Hari Dass and D.J. Smit.

References

1. B. de Wit and H. Nicolai. Nucl. Phys. B231 (1984) 506.
2. M.J. Duff, B.E.W. Nilsson, C.N. Pope and N.P. Warner, Phys. Lett. 149B (1984) 90.
3. E. Cremmer and B. Julia, Nucl. Phys. B159 (1979) 141.
4. M.J. Duff and C.N. Pope, Nucl. Phys. B255 (1985) 355.
   See also, M.J. Duff, in "Supersymmetry", proc. of the NATO Advanced Study Institute, Bonn, 1984, eds. K. Dietz, R. Flume, G. von Gehlen and V. Rittenberg (Plenum, 1985).
5. M.A. Awada, B.E.W. Nilsson and C.N. Pope, Phys. Rev. D29 (1984) 334.
6. H. Nicolai, in proc. of the 1985 Les Houches Summer School, preprint CERN-TH. 4290/85, North-Holland, to published.
7. B. de Wit and H. Nicolai, preprint CERN-TH. 4359/86.
8. H. Nicolai and B. de Wit, Phys. Lett. 155B (1985) 47; in proc. of the Cambridge Workshop on Applications of Supersymmetry, June 1985, preprint CERN-TH. 4291/85, Cambridge Univ. Press, to be published. H.Nicolai, in Recent Developments in Quantum Field Theory (North-Holland, 1985).
9. B. de Wit and H. Nicolai, Nucl. Phys. B208 (1982) 323.
10. B. de Wit, H. Nicolai and N.P. Warner, Nucl. Phys.B255 (1985) 29.

11. P. Candelas, G.T. Horowitz, A. Strominger and E. Witten, Nucl. Phys. B258 (1985) 46.

12. P. Candelas and D.J. Raine, Nucl. Phys. B248 (1984) 415.

13. P. Candelas, Nucl. Phys. B256 (1985) 717.

14. B. de Wit, D.J. Smit and N.D. Hari Dass, NIKHEF-H preprint, in preparation.

15. P.S. Howe, H. Nicolai and A. Van Proeyen, Phys. Lett. 124B (1983) 45.

16. A.H. Chemseddine, Nucl. Phys. B185 (1981) 403; Phys. Rev. D24 (1981) 3065.

17. E. Bergshoeff, M. de Roo, B. de Wit and P. van Nieuwenhuizen, Nucl. Phys. B195 (1982) 97; E. Bergshoeff, M. de Roo and B. de Wit, Nucl. Phys. B217 (1983) 489.

18. G.F. Chapline and N.S. Manton, Phys. Lett. 120B (1983) 105.

19. R. Haag, J. Lopuszanski and M.F. Sohnius, Nucl. Phys. B88 (1975) 61.

20. M.B. Green and J.H. Schwarz, Phys. Lett. 149B (1984) 117.

21. D.Z. Freedman, G. Gibbons and P.C. West, Phys. Lett. 124B (1983) 91.

22. B. de Wit and H. Nicolai, Phys. Lett. 148B (1984) 60.

23. P. van Nieuwenhuizen and N.P. Warner, Comm. Math. Phys. 99 (1985) 141.

RECENT WORK IN ANTI DE SITTER SUPERSYMMETRY

Daniel Z. Freedman
Dept. of Mathematics and Center for Theoretical Physics,
Laboratory for Nuclear Science, Massachusetts Institute of Technology, Cambridge, MA 02139, USA
and
Laboratoire de Physique Théorique de l'Ecole Normale Supérieure, Paris, France

Abstract. Recent results on supersymmetry in a fixed AdS background geometry are summarized. These results include i) required modification of the generators of the O(3,2) isometry group in the AdS Wess-Zumino model, ii) the one-loop renormalization structure of this model, showing that the special "naturalness" properties of flat space supersymmetry do not extend to AdS, and iii) a non-perturbative Lehmann spectral representation. Open problems suggested by recent work are emphasized.

I. INTRODUCTION

The original motivation for work in AdS supersymmetry came from the vacuum stability problem in gauged extended supergravity and Kaluza-Klein theories, and we have made considerable progress on this problem in the past few years (Breitenlohner & Freedman, 1982; Gibbons et al., 1983; Boucher, 1984). There was also the hope to illuminate the cosmological constant problem, and on this question there has been little progress. The solution, if it exists at all within AdS quantum field theory, perhaps lies in the unfathomable "deep quantum limit", rather than within the semi-classical perturbative techniques that have been used so far. For this and for other reasons the subject of AdS supersymmetry no longer seems as relevant to theoretical physics as it did three years ago, and I myself have decided that it is time to study something else. It is a decision made with some reluctance, because I find that the subject has intellectually attractive aspects. Glaring problems present themselves, but when you look more closely a solution is found. There is a satisfying coherence and harmony in this study of quantum field theory in a context intermediate in complication between flat space and quantized gravity. Further there are some interesting, tractable open problems, and it is one goal of this report to describe them in the hope that some physicists with fresh enthusiasm will study them. It would not surprise me if the subject regains its relevance in a few years time.

The results discussed at the Torino Meeting included three topics. The first of these was the energy crisis in the AdS Wess-Zumino model formulated by (Ivanov & Sorin 1980). In this model the O(3,2) generators obtained by the Noether method or by the Killing charge method (Abbott & Deser 1982) do not have manifest positive energy and vanishing vacuum expectation value as required by the OSp(1,4) supersymmetry algebra. This problem was first solved in the two-dimensional case (Bardeen & Freedman 1985) and more recently in four dimensions (Burges et al. 1985). It turns out that the Noether generators must be modified by adding surface terms, and the improved generators do enjoy the required properties.

Some features of the Pauli-Villars regulated perturbation calculations used to test the improved generators suggested that the non-renormalization theorems of flat-space supersymmetry are modified in AdS and detailed one-loop calculations were studied to ascertain the situation. The results (Düsedau & Freedman 1985a) showed that the renormalization of the 2- and 3-point functions is unmodified by the curved background, but there is a linearly divergent vacuum expectation value of the physical scalar field $\langle A \rangle$ while $\langle F \rangle = 0$. This supersymmetric divergence is cancelled by adding a counter term to the superpotential which is linear in A and which is not present in the flat-space theory. This counter term can be understood as a $d^4\theta$ superspace integral in curved superspace (Grisaru & Stelle 1985) so that the non-renormalization theorems in the sense of no $d^2\theta$ terms still hold. What breaks down in the curved background is the special "naturalness" property of flat-space supersymmetry which implies that the classical superpotential is not renormalized. At the moment the violation of the naturalness property occurs only in the 1-point function and is not very drastic. However dimensional arguments allow modifications of the relation between mass and wave function renormalization which did not actually occur in 1-loop order. One of the interesting open questions mentioned in the introduction is to study the situation in higher loop order.

The last result discussed does not require supersymmetry; it is valid for a general field theory in an AdS background. The O(3,2) invariance properties of such field theories are used to derive a non-perturbative Lehmann spectral representation for two-point functions of scalar operators (Düsedau & Freedman 1985b). Using a suitable integral transform, one can derive a dispersion relation in a variable which is the eigenvalue of the quadratic Casimir operator of O(3,2). The "mass" of a particle can be defined non-perturbatively as the location of a pole in the dispersion relation.

II. $(AdS)_4$ GEOMETRY

Anti de Sitter space can be realized on the hyperboloid $\eta_{AB}\, y^A y^B = a^{-2}$ embedded in $\mathbb{R}^5$ with Cartesian coordinates y^A, $A = 0,1,2,3,4$ and flat metric $\eta_{AB} = (+\,---\,+)$. One can introduce intrinsic coordinates t, ρ, θ, φ, collectively denoted by x^μ, by an explicit formula $y^A(x^\mu)$ such that the induced metric takes the form

$$ds^2 = \frac{1}{a^2 \cos^2\rho}\left[dt^2 - d\rho^2 - \sin^2\rho\,(d\theta^2 + \sin^2\theta\, d\varphi^2)\right] \tag{1}$$

This is a maximally symmetric metric with isometry group O(3,2) and scalar curvature $R = 12\, a^2$.

Infinitesimal O(3,2) transformations are realized by Killing vectors

$$K_{AB} = y_A \frac{\partial}{\partial y^B} - y_B \frac{\partial}{\partial y^A} \tag{2}$$

To formulate supersymmetry one needs Killing spinors $\varepsilon(x)$ which satisfy the equations

$$\left(D_\mu + \tfrac{1}{2} i a\, \gamma_\mu\right)\varepsilon(x) = 0 \tag{3}$$

There are four independent solutions which can be written as $\varepsilon(x) = S(x)\xi$ where ξ is a constant Majorana spinor and the matrix S(x) is an explicitly known matrix in the fundamental spinor representation of SO(3,2). Two key properties of S(x) are

$$\bar{S}(x)\, S(x) = \mathbb{1}$$

$$\bar{S}(x)\gamma^\mu S(x)\partial_\mu = \gamma^{a4} K_{a4} + i\sigma^{ab} K_{ab} \equiv i\,\Gamma^{AB} K_{AB} \tag{4}$$

where $a, b = 0,1,2,3$, and the K_{AB} are the Killing vectors expressed in intrinsic coordinates. The 10 matrices Γ^{AB} generate the spinor representation. For more details of $(AdS)_4$ geometry, see (Breitenlohner & Freedman 1982).

III. The Wess-Zumino Model and its Conserved Currents

The chiral multiplet in $(AdS)_4$ consists of component fields $z(x), \psi(x), F(x)$ with transformation rules (with $L, R = \frac{1}{2}(\mathbb{1} \mp \gamma_5)$ and with Killing spinor parameter ε)

$$\delta z = \bar{\varepsilon} L \psi$$
$$\delta L\psi = L(-i\not\partial z + F)\varepsilon$$
$$\delta F = -\bar{\varepsilon}(i\not{D} + a)L\psi \tag{5}$$

For a general superpotential W(z) the following kinetic and interaction Lagrangians each transform into a total derivative under (5).

$$\mathcal{L}_{KIN} = \partial^\mu \bar{z}\, \partial_\mu z + \tfrac{1}{2} i \bar{\Psi} \not{D} \Psi + \bar{F} F + a(\bar{z} F + z \bar{F}) + 3a^2 \bar{z} z$$

$$\mathcal{L}_{INT} = F W' + \bar{F} \bar{W}' + 3a(W + \bar{W}) - \tfrac{1}{2} \bar{\Psi} (L W'' + R \bar{W}'') \Psi \tag{6}$$

Note that the Lagrangian depends on W(z) as well as its derivatives W'(z) and W"(z). This leads to the energy crisis.

Expressions for the generators of O(3,2) transformations and for the supersymmetry charge may be obtained by the Noether method which yields

$$\bar{M}_{AB} = \int d\Sigma_\nu\, K^\mu_{AB}\, T_\mu{}^\nu \tag{7}$$

$$Q_\alpha = \int d\Sigma_\nu\, \bar{S}(x)_{\alpha\beta}\, J^\nu_\beta \tag{8}$$

where $d\Sigma_\nu$ is the volume element of a spacelike surface, and the stress tensor and super current are

$$T_{\mu\nu} = \partial_\mu \bar{z} \partial_\nu z + \partial_\nu \bar{z} \partial_\mu z - g_{\mu\nu}(\partial\bar{z}\partial z - V(z,\bar{z})) + \frac{i}{4} \bar{\Psi} (\gamma_\mu D_\nu + \gamma_\nu D_\mu) \Psi \tag{9}$$

$$V(z,\bar{z}) = \bar{F} F - 3a(a \bar{z} z + W + \bar{W}) \cdot$$

$$F = -\bar{W}'(\bar{z}) - a z \tag{10}$$

$$J^\mu = L(\not{\partial}\bar{z} - iF)\gamma^\mu \Psi + R(\not{\partial} z - i\bar{F})\gamma^\mu \Psi \tag{11}$$

IV. The Energy Crisis

Repeated supersymmetry transformations (5) give O(3,2) transformations and the resulting algebraic structure is the superalgebra OSp(1,4) with structure relation

$$\{Q_\alpha, \bar{Q}_\beta\} = (\gamma^a M_{a4} + i\sigma^{ab} M_{ab})_{\alpha\beta} \tag{12}$$

The time displacement is M_{04} and by a trace of (12) with the matrix γ^0 one finds that M_{04} is a positive operator whose vacuum expectation value vanishes in a supersymmetric vacuum. However since the potential in (10) has indefinite sign at a critical point, even in a supersymmetric state, F=0, the Noether energy $\bar{M}_{04}$ of (7) apparently does not satisfy the requirements deduced from the algebra (12). This is the energy crisis.

Some readers may be puzzled as to why this crisis was not already resolved by the early work on the vacuum stability problem which did establish stability in gauged supergravity with potentials which were unbounded below. One answer is that the previous studies were essentially classical and did not really address the question whether $\langle 0 | M_{AB} | 0 \rangle =$ 0 in the quantum theory when $Q_\alpha | 0 \rangle = 0$. Another more subtle issue is the boundary conditions at spatial infinity (that is as $\rho \to \frac{1}{2}\pi$ in the coordinates of (1)) which must be imposed on fluctuation fields (Avis et al. 1978). I actually suspect that the generalized Witten positive energy argument of Gibbons et al. (1983) must be modified to incorporate these boundary conditions more accurately, since there is an unresolved discrepancy between the boundary behavior assumed in the work above and that which occurred in the linearized analysis of Breitenlohner & Freedman (1982). This modification is another open problem.

The boundary conditions are required to ensure that formally conserved quantities such as the Noether charges (7-8) are actually conserved in time; there must be no leakage of the conserved quantities through spatial infinity. The boundary conditions are quite well understood for free fields in $(AdS)_4$. For a free scalar field of Lagrangian mass m^2, there are solutions of the wave equation with two distinct boundary behaviors, namely

$$\varphi(x) \sim (\cos\rho)^{\lambda_\pm}$$

$$\lambda_\pm = \frac{3}{2} \pm \sqrt{\frac{9}{4} + \frac{m^2}{a^2}} \qquad (13)$$

The upper and lower signs correspond to regular and irregular boundary conditions respectively. There is a group theoretic interpretation for $\lambda_\pm$; it is the lowest weight of the O(3,2) irrep. in which the field $\varphi(x)$ transforms. Specifically, $\lambda_\pm$ is the lowest eigenvalue of the energy operator M_{04}; see Fronsdal (1976), Heidenreich (1982) or Nicolai (1985). In the mass range $-9a^2 < 4m^2 < -5a^2$, there is a choice of regular or irregular boundary conditions. However in a supersymmetric theory such as the Wess-Zumino model, where $z(x) = (A(x) + i\,B(x))/\sqrt{2}$, if the scalar A(x) is quantized with regular boundary conditions, then one must choose irregular boundary conditions for B(x), and vice versa. Otherwise the supercharge Q_α is not conserved; see Breitenlohner & Freedman (1982). If $4m^2 \geq -5a^2$, then only the regular boundary conditions lead to unitary representations of O(3,2) or OSp(1,4).

In the free massive Wess-Zumino model with Lagrangian (6) and superpotential $W(z) = \frac{1}{2}\mu z^2$, one finds after elimination of auxiliary fields that A(x) and B(x) have different Lagrangian masses. Further analysis of the boundary conditions shows that, if $\mu > -\frac{1}{2}a$, one can take $\lambda_A = 1 + \mu/a$ and $\lambda_B = 2 + \mu/a$. If $\mu > \frac{1}{2}a$, then both A and B have regular boundary conditions, but if $\mu < \frac{1}{2}a$ then A has irregular boundary conditions.

In the recent work described here we must choose the proper boundary conditions for interacting fields. The old integral representation for solutions of nonlinear field equations of Yang and Feldman (1950) suggests that the behavior at spatial infinity is still determined by the physical mass parameter of the Lagrangian. Thus we choose $A(x) \sim (\cos\rho)^{\lambda_A}$ and $B(x) \sim (\cos\rho)^{\lambda_B}$ in the following.

V. Resolution of the Energy Crisis

The previous discussion of boundary conditions shows that it is important to check that the Noether method charges (7-8) are actually conserved in time. It turns out that, for more detail see Burges et al. (1985), that the supercharge Q_α is conserved for both regular and irregular boundary conditions, although delicate cancellations are necessary to establish this in the irregular case. The O(3,2) generators, however, are conserved only for regular boundary conditions (i.e. $\mu > \frac{1}{2}a$). Thus both flux conservation and failure of manifest positive energy suggest that modification of the $\overline{M}_{AB}$ is required.

Since the supercharge Q_α is conserved, it is quite natural to find the generators M_{AB}, as they occur in the algebra (12), by computing the anti-commutator $\{Q_\alpha, \overline{Q}_\beta\}$ from (8) using either canonical commutation relations or by computing the supersymmetry transform δJ^ν. After dropping surface terms which actually vanish at infinity, one finds the improved O(3,2) generators

$$M_{AB} = \int d\Sigma_\nu \left\{ K^\nu_{AB} T^\mu_{\ \nu} - a^{-1} D_\nu \left[(D^\nu K^\mu_{\ AB})(a\bar{z}z + W + \overline{W}) \right] \right\} \quad (14)$$

One can check that these are actually conserved, due to cancellation between the Noether term and the new term for irregular boundary conditions. When only regular boundary conditions are needed the surface term actually vanishes. It could be discarded, but it is useful to keep it in order that M_{AB} satisfy the required properties manifestly.

I now briefly summarize the work of (Burges et al., 1985) which showed that the improved generators M_{AB} do satisfy the properties which follow from the algebra (12). Actually we proved stronger results about the "densities" of these generators. Namely the energy density of M_{04} is positive and vanishes only if F = 0, and that the densities of all 10 have vanishing vacuum expectation value to all orders in perturbation theory in a supersymmetric vacuum.

Positive energy density was proved by examining M_{04} on a constant time surface in the coordinates of (1). One finds

$$M_{04} = a^2 \int d^3x \sqrt{-g}\, \cos^2\rho \Big[\partial_0 \bar{z}\, \partial_0 z + \tfrac{1}{2} i \bar{\Psi} \gamma_0 D_0 \Psi + (\sin^{-2}\rho)\big(\partial_\theta \bar{z}\, \partial_\theta z + \sin^{-2}\theta\, \partial_\varphi \bar{z}\, \partial_\varphi z\big) + (\partial_\rho \bar{z} - a^{-1}\tan\rho\, \bar{F})(\partial_\rho z - a^{-1}\tan\rho\, F) + a^{-2} \bar{F} F \Big] \tag{15}$$

The bosonic terms of the integrand are manifestly positive and vanish only if $\partial_\mu z$ = 0 and F=0, i.e. when the classical vacuum state is supersymmetric. Note that the result was established by a non-covariant technique. It is desireable to establish positivity of the improved energy density for any spacelike surface and any coordinatization of $(AdS)_4$. This is another open problem.

The next step is to show that $\langle M_{AB} \rangle$ = 0 to arbitrary order in perturbation theory. To do this we take the trace of (14) and use the general relation $\Box K^\mu = -3a^2 K^\mu$ for the Killing vectors of an Einstein space, together with $\langle T^{\mu\nu} \rangle = \frac{1}{4} g^{\mu\nu} \langle T_\alpha{}^\alpha \rangle$ and $\langle \partial_\mu (a \bar{z} z + W + \bar{W}) \rangle$ = 0 which are valid because of the high symmetry of $(AdS)_4$. This leads to

$$\langle M_{AB} \rangle = \frac{1}{4} \int d\Sigma_\mu K^\mu{}_{AB} \langle T_\nu{}^\nu + 12a\,(a\bar{z}z + W + \bar{W}) \rangle \tag{16}$$

The local bosonic operator which appears in (16) can be shown to be the supersymmetric transform of a fermionic operator $\mathcal{F}_\alpha(x)$, and we can write, schematically,

$$\langle M_{AB} \rangle = \frac{-i}{4} \int d\Sigma_\mu K^\mu{}_{AB} \langle \{ \mathcal{F}_\alpha(x), Q_\beta \} \rangle \bar{S}_{\beta\alpha}(x) \tag{17}$$

so that the right-side vanishes if $Q_\beta |0\rangle$ = 0.

We hope that interested readers will consult Burges et al. (1985) for more details on the resolution of the energy crisis. These include a parallel treatment which starts with the conformally improved supercuurent and leads to charges M^c_{AB} which involve the conformal stress tensor plus an additional surface term. There are also Pauli-Villars regulated calculations to test the formal supersymmetric Ward identities used to prove $\langle M_{AB} \rangle$ = 0, and a discussion of the relation between $\langle M^c_{AB} \rangle$ = 0 and the standard curved space trace anomaly.

The most interesting open question suggested by this work is the relation of the fixed background model considered here to the chiral multiplet coupled to N=1 supergravity where the AdS background arises dynamically. It would be useful to know whether or how the surface terms in the M_{AB} of (14) arise in supergravity. I suspect that they arise in the Witten positive energy argument of Gibbons et al. (1983) when the boundary conditions are carefully considered.

VI. Renormalization in AdS Supersymmetry

I turn now to the second major topic discussed at the meeting namely the 1-loop renormalization structure of the Wess-Zumino model (6) with superpotential

$$W(z) = \frac{1}{2}\mu z^2 + \frac{1}{3\sqrt{2}}\lambda z^3 \tag{18}$$

I will summarize the work of Düsedau and Freedman (1985a).

Careful calculation requires a method of regularization which maintains supersymmetry, and we chose the Pauli-Villars method. We used a set of regular multiplets z_i, Ψ_i, F_i with kinetic coupling and mass parameters c_i and μ_i. The physical field, i=0, has $c_0 = 1$ and $\mu_0 = \mu$. Sum rules

$$\sum_i c_i \mu_i^p = 0 \qquad p = 0, 1, 2 \tag{19}$$

are imposed to regulate divergent expressions and the limit $\mu_i \to \infty$, $i \geq 1$ is taken after calculation. The regularized superpotential is

$$W(z_i) = \frac{1}{2}\sum_i c_i^{-1}\mu_i z_i^2 + \frac{1}{3\sqrt{2}}\lambda\Big(\sum_i z_i\Big)^3 \tag{20}$$

and the regularized form of (6) turns out to be, after elimination of auxiliary fields, via $F_i = -a z_i - c_i \overline{W}_{,i}$,

$$\begin{aligned}\mathcal{L} = {} & \frac{1}{2}\sum_i c_i^{-1}\big[\partial_\mu A_i \partial^\mu A_i + \partial_\mu B_i \partial^\mu B_i + \overline{\Psi}_i (i\not{D} - \mu_i)\psi_i \\ & \qquad - (\mu_i^2 - a\mu_i - 2a^2) A_i^2 - (\mu_i^2 + a\mu_i - 2a^2) B_i^2\big] \\ & - \frac{1}{2}\lambda\Big[\Big(\sum_i \overline{\Psi}_i\Big)\sum_j (A_j - i\gamma_5 B_j)\Big(\sum_k \psi_k\Big) + \Big(\sum_i \mu_i A_i\Big)\Big[\Big(\sum_j A_j\Big)^2 - \Big(\sum_j B_j\Big)^2\Big] \\ & \qquad + 2\Big(\sum_i \mu_i B_i\Big)\Big(\sum_j B_j\Big)\Big(\sum_k A_k\Big)\Big]\end{aligned} \tag{21}$$

This Lagrangian is supersymmetric for $(AdS)_4$ and, with a=0, for flat space. The surprising feature of this Lagrangian is that there are no quartic terms, because they appear in (21) with coefficient $\sum_i c_i$ which vanishes by (19). Thus quartic vertices do not explicitly occur in any calculation, although their physical effects are properly reproduced by regulator field exchanges as $\mu_i \to \infty$.

Propagators for spinless fields which satisfy the AdS boundary conditions can be expressed as hypergeometric functions (Burgess & Lütken 1985; Inami & Ooguri 1985; Burges et al. 1985) which turn out to be derivatives of Legendre functions $Q_\nu(1-u)$ with respect to the chordal distance variable $u(x,x') = \frac{1}{2}a^2(y^A - y'^A)^2$ on the AdS hyperboloid. For the supersymmetric case we have

$$\langle A(x)\ A(x')\rangle = \frac{a^2}{4\pi^2}\frac{d}{du}Q_{\lambda_A-2}(1-u+i\varepsilon)$$

$$\langle B(x)\ B(x')\rangle = \frac{a^2}{4\pi^2}\frac{d}{du}Q_{\lambda_B-2}(1-u+i\varepsilon) \tag{22}$$

There is a supersymmetric Ward identity which relates the spinor propagator to either $\langle A(x)A(x')\rangle$ or $\langle B(x)B(x')\rangle$, and we can take this as the definition of the spinor propagator which then satisfies the correct boundary conditions. The propagator is

$$\langle\psi(x)\ \bar{\psi}(x')\rangle = [(i\gamma_x + \mu + a)\langle A(x)\,A(x')\rangle]\,S(x)\,\bar{S}(x')$$

$$= [(i\gamma_x + \mu - a)\langle B(x)\,B(x')\rangle]\,\gamma_5\,S(x)\,\bar{S}(x')\,\gamma_5 \tag{23}$$

where S(x) is the Killing spinor matrix discussed below (3). The expressions (23) obviously have the correct flat space limit. It is curious that for a massless spinor ($\mu = 0$) the propagator does not satisfy the properties of anti-commutativity with γ_5 required by the chiral invariance of the Lagrangian. The reason for this is the AdS boundary conditions which imply that the axial charge is not conserved (Fronsdal 1976) although there is a locally conserved axial current. Thus chiral spinors do not exist in $(AdS)_4$. One should note that (23) gives the correct spinor propagator even for non-supersymmetric theories in $(AdS)_4$. It is easy to show directly that it satisfies the correct differential equation, and the quantity $S(x)S(x')$ can be expressed in terms of more traditional geometrical objects such as the bispinor of parallel transport (Allen & Lütken 1985).

The potential of the regularized Lagrangian (21) has a classical stationary point at $A_i = B_i = 0$ which also satisfies $F_i = 0$. It is the radiative corrections to this classical supersymmetric phase that we investigated.

Dimensional analysis of the general form of the supersymmetric action (6) strongly constrains the possible modifications of the renormalization structure due to the curved background. In particular the dimensionless wave function and coupling constant renormalization constants Z and Z_λ are unchanged from their flat space values. However one might expect that the mass renormalization for the $(AdS)_4$ model is obtained by the replacement in (6)

$$\mu \to Z_\mu \mu + Z' a \tag{24}$$

where $Z_\mu = Z^{-1}$, as required by the flat-space non-renormalization theorem, and Z' is a possible new logarithmically divergent term. One might also expect that the superpotential W(z) of (20) is modified by the additive counter term

$$\Delta W(z) = V z \tag{25}$$

where the degree of divergence of V as the regulator masses $\mu_i \to \infty$ is at most linear with V proportional to a.

To determine whether these possible modifications actually occurred in 1-loop order, we calculated the regulated 1-point and 2-point functions for the Lagrangian (21). It is not necessary to calculate 3-point functions because no modification of the flat space result is allowed by the previous argument.

Let us first discuss the results for 1-point functions which have the highest degree of divergence. Truncated 1-point functions $\langle A_i \rangle_{Tr}$ have contributions from scalar, pseudoscalar, and spinor tadpoles. These can be computed from the Dyson-Wick expansion of the interaction part of (21), and the result is

$$\begin{aligned}\langle A_i \rangle_{Tr} &= -\tfrac{1}{2}\lambda \sum_j c_j \{(\mu_i + 2\mu_j)\langle A_j^2\rangle - (\mu_i - 2\mu_j)\langle B_j^2\rangle + \langle \bar{\Psi}_j \psi_j \rangle\} \\ &= -\tfrac{1}{2}\lambda(\mu_i - 2a)\sum_j \{\langle A_j^2\rangle - \langle B_j^2\rangle\} \\ &= \frac{a\lambda}{8\pi^2}(\mu_i - 2a)\sum_j c_j \mu_j \psi(\mu_j/a)\end{aligned} \tag{26}$$

In the first line the regulated sum of zero separation propagators appears and is well defined. To obtain the second line the trace of the Ward identity (23) is used. In the third line Euler's function $\psi(z)$ appears because the short distance expansion of the propagator (22) is used. The singular terms in the expansion cancel due to the sum rules (19). One can easily see that $\langle A_i \rangle_{Tr}$ diverges linearly as the regulator masses $\mu_i \to \infty$.

Separate computations show that $\langle B_i \rangle = 0$, essentially because of parity conservation, and, more important, that $\langle F_i \rangle = 0$. Thus the supersymmetry of the classical vacuum is preserved by the 1-loop radiative corrections. The results show that $\langle A_i \rangle$ is non vanishing (and ultraviolet divergent) but $\langle F_i \rangle$ vanishes. Since no symmetry of the initial Lagrangian is violated by these 1-loop vacuum expectation values, we attempt to cancel the divergence by adding the linear counter term $\Delta W = v \sum_i z_i$ to the regulated superpotential (20). After eliminating auxiliary fields one finds that the interaction Lagrangian (21) changes by $\Delta \mathcal{L} = -\sqrt{2} v \sum (\mu_i - 2a) A_i$. Comparison with (26) shows that the divergence is cancelled by the simple choice

$$v = -\frac{\lambda}{2\sqrt{2}} \sum_j \left[\langle A_j^2 \rangle - \langle B_j^2 \rangle \right] \tag{27}$$

Both the cutoff dependent and finite parts of $\langle A_i \rangle$ are cancelled, as is required for a proper operator interpretation of the theory (Glaser et al. 1957).

The philosophy underlying the procedure just discussed is simply the idea of "naturalness" in non-supersymmetric flat space field theories. We found a divergence which was consistent with the symmetries of the Lagrangian with couplings μ and λ, and we cancelled it by adding a term which was also consistent with those symmetries and could have been put in at the start. Thus the modified view of naturalness in flat space supersymmetry, where one does not have to take the most general superpotential consistent with symmetries, does not extend to anti de Sitter space.

To complete the study of the 1-loop renormalization, we computed the boson and fermion 2-point functions using a modified form of the adiabatic momentum space technique of Bunch & Parker (1979). The work may be (all too briefly) summarized by giving the results for the 1-loop contributions to the renormalization constants which are

$$\delta Z = -\delta Z_\mu = -\frac{\lambda^2}{32\pi^2} \sum_{j,k} c_j c_k \frac{\mu_j^2 + \mu_k^2}{\mu_j^2 \mu_k^2} \ln\left(\mu_j^2 / \mu_k^2\right)$$

$$\delta Z' = 0 \tag{28}$$

The first result, which is valid in both flat space and $(AdS)_4$ reflects the standard nonrenormalization theorem which requires $\delta Z = -\delta Z_\mu$. However the expression for the renormalization constants, using Pauli-Villars regularization, seems to be new.

The curious point is that the possible curved space modification of the mass renormalization allowed by dimensional arguments in (24) is actually absent in 1-loop order, and we again state that it is an interesting open question to determine what happens in higher orders of perturbation theory. It might be best to deduce a general result using either the form of superfield perturbation theory for $(AdS)_4$ supersymmetry (Gates et al. 1983) or functional techniques similar to those which were first used to establish the general renormalization structure of the flat-space model (Iliopoulos & Zumino 1974; Tsao 1974).

The result that the 1-loop mass renormalization is not modified in the $(AdS)_4$ Wess-Zumino model was also obtained by Belluci & Gonzalez (1985a) who used a form of dimensional regularization. In a later preprint (Belluci & Gonzalez 1985b), these authors adopted Pauli-Villars regularization and studied the 1-point function. Due to a serious error, they incorrectly concluded that supersymmetry was spontaneously broken by 1-loop radiative corrections while non-renormalization theorems were unmodified.

An interesting open problem is suggested by the non-existence of chiral spinors in $(AdS)_4$, as discussed above. Because of this the particles of the massless ($\mu = 0$) Wess-Zumino model should develop a mass in 1-loop order. Since the mass renormalization vanishes if $\mu = 0$, the mass shift should be calculable. One way to approach the problem is to use the definition of particle mass given by the Lehmann representation which we now discuss.

VII. Lehmann Spectral Representation

The standard Lehmann representation is a consequence of the Poincaré invariance of flat space quantum field theory. It is very useful because it incorporates non-perturbative information. The analogous representation has recently been derived (Düsedau and Freedman 1985b) for the 2-point functions of scalar operators in a field theory in an $(AdS)_4$ background, and I now discuss the results.

As one might expect the main tool in the derivation is O(3,2) invariance. Specifically the generators M_{AB} of this group act on any scalar operator $\varphi(x)$, whether elementary or composite, as

$$[M_{AB}, \varphi(x)] = -i K_{AB} \varphi(x) \tag{29}$$

where K_{AB} is a Killing vector (2) expressed as a differential operator in an intrinsic coordinate system such as (1). We also assume that there is a unique invariant vacuum state $|0\rangle$, and that excited states are classified in unitary positive energy irreps. (λ, s) where λ is the lowest eigenvalue of the energy operator, and s is the spin of the unique state with energy λ. See Fronsdal (1975) or Nicolai (1985) for more information about these representations. Thus an excited state is denoted by $|(\lambda,s)\, \omega, j, m\, (\alpha)\rangle$ where ω, j, m are energy and angular momentum labels for the basis states of an irrep., and (α) denotes additional O(3,2) invariant quantum numbers.

To derive the Lehmann representation we first study the "wave functions" $\langle 0|\phi(x)|(\lambda,s)\omega,j,m(\alpha)\rangle$. By repeated use of (29), and its generalization for the Casimir operator $\frac{1}{2}M_{AB}M^{AB}$, one can deduce that only irreps. $(\lambda,0)$ have non-vanishing wave functions, since $\varphi(x)$ is a scalar operator, and that these wave functions can be written as

$$\langle 0|\varphi(x)|(\lambda,0)\ \omega,j,m(\alpha)\rangle = N(\phi,\lambda,\alpha)\,\phi^{\lambda}_{\omega j m}(x) \tag{30}$$

where $\phi^{\lambda}_{\omega,m}(x)$ is a normalized mode function for free scalar fields in the $(\lambda,0)$ irrep., and $N(\varphi,\lambda,\alpha)$ is a "normalization constant" whose value reflects the dynamics of the interacting field theory.

Next we study the 2-point function $\langle 0|T\,\phi(x)\,\phi(x')|0\rangle$. After expansion in intermediate states one finds that the sum over mode functions for the irrep. $(\lambda,0)$ exactly gives the free field propagator

$$i\Delta_F(x,x',\lambda) = \frac{a^2}{4\pi^2}\frac{d}{du}Q_{\lambda-2}(1-u+i\varepsilon) \tag{31}$$

and we can write the Lehmann representation as

$$\langle 0|T\,\phi(x)\,\phi(x')|0\rangle = \sum_{\lambda}\rho(\lambda,\phi)\,i\Delta_F(x,x',\lambda) \tag{32}$$

$$\rho(\lambda,\phi) = \sum_{(\alpha)}|N(\phi,\lambda,\alpha)|^2 \tag{33}$$

Thus, just as in the flat space case, the space-time dependence is expressed by known free field propagators and the dynamics by the weight function $\rho(\lambda,\phi)$.

The standard Lehmann representation is most useful in momentum space, where the analyticity properties of 2-point functions are conveniently displayed. The Fourier transform is not natural in AdS since ordinary "plane waves" are not eigenfunctions of the wave operator. However there is an eigenfunction transform involving the appropriate generalization of "plane waves", which has been described for the hyperboloid H_4 by Vilenkin (1968), and this integral transform can be adapted to $(AdS)_4$.

For a function of the single variable u or, equivalently $z = 1 - u$, such as

$$F(z) \equiv \langle 0|T\,\phi(x)\,\phi(x')|0\rangle \tag{34}$$

the integral transform and its inverse are given by

$$F(z) = \frac{i}{2}\int_{-\frac{1}{2}-i\infty}^{-\frac{1}{2}+i\infty} d\sigma\,(\sigma^2-1)\,\sigma\,\operatorname{ctg}\pi\sigma\,\hat{F}(\sigma)\,P_{\sigma}^{-1}(z)\,(z^2-1)^{-1/2} \tag{35}$$

$$\hat{F}(\sigma) = \int_1^{\infty} dz\,F(z)\,(z^2-1)^{1/2}\,P_{\sigma}^{-1}(z) \tag{36}$$

The associated Legendre function $P_\sigma^{-1}(z)$ appears, and the integral over z in (36) extends over the region where the points x, x' are spacelike separated. See (Düsedau & Freedman 1985b) for discussion of the mathematical conditions for the validity of this transform. The transform can be applied to the free-field propagator (31) and one obtains

$$\lambda \Delta_F(x,x',\lambda) = \frac{ia^2}{8\pi^2}\int_{-\frac{1}{2}-i\infty}^{-\frac{1}{2}+i\infty} d\sigma (\sigma^2-1)\,\sigma \operatorname{ctg}\pi\sigma \,\frac{1}{\lambda(\lambda-3)-\sigma(\sigma+1)+2}\,\frac{P_\sigma^{-1}(z)}{(z^2-1)^{1/2}} \tag{37}$$

The transform of the free propagator is thus a pure pole at $\lambda(\lambda-3)$ which is just the eigenvalue of the quadratic Casimir operator of O(3,2). This simple result shows that the integral transform used is a natural spectral representation for AdS.

If we use (35) on the left-side of the Lehmann representation (32) and then use (37) on the right-side, we can equate the transforms and deduce the result

$$\hat{F}(\sigma) = \frac{a^2}{4\pi^2}\sum_\lambda \frac{\rho(\lambda,\phi)}{\lambda(\lambda-3)-\sigma(\sigma+1)+2} \tag{38}$$

Thus the transform of the exact 2-point function satisfies a dispersion relation, which is entirely analogous to the situation in flat space. One difference is that the spectrum of λ is discrete. In a field theory such as $\lambda\varphi^4$ theory in AdS one expects that the lowest contributing value corresponds to the 1-particle intermediate states, and there will be additional contributions at 2λ, $2\lambda+2$,... from 2-particle states and 3λ, $3\lambda+1$,... from 3-particles states, etc.

For the 2-point function of a canonical scalar field theory one can write a Lehmann sum rule and define the physical mass as $m^2_{phys} = \lambda(\lambda-3)$ where λ is the location of the lowest pole in (38). One can use this for perturbative studies of mass shifts as in the chirality problem discussed at the end of Sec. VI, although it may be more convenient to work with the inverse propagator, whose transform in (35) is $\hat{F}(\sigma)^{-1}$, and look for the shift in the zero of this function due to radiative corrections.

The first discussion of the Lehmann spectral representation for a theory where the invariance group was not the Poincaré group occurred in the O(2,1) invariant quantization of the Liouville theory (d'Hoker et al. 1983). This served as a simpler model for the treatment above. A spectral representation for two-point functions of the stress energy tensor in de Sitter space was used by Mottola (1984).

The Lehmann representation (32) and the corresponding dispersion relation (38) may simply be the expected consequence of micro-causality and O(3,2) invariance and signify little more. Nevertheless the close parallel between these results in AdS and their flat space analogues leads me to speculate that AdS quantum field theory can be developed considerably further. One can formally define 1PI vertex functions and truncated Green's functions in AdS field theory using functional methods and combinatorics. One should also be able to define "on shell" amplitudes by integrating products of a truncated Green's function and the appropriate free field mode functions. I conjecture that these amplitudes share some of the properties of ordinary scattering amplitudes such as independence of the choice of interpolating field and gauge independence (in a gauge field theory). I suggest that these questions be studied as an open problem, although I cannot really foresee what application they will have. Perhaps AdS field theory can be useful as an infrared regularization of non-Abelian gauge theories.

REFERENCES

Abbott, L.F. & Deser, S. (1982). Stability of gravity with a cosmological constant. Nucl.Phys. B159, 76-96.

Allen, B. & Lütken, C.A. (1985). Private communication.

Avis, S.J., Isham, C.J. & Storey, D. (1978). Quantum field theory in Anti de Sitter spacetime. Phys.Rev. D18, 3565-3576.

Bardeen, W.A. & Freedman, D.Z. (1985). On the energy crisis in Anti de Sitter Supersymmetry. Nucl.Phys. B253, 635-649.

Bellucci, S. & Gonzalez, J. (1985a). One-loop order renormalization of the massless Wess-Zumino model in Anti de Sitter space. Brandeis preprint BRX-TH-185.
One-loop order renormalization of the massive Wess-Zumino model in Anti de Sitter space. Brandeis preprint BRX-TH-187.

Bellucci, S. & Gonzalez, J. (1985b). Radiative breaking of supersymmetry in the Wess-Zumino model in Anti de Sitter space. Brandeis preprint BRX-TH-189.

Boucher, W. (1984). Positive energy without supersymmetry. Nucl.Phys. B242, 282-296.

Breitenlohner, P. & Freedman, D.Z. (1982). Stability in gauged extended supergravity. Ann.Phys. 144, 249-276.

Bunch, T.S. & Parker, L. (1979). Feynman propagator in curved spacetime: A momentum-space representation. Phys.Rev. D20, 2499-2510.

Burges, C.J., Freedman, D.Z., Davis, S. & Gibbons, G.W. (1985). Supersymmetry in Anti de Sitter space. To appear in Ann.Phys.

Burgess, C.P. & Lütken, C.A. (1985). Propagators and effective potentials in Anti de Sitter space. Phys.Lett. 153B, 137-141.

D'Hoker, E., Freedman, D.Z. & Jackiw, R. (1983). SO(2,1) invariant quantization of the Liouville theory. Phys.Rev. D28, 2583-2598.

Düsedau, D.W. & Freedman, D.Z. (1985a). Renormalization in Anti de Sitter

Supersymmetry. M.I.T. preprint CTP-1291.
Düsedau, D.W. & Freedman, D.Z. (1985b). Lehmann spectral representation for Anti de Sitter quantum field theory. M.I.T. preprint CTP - 1290.
Fronsdal, C. (1975). Elementary particles in a curved space. IV. Massless particles. Phys.Rev. D12, 3819-3830.
Gates, S.J., Grisaru, M.T., Roček, M. & Siegel, W. (1983). Superspace: one thousand and one lessons in supersymmetry. Reading, Mass.: Benjamin-Cummings.
Gibbons, G.W., Hull, C.M. & Warner, N.P. (1983). The stability of gauged supergravity. Nucl.Phys. B218, 173-190.
Glaser, V., Lehmann, H. & Zimmermann, W. (1957). Field Operators and Retarded Functions, Nuovo Cimento 6, 1122-1128.
Grisaru, M.T. & Stelle, K. (1985). Private communication.
Heidenreich, W. (1982). All linear unitary irreducible representations of de Sitter supersymmetry with positive energy. Phys.Lett. 110B, 461-464.
Iliopoulos, J. & Zumino, B. (1984). Broken supergauge symmetry and renormalization. Nucl.Phys. B76, 310-322.
Inami, T. & Ooguri, H. (1985). One-loop effective potential in Anti de Sitter space. Progr.Theor.Phys. 73, 1051-1054.
Ivanov, E.A. & Sorin, E.A. (1980). Superfield Formulation of OSp(1,4) supersymmetry. J.Phys. A: Math.Gen. 13, 1159-1188.
Mottola, E. (1984). Particle creation in de Sitter space. UCSB preprint NSF ITP 84-123.
Nicolai, H. (1985). Representations of supersymmetry in Anti de Sitter space. In Supersymmetry and Supergravity '84, eds. B. de Wit, P. Van Nieuwenhuizen, pp.368-399. Singapore: World Scientific.
Tsao, H.S. (1974). Supergauge Symmetry, broken chiral symmetry and renormalization. Phys.Lett. 53B, 381-383.
Vilenkin, N. (1968). Special functions and the theory of group representations. Providence, RI: American Mathematical Society.
Yang, C.N. & Feldman, D. (1950). S-matrix in the Heisenberg Representation. Phys.Rev. 79, 972-978.

ADDED NOTES:

The fact that the modified mass counterterm aZ' in (24) does not actually appear in one-loop order is probably a consequence of the R-invariance of the massless model. I thank K. Stelle for pointing this out.

A revised and retitled version of (Bellucci and Gonzalez 1985b) has appeared in which the error mentioned at the end of Section VI above is corrected. Results on one-loop divergences now agree with (Dusedau and Freedman, 1985a).

EXTERIOR CANONICAL FORMALISM FOR UNIFIED GRAVITY

J.E.NELSON* AND T.REGGE

Istituto di Fisica Teorica dell'Universita' di Torino

Corso M. d'Azeglio 46, 10125 Torino, Italy

Institute for Scientific Interchange (ISI)

Villa Gualino, 10131 Torino, Italy

We apply a previously developed covariant canonical formalism (CCF) to gravity and supergravity. In this CCF there is no preferred time direction and the exterior derivative is given through a first class Hamiltonian using a form/superform bracket. We establish the correspondence between the CCF and the canonical vierbein formalism (CVF), and relate the form Hamiltonian in CCF to a compound Poincare'/superPoincare' multiplet of the primary constraints in CVF. This multiplet generates Lie derivatives which agree with those constructed in the CCF, using an integral relationship between the form/superform brackets and Poisson brackets.

1. INTRODUCTION

We discuss a piece of work which has been developed in the last two years in Turin in collaboration with A.D'Adda and

A.Lerda [1-4]. The first part concerns a canonical formalism (CCF)[5], for the exterior calculus on differential forms, well suited to the group manifold approach [6] applied to 4-dimensional pure gravity and to N=1 supergravity. This formalism has no preferred time direction, in contrast to the more usual canonical vierbein formalism (CVF)[7-12] which implies a preferred time variable t and makes use of a form/superform bracket (related to the Poisson bracket of classical mechanics) and in the case of supergravity, fermi gradings and compensating phases [6].

In both theories, the Hamiltonian is found to be first class and constructed out of the primary constraints. There are no secondary constraints. The form/superform bracket is fundamental in that it can be used to implement the definitions of the exterior derivatives,of the contraction operator (itself a derivative of degree -1), and therefore of the Lie derivatives which control the evolution of the system.

In the second part of our work we establish the correspondence, for both gravity and N=1 supergravity, between the CCF and the CVF, and relate our form/superform bracket to the conventional Poisson bracket. We wrote the extended Hamiltonian of CVF as a compound Poincare'/superPoincare' multiplet of constraints and calculated the constraint algebra. This Hamiltonian generates Lie derivatives which agree with those constructed in CCF.

Throughout we use systematically the differential calculus on forms and a compound Poincare' notation, which is extended to include spinor components in the case of supergravity. In both parts of this research the results for the two theories, namely pure gravity and supergravity, are identical, and differ only in the range of the compound indices and the use of compensating phases.

We expect that most of our compound formulae can be readily generalized and applied to a broader class of group manifold theories and in higher dimensions.

2. EXTERIOR CALCULUS IN CANONICAL FORM

The action for N=1 supergravity in 4 dimensions is

$$S = \int L$$

$$L = \tfrac{1}{2} R^{ab}\wedge V^c\wedge V^d \, \varepsilon_{abcd}$$
$$+ iAD\bar{\psi}\wedge\gamma_5\gamma_a\psi\wedge V^a$$
$$= \omega^{ab}\wedge dV^c\wedge V^d \, \varepsilon_{abcd}$$
$$- \tfrac{1}{2} \omega^{at}\wedge\omega_t{}^b\wedge V^c\wedge V^d \varepsilon_{abcd}$$
$$+ iAD\bar{\psi}\wedge\gamma_5\gamma_a\psi\wedge V^a$$
$$+ \text{ a total derivative} \qquad (2.1)$$

with $R^{ab} = d\omega^{ab} - \omega^{at}\wedge\omega_t{}^b$

$$R^a = dV^a - \omega^{ab}\wedge V_b - \tfrac{1}{4}A\bar{\psi}\,\gamma^a\wedge\psi$$
$$R^\alpha = d\bar{\psi}^\alpha - \tfrac{1}{2}\,(\bar{\psi}\sigma_{ab})^\alpha\wedge\omega^{ab} \qquad (2.2)$$

The field variables are V^a, ω^{ab} (Bose) and $\bar{\psi}^\alpha$ (Fermi, Majorana, anticommuting components) and are 1-forms. We use $\bar{\psi} = \psi^\dagger\gamma_0$, and our gamma matrices are real as in [8,9]. We define

$$\sigma_{ab} = \tfrac{1}{4}[\gamma_a,\gamma_b] \text{ and}$$
$$\gamma_5 = i\gamma_0\gamma_1\gamma_2\gamma_3$$

The parameter A in (2.1) is a real number, and setting A = 0 corresponds everywhere to the restriction to the case of pure gravity.

In the group manifold approach one necessarily uses a first-order formalism. From now on we will also use a compound notation. We write our 1-form field variables V^a, ω^{ab}, $\bar{\psi}^\alpha$ as q^A where $A = (a,ab,\alpha)$ and $a,b,\alpha.... = 0,1,2,3$ with the summation convention

$$M^A L_A = M^a L_a + \tfrac{1}{2} M^{ab} L_{ab} + \bar{M}^\alpha L_\alpha$$

The curvatures (2.2) can therefore be written compoundly as

$$R^A = dq^A - \tfrac{1}{2} q^C \wedge q^B C_{BC}{}^A \qquad (2.3)$$

where $C_{AB}{}^C$ are the superPoincare' structure constants given in [3], or the Poincare' structure constants given in [1,2] for the case A = 0.

The momenta canonical to V^a, ω^{ab}, $\bar{\psi}^\alpha$ are defined as the variation of (2.1) with respect to dV^a, $d\omega^{ab}$, $d\bar{\psi}^\alpha$ and are denoted π_a, π_{ab}, π_α respectively. We find the set of primary constraints (2-forms)

$$\Phi_a = \pi_a - \omega^{cd} \wedge V^b \varepsilon_{abcd} \sim 0$$

$$\Phi_{ab} = \pi_{ab} \sim 0$$

$$\Phi_\alpha = \pi_\alpha - iA(\gamma_5 \gamma_a \psi)_\alpha \wedge V^a \sim 0$$

or written compoundly

$$\Phi_A = \pi_A - \tfrac{1}{2} q^B \wedge q^C M_{CBA} \sim 0 \qquad (2.4)$$

where $M_{ABC} = (-1)^{AB+1} M_{BAC}$ (for Fermi gradings

A,B) and has components

$$M_{a\ bc\ d} = -2\,\varepsilon_{abcd}$$

$$M_{a\alpha\beta} = iA(\gamma_5\gamma_a C)_{\beta\alpha}$$

with all others zero, and C, the charge conjugation matrix, is

$$(C)_{\alpha\beta} = (\gamma_0)_{\alpha\beta}$$

The canonical Hamiltonian (a Bose four-form) is

$$H = dq^{A}\wedge\pi_A - L + \Lambda^{A}\wedge\Phi_A$$

and we have added on the primary constraints Φ_A with 2-form Lagrange multipliers Λ^A. It is

$$H = \Lambda^{A}\wedge\Phi_A - q^{D}\wedge q^{C}\wedge q^{B}\wedge q^{A} C_{AB}{}^{T} Q_{CTD}(-1)^{CT}/8 \qquad (2.5)$$

where

$$Q_{ABC} = (-1)^{BC} Q_{ACB}$$

$$= (-1)^{AB} M_{BAC} + (-1)^{C} M_{CAB}$$

with A+B+C even, and A,B,C denote the Fermi gradings of the corresponding forms.

The "superform bracket" is an extension of the "form bracket" of [1,2] with the properties:

$$(q^A,\pi_B) = \delta_B{}^A(-1)^A$$

$$(A,B) = (-1)^{AB+ab}(B,A)$$

$$(A,BC) = (A,B)C + (-1)^{b(a+1)+AB}B(A,C) \qquad (2.6)$$

$$(AB,C) = (-1)^{BC+bc}(A,C)B + (-1)^{a}A(B,C)$$

$$(-1)^{CB+c(b+1)}(C,(A,B))+(-1)^{AC+a(c+1)}(A,(B,C))+(-1)^{BA+b(a+1)}(B,(C,A)) = 0$$

for form gradings a,b,c and Fermi gradings A,B,C. For Bose fields i.e. zero Fermi gradings the superform bracket of

(2.6) reduces to the form bracket of pure gravity. We use the definition

$$dF = \partial F + (H,F) \tag{2.7}$$

for functionals F of q^A, π_A, where the operator ∂ acts non-trivially on external fields only(the components $\Lambda_{AB}{}^C$ of Λ^C). Using (2.6) and (2.7) we find that

$$dq^A = \Lambda^A = R^A - C^A$$

$$= -\tfrac{1}{2}q^{B}\wedge q^C \;(R_{CB}{}^A - C_{CB}{}^A)$$

which follows from (2.3)and

$$d\Phi_A = (H,\Phi_A) = -q^{B}\wedge\Lambda_{BA}{}^{C}\wedge\Phi_C$$

so that H given by (2.5) is first class, and there are no secondary constraints. We also find the Bianchi identities for the curvatures:

$$DR^A = dR^A + R^{B}\wedge\; q^C\; C_{CB}{}^A = 0$$

We define the contraction operator on a generic functional F of q^A, π_B and external fields as

$$\iota_B F = (\pi_B,F) + (-1)^T M_{PBT}q^{P}\wedge(q^T,F) \tag{2.8}$$

and introduce

$$\iota_\varepsilon = \varepsilon^A\iota_A$$

for external fields ε^A. From (2.7) and (2.8) the Lie derivative

$$\pounds_\varepsilon = \iota_\varepsilon d + d\iota_\varepsilon \tag{2.9}$$

satisfies

$$\pounds_\varepsilon\pounds_\eta - \pounds_\eta\pounds_\varepsilon \sim \pounds_{(\varepsilon,\eta)}$$

where

$$\{\varepsilon,\eta\}^A = \varepsilon^C\eta^B\Lambda_{BC}{}^A + \varepsilon^B\eta_B{}^A - \eta^B\varepsilon_B{}^A$$

We see therefore that our extended compound notation and our form/superform bracket provide us with a canonical structure (CCF) for the exterior calculus on forms, applied to pure gravity and N=1 supergravity.

3. RELATION BETWEEN CVF AND CCF

The above formalism (CCF) can be related to the more familiar component formalism (CVF). For pure gravity we need to distinguish between first order CVF given in [9] and second order CVF as developed in [7,8]. (In [8] the theory is developed for the coupling to spin ½ fields. To retrieve pure gravity it is enough to set the spin ½ field equal to zero.) These relationships are discussed in detail in [2]. Here we give only details that are essential to the construction of the Hamiltonian multiplet of constraints and the calculation of the constraint algebra. For both theories, this multiplet and constraint algebra are identical, differing only in the range of the compound indices, and coincide*, when expanded in components ,with those of [8] and [12,13] respectively.

We now clarify our notation with respect to those of [8] and [12]. The Lagrangian of [8], in the group manifold approach, is identical to ours, (2.1), with A = 0, whereas the

---

*We found one term in the supersymmetry generator which was missing in [12].

Lagrangian of Pilati [12] is equal to one half of ours with $A = -1$. This factor of one half causes some confusion that can however be handled. By Σ we mean a member of the family of t = constant space-like hypersurfaces in space-time, and denote the restriction to Σ of the functions and variables V^a, ω^{ab}, Φ_A as L^a, $-B^{ab}$, Ψ_A.

The authors in both [8] and [12] used $-\omega^{ab}$ as the 4-dimensional spin connection and hence the opposite sign. We denote the spin connection on Σ by Ω^{ab} and have the relationship between the components of B^{ab} and Ω^{ab}

$$B_{jab} = \Omega_{jab} + (n_b L_a{}^i - n_a L_b{}^i)\, K_{ji} \tag{3.1}$$

where $i,j.. = 1,2,3$ and K_{ji} is the extrinsic curvature of Σ, given by

$$K_{ji} = (2N^{\perp})^{-1}(-\dot{g}_{ij} + N_{i|j} + N_{j|i}) - C_{ij\perp} \tag{3.2}$$

In (3.2) $C_{\lambda\mu\nu}$ is the contorsion tensor, related to the torsion $S_{\lambda\mu\nu}$ by

$$C_{\lambda\mu\nu} = S_{\lambda\mu\nu} + S_{\nu\mu\lambda} + S_{\nu\lambda\mu}$$

with $S_{\mu\nu\lambda} = -\ \tfrac{1}{4}A\bar{\Psi}_\mu\gamma_\lambda\Psi_\nu$. For pure gravity, i.e. for $A = 0$ both the torsion S and contorsion C are zero, and $K_{ij} = K_{ji}$. In the sequel Π_a denotes the form with components equal to the vierbein momenta of [8] and twice the vierbein momenta of [12]. We establish the relationship between our momenta π_a and Π_a

$$\Pi_a = \pi_a + \varepsilon_{abcd}\,\Omega^{cd}\wedge L^b - n_a C^{jik} g_{jk}\,\Sigma_i$$
$$= \pi_a + \varepsilon_{abcd}\,\overset{\circ}{\Omega}{}^{cd}\wedge L^b \tag{3.3}$$

where $\overset{\circ}{\Omega}{}^{cd}$ is the spin connection on Σ for pure gravity, a functional of the vierbein L_{ai} only and

$$\Sigma_i = \varepsilon_{ijk}\, dx^j\wedge dx^k.$$

The Lagrangians in [8] and [12] differ from ours (see equation (2.1)) by a total divergence and hence equation (3.3). This allows us to write the restriction to Σ of the constraints Φ_a, (equation (2.4)), denoted Ψ_a, as

$$\Psi_a = \Pi_a + (L_a{}^k g^{ij} - L^{ai} g^{jk}) K_{kj}\Sigma_i + n_a C^{jik} g_{jk}\,\Sigma_i \tag{3.4}$$

The constraint Ψ_{ab} and the momenta π_{ab} are now set strongly zero, and we consider the B^{ab} as functionals of fields and momenta, as given by equation (3.1).

Using (3.1-3.4) we can show that the total Hamiltonians of [8] and [12]

$$H = \int (L^a{}_0 H_a + \tfrac{1}{2}\omega_0{}^{ab} H_{ab} + \psi_0{}^{\alpha} H_{\alpha})\, d^3x \tag{3.5}$$

can be written as

$$H = \int q_0{}^A\, H_A d^3x$$

where

$$H_A d^3x = -D_P \Psi_A + R^B\wedge L^C\, Q_{CBA} \tag{3.6}$$

In (3.6) D_P denotes the Poincare' covariant derivative

$$D_P W_A = {}^3dW_A - q^B C_{BA}{}^{D\wedge} W_D$$

and R^C the restrictions to Σ of the curvatures.

For pure gravity the third term of (3.5) is clearly not present, and the Hamiltonian agrees with that of [8]. For N = 1 supergravity however we find*

$$H_a = 2\,(-n_a H_{\times\lrcorner} + L_a{}^i H_{\times i})$$
$$= 2\,H_{a(PILATI)} \tag{3.7a}$$

$$H_{ab} = 2\,J_{ab(PILATI)} \tag{3.7b}$$

$$H_\alpha = 2\,S_{\alpha(PILATI)} +$$
$$i\varepsilon^{ijk}\gamma_5\gamma_\lrcorner\psi_k\bar{\psi}_i\gamma_\lrcorner\psi_j g^{-1/2}/4 \tag{3.7c}$$

The original factor of 2 between our Lagrangian and that of Pilati is responsible for the overall factors of 2 in equation (3.7). In (3.7c) we have found a term, not present in [12], but present in the CVF calculation. The Hamiltonian given by (3.5) and (3.6) generates the time evolution of a generic functional of the q^A and π_A by

$$[H, F(y)] \sim \dot{F}(y) \tag{3.8}$$

In (3.8) [,] denotes the Poisson brackets of two forms, computed by expanding them in their bases, taking the bracket of the components, and reassembling the forms. Equation (3.8) is a particular case of the CVF Lie derivative defined by

---

*For typographical reasons we use $H_\times$ to denote $\widetilde{H}$ of[8,12].

$$\pounds_\varepsilon O(x) = \varepsilon^A \partial_A O(x) + \int \varepsilon^A(y)[H_A(y), O(x)]d^3y \qquad (3.9)$$

whose commutator is

$$\pounds_\varepsilon \pounds_\eta - \pounds_\eta \pounds_\varepsilon = \pounds_{(\varepsilon,\eta)}$$

When $\varepsilon^A = q^A_0$ equations (3.8) and (3.9) coincide. The CCF Lie derivative given by (2.9) and the CVF Lie derivative given by (3.9) also coincide when we use the following integral relationship between form and Poisson brackets.

$$(-1)^{a+1}\int_\Sigma \alpha\wedge(A,B)\wedge\beta = \int\int_{\Sigma\times\Sigma} \alpha(x)\wedge[A(x),B(y)]\wedge\beta(y)$$

for test forms α, β of degree 3-a,3-b respectively.

Finally, the constraint algebra of the CVF, as in [8] and [13], can now be written as

$$[H_A(x),\ H_B(y)] = \Lambda_{AB}{}^C\ H_C\ \delta^3(x,y)$$

with $\Lambda_{BC}{}^A = R_{BC}{}^A - C_{BC}{}^A$

As a final comment, we note that the CCF can be extended to non-trivial free differential algebras (algebras in which not all the fields q^A are 1-forms), for example the 6-dimensional theory of supergravity [6],[14].

REFERENCES

[1] A.D'Adda, J.E.Nelson and T.Regge, Ann.Phys.165 (1985) 384

[2] J.E.Nelson and T.Regge, Ann.Phys.166 (1986)

[3] A.Lerda, J.E.Nelson and T.Regge, Phys.Lett.161B (1985) 294

[4] A.Lerda, J.E.Nelson and T.Regge, Phys.Lett.161B (1985) 297

[5] P.A.M.Dirac, Canad.J.Math.2 (1950) 129;"Lectures on Quantum Mechanics" Academic Press, New York 1965; A.Hanson, T.Regge and C.Teitelboim "Constrained Hamiltonian Systems", Accademia Nazionale dei Lincei, Roma 1976.

[6] T.Regge, "The Group Manifold Approach to Unified Gravity" -Lectures presented at the Les Houches Summer School, Session XV (1983),(North Holland 1984);

L.Castellani, R.D'Auria and P.Fre' "Seven lectures on the Group Manifold Approach to Supergravity " in 'Supersymmetry and Supergravity', Proc. of Karpacz Winter School (1983), (World Scientific 1983).

[7] P.A.M.Dirac, "Recent Developments in General Relativity", Pergamon, New York 1962.

[8] J.E.Nelson and C.Teitelboim, Phys.Lett.69B (1977) 81; Ann.Phys.116 (1978) 1.

[9] L.Castellani, P.van Nieuwenhuizen and M.Pilati, Phys.Rev.D26 2 (1982) 352.

[10] S.Deser and C.J.Isham, Phys.Rev.D14 (1976) 2505.

[11] S.Deser, J.H.Kay and K.S.Stelle,Phys.Rev.D16 (1977) 2448.

[12] M.Pilati, Nucl.Phys. B132 (1978) 138.

[13] C.Teitelboim, Phys.Rev.Lett.38 (1977) 1106.

[14] R.D'Auria, P.Fre' and T.Regge, Phys.Lett.128B (1983) 44.

KALUZA-KLEIN APPROACH TO THE HETEROTIC STRING

M.J. Duff*) and B.E.W. Nilsson+)

CERN - Geneva

and

C.N. Pope

Blackett Laboratory, Imperial College
London SW7 2BZ, U.K.

A B S T R A C T

Both the gravitational and gauge field sectors of the $d = 10$ superstring, including the Chern-Simons terms, may be obtained from the purely gravitational bosonic string in $d = 10 + \dim G$ via a spontaneous compactification on the group manifold G with $\langle H_{mnp} \rangle$ the parallelizing torsion. The Kaluza-Klein relation between the coupling constants $\alpha' g^2 \sim \kappa^2$ is that of the heterotic string. Consistency of the Kaluza-Klein ansatz is essential and reduces the symmetry from $G_L \times G_R$ to G_L.

---

*) On leave of absence from the Blackett Laboratory, Imperial College, London SW7 2BZ, United Kingdom.

+) On leave of absence from the Institute of Theoretical Physics, S-41296 Goteborg, Sweden.

The most appealing feature of Kaluza-Klein theories is that what we perceive to be internal symmetries are really space-time symmetries in a higher dimension. In particular, it is not necessary to postulate the separate existence of Yang-Mills fields, they are an automatic consequence of gravity [1].

It is ironic therefore that the recent spectacular successes of superstrings [2,3,4] seem to ignore this beautiful concept. For although formulated in 10 rather than 4 dimensions, the Yang-Mills fields of $E_8 \times E_8$ or SO(32) are present as primary fields already in the d=10 Lagrangian. Moreover, the favoured compactification to d=4 [4] yields no Kaluza-Klein gauge fields since Calabi-Yau manifolds have no continuous symmetries.

In this paper we show, in fact, that Kaluza-Klein can provide an explanation for the gauge fields of the heterotic string [3] , as might have been suspected from the Kaluza-Klein like relation between the Yang-Mills coupling constant g ,the gravitational constant κ and the slope parameter α' , namely $\alpha' g^2 \sim \kappa^2$. (This is to be contrasted with the corresponding relation $g^2 \sim \kappa \alpha'$ of the open string [2] which appears to admit no such K-K interpretation). The Yang-Mills gauge group will, as usual, be a subgroup of the higher dimensional general coordinate group. Although the theory compactifies on a group manifold G, however, our Kaluza-Klein ansatz will involve only the gauge bosons of G_L and not those of the full isometry group $G_L \times G_R$. As discussed elsewhere [1,5,6,7,8] this is essential for the consistency of the Kaluza-Klein ansatz. Surprisingly, consistency is achieved without the introduction of Kaluza-Klein scalars.

Our starting point is the off-shell effective action Γ of Fradkin and Tseytlin [9] for the infinite set of local fields corresponding to the modes of a free bosonic string spectrum. Γ is the generating functional of all possible off-shell scattering amplitudes on arbitrary backgrounds. In the case of closed oriented strings Γ is defined (in Euclidean signature) by

$$\Gamma[\hat{\phi}(x), \hat{g}_{MN}(x), \hat{B}_{MN}(x), \cdots] = \sum_{\chi=2,0,-2\ldots} e^{\sigma\chi} \int [d\gamma_{ab}][dx^M]\, e^{-S} \quad (1)$$

where σ is a dimensionless parameter and where

$$S = \int_{M_2} \frac{d^2z}{4\pi\alpha'} \Big\{ \sqrt{\gamma}\, \gamma^{ab} \partial_a x^M \partial_b x^N \hat{g}_{MN}(x) + \alpha' \sqrt{\gamma}\, R(\gamma)\, \hat{\phi}(x) + \varepsilon^{ab} \partial_a x^M \partial_b x^N \hat{B}_{MN}(x) + \cdots \Big\} \quad (2)$$

Here $x^M(z)$ defines the embedding of the two dimensional string worldsheet M_2 in a spacetime M_d (M,N=1,...,d); z^a (a=1,2) are the coordinates on M_2; $R(\gamma)$ is the curvature scalar of the metric γ_{ab}. The graviton $\hat{g}_{MN}(x)$, the antisymmetric tensor $\hat{B}_{MN}(x)$ and the dilaton $\hat{\phi}(x)$ correspond to the massless modes of the bosonic string spectrum. The dots refer to terms describing the higher spin massive modes and the scalar 'tachyon'. In writing the action (2), and in the remainder of the paper, we revert to Minkowski signature for both the worldsheet and for spacetime.

For string consistency, the two dimensional theory must be conformally invariant and hence the two-dimensional worldsheet stress tensor must be traceless. The general structure of the trace is given by

$$\Theta_a{}^a = \beta^{\phi} \sqrt{\gamma}\, R(\gamma) + \beta^{g}_{MN} \sqrt{\gamma}\, \gamma^{ab} \partial_a x^M \partial_b x^N + \beta^{B}_{MN}\, \varepsilon^{ab} \partial_a x^M \partial_b x^N \quad (3)$$

where β^{ϕ}, β^{g} and β^{B} are local functionals of $\hat{\phi}(x)$, $\hat{g}_{MN}(x)$ and $\hat{B}_{MN}(x)$. As noted by Polyakov [10] the first term is just the two-dimensional gravitational trace anomaly [11] while the second and third terms are present even in flat space and arise from the self-interaction. The β functions in the tree approximation ($\chi=2$) have been calculated by Callan et al. [12]. They find

$$16\pi^2\beta^{\phi} = \frac{d-26}{3\alpha'} + \left[4(\partial\hat{\phi})^2 - 4\hat{\Box}\hat{\phi} - \hat{R} + \frac{1}{12}\hat{H}^2\right] + O(\alpha') \quad (4)$$

$$\beta^{g}_{MN} = \hat{R}_{MN} - \frac{1}{4}\hat{H}_{M}{}^{PQ}\hat{H}_{NPQ} + 2\hat{\nabla}_M\partial_N\hat{\phi} + O(\alpha') \quad (5)$$

$$\beta^{B}_{MN} = \hat{\nabla}_P\hat{H}^{P}{}_{MN} - 2\hat{H}^{P}{}_{MN}\partial_P\hat{\phi} + O(\alpha') \quad (6)$$

where $\hat{H}_{MNP} = 3\partial_{[M}\hat{B}_{NP]}$ and satisfies

$$\partial_{[Q}\hat{H}_{MNP]} = 0 \quad (7)$$

The absence of a trace anomaly, i.e. the vanishing of β^{ϕ}, β^{g} and β^{B}, is therefore just equivalent to the Einstein-matter field equations obtained from the effective action

$$\Gamma \sim \int d^dx\sqrt{-\hat{g}}\,e^{-2\hat{\phi}}\left[\frac{d-26}{3\alpha'} - \hat{R} - 4(\partial\hat{\phi})^2 + \frac{1}{12}\hat{H}^2 + O(\alpha')\right] \quad (8)$$

One obvious solution of the field equations corresponds to

$\langle\hat{\phi}\rangle$ = constant, $\langle\hat{g}_{MN}\rangle = \eta_{MN}$ and $\langle\hat{H}_{MNP}\rangle = 0$, but this is valid only for d=26. For $d > 26$, the cosmological term in (4) obliges us to look for solutions in which some of the dimensions are compactified and we can now follow the traditional Kaluza-Klein interpretation. Accordingly we split the indices $x^M = (x^\mu, y^m)$ where x^μ (μ=1,...,n) refers to spacetime and y^m (m=1,...,k) refers to the extra dimensions. One solution which suggests itself corresponds to the case where the extra dimensions are a group manifold G with k= dim G. In this case

$$\langle \hat{g}_{mn}(x,y)\rangle = g_{mn}(y) = \frac{1}{C_A} c_{mpq} c_n{}^{pq} \tag{9}$$

where c_{mnp} are the structure constants of the group, converted to world indices, and C_A is the second-order Casimir in the adjoint representation. This will indeed be a solution provided

$$\langle \hat{\phi}(x,y)\rangle = \text{constant} \tag{10}$$

and

$$\langle \hat{H}_{mnp}(x,y)\rangle = H_{mnp}(y) = m\, C_{mnp} \tag{11}$$

where m is a constant with dimensions of mass. The left-invariant Killing vector fields $K^i{}_m(y)$ are normalized by

$$g^{mn} K^i{}_m K^j{}_n = \delta^{ij} \tag{12}$$

and satisfy

$$[K^i, K^j] = m\, c^{ij}{}_k K^k \tag{13}$$

where $K^i = K^{im}\partial_m$. With these normalizations H_{mnp} corresponds to the parallelizing torsion[1]. The curvature is given by

$$\langle R_{mnpq} \rangle = \frac{m^2}{4} C_{mn}{}^{\ell} C_{pq\ell} \tag{14}$$

$$\langle R_{mn} \rangle = \frac{m^2}{4} C_A \langle g_{mn} \rangle \tag{15}$$

$$\langle R_K \rangle = \frac{m^2}{4} C_A R \tag{16}$$

where R_k is the curvature scalar of the k-dimensional group manifold. Substituting (9)-(11) into the field equations (4)-(7) we see that (5)-(7) are identically satisfied while (4) yields

$$d - 26 = k\alpha' m^2 C_A/2 + O(\alpha'^2) \tag{17}$$

Our aim is to obtain the bosonic sector of the heterotic string in ten dimensions. Neglecting the higher derivative terms, this corresponds to the bosonic sector of the Chapline-Manton [14,15] N=1 supergravity with fields $\phi(x)$, $g_{\mu\nu}(x)$, $H_{\mu\nu\rho}(x)$ coupled to Yang-Mills fields $A^i{}_\mu(x)$ with $G = E_8 \times E_8$ or SO(32). This means that d=506 and k=496. We shall now write down a Kaluza-Klein ansatz which achieves this goal, but for the sake of generality we allow G and the dimensions n and d to be arbitrary. For the scalar, we write

$$\hat{\phi}(x,y) = \text{constant} + \phi(x) \tag{18}$$

For the metric ansatz, we employ the 1-forms $\hat{e}^A = \hat{e}^A{}_M\, dx^M$ for which

$$\hat{g}_{MN} = e_M{}^A e_N{}^B \eta_{AB} \tag{19}$$

and write

$$\hat{e}^\alpha(x,y) = e^\alpha{}_\mu(x)\, dx^\mu \tag{20}$$

$$\hat{e}^a(x,y) = e^a{}_m(y)\, dy^m - m^{-1} K^{ia}(y)\, A^i{}_\mu(x)\, dx^\mu \tag{21}$$

The quantities $\phi(x)$, $e^\alpha{}_\mu(x)$ and $A^i{}_\mu(x)$ will be interpreted as the scalar, vielbein and Yang-Mills gauge potential in n-dimensional spacetime. We must also make an ansatz for $\hat{H}_{MNP}(x,y)$. Experience with antisymmetric tensors in d=11 supergravity [1,16] suggests that this must also involve the Yang-Mills fields as well as the d=10 three-form $H_{\mu\nu\rho}(x)$. Accordingly, we write

$$\hat{H}_{\alpha\beta\gamma}(x,y) = H_{\alpha\beta\gamma}(x) \tag{22}$$

$$\hat{H}_{\alpha\beta c}(x,y) = -m^{-1} F^i{}_{\alpha\beta}(x)\, K^i{}_c(y) \tag{23}$$

$$\hat{H}_{\alpha bc}(x,y) = 0 \tag{24}$$

$$\hat{H}_{abc}(x,y) = m\, c_{abc} \tag{25}$$

where $F^i{}_{\alpha\beta}(x)$ is the Yang-Mills field strength

$$F^i{}_{\mu\nu} = \partial_\mu A^i{}_\nu - \partial_\nu A^i{}_\mu + c^i{}_{jk} A^j{}_\mu A^k{}_\nu \tag{26}$$

It is important to note that (α, a) are tangent space indices and (μ, m) are world indices and that $c_{abc} = c_{ijk}\, K^i{}_a K^j{}_b K^k{}_c$.

Several comments are now in order. That equations (18) to (25) correspond to the correct Kaluza-Klein ansatz can only be justified a posteriori. First they must be mathematically consistent and, as emphasized elsewhere [1,5,6,7,8] this is a highly non-trivial requirement. A 'consistent' ansatz is one for which all solutions of the n-dimensional theory are solutions of the original d-dimensional theory. For a generic Kaluza-Klein theory with homogeneous extra dimensions, a consistent truncation is achieved by an ansatz which retains all those fields and only those fields invariant under a transitively acting subgroup of the isometry group [6] . In particular, for group manifolds the gauge bosons are only those of G_L and not those of the full isometry group $G_L \times G_R$ [5] . In certain cases, it may not be necessary to retain <u>all</u> the G_R singlets. Indeed, surprisingly, our ansatz (18)-(25) is consistent in spite of omitting the usual Kaluza-Klein scalars. Secondly, equations (18) to (25) should be physically consistent in that we recover the desired n-dimensional theory. In our case that means, for n=10, that we recover the low energy limit of the heterotic string [3] , i.e. the Chapline-Manton N=1 supergravity coupled to Yang-Mills with gauge group G including, of course, the correct Yang-Mills Chern-Simons terms [14,15] .

Note that we have not specified the ground-state values of the spacetime fields $\phi(x)$, $g_{\mu\nu}(x)$, $H_{\mu\nu\rho}(x)$ or $A^i{}_\mu(x)$. Since, by definition, a 'consistent' ansatz is one for which all solutions of the n-dimensional field equations are solutions of the original d-dimensional field equations, it is not necessary at this stage to single out any preferred configuration such as $g_{\mu\nu} = \eta_{\mu\nu}$, $\phi = H_{\mu\nu\rho} = A^i{}_\mu = 0$. (For example, there can be other solutions such as Calabi-Yau manifolds.)

We now substitute the ansatze (18) to (25) into the field equations (4) to (7). Most of the relevant calculations may be found in Appendix A of [1] . The $\beta^g_{\mu\nu}$ equation becomes

$$R_{\mu\nu} - \frac{1}{4} H_\mu{}^{\rho\sigma} H_{\nu\rho\sigma} + 2 \nabla_\mu \partial_\nu \phi - m^{-2} F^i{}_{\mu\rho} F^i{}_\nu{}^\rho = 0 \qquad (27)$$

The $\beta^g_{\mu n}$ equation becomes

$$D_\rho F^i{}_\mu{}^\rho - \frac{m^{-2}}{2} H_\mu{}^{\rho\sigma} F^i{}_{\rho\sigma} - 2 F^i{}_\mu{}^\rho \partial_\rho \phi = 0 \qquad (28)$$

where D_ρ is the Yang-Mills covariant derivative. Note that since our ansatz is invariant under the transitive action of G_R, all the y dependence of the ansatz disappears from the n-dimensional equations. Note also that half the Yang-Mills contribution to the energy-momentum tensor comes from the standard metric ansatz and an equal contribution from the antisymmetric tensor ansatz (2). The β^g_{mn} equation is satisfied identically. (This is the equation in which one might have expected to encounter inconsistencies by omitting the Kaluza-

Klein scalars $S^{ij}(x)$ since in a generic theory one obtains an equation of the form $\Box S^{ij} \sim F^{i}{}_{\mu\nu} F^{j\,\mu\nu}$. In our case, however, the inclusion of the Yang-Mills fields in the ansatz (23) for $\hat{H}_{\alpha\beta c}$ exactly cancels this right hand side and hence S^{ij} may consistently be set to zero).

The $\beta^{B}_{\mu\nu}$ equation becomes

$$\nabla_{\rho} H^{\rho}{}_{\mu\nu} - 2 H^{\rho}{}_{\mu\nu} \partial_{\rho} \phi = 0 \tag{29}$$

the $\beta^{B}_{\mu n}$ equation simply reproduces (28); while the β^{B}_{mn} equation is satisfied identically.

Next we consider the β^{ϕ} equation, which becomes

$$4 \partial_{\mu}\phi \partial^{\mu}\phi - 4\Box\phi - R + \frac{1}{12} H_{\mu\nu\rho} H^{\mu\nu\rho} + \frac{m^{-2}}{2} F^{i}{}_{\mu\nu} F^{i\,\mu\nu} = 0 \tag{30}$$

after using (17). Finally, equation (7) becomes

$$\partial_{[\mu} H_{\nu\rho\sigma]} + \frac{3m^{-2}}{2} F^{i}{}_{[\mu\nu} F^{i}{}_{\rho\sigma]} = 0 \tag{31}$$

or, in terms of differential forms,

$$dH + m^{-2} F^{i} \wedge F^{i} = 0 \tag{32}$$

In the case n=10, the reader will no doubt recognize equations (27)-(32) as the bosonic sector of the Chapline-Manton N=1 supergravity with coupling constant κ coupled to N=1 Yang-Mills with coupling constant g, where we make the identification

$$g^2 = \kappa^2 m^2 \tag{33}$$

It is perhaps remarkable that in (32) we correctly reproduce the Yang-Mills Chern-Simons term in n dimensions even though there was no such term in d dimensions. (The appearance of the $F^i \wedge F^i$ term on reducing from d to n dimensions is known in the mathematical literature as a transgression).

Equation (33) is of the standard Kaluza-Klein form since m^{-1} is just the radius of the extra dimensions. As in d=11 supergravity, however, the numerical coefficient will differ from that obtained in pure gravity Kaluza-Klein theories [17] owing to the Yang-Mills content of the antisymmetric tensor ansatz [16] .

If, on the other hand, we compare equations (27)-(32) with the low energy limit ($\lesssim$ 2 derivatives) of the heterotic string [12] , we find

$$m^2 \alpha' = 1 \tag{34}$$

Hence we recover the well-known relation [3] between κ , g and α' of the heterotic string

$$\alpha' g^2 = \kappa^2 \tag{35}$$

One major difference from generic Kaluza-Klein compactifications is that we can obtain an n-dimensional theory with zero cosmological constant. The particular $\hat{\phi}$ dependence of the action (8) ensures that the Einstein equation (5) contains no cosmological term, while the cosmological term in the d-dimensional $\hat{\phi}$ equation cancels out in the n-dimensional ϕ equation.

By ignoring the $O(\alpha')$ terms in the d-dimensional equations we have obtained in this paper the Chapline-Manton 10 dimensional equations including the Yang-Mills Chern-Simons term. In a subsequent paper we shall show that including the $\alpha' \hat{R}_{MNPQ} \hat{R}^{MNPQ} + \ldots$ terms in d dimensions, we can obtain the modified field equations of Candelas et al. [4] in 10 dimensions with their $\alpha'(R_{\mu\nu\rho\sigma} R^{\mu\nu\rho\sigma} - F^i{}_{\mu\nu} F^{i\mu\nu})$ terms. This Lagrangian will contain not only the Chapline-Manton Yang-Mills Chern-Simons term but also the Green-Schwarz [2] gravitational Chern-Simons term ensuring both Yang-Mills and gravitational anomaly cancellations in the case G = $E_8 \times E_8$ or SO(32); i.e. equation (32) becomes

$$dH + \alpha'(F^i \wedge F^i - R^{ab} \wedge R^{ab}) = 0 \tag{36}$$

One is then in a position to seek a second compactification from n=10 to n=4 of the kind discussed by Candelas et al. [4] . If so desired, of course, one could descend directly from d=506 to n=4 in which case the embedding of the Yang-Mills potential in the spin connection [18] employed by Candelas et al. [4] could be reinterpreted as a purely gravitational phenomenon. Since, in our compactification from d=506 to n=10 the $\hat{H}_{mnp}$ background was non-trivial, moreover, there seems to be no compelling reason to insist that it be zero in going from n=10 to n=4. An interesting question is whether one can attach physical significance to compactifications which go directly from d=506 to n=4 and which do not admit any intermediate n=10 interpretation. The answer to this question would seem to require a better understanding of how the fermions are incorporated.

It is interesting to note that whereas in d dimensions the α' expansion corresponds to an expansion in numbers of derivatives (i.e. zero slope limit = low energy limit), in n dimensions this correspondence breaks down. For example, in the present paper we obtain $\alpha' F^i{}_{\mu\nu} F^{i\mu\nu}$ terms in n dimensions even though we ignored $\alpha' \hat{R}_{MNPQ} \hat{R}^{MNPQ}$ terms in d dimensions. In fact we conjecture that the group manifold with parallelizing torsion solves the field equations to all orders in α' and that the Kaluza-Klein interpretation applies to the full string theory and not merely to its field theory limit.

All this, of course, is entirely consistent with previous work on strings in curved space and non-linear σ models [19-24]. The Lagrangian (2) with the fields set equal to their background values, i.e. $\hat{\phi}$ = constant; $\hat{g}_{mn}$ the metric on the group manifold; and $\hat{H}_{mnp}$ the parallelizing torsion is nothing but the non-linear σ model with Wess-Zumino term. According to Witten [20], this model has vanishing β functions provided $\alpha' m^2 = 1$, which is just equation (34). Furthermore, the critical dimension of these theories is given by [23,24]

$$d - 26 = \frac{k C_A}{2 + C_A} \qquad (37)$$

(e.g. with G = SO(32), k=496, C_A = 60 and hence d=506 and n=10). On the other hand, our equation (17) yields

$$d - 26 = \frac{k C_A}{2} + \ldots \tag{38}$$

on using (34). This suggests that solving the β^{ϕ} equation to all orders in α' would yield the exact equation (37).

Note, however, that the ordinary bosonic string formulated on the group manifold G would yield states belonging to representations of the full isometry group $G_L \times G_R$ in contrast to the heterotic string where the states belong only to representations of G_L. In our Kaluza-Klein formulation this is taken care of by the consistency of the truncation in the Kaluza-Klein ansatz which demands that we retain only G_R singlets.

In this Kaluza-Klein picture, the $E_8 \times E_8$ or SO(32) symmetry of the heterotic string is a subgroup of the general coordinate group in d=506. This is to be contrasted with the Frenkel-Kac-Goddard-Olive [25] approach where the gauge symmetry emerges from compactification of the d=26 string on the maximal torus T^{16} [26,3,27]. No doubt the two approaches are ultimately equivalent and it will be interesting to establish the appropriate dictionary. In particular one can ask whether the n=10 fermions can be incorporated in any natural way into the d=506 bosonic theory. For example, one has the feeling that the gravitational, Yang-Mills and mixed anomaly cancellations of Green and Schwarz in n=10 [2] may have a simpler pure gravity interpretation in d=506.

ACKNOWLEDGEMENTS

We are grateful to C. Isham, D. Olive, M. Perry and E. Sezgin for discussions, and to the organizers and participants of the Cambridge Workshop on Supersymmetry and its Applications, June-July 1985.

FOOTNOTES

(1) Setting $H_{mnp} \sim c_{mnp}$ was first attempted (albeit unsuccessfully) in [13] .

(2) Similarly in d=11 supergravity [5] one quarter of the Yang-Mills stress tensor comes from the standard metric ansatz and three quarters from the antisymmetric tensor ansatz.

REFERENCES

1. For a recent review, see M.J. Duff, B.E.W. Nilsson and C.N. Pope "Kaluza-Klein Supergravity", Phys. Rep. (to appear)
2. M.B. Green and J.H. Schwarz, Phys. Lett. 149B, 117 (1984)
3. D.J. Gross, J.A. Harvey, E. Martinec and R. Rohm, Phys. Rev. Lett. 54, 502 (1985)
4. P. Candelas, G.T. Horowitz, A. Strominger and E. Witten, UCSB-ITP preprint NSF-ITP-84-170
5. M.J. Duff, B.E.W Nilsson, C.N. Pope and N.P. Warner, Phys. Lett. 149B, 90 (1984)
6. M.J. Duff and C.N. Pope, UCSB-ITP preprint NSF-ITP-84-166, to appear in Nucl. Phys. B
7. C.N. Pope, Class. and Qu. Grav. 2, L77 (1985)
8. M.J. Duff, Proceedings of the Jerusalem Winter School on "Physics in Higher Dimensions", Jerusalem, December 1984, Eds T. Piran and S. Weinberg; Proceedings of the Kyoto Summer School "Quantum Gravity and Cosmology", Kyoto, May 1985, Eds T. Inami and H. Sato
9. E.S. Fradkin and A.A. Tseytlin, Lebedev Inst. preprint N261 (1984); A.A. Tseytlin, Lebedev Inst. preprint N153 (1985)
10. A.M. Polyakov, Phys. Lett. 103B, 207 (1981)
11. D.M. Capper and M.J. Duff, Nuovo Cim. 23A, 173 (1974); S. Deser, M.J. Duff and C.J. Isham, Nucl. Phys. B114, 29 (1976); M.J. Duff, Nucl. Phys. B125, 334 (1977)
12. C.G. Callan, D. Friedan, E.J. Martinec and M.J. Perry, Princeton University preprint (1985)
13. M.J. Duff, P.K. Townsend and P. van Nieuwenhuizen, Phys. Lett. 122B, 232 (1983)
14. E. Bergshoeff, M. de Roo, B. de Wit and P. van Nieuwenhuizen, Nucl. Phys. B195, 97 (1982)
15. G.F. Chapline and N.S. Manton, Phys. Lett. 120B, 105 (1983)
16. M.J. Duff and C.N. Pope, in "Supersymmetry and Supergravity 82" Eds S. Ferrara, J.G. Taylor and P. van Nieuwenhuizen (World Scientific Publishing, 1983)
17. S. Weinberg, Phys. Lett. 124B, 265 (1983)

18. J.M. Charap and M.J. Duff, Phys. Lett. 69B, 445 (1977); See also F. Wilczek in "Quark Confinement and Field Theory", Eds Stump and Weingarten (Wiley Interscience, NY 1977)

19. C. Lovelace, Phys. Lett. 135B, 75 (1984)

20. E. Witten, Comm. Math. Phys. 92, 455 (1984)

21. T. Curtwright and C. Zachos, Phys. Rev. Lett. 53, 1799 (1984)

22. P. Howe and G. Sierra, Phys. Lett. 148B, 451 (1984)

23. D. Nemeschansky and S. Yankielowicz, Phys. Rev. Lett. 54, 620 (1984)

24. E. Bergshoeff, S. Randjbar-Daemi, Abdus Salam, H. Sarmadi and E. Sezgin, ICTP Trieste preprint IC/85/51

25. I. Frenkel and V.G. Kac, Inv. Math. 62, 23 (1980); P. Goddard and D. Olive, in "Workshop on Vertex Operators in Mathematics and Physics", Berkeley (1983)

26. P.G.O. Freund, Phys. Lett. 151B, 387 (1985)

27. F. Englert and A. Neveu, CERN preprint TH 4168 (1985)

KALUZA-KLEIN SPECTRA ON A CONTORTED VACUUM *

Claudio A. Orzalesi**

Dipartimento di Fisica dell'Università, 43100 Parma, and

Ambasciata d'Italia, 1601 Fuller St.NW, Washington DC 20009

1. INTRODUCTION

Modern approaches[1] to Kaluza-Klein (KK) theories[2,3] rely on the phenomenon of spontaneous compactification, whereby the physical ground-state metric corresponds to a space M_4xV_K , where M_4 denotes the ordinary Minkowski spacetime and V_K is a K-dimensional compact internal space of very small size. Once spontaneous compactification sets in, the zero-mass fluctuations of the metric should yield the effective low-energy approximation in quantum path integrals. Such zero-modes contain the Einstein metric (describing ordinary gravity), Yang-Mills potentials and possibly also Jordan-Thiry scalars.

According to the common lore, the gauge group G_{YM} of these YM potentials must coincide with the symmetry group G_{INT} of the internal space V_K. However, we shall present examples[4] where G_{YM} is a proper subgroup of G_{INT}. This point deserves further comment, because there exists a "general argument" as well as a "direct proof" in support of the common lore.

The general argument is that a global symmetry of the background becomes a local gauge symmetry for the local fluctuations of the metric, hence such fluctuations must contain the YM potentials of G_{INT}. Here, the loophole is that these fluctuations will be "massless" in 4+K dimensions, but may still correspond to massive fields in 4 dimensions.

The direct proof goes as follows: let $K_i = K_i^m \partial_m$ be the

*. Supported in part by Istituto Nazionale di Fisica Nucleare and by Consiglio Nazionale delle ricerche.

**. On Leave of absence from Dipartimento di Fisica, dell'Università, 43100 Parma.

Killing vectors of V_K, and consider the KK ansatz for the metric*

$$\gamma_{\mu\nu} = g_{\mu\nu}(x), \quad \gamma_{\mu m} = A_\mu^{\ i}(x)K_{im}(y), \quad \gamma_{mn} = g_{mn}(y), \tag{1.1}$$

where $g_{mn}(y)$ is a G_{INT}-invariant metric on V_K. Then, compute the Riemann scalar $\overset{\circ}{R}$ for this metric and take the average of $\overset{\circ}{R}$ over V_K. The resulting effective four-dimensional Lagrangian contains the YM Lagrangian for the potentials A^i_μ with G_{INT} as gauge group**.

This proof is faulty because $\overset{\circ}{R}$ is not the only relevant piece of the Lagrangian in 4+K dimensions. As a matter of fact, for nontrivial V_K the ansatz (1.1) does not correspond to zero-modes of the homogeneous linearized Einstein equations, $\overset{\circ}{R}_{MN}=0$, which would determine the small fluctuations if $\overset{\circ}{R}$ were the Lagrangian. In order to obtain zero-modes with Killing vectors, as in (1.1), the field equations must contain additional terms. Additional terms are also needed in order that the bacground metric of $M_4 x V_K$,

$$\overline{\gamma}_{\mu\nu} = \eta_{\mu\nu}, \quad \overline{\gamma}_{\mu m} = 0, \quad \overline{\gamma}_{mn} = g_{mn}(y) \ , \tag{1.2}$$

be a solution of the field equations, see ref.[5].

We now summarize the features of the models to be discussed. The models are based on the Einstein-Cartan theory[6] of gravity in D=4+K dimensions.

To achieve spontaneous compactification with vanishing cosmological constant (in 4+K and in 4 dimensions) we rely[7] on nontrivial values of torsion in the ground state. In the models

---

*. Notation: Greek, lower-case Latin and upper-case Latin indices refer respectively to M_4, to V_K and to the total (4+K) dimensional space. On $M_4 x V_K$ we label points X as (x,y), with coordinates $(X^M)=(x^\mu,y^m)$. The metric in 4+K dimensions is denoted by $\underset{\sim}{\gamma}$.

**. This calculation of the dimensionally reduced Lagrangian can similarly be carried out for the more general KK ansatz[3], where Jordan-Thiry scalars are included in the fluctuations of the internal components of the metric.

considered, V_K is the space of a compact nonabelian group G and $g_{mn}(y)$ is the Killing-Cartan metric on G. The symmetry group of the background metric is the Poincarè group (in four dimensions) times GxG. For the background torsion, we take a Poincarè invariant "parallelizing" torsion on G, which gives vanishing Riemann-Cartan curvature on G and has GxG as its invariance group.

In our models, matter is described by a Dirac field minimally coupled to gravity in D=4+K dimensions, and the background parallelizing torsion leads to zero-modes for the Dirac field[5]. These Dirac zero-modes fall[5] into the (singlet, spinor) representation og GxG, hence they decouple from half of the YM fields of GxG. Consistently with this result, we find[4] that the decoupled YM potentials belong to massive modes of the metric. Therefore, the YM gauge group of the zero-modes is G rather than the "expected" GxG.

A zero-mode truncation of the theory can consistently be made for the small fluctuations. This truncation leads in four dimensions to an effective theory of Einstein gravity, Jordan-Thiry scalars and YM fields (with gauge group G) minimally coupled to a Dirac field in the spinor representation of G.

The zero-modes of the metric on the contorted vacuum reproduce the old-fashioned KK ansatz[3], which implements[3,7] a requirement of G-invariance for the metric on M_4xG. However, now we have a physical rationale for this ansatz, because it indeed yields the zero modes, while the older approaches offered no explanation for why the KK ansatz selected physically relevant modes. Furthermore, our approach avoids the inconsistencies[5,7] which in older approaches arose from the cosmological constant induced by the Riemannian curvature of the internal space.

It should also be recalled that the torsion compactification mechanism used here relies[5] on the occurrence of a spin condensate in vacuo. Since the spin density is a fermion bilinear, this mechanism is quantum-mechanical and could originate from the quantum dynamics of torsion[8].

2. FIELD EQUATIONS

As in ref.[5], we deal with the Einstein-Cartan (EC) theory[6] in D=4+K dimensions and we use a first-order formulation, where the metric $\underset{\sim}{\gamma}$ and the linear connection $\underset{\sim}{\omega}$ are treated as independent variables. We use capital Latin indices to label components in an arbitrary coframe $\theta^A = \theta^A{}_M dx^M$ or dual frame $\partial_A = \theta_A{}^M \partial_M$, A=1,...,D, so that, e.g.,

$$\underset{\sim}{\gamma} = \gamma_{AB}\,\theta^A\theta^B = \gamma_{MN}\,dx^M dx^N\ . \tag{2.1}$$

For consistency, since we do include spinor fields we must require that the connection $\underset{\sim}{\omega}$ be metric: $\nabla\underset{\sim}{\gamma} = 0$. The metricity constraint is equivalent to the following relation between the connection, the metric and the torsion $\underset{\sim}{T}$:

$$\omega_{CA}{}^B = \overset{\circ}{\omega}_{CA}{}^B + K_{CA}{}^B\ , \tag{2.2}$$

where $\overset{\circ}{\underset{\sim}{\omega}}$ is the Levi-Civita connection of $\underset{\sim}{\gamma}$ and the contorsion $\underset{\sim}{K}$ is defined by

$$K_{CA}{}^B = T_{CA}{}^B/2 - \gamma^{BD}\gamma_{E(A}T_{C)D}{}^E\ . \tag{2.3}$$

It should be kept in mind that the torsion is in principle independent of the metric, while the contorsion depends on $\underset{\sim}{\gamma}$, as indicated in (2.3). Therefore, the independent variables to be varied in the action principle are the metric and the torsion*.

The EC action is given by

$$\mathcal{A} = \mathcal{A}_{Grav} + \mathcal{A}_{Mat} = \int (R/2\hat{\kappa} + \mathcal{L}_{Mat})\,|\gamma|^{1/2}\,\theta^1\wedge\cdots\wedge\theta^D\ , \tag{2.4}$$

where we choose the Dirac action for matter,

$$\mathcal{L}_{Mat} = \mathcal{L}_D = (\overline{\Psi}\Gamma^A\nabla_A\Psi - (\nabla_A\overline{\Psi})\Gamma^A\Psi)/2\ , \tag{2.5}$$

and R is the Riemann-Cartan curvature scalar of $\underset{\sim}{\omega}$:

$$d\omega_A{}^B + \omega_C{}^B\wedge\omega_A{}^C = R_{CDA}{}^B\,\theta^C\wedge\theta^D/2,\quad R_{AB} = R_{CAB}{}^C,\quad R = \gamma^{AB}R_{AB}. \tag{2.6}$$

---

*. The procedure usually followed[8] is that of varying the metric and the connection or the contorsion; this becomes rather clumsy when metricity is required. By expressing the action in terms of the metric and the torsion and by varying these variables, metricity is automatically implemented throughout, because now the contorsion is a dependent variable defined by (2.3).

To simplify matters, we can use a traceless torsion; this is not an additional assumption, because it will be found below that, for a Dirac field, the torsion is totally antisymmetric. With a traceless torsion, one finds the following relation between the Riemannian and non-Riemannian quantities:

$$R_{AB} = \mathring{R}_{AB} + Z_{AB} + \nabla_C K_{AB}{}^C , \quad R = \mathring{R} + Z , \tag{2.7}$$

where

$$Z_{AB} = K_{CA}{}^D K_{DB}{}^C = Z_{BA} , \quad Z = \gamma^{AB} Z_{AB} . \tag{2.8}$$

It follows from (2.7-8) that the EC gravitational Lagrangian differs from the usual Einstein Lagrangian by the additional term $Z/2\hat{\varkappa}$, which is quadratic in the torsion.

The field equations are

$$\delta \mathcal{A} / \delta \gamma_{AB} = 0 \qquad (T_{AB}{}^C \text{ and } \Psi \text{ fixed}), \tag{2.9a}$$

$$\delta \mathcal{A} / \delta T_{AB}{}^C = 0 \qquad (\gamma_{AB} \text{ and } \Psi \text{ fixed}), \tag{2.9b}$$

$$\delta \mathcal{A} / \delta \Psi = 0 \qquad (\gamma_{AB} \text{ and } T_{AB}{}^C \text{ fixed}). \tag{2.9c}$$

We define the Rosenfeld energy-momentum density T_{AB} and the spin density $\sigma^{AB}{}_C$ by

$$|\gamma|^{1/2}\, T^{AB} = \delta \mathcal{A}_{Mat} / \delta \gamma_{AB} , \tag{2.10}$$

$$|\gamma|^{1/2} \sigma^{AB}{}_C = -2 \delta \mathcal{A}_{Mat} / \delta T_{AB}{}^C . \tag{2.11}$$

For the case (2.4-6) at hand one finds [4]

$$T_{AB} = [(\nabla_A \bar{\Psi}) \Gamma_B \Psi - \bar{\Psi} \Gamma_B \nabla_A \Psi]/2 + \gamma_{AB} \mathcal{L}_D - \nabla^C (\bar{\Psi} \Gamma_{CBA} \Psi)/4 , \tag{2.12}$$

$$\sigma^{AB}{}_C = \bar{\Psi} \Gamma^{AB}{}_C \Psi / 4 , \tag{2.13}$$

and the field equations*

$$\mathring{G}_{AB} + Z_{AB} - Z \gamma_{AB}/2 = \hat{\varkappa} T_{AB} , \tag{2.14}$$

---

* . The antisymmetric components of T_{AB} in equation (2.12) vanish identically on-shell; therefore, the antisymmetric components of (2.14) are identities on-shell. Such identities express the local Lorentz invariance of the EC action, see ref.[4] for details.

$$K_{CBA} = \hat{\varkappa}\sigma_{ABC} \ , \tag{2.15}$$

$$\not{\nabla}\Psi = 0. \tag{2.16}$$

For our purposes, it is convenient to rewrite the field equations in terms of the generalized Einstein tensor G_{AB},

$$G_{AB} \equiv R_{AB} - R\gamma_{AB}/2 = \overset{\circ}{G}_{AB} + Z_{AB} - Z\gamma_{AB}/2 + \nabla_C K_{AB}{}^C, \tag{2.17}$$

and of the canonical energy-momentum density Σ_{AB},

$$\Sigma_{AB} = [(\nabla_A\bar{\Psi})\Gamma_B\Psi - \bar{\Psi}\Gamma_B\nabla_A\Psi]/2 + \gamma_{AB}\mathcal{L}_D \ . \tag{2.18}$$

From the explicit expression of the generalized Einstein tensor, one sees that eq.(2.14) can be rewritten in the form

$$G_{AB} = \hat{\varkappa}\Sigma_{AB} + \nabla_C(K_{AB}{}^C + \hat{\varkappa}\sigma_{AB}{}^C) \ . \tag{2.19}$$

Keeping into account the "algebraic" Cartan field equation (2.15), one obtains the simpler field equation

$$G_{AB} = \hat{\varkappa}\Sigma_{AB} \ , \tag{2.20}$$

which, together with (2.15-16), describes the classical theory.

It should be kept in mind that the generalised Einstein equation (2.20) is equivalent to (2.14) only insofar as the "conservation law"

$$\nabla_C(K_{AB}{}^C + \hat{\varkappa}\sigma_{AB}{}^C) = 0 \tag{2.21}$$

is verified. Now, eq.(2.21) is identically satisfied when the Cartan field equation (2.15) holds; in this case, the torsion orginates from the matter fields and is not simply an "external" field.

3. SPIN-TORSION COMPACTIFICATION

In a quantum theory, the classical field equations can be used in the semiclassical <u>Ehrenfest approximation</u> as equations for expectation values of quantum operators. In this approximation, the classical action is adopted and the expectation values of operator products are approximated by corresponding products of expectation values.

Quantum effects limit the validity of the Ehrenfest approximation, because loop contributions can change the form of the effective action and because the expectation value of an operator product can differ from the product of expectation values.

These limitations bear particularly on the role of torsion in a quantised EC theory. Classically, the action contains only quadratic and linear terms in the torsion; these lead to the algebraic field equation (2.15), which implies that torsion vanishes outside matter and propagates only "convectively". At the quantum level, the effective action for torsion is non-polynomial and leads to torsion propagation also in vacuo, through virtual loop effects[8].

It is also clear that fermion condensates can lead to a failure of the Ehrenfest theorem for expectation values of fermion bilinears. In particular, fermion fields must vanish in a Poincarè invariant vacuum, while Poincarè invariant expectation values of products of fermion fields need not vanish. When this happens for the internal space components of the fermion spin density in vacuo, one obtains a nontrivial background value for the torsion.

If a satisfactory quantum theory were available, ground state values of the metric and of the torsion would in principle be computable from the minima of the effective quantum potential. Lacking a full quantum theory, we proceed on the basis of an ansatz and we verify its consistency with the classical field equations. To make sure that the ansatz indeed corresponds to a ground state, we should also verify that there are no instabilities. This will be checked within a semiclassical approximation, because it will be found that the mass spectrum of the small fluctuations does not contain negative-energy modes.

The ansatz is as follows[5]:

(i) the ground-state metric is

$$\langle \underset{\sim}{\gamma} \rangle_0 \equiv \bar{\underset{\sim}{\gamma}} = \eta_{\mu\nu} dx^\mu dx^\nu + \delta_{ab} R^a R^b \,, \tag{3.1}$$

where R^a are right-invariant Maurer-Cartan forms on G normalized by the communtation rules

$$[R_a, R_b] = -mf_{ab}{}^c R_c \,, \tag{3.2}$$

where $f_{ab}{}^c$ are the canonical structure constants of G and m^{-1} is the linear size of G;

(ii) in the frame (∂_μ , R_a), the ground-state torsion has the components

$$\langle T_{AB}{}^C\rangle_0 = \bar{T}_{AB}{}^C \text{ with } \bar{T}_{ab}{}^c = mf_{ab}{}^c \text{ and } \bar{T}_{AB}{}^C = 0 \text{ otherwise.} \tag{3.3}$$

It is easily shown that the background linear connection vanishes in the frame (∂_μ, R_a), hence it has zero Riemann-Cartan curvature:

$$\bar{\underset{\sim}{\omega}} = \overset{\circ}{\underset{\sim}{\omega}}[\bar{\underset{\sim}{\gamma}}] + \underset{\sim}{\bar{T}}/2 \,,\; \overset{\circ}{\omega}[\bar{\underset{\sim}{\gamma}}]_{ab}{}^c = -mf_{ab}{}^c/2 \,,\; R_{ABC}{}^D[\bar{\omega}] = 0. \tag{3.4}$$

Therefore, the contorted background is a solution of the classical homogeneous field equations:

$$R_{AB}[\bar{\omega}] = 0 \,, \qquad \langle \Sigma_{AB}\rangle_0 = 0, \tag{3.5}$$

and it is consistent to assume that the ground state has the geometry of $M_4 x G$.

4. SMALL FLUCTUATIONS

As we have already enphasized, when one takes into account the fact that the torsion is generated by matter, the relevant field equation is the generalised Einstein equation (2.20), and only its symmetric components are dynamically relevant. In analogy with the procedure usually adopted in the Einstein theory, we shall now proceed to linearize this equation in vacuo, so as to determine the small fluctuations of the metric near the background (3.1-3).

The equation to be linearized is $G_{(AB)} = 0$, or equivalently

$$R_{(AB)} = 0 \,. \tag{4.1}$$

In the linearization of (4.1), we shall keep the torsion fixed to its background value, because we expect that the fluctuations of torsion should be of order $\varkappa$, although the background torsion has large internal components.

First, we need to generalise to our case two relations which are well known in the Riemannian theory. When the metric differs by an infinitesimal fluctuation $\underset{\sim}{h}$ from its background value,

$$\underset{\sim}{\gamma} = \underset{\sim}{\bar{\gamma}} + \underset{\sim}{h}, \qquad (4.2)$$

and the torsion is kept fixed to its background value, one finds that:

(a) the linear connection differs from its background value by the infinitesimal amount

$$\delta\omega_{AB}{}^{C} = \nabla_{(A}h_{B)}{}^{C} - \nabla^{C}h_{AB}/2 , \qquad (4.3)$$

where ∇_A denotes the covariant derivative with the background connection and the indices are raised with the background metric;
(b) the corresponding variation of the generalized Ricci tensor, $\delta R_{AB} = R_{AB}[\underset{\sim}{\bar{\gamma}}+\underset{\sim}{h}] - R_{AB}[\underset{\sim}{\bar{\gamma}}]$, is given by the generalized Palatini identity

$$\delta R_{AB} = 2\nabla_{[D}\delta\omega_{A]B}{}^{D} + \delta_{DB}{}^{C}\,\bar{T}_{CA}{}^{D} . \qquad (4.4)$$

Keeping in mind that R_{AB} vanishes in the background, one finds

$$R_{(AB)}[\underset{\sim}{\bar{\gamma}}+\underset{\sim}{h}] = -\Delta h_{AB}/2 + \nabla_{(A}[\nabla^{D}h_{B)D} - \nabla_{B)}h/2] , \qquad (4.5)$$

where Δ is the background generalized Laplacian $\bar{\gamma}^{AB}\nabla_A\nabla_B$ and $h = \bar{\gamma}^{AB}h_{AB}$. Now we observe that, in our background,

$$\nabla^{D}h_{BD} - \nabla_{B}h/2 = \mathring{\nabla}^{D}h_{BD} - \mathring{\nabla}_{B}h/2 , \qquad (4.6)$$

and that, with suitable coordinates, we can satisfy the harmonic gauge condition

$$\mathring{\nabla}^{D}h_{BD} - \mathring{\nabla}_{B}h/2 = 0 . \qquad (4.7)$$

We conclude that the linearized field equation for the fluctuations in the harmonic gauge is simply

$$\Delta h_{AB} = 0 . \qquad (4.8)$$

Since in the frame (∂_μ, R_a) the background connection vanishes, the generalised Laplacian is simply $\Box + \delta^{ab}R_aR_b$, and the equation becomes

$$(\Box + \delta^{ab}R_aR_b)h_{AB} = 0. \tag{4.9}$$

We see that the generalized Laplacian $\delta^{ab}R_aR_b$ of the background internal space G becomes a mass operator from the four-dimensional viewpoint.

To determine the spectrum of the fluctuations, we can refer the metric to the frame (∂_μ, R_a) and expand the components in harmonics on G, using the complete set of unirreps $U^{(\lambda)}$ of G:

$$h_{AB}(x,y)=\sum_\lambda h_{AB}{}^{(\lambda)}(x,y),\ h_{AB}{}^{(\lambda)}=\sum_{i,j} h_{AB}{}^{(\lambda)}{}_{ij}(x)U^{(\lambda)}{}_{ji}(y). \tag{4.10}$$

On the unirreps, the R_a's act as generators of left translations:

$$R_aU^{(\lambda)}{}_{ji}(y) = R_a{}^n\partial_nU^{(\lambda)}{}_{ji}(y) = m\sum_k T^{(\lambda)}{}_{ajk}\ U^{(\lambda)}{}_{ki}(y), \tag{4.11}$$

where the $T^{(\lambda)}{}_a$ are the (matrix) generators of $U^{(\lambda)}$ with the canonical commutation rules

$$[T^{(\lambda)}{}_a, T^{(\lambda)}{}_b] = f_{ab}{}^c\ T^{(\lambda)}{}_c\ . \tag{4.12}$$

It follows that the harmonics of h_{AB} satisfy the equation

$$[\Box - m^2C(\lambda)]\ h_{AB}{}^{(\lambda)}{}_{ij}(x) = 0, \tag{4.13}$$

where $C(\lambda)$ is the quadratic Casimir invariant of $U^{(\lambda)}$ and is positive semi-definite, with $C(\lambda)=0$ if and only if $U^{(\lambda)}$ is the singlet ($\lambda=0$).

We conclude that, in the frame (∂_μ, R_a), the zero modes of the fluctuations have the general form

$$h_{\mu\nu}{}^{(0)} = h_{\mu\nu}{}^{(0)}(x),\ h_{\mu a}{}^{(0)} = A_{\mu a}(x),\ h_{ab}{}^{(0)} = h_{ab}{}^{(0)}(x). \tag{4.14}$$

The spectral analysis for the Dirac field can be similarly carried out, see ref.[5]. The zero-modes are easily found, because in the frame (∂_μ, R_a) the Dirac equation on the background becomes simply

$$(\gamma^\mu\partial_\mu + \Gamma^aR_a)\Psi = 0\ . \tag{4.15}$$

Now, it is clear that the solutions of the equation

$$R_a\Psi^{(0)} = 0 \tag{4.16}$$

are certainly zero-modes, and it can be shown that there are no other zero-modes. Once again, also for the Dirac zero-modes we see that the corresponding harmonic in the frame (∂_μ, R_a) is independent of y: $\Psi^{(0)} = \psi(x)$.

Now we recall that G acts naturally on itself as the group G_L of left translations and as the group G_R of right translations, respectively with generators R_a and L_a, where

$$R_a = U_a{}^b L_b \tag{4.17}$$

and U is the adjoint representation of G. The frame R_a if G_R-invariant and the frame L_a is G_L-invariant.

Since the zero-modes are y-independent in the G_R-invariant frame, they are singlets of G_R. To obtain the quantum numbers relative to G_L, we may simply regard (4.17) as a frame transformation to go over to the frame (∂_μ, L_a) and use the tensorial transformation laws for the components of the metric and the spinorial transformation for the spinor components. We therefore see that the components of the metric zero-modes in the frame (∂_μ, L_a) are given by

$$\begin{aligned} \hat{h}_{\mu\nu}{}^{(0)} &= h_{\mu\nu}{}^{(0)}(x), \quad \hat{h}_{\mu a}{}^{(0)} = A_{\mu b}(x) U_a{}^b(y), \\ \hat{h}_{ab}{}^{(0)} &= h_{cd}{}^{(0)}(x) U_a{}^c(y) U_b{}^d(y). \end{aligned} \tag{4.18}$$

Therefore, $h_{\mu\nu}{}^{(0)}$ is a singlet of G_L, $h_{\mu a}{}^{(0)}$ belongs to the adjoint representation of G_L and $h_{ab}{}^{(0)}$ belongs to the symmetric product of two adjoint representations of G_L. Similarly, the Dirac zero modes belong to the spinor representation of G_L.

To summarize, the zero modes of the mixed components of the metric belong to the (singlet,adjoint) representation of $G_R x G_L$ and the Dirac zero modes to the representation (singlet,spinor). Therefore, the metric zero modes contain the YM potentials of G_L, while the YM potentials of G_R are contained in the harmonic corresponding to the representation (adjoint,singlet) of $G_R x G_L$; this harmonic is massive in four dimensions, with $(\text{mass})^2$ given by the quadratic Casimir of the adjoint representation of G.

This result has been obtained in spite of the fact that the

background metric and torsion have $G_R x G_L$ as symmetry group*.

Finally, it should be observed that the general G_R-invariant metric on M_4xG is one which has y-independent components in a G_R-invariant frame. Therefore, eq.(4.14) gives the general form of fluctuations which are G_R-invariant. Now, the requirement of G_R-invariance is precisely the one used in the old-fashioned approaches to KK theories[3,7]. Hence, as we anticipated in the Introduction, the metric zero modes reproduce precisely the old-fashioned KK ansatz, including the Jordan-Thiry fields**. However, our approach based on spin-torsion compactification does not suffer from the inconsistencies[5,7], which in the older approach arose from the cosmological constant induced by the Riemannian curvature of G.

---

*. Indeed, the Killing-Cartan metric on G is bi-invariant, and also the parallelizing torsion $mf_{ab}{}^c$ is invariant under $G_R x G_L$. There is, however, a discrete symmetry which is broken by the background torsion: the Killing-Cartan metric is invariant under the interchance of G_R with G_L, which amounts to the invariance of the components when the reference frame is changed from the L_a's to the R_a's, while under this interchange the components of the parallelizing torsion change sign.

**. To be precise, the KK ansatz is reproduced only in the linear approximation, because we have used the equations for small fluctuations.

REFERENCES

1. E.Witten, Nucl.Phys. B108, 409 (1981);
A.Salam and J. Strathdee, Ann.Phys.(N.Y.) 141, 321 (1983).
2. Th.Kaluza, Sitzber.Z.Akad.Wiss. 966 (1921);
O.Klein, Z.Phys. 37, 895 (1925);
P.Jordan, Astron.Nachr. 276, 193 (1948);
Y.R.Thiry, C.R.Acad.Sci.Paris 226, 216 (1948);
A.Lichnerowicz, Théories Relativ.de la Gravitation et de l'Electromagnetisme, Masson, Paris 1953.
3. B.S.De Witt, in Relativity, Groups and Topology, Ed.by B.S.and C.De Witt, Gordon and Breach, New York 1964;
Y.M.Cho, J.Math.Phys. 16,2029 (1974);
Y.M.Cho and P.G.O.Freund, Phys.Rev. D12, 1711 (1975);
C.A.Orzalesi, Fortschr.f.Phys. 29, 413 (1981).
4. R.Camporesi, C.Destri, G.Melegari and C.A.Orzalesi, Class. Quantum Grav. 2, 461 (1985).
5. C.Destri, C.A.Orzalesi and P.Rossi, Ann.Phys.(N.Y.) 147, 321 (1983).
6. A.Trautman, Symp.Math. 12, 139 (1973);
F.W.v.Hehl, P.V.d.Heyde, G.D.Kerlick and J.M.Nester, Revs. Mod.Phys. 48, 393 (1975).
7. C.A.Orzalesi and M.Pauri, Phys.Lett. B107, 186 (1981);
C.A.Orzalesi, in Proc.Conf.on Differential Geom.Methods in Mathematical Phys.,Trieste 1981, Ed.by G.Denardo and H.D. Doebner, World Scient.Singapore 1983.
8. C.A.Orzalesi, Phys.Lett. B140, 39 (1984);
C.A.Orzalesi and G.Venturi, Phys.Lett. B139, 357 (1984).

ON THE EFFECTIVE GAUGE GROUP FROM G/H SPONTANEOUS COMPACTIFICATION

A Jadczyk*
CERN, Geneva

ABSTRACT

We discuss two schemes of dimensional reduction: G-invariant and non-G-invariant one. The first gives rise to a consistent truncation with N(H)/H gauge bosons, while the second leads to the effective gauge group $G_{eff} \overset{loc}{\approx} N(H)/H \text{ Aut } G$.

* Permanent address: Inst. Theor. Physics, University of Wrocław, Cybulskiego 36, 50-205 Wrocław, Poland.

1. Introduction

The idea of spontaneous compactification is both simple and attractive. One starts here with a field theory in d=m+s dimensions. Let {F} denote the set of primitive fields of the theory. Usually {F} contains metric tensor or vielbein field, but there are known interesting models in which these fields are composite. After contemplating the Lagrangian and field equations for {F}, and after convincing oneself that the set {F} is reasonably complete (in particular one has to take care of the anomalies) one looks for a "compactifying ground-state solution" $\{F_o\}$.$\{F_o\}$is expected to be a highly symmetric solution of the field equations and it should be stable in an appropriate sense. The symmetry of $\{F_o\}$ need not (and in many cases will not) be the highest possible one-provided that the stability of $\{F_o\}$ is assured. If $\{F_o\}$ gives rise to a splitting of the m+s dimensional world - we shall denote it by E - into a product E=M×S of m and s dimensions, with S compact, then one says that a spontaneous compactification is taking place. Such a phenomenon may, for instance, easily occur in theories containing in their "menu" antisymmetric tensor fields which acquire nonzero vacuum expectation values. For example, if $F_{AB...C}$ is such a field, and if dF=0, then one has a natural definition of the internal directions X^A as those satisfying $X^A F_{AB...C}=0$. The details of the mechanism of spontaneous compactification may, however, be model-dependend to a great extent. Therefore if we want to draw some *model-independent* conclusions, the natural thing to look at is the symmetry group of the ground state. We shall assume that this group splits in a natural way into a product $G_E=G_M \times G_S$ of "spacetime" symmetry group G_M and "internal" symmetry group G_S. Several comments are to be made at this point. First, we have taken "spacetime" into the quotation marks. Reason for this is the following: one should not be prejudiced and think that a spontaneous compactification occurs necessarily in one step. To the contrary, a realistic scenario may proceed in several steps. For instance, if we think of pure Einstein's gravity as the primitive field, then the first compactification will produce (via the Kaluza-Klein mechanism) Yang-Mills fields, while the second compactification may occur around a nontrivial Einstein-Yang-Mills background to produce chiral fermions and spontaneous symmetry breaking via the Higgs potential resulting from the Yang-Mills part of the effective "after-first-step" Lagrangian. One can also think of three or more consecutive steps (M. Duff [1] considers a possibility of such a more-than-one-step scenario for the field theory limit of strings). What is important to notice here is that each step taken separate-

ly can satisfy the appropriate criterium of stability without the final result being a stable compactification.
Next remark concerns the group G_S. It is convenient to assume that G_S is a compact Lie group. On the other hand some non-compact groups may well occur here. There is no reason for G_S to be simple either ... . Therefore, when in the following we shall consider mainly the case of G_S compact and/or simple, it will be only for the sake of convenience. Since in the following G_M will not be discussed we shall denote G_S simply by G. And when we will talk of G being the "isometry group of the vacuum", what we will really mean is that "G is some natural part of the symmetry group of $\{F_o\}$ ".
Another remark: we say "isometry group", but what if $\{F_o\}$ has no isometries at all? What if S-the internal space and/or M - the spacetime, is a complicated manifold which is not a homogeneous space and which has no Killing vectors? The answer to this question is not a straightforward one. First remark is that, as we will see, there are interesting manifolds which are not homogeneous spaces but which are "born out" of homogeneous spaces - they are "Kaluza-Klein projections" of homogeneous spaces. These manifolds are the *double cosets* (or parts of double cosets). They carry a finite - parameter family of natural metrics inherited from their homogeneous parents, although they need not to have Killing vectors. (It is also remarkable that, apparently, some of these double-coset manifolds may carry naturally exotic differentiable structures). Therefore, even if S has no isometries, it may happen that there is a still higher dimensional theory, in some $\bar{E}$ containing E, in which the usual Kaluza-Klein mechanism involving isometries works, and such that E is obtained by a projection from $\bar{E}$. We will discuss some of the relevent geometrical constructions later on.
Second remark is that if S has no group acting on it, and if there is no way of using some trick (like that of embedding E into $\bar{E}$ where some G can act), then it becomes really a problem. And this because after spontaneous compactification it is necessary to perform "truncation" and "dimensional reduction", and the only known way of doing that seems to be via "harmonic expansion". But there is no harmonic expansion if there is no group action! This brings us to the next important concepts: harmonic expansion, truncation and dimensional reduction.

Suppose $\{F_o\}$ is given, is stable, and the d=m+s dimensional world splits into M×S. Let also G be the isometry group of $\{F_o\}$, with S acting transitively on S. The next thing to do is to analyse the fluctuations $F_o+\delta F$ around F_o. However, most of these fluctuations will completely destroy the product structure M×S - they will hopelessly mix the "internal" with the "external". If we want to have an effective description of phenomena from the point of view of the m-dimensional base manifold M, then we have to use: harmonic expansion, dimensional reduction and, eventually, some or other truncation. The first step - harmonic expansion - has as its aim to select a basis in the space of all fluctuations δF, a basis δF_n such that every δF can be expressed as a series $\delta F = c_n \delta F_n$, and every δF_n can be interpreted as a finite-component field on M. The number of components of δF_n will, in general, increase with n and will be related to the consecutive dimensions of irreducible representations of G. If this process of harmonic expansion is induced by geometry, then there is a good chance that it will not destroy the gauge invariance of the theory. Such a geometrical scheme of harmonic expansion has been proposed in Ref. [2].

The next step, if possible, consists of truncation of the infinite tower of fields so as to get a theory on M with only a finite number of fields. Here one may wish to truncate the theory in such a way that masses of the Planck order do not appear in the effective m-dimensional theory and the truncated theory deals with the massless modes done. The important point which should be observed here is that *neither the procedure of harmonic expansion nor that of truncation is, in general, unique*. This fact is neither "good" nor "bad" - it is a reality one has to live in. A simple receipt for truncation can be given: let G be a subgroup of the isometry group of S, and suppose G is transitive on S. Then consider only those δF's which are G-invariant - this defines a truncation scheme. We get an effective theory on M with only finite number of fields and no Planck masses in the effective mass spectrum. A source of nonuniqueness is clearly seen: there may be more than one choice of G, and the smaller G is, the reacher is the spectrum of the effective m-dimensional theory. A well known example comes from 11-dimensional supergravity compactified on $S=S^7$ - the seven-sphere. If we take for G the group SO(8) then *the SO(8)-invariant ansatz gives no gauge fields at all*. On the other hand the group $U(2;\mathbb{H})$ is a subgroup of SO(8) which is also transitive on S^7, and the $U(2;\mathbb{H})$-invariant ansatz produces gauge bosons of SU(2). Of course, this second ansatz is more natural for the "squashed" ground state rather than the "round" one; nevertheless there are no good reasons why should it not be applied in the latter case too. Here it is important to stress that the G-invariant ansatz is, as a rule, consistent.

Before discussing briefly this problem of the consistency let us first try to make it more precise what is meant by the term "ansatz". Once a stable ground state solution $\{F_o\}$ of the classical field equation in m+s dimensions has been selected, and once a spontaneous compactifiction $E \to M\times S$ induced by $\{F_o\}$ has taken place, then we have still to decide on the form of the fluctuations δF-s which will define the effective quantum theory in m dimensions. Usually one selects some finite set of fields $\{f_i\}$ on M and considers δF's as build out of the δf_i-s. In the following we will not split F into F_o and δF but, instead, we will describe how are the field configurations F in E build out of the field configurations f in M. This is called "ansatz", and one should not confuse this "ansatz" with a method of finding topologically interesting solutions of the field equations. Here we are not that much interested in *solutions* but rather in restricting the space of *field configurations* (which defines a domain of the functional integral). For instance, what we call "G-invariant ansatz" is defined as follows: assuming G is a symmetry group of the ground state $\{F_o\}$, we consider only these field configurations {F} in E which are G-invariant (G-singlets). One then finds that every such F can be expressed in terms of a certain number of fields f on M. Solving out these constraints of G-invariance in terms of f's gives then the explicit form of the ansatz. Now, consider the problem of a "consistency" of a given ansatz. Suppose that we have given an explicit expression for F[f], where {f} are fields on M. It is then an easy matter to put F[f] into the action $A_E=\int_{M\times S}L[F[f]]$, to integrate it over S, and to obtain in this way an "effective action" A_M for the fields {f} on M. However, there is no whatsoever guarantee that the field theory on M obtained in this way will be *consistent* with the original one. The requirement of *consistency* is similar to that of *stability*. A truncation obtained by an ansatz F[f] is called *consistent* if the extrema $\{f_o\}$ of the effective action A_M determine extrema $F[f_o]$ of the original action. Or, even simpler, if every solution of the reduced theory is a solution of the original one (see [3-4]). There are many ansatze which are inconsistent. The G-invariant ansatz, which will be discussed in more details later, can be shown to be consistent [3,5]. On the other hand the most popular not-G-invariant ansatz involving Killing vectors is, in general, inconsistent. However, it is to be noticed, that an ansatz which is inconsistent with one set of fields may well become consistent with another. This situation apparently happens with the "Killing-vectors-ansatz" used in 11 dimensional supergravity (see the discussion in [3]).

2. The G-invariant ansatz.

We will now discuss the geometrical "milieu" of the G-invariant ansatz. The fact that we will strictly adhere to the pure geometrical aspects will make much of the discussion *model-independent*. The drawback of using geometrical methods in the particular case which interests us, but as well in any other case, is that the results depend on satisfying the assumptions, and the assumptions may happen to be too restrictive to accomodate some interesting models. After discussing the G-invariant scheme we will later on weaken our assumptions. But it must be understood that even these weaker assumptions are arbitrarily imposed - they seem to constitute a natural description of the to-day's models, but tomorrow... . But even if this is going to happen the "*tool*" of the G-invariant scheme will remain to be useful.

Observe that transformation properties and dynamics of gauge fields are most naturally expressed in geometrical terms when gauge fields are represented by connections on principal fibre bundles. Therefore if in a theory of a Kaluza-Klein type one believes that the dimensional reduction scheme leads to gauge fields with a certain gauge group G, then the natural question to ask is: "where is the principal bundle on which the gauge field is supposed to live?". Answering this question is not a problem *if one starts with assuming* that th e universe is a principal bundle to begin with... However, such a position seems to be not quite what one wants; and indeed there exists a more natural and more general framework. This is the framework of G-invariant dimensional reduction. This framework is conceptually simple and it has a nice geometrical interpretation. It is well adopted for harmonic expansion and for reduction of all kinds of geometrical objects and matter fields. Last but not least, it leads as a rule to a consistent truncation of massive modes. The methods has many advantages but, at the same time, it is certainly not the key to all the enigmas of the Universe. It should be considered rather as a *powerfull and convenient mathematical tool*, which it is good to have at hand when it is needed. After describing first this tool of G-invariant dimensional reduction, we will next consider a more general setting, covering it, and we will see how can this universal tool be applied to produce another scheme of dimensional reduction which is more subtle than the G-invariant one.

Let us begin with the following remark: our tool will work and will do its job *whenever there is a group acting on some manifold*. It need not be in the context of dimensional reduction, for instance one can look for solutions of some field equations having certain symmetry, or one can think of the infinite dimensional group of (x-dependent) gauge transformations or diffeomerphisms acting on an infinite dimensional manifold of field configurations... It is therefore for convenience and in order to have some concrete picture in mind that, while describing this tool, we will use a terminology which is adapted to the problems of dimensional reduction in Kaluza-Klein theories.

Let therefore G be a group acting on a manifold E. To make things regular and easy we will assume that G is a compact Lie group which acts smoothly from the right on a d-dimensional smooth manifold E. Given $y \in E$ we denote by G_y the stabilizer of y. The manifold E decomposes now into several *strata* according to the type of the stabilizer. We choose one of these strata and call it E in the following. All the stabilizers G_y, $y \in E$ are now conjugated to a standard one, say H. We now define M to be the space of orbits: $M=E/G$, so that locally $E=M\times(H\backslash G)$ (we write $H\backslash G$ and not G/H since we have chosen *right* action of G). Thinking of some dynamical theory with gauge fields as an output it is now natural to ask: "what principal fibre bundle over M can be seen in the structure we have?". The answer reads: the only potentially nontrivial principal fibre bundle over M which can be constructed out of the ingredients we have put into the game is a principal bundle P with structure group N(H)/H, N(H) being the normalizer of H in G. P is constructed as a subset of E

$$P = \{y \in E : G_y = H\}$$

The point to be stressed is that no nontrivial fiber bundle with structure group G can be seen emerging. This was the surprising result of [6] , where we have found N(H)/H as the effective gauge group, instead of the expected G. It was also shown that what is geometrically allowed and natural is also dynamically available i.e. one really gets an N(H)/H gauge field and its Lagrangian from dimensional reduction of G-invariant metric and Einstein-Hilbert action on E.

The effective gauge group from G-invariant dimensional reduction is therefore $G_{eff}=N(H)/H$. In many cases this group N(H)/H can be considered as the biggest subgroup K of G such that $H\times K \subset G$. Here "in many" does not mean "in all"! One must be particularily carefull if G is non-compact or non-semisimple. For instance, if G is the Poincaré group and H is the translation group then N(H)/H is the Lorentz group which is not the *direct* factor of the translation group in G.

In [6] the following result has been proven: there is 1-1 correspondence between G-invariant metrics g_E on E and triples (g_M, A, ϕ) of fields on M, whre g_M is a metric on M, A is a gauge field with gauge group N(H)/H and ϕ is a multiplet of scalar fields. The metric g_M induced by g_E is called "the Kaluza-Klein projection of g_E". The following comments can be given to this result:

i) The projection $E \to M$ is an example of what is called in the mathematical literature a "Riemannian submersion". There are many results in the mathematical papers dealing with what is called "totally geodesic" case. The case we have to deal with is not of that kind unless the scalar fields are switched off. (There are also mathematical results dealing with the case of gauge fields switched off).

ii) The results and the formulae of [6] form up a *tool*. It can be applied whenever one has a group acting on some manifold, and whenever one is interested in geometrical objects invariant under this group action. A general theory of dimensional reduction of geometrical objects has been initiated in [2].

iii) It is convenient to introduce a concept of a "dimensionally reducible geometrical object". This is an object on E which can be also interpreted as a finite-component field on M. As a rule all objects on E which are G-invariant (G-singlets) are dimensionally reducible. But also objects whose values transform under a finite dimensional representation of G are dimensionally reducible. Sometimes it may be, however, convenient to consider objects transforming under an infinite dimensional representation of G as dimensionally reducible too.

iv) The assumption of the global action of G on E is used in the process of harmonic expansion of fields. As we shall discuss it later it is not necessary to assume that much for the harmonic expansion scheme to work.

3. The non-G-invariant scheme of dimensional reduction.

Let us start with an example which will illustrate the idea of "non-invariant" dimensional reduction. The example will at the same time introduce the concept of a double coset, the concept which may prove to be usefull for building model manifolds with interesting geometrical properties.
Consider a homogeneous space $H\backslash G$, on this space there is a finite parameter family of G-invariant metrics. Indeed, since G acts transitively on the coset

$H\backslash G$, a G-invariant metric on H G is completely determined by knowing it at one point; the number of G-invariant metrics on $H\backslash G$ is therefore equal to the number of AdH invariant scalar products at the origin of the coset space. Let now K be another closed Lie subgroup of G, we can form then the double coset space $H\backslash G/K$. The (right) action of K on $H\backslash G$ will have, in general, more than one orbit type. In such cases we can restrict ourselves to an open dense submanifold of $H\backslash G$ which constitutes the principal stratum of K-action on $H\backslash G$. With this understanding $H\backslash G/K$ becomes a manifold. Observe the analogy: $E \sim H\backslash G$, $M \sim H\backslash G/K = E/K$. Every G-invariant metric on $H\backslash G$ is now, a fortiori, K-invariant and therefore, according to the G-invariant scheme of dimensional reduction, determines its Kaluza-Klein projection on $H\backslash G/K$. In this way we obtain a finite-parameter family of metrics on $H\backslash G/K$ which, in general, have no isometries at all. Observe that the group which survives the double quotient and still acts on $H\backslash G/K$ is $N(H)/H \times N(K)/K$ (it is however not automatically guaranted that the action of this group on $H\backslash G/K$ is effective). It may be instructive to consider a concrete example. Let us therefore take for G the group $U(2;\mathbb{H})$, and for H and K the following two subgroups of $U(2;\mathbb{H})$, each isomorphic to $U(1;\mathbb{H}) \sim SU(2)$:

$$H = \left\{ \begin{pmatrix} q & 0 \\ 0 & q \end{pmatrix} \quad : q \in \mathbb{H}\,, \quad q \neq 0 \right\}$$

$$K = \left\{ \begin{pmatrix} q & 0 \\ 0 & 1 \end{pmatrix} \quad : q \in \mathbb{H}\,, \quad q \neq 0 \right\}$$

The coset G/K is isomorphic to a seven-sphere S^7. The coset $H\backslash G/K$ is S^4 and $G/K \to H\backslash G/K$ is nothing but the Hopf fibration of S^7. Observe that the residual group which still acts on S^4 is $N(H)/H \times N(K)/K = O(2) \times SU(2)$.

Remark. The groups H and K are both naturally isomorphic to $U(1;\mathbb{H})$, therefore we can take first a partial quotient of G by the "diagonal" $U(1;\mathbb{H})$. The resulting manifold of orbits of the diagonal $U(1;\mathbb{H})$ acting (on both sides) on $U(2;\mathbb{H})$ is diffeomorphic to an exotic seven-sphere Σ^7. The group $O(2)\times SU(2)$ acts therefore on Σ^7. The principal stratum of this action projects onto an open dense subset of S^4. It would be interesting to know whether there is any relation between this construction and exotic $\mathbb{R}^4$ recently investigated (see [7] for Σ^7 and [8] for a review on exotic $\mathbb{R}^4$-s).

After discussing the double-coset example let us discuss a similar construction which will generate a class of non-G-invariant, dimensionally reducible, metrics on a G-space E. Let therefore E, G and M be as in the discussion of the G-invariant ansatz. We already know the form of the most general G-invariant metric on E. We will enlarge now this class of metrics so as to include some of dimensionally reducible non-G-invariant ones. To this end we will use the following receipt: first, replace E with a bigger space $\bar{E}=E\times G$. Then, on $\bar{E}$ we have right action of the group G×G: $(y,a)(b,c)=(yb,c^{-1}a)$, and E is isomorphic to the quotient $\bar{E}/G^d$, where G^d is the diagonal of G×G. Indeed, the isomorphism of $\bar{E}/G^d$ onto E is given by $(y,a)\to ya$. Observe that, in fact, we have action on $\bar{E}$ of the product (G×G)×G, the last factor being the right action of G on itself; it goes to the quotient $E=\bar{E}/G^d$ to coincide with the right action of G on E. Now, consider the class of all G×G-invariant metrics on $\bar{E}$. Since G×G acts on $\bar{E}$ with the stability group $\bar{H}=H\times id$, it follows (by application of the tool of G-invariant dimensional reduction, with $G\to G\times G$) that these metrics can be described in terms of fields on M, and that they give rise to gauge fields of $N(\bar{H})/\bar{H}=N(H)/H\times G$. But each of these metrics, being G×G invariant is, a fortiori, G^d-invariant, and therefore it defines, by the Kaluza-Klein projection, a metric on E.The class of metrics on E induced that way contains the class of G-invariant ones, as a subclass. But it contains much more: it contains also those metrics on E which give rise to gauge bosons of G, which degrees of freedom are not contained in the G-invariant ansatz. The receipt given above may seem to be unnatural, this is not, however, so. We will describe now a framework for dimensional reduction which does not require the assumption of a global G-action. And in this framework the receipt above will find its natural place. But before discussing the technical side of the extended framework, let us first analyse the following simple illustrative example: the two-torus contra the Klein-bottle. Both are S^1 fibrations over S^1. Both carry a flat riemannian metric which correspond to the vacuum configuration $\{F_o\}$ discussed at the beginning. Both are candidates for E with $M=S^1$, $G=U(1)$, H trivial. But the internal U(1) acts globally on the two-torus but does not act globally on the Klein bottle. It is this Klein bottle example which is an archetype for the extended model. This model can be defined by the following axioms

1. There are two fibrations $\mathbb{G}$ and E over M.
2. The fibers G_x ($x\in M$) are groups which act transitively (from the right) on the fibers E_x of E.

3. There is an open covering (U_α) of M and, for each α, there are maps

$$\phi_\alpha : \pi_E^{-1}(U_\alpha) \to U_\alpha \times (H\backslash G)$$

$$\psi_\alpha : \pi_{\mathbb{G}} (U\ ) \to U_\alpha \times G$$

(ϕ_α and ψ_α are assumed to be diffeomorphisms and are called local trivializations of E and $\mathbb{G}$ respectively; G is a (compact) Lie group, and H is a closed Lie subgroup of G) such that ϕ_α restricts to group isomorphisms $G_x \to G$ on the fibers, and ψ_α satisfies

$$\psi_\alpha(ya) = \psi_\alpha(y)\phi_\alpha(a)$$

for all $y \in E_x$, $a \in G_x$, $x \in U_\alpha$.

The Klein bottle example is a particular case of such a structure. The group G is U(1) here, and the bundle of groups $\mathbb{G}$ coincides with E i.e. with the Klein bottle itself, in this case. The model of a global G-action considered in Sect.2. is a particular case of the above situation corresponding to the case of $\mathbb{G}$ being the global product $\mathbb{G}=M\times G$.

The important question to be answered reads as follows: what is the natural class of metrics on E? We will answer this question later on, where we will see that the effective gauge group, resulting from the class of metrics we will describe, consists of gauge bosons of the group $G_{eff} \approx N(H)/H \times G$ (modulo the common central factors). Here, anticipating the final result, we will first concentrate ourselves on the group theoretical structure arising from the discussed scheme .

Let us start with giving the precise definition of the group G_{eff}. To construct this group we will have to introduce the groups AutG and $\mathrm{Aut}_H G$. The group AutG of all automorphisms of G is a Lie group. This group, however, need not be compact even if G is such. Indeed, the group of automorphisms of the torus U(1)×U(1) contains the non-compact group SL(2,Z) (the map $(u,v) \to$ $\to (u^m v^n, u^k v^l)$, with $u,v \in U(1)$, is 1-1 if and only if $ml-nk=\pm 1$). If H is a Lie subgroup of G then $\mathrm{Aut}_H G$ will denote the subgroup of AutG consisting of those automorphisms ϕ of G for which $\phi(H)$ is conjugated to H:

$$\mathrm{Aut}_H G = \{\phi \in \mathrm{Aut}\, G : \exists a \in G,\ \ \phi(H) = aHa^{-1}\} .$$

Recall that an automorphism ϕ of G is called inner if there exists $a \in G$ such that $\phi(b)=aba^{-1}$ for all $b \in G$. The group of all inner automorphisms of G is an invariant subgroup of Aut G . Observe that all inner automorphisms belong automatically to $\mathrm{Aut}_H G$. Later we will be interested in the Lie algebra of $\mathrm{Aut}_H G$. Locally (i.e. in a neighborhood of the identity) the group $\mathrm{Aut}_H G$ is isomorphic to $G/Z(G)$, $Z(G)$ being the center of G. On the Lie algebra level therefore we have $\mathrm{Lie}(\mathrm{Aut}_H G)=\mathrm{Lie}(G) - \mathrm{Lie}(Z(G))$.

We will describe now the structure of the group G_{eff} - the effective gauge group arising from the non-invariant scheme. We will first describe the construction of G_{eff}, and only later justify it. The first step is to build the semidirect product $\bar{G}=G \otimes \mathrm{Aut}_H G$. The group $\bar{G}$ consists of pairs (a,ϕ), $a \in G, \phi \in \mathrm{Aut}_H G$, with the semidirect product multiplication law:

$$(a,\phi)(a',\phi') = (a\phi(a'),\phi\phi').$$

The group G_{eff} is then defined as

$$G_{eff} = N(\bar{H})/\bar{H},$$

where $\bar{H}=H \otimes \mathrm{id}$ is the subgroup of $\bar{G}$ which is isomorphic to H, and $N(\bar{H})$ is the normalizer of $\bar{H}$ in $\bar{G}$.

<u>Remark</u>. A similar construction of the effective gauge group appeared in studying symmetric Yang-Mills fields [9], with the difference that $\bar{G}$ there was the direct product $\bar{G}=G\times R$ of G and the initial gauge group R, and $\bar{H}$ was the diagonal $H\times\lambda(H)$, $\lambda : H \to R$ being a group homomorphism characterizing the action of G on a principal bundle carrying the initial gauge fields.

Some relevant information concerning the group G_{eff} is contained in the following diagram whose rows and columns are exact

$$\begin{array}{ccccccccc}
 & & 1 & & 1 & & 1 & & \\
 & & \downarrow & & \downarrow & & \downarrow & & \\
1 & \to & H & \to & N(H) & \to & N(H)/H & \to & 1 \\
 & & \downarrow & & \downarrow & & \downarrow & & \\
1 & \to & \bar{H} & \to & N(\bar{H}) & \to & N(\bar{H})/\bar{H} & \to & 1 \\
 & & \downarrow & & \downarrow & & \downarrow & & \\
 & & 1 & \to & \mathrm{Aut}_H G & \to & \mathrm{Aut}_H G & \to & 1
\end{array}$$

of particular interest is the last column which tells us that $G_{eff}=N(\bar{H})/\bar{H}$ is an extension of $Aut_H G$ by $N(H)/H$. In the compact case we therefore locally have $N(\bar{H})/\bar{H} \overset{loc}{\sim} N(H)/H \times Aut_H G \overset{loc}{\sim} (N(H)/H) \times (G/Z(G))$. To describe the Lie algebra of G_{eff} it is convenient to decompose the Lie algebra of G as follows (we assume that the action of G on G/H is effective, what implies that $H \cap Z(G)$ is trivial)

$$Lie(G) = Lie(H) + S ,$$

$$S = K + L ,$$

$$K = Z + K_1$$

where S is a reductive complement of Lie(H) in Lie(G), K is the subset of S consisting of H - singlets of the adjoint representation, Z is the Lie algebra of the center Z(G) of G, K_1 is a complement of Z in K , and L is a reductive complement of K in S.
The Lie algebra of G_{eff} is then given by

$$Lie(G_{eff}) = Z + K_1 + H + K_1 + L .$$

Observe that K_1 , which is the Lie algebra of $N(H)/(H\times Z)$, enters twice. For instance, if G is simple and H is trivial, then the effective gauge group is G×G. At this place it is to be stressed again that the extended, non-G-invariant scheme of dimensional reduction will, in general, lead to an *inconsistent* ansatz, unless one retains *all* the modes in the harmonic expansion of fields. Before, however, commenting on the problem of a harmonic expansion in the absence of global G-action, let us first show the relation of G_{eff} defined above to the extended scheme based on a pair $(E,\mathbb{G})$, $\mathbb{G}$ being a bundle of groups. The axioms for $(E,\mathbb{G})$ deal with local trivializations ϕ_α, ψ_α so that, on the intersections of the domains $U_\alpha \cap U_\beta$, we get transition functions $\phi_\alpha \circ \phi_\beta^{-1}$ and $\psi_\alpha \circ \psi_\beta^{-1}$. The function $\phi_\alpha \circ \phi_\beta^{-1}$ take value in the group $Aut_H G$ - the group we already know. The functions $\psi_\alpha \circ \psi_\beta^{-1}$ take value in another group - the group Taut(H\G) of *twisted automorphisms* of (H\G). The group N(H)/H can be identified with the group of automorphisms Aut(H\G) of the homogeneous space H\G. Recall that an automorphism of H\G is a map ψ: H\G → →H\G such that $\psi([a]b)=\psi([a])b$ for all $a,b \in G$. To every $n \in N(H)$ there corresponds an automorphism $\psi_n:[a]\to[na]$ of H\G, and the map $n \to \psi_n$ defines an isomorphism between N(H)/H and Aut(H\G). The group Taut(H\G) of twisted automorphisms of H\G contains Aut(H\G), and is defined as follows

Definition A twisted automorphism of $H\backslash G$ is a pair of diffeomorphisms $\phi : G \to G$, $\psi : H\backslash G \to H\backslash G$, where ϕ is an automorphism of G, and ψ satisfies

$$\psi([a]b) = \psi([a])\phi(b)$$

for all $a,b \in G$. The set of all twisted automorphisms $\{(\phi,\psi)\}$ of $H\backslash G$ is a group under the composition of maps, and is denoted by $Taut(H\backslash G)$.

Remark. We always assume that $H\backslash G$ is an *effective* homogeneous space. In this case the map ϕ in the formula above is completely determind by ψ.

Let us now show that $G_{eff}=N(\bar{H})/\bar{H}$ is indeed isomorphic to $Taut(H\backslash G)$. First of all observe that $(a,\phi) \in N(\bar{H})$ if and only if $\phi(H)=a^{-1}Ha$. Now, given $(\phi,\psi) \in Taut(H\backslash G)$ let $[a]=\psi([e])$. Then $[a]=\psi([e])=\psi([e]h)=\psi([e])\phi(h)=[a]\phi(h)$, and so $(\phi,a) \in N(\bar{H})$. It is easy to see that this gives rise to the required isomorphism.

Transition functions allow one to construct a principal bundle. Therefore, as a corollary to the above considerations we find: given an extended Kaluza-Klein scheme $(E,\mathbb{G})$ one can automatically construct two principal bundles: a principal bundle Q with structure group G_{eff}, and another principal bundle, with structure group $Aut_H G$. In fact the second bundle is the quotient of Q by $N(H)/H$ which is an invariant subgroup of G_{eff}. The group bundle $\mathbb{G}$ is a bundle associated to this quotient bundle. Knowing the above structure one can easily distinguish now a class of dimensionally reducible metrics on E which give rise to G_{eff} gauge boson on M. Namely, having the principal bundle Q over M, with structure group $G_{eff}=N(\bar{H})/\bar{H}$, we can construct an associated bundle $\bar{E}$ with fiber $\bar{H}\backslash\bar{G}$. The group $\bar{G}$ acts now globally on $\bar{E}$ from the right, and E can be identified with a quotient of $\bar{E}$ by $Aut_H G$, which is a subgroup of $\bar{G}$. The class of metrics on E which interests us is now defined as the Kaluza-Klein projections of $\bar{G}$-invariant metrics on $\bar{E}$. Indeed, every $\bar{G}$-invariant metric on $\bar{E}$ is, a fortiori, $Aut_H G$-invariant, and therefore projects onto $E=\bar{E}/Aut_H G$ by the Kaluza-Klein projection. This ansatz produces automatically gauge bosons of G_{eff} on H.

Let us finally comment on the problem of harmonic expansion in the absence of global G-action. A particular example we may keep in minds is that of harmonic expansion of fields defined on the Klein bottle. The important point to observe in this connection is that the harmonic expansion scheme is well defined provided the bundle $Q/(N(H)/H)$ is equipped with a *flat connection*. It does not mean it has to be trivial - by a *flat* connection we mean a connection with zero curvature but possibly nontrivial global holonomy group. (The ground state metric should be a natural source of such a connection). This flat connection will distinguish a class of trivializations of the associated group bundle $\mathbb{G}$, which will be related one to another by *constant* transition functions. Or, in other words, there is a restricted class of *local* G-actions of G on E, related one to another by constant automorphisms of G. And, as one can easily see, the method of harmonic expansion developed in Ref. [2] , can be applied to each local G-action with the results being independent of the choice of a local G-actions in the restricted class.

REFERENCES

1) M.J. Duff, "Recent Results in Extra Dimensions", CERN preprint TH.4243/85 see also references there.

2) R. Coquereaux and A. Jadczyk, "Harmonic expansion and dimensional reduction in G/H Kaluza-Klein theories", Class.Quantum Grav.2(1985)

3) M.J. Duff and C.N.Pope, "Consistent Truncations in Kaluza-Klein Theories", Nucl.Phys.B255,(1985)355-364

4) M.J. Duff, B.E.W. Nilsson, C.N. Pope and N.P. Warner, "On the consistency of the Kaluza-Klein Ansatz", Phys.Lett.149B(1984)90-94

5) R. Coquereaux and A. Jadczyk, "Consistency of the G-invariant Kaluza--Klein Scheme", CERN preprint TH.4305/85

6) R. Coquereaux and A. Jadczyk, "Geometry of Multidimensional Universes", Commun.Math.Phys.80(1983)79-100

7) D. Gromoll and W. Meyer, "An Exotic Sphere with Nonnegative Sectional Curvature", Ann of Math.100(1974)401-406

8) D. Freed and K.K. Uhlenbeck, "Instantons and Four-Manifolds", Springer-Verlag, New York-Berlin-Heidelberg-Tokyo, 1984

9) R. Coquereaux and A. Jadczyk, "Symmetries of Einstein-Yang-Mills Fields and Dimensional Reduction", Commun.Math.Phys.98(1985)79-104.

Part Two: Matter Coupling and Supersymmetry Breaking

N=2 Matter couplings in d=4 and 6 from Superconformal Tensor Calculus

A. Van Proeyen *

Istituto di Fisica Teorica - Università di Torino

Abstract

A review is given on the possible matter couplings in N=2, d=4 and d=6 and their construction using the superconformal tensor calculus. Especially the sector containing the scalars will be explained, giving rise to Kähler or quaternionic manifolds. In d=6 the treatment of a self dual tensor is crucial for the derivations of actions.

* On leave of absence from K.U. Leuven, Celestynenlaan 200 D, B - 3030 Leuven, Belgium.
Bevoegdverklaard navorser N.F.W.O., Belgium.

1. - Introduction.

I review mainly the work contained in refs. 1) and 2), where complete actions have been derived for the coupling of matter multiplets to N=2 supergravity in d=4 and d=6. I will treat the two theories parallel in the beginning, but when constructing the actions there are important differences. The importance of the self dual tensors will then manifest itself in d=6. Real self dual tensors can exist in dimensions d=2 + 4n. The treatment of self dual tensors is thus also important in the construction of 10-dimensional theories.

We use the superconformal tensor calculus[3),4)] to construct general matter couplings. For reviews explaining details we refer to ref. 5). The essential method consist of constructing Poincaré gravity theories by first starting from the larger conformal group. E.g. for a scalar field the conformal invariant action is[*]

$$\mathcal{L} = \sqrt{g}\left(\frac{1}{2}(\partial_\mu A)^2 - \frac{1}{12} R A^2\right) \tag{1.1}$$

The scalar A will not represent a dynamical degree of freedom, but compensates for the dilatational invariance of (1.1)

$$\begin{aligned} \delta_D A &= \Lambda_D A \\ \delta_D g_{\mu\nu} &= -2\Lambda_D g_{\mu\nu} \end{aligned} \tag{1.2}$$

In the dilatational-gauge

$$D\text{-gauge}: \quad A = \sqrt{6} \tag{1.3}$$

the action (1.1) is reduced to Poincaré gravity. Schematically we depict this procedure in fig. 1.

----

* For conventions we refer to 1,2) and 5).

Fig. 1: Gravity from conformal gravity

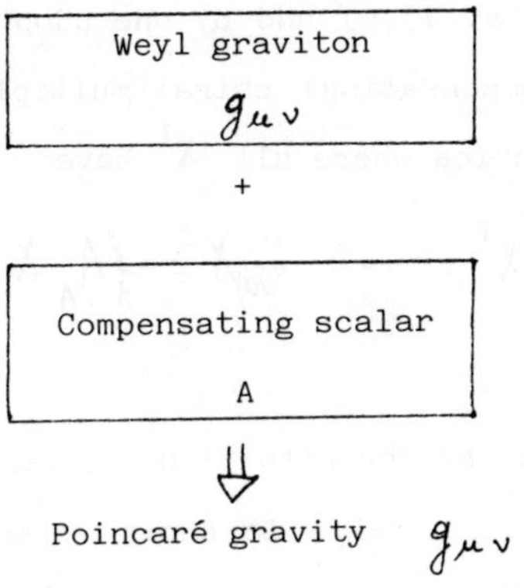

2. - The example of N=1 supergravity.

In more complicated theories as supersymmetry it pays off to start with as much symmetry possible. In this way the construction of N=1 matter couplings in d=4 has been facilitated[6]. One starts from the superconformal group[7] which contains the 15-parameter conformal group (General coordinate transformations GCT, dilatations D, Lorentz rotations M_{ab} and special conformal transformations K_a). When enlarging it with a supersymmetry (Q), one needs an extra supersymmetry S and moreover a U(1) in order to satisfy the Jacobi identities. This group is gauged[3),8)] by a "Weyl multiplet" (e_μ^a, ψ_μ, A_μ). The gauge fields of M_{ab}, S and K_a are combinations of the mentionned ones as is well known for ω_μ^{ab}, gauge field of M^{ab}. In principle we should include b_μ, gauge field of D, in the Weyl multiplet. We take however always $b_\mu = 0$ as K-gauge choice. This was already implicit in the gravity example above, and we will not mention this further on. The independent remaining gauge fields are thus those of GCT, Q and U(1). Then we introduce matter multiplets and for simplicity I will restrict myself here to chiral multiplets (X^I, Ω^I, h^I). I will choose here the Weyl weight of all chiral multiplets equal to one. The Weyl weight of X^I is a free parameter and different choices amount to field redefinitions. In refs. 1),5) and 6) one adopted zero Weyl weight for all but one (the compensating) chiral multiplet, but here we prefer the more symmetric choice where all X^I have

$$\delta_D X^I = \Lambda_D X^I \quad \rightarrow \quad \delta_{U(1)} X^I = -\frac{i}{3}\Lambda_A X^I \tag{2.1}$$

This is inspired by the situation in N=2, but this formulation has also advantages for N=1. It shows some symmetries explicit, as e.g. the SU(1,1) in the flat potential model[9). One can also gauge more general groups, which have the compensating and the other chiral multiplets in one representation. In this way one obtains gaugings of transformations which are non-linear in the independent fields as in ref. 10). It is well known that the kinetic part of the actions for these multiplets are obtained from $\int d^4\theta$ of a real superfield. (I will also neglect the superpotential $\int d^2\theta\, g(X)$ here). This can now be made superconformal invariant. One needs then a real multiplet $N(X^I, \bar{X}^I)$ (multiplets are identified by their first component). To get a superconformal invariant action from N, this function should be of "Weyl weight 2", and U(1) invariant, i.e.

$$\left.\begin{aligned} 2N(X,\bar{X}) &= N_I X^I + N_{\bar{I}} \bar{X}^I \\ 0 &= N_I X^I - N_{\bar{I}} \bar{X}^I \end{aligned}\right\} N = X^I N_{I\bar{J}} \bar{X}^J \tag{2.2}$$

As explained in 5),6) the bosonic part of the action which one obtains (omitting the trivial auxiliary fields h^I)

$$\mathcal{L}_B = e N_{I\bar{J}} \mathcal{D}_\mu X^I \mathcal{D}^\mu \bar{X}^J - \frac{1}{6} e R X^I N_{I\bar{J}} \bar{X}^J \tag{2.3}$$

$\mathcal{D}_\mu X^I$ is covariant w.r.t. the U(1) in the superconformal group. Its charge is proportional to the Weyl weight. As we took these all equal, we have

$$\mathcal{D}_\mu X^I = \partial_\mu X^I + \frac{i}{3} A_\mu X^I \tag{2.4}$$

Eq. (2.3) is similar to (1.1) and the D-gauge choice can then also be taken as

$$\text{D-gauge}: \quad X^I N_{I\bar{J}} \bar{X}^J = 3 \tag{2.5}$$

In this way there is no mixing between $g_{\mu\nu}$ and scalars in the kinetic part (In other words, the Einstein term gets its standard form). This avoids later rescalings. The condition (2.5) eliminates one of the components in terms of the others. In fact this is only one real component, e.g. the modulus of X^0, but the U(1) invariance makes its phase also a gauge degree of freedom. The independent component will be called

$$z^I = \frac{X^I}{X^0} = (1, z^A) \qquad \begin{array}{l} I = 0, 1, \ldots, n \\ A = 1, \ldots, n \end{array} \tag{2.6}$$

These Z^A are the Weyl weight zero, and U(1) neutral fields used in 6).

The A_μ field is auxiliary, its field equation is

$$i\,(X^I \overleftrightarrow{\mathcal{D}}_\mu \bar{X}^J)\, N_{I\bar{J}} = 0 \tag{2.7}$$

Using (2.5-7) in (2.3) one obtains

$$\mathcal{L}_B = -\frac{1}{2} eR + e\, \mathcal{G}_{A\bar{B}}\, \partial_\mu z^A\, \partial_\mu \bar{z}^B \tag{2.8}$$

The kinetic terms of the scalars define a metric on the n-dimensional manifold in which the scalars take values. As there are only $\partial_\mu z\, \partial^\mu \bar{z}$ and no $\partial_\mu z\, \partial^\mu z$ or $\partial_\mu \bar{z}\, \partial_\mu \bar{z}$ terms, the metric is said to be "hermitian" (see e.g. the review in ref. 11). If one has found such a parametrization the manifold is hermitian and one can check whether it is Kählerian. On Kähler manifolds the hermitian metric is the second derivative of a scalar function $G(z,\bar{z})$

$$\mathcal{G}_{A\bar{B}} = \frac{\partial}{\partial z^A} \frac{\partial}{\partial \bar{z}^B} \mathcal{G}(z,\bar{z}) \tag{2.9}$$

In the N=1 matter couplings this is indeed the case with

$$\mathcal{G}(z,\bar{z}) = 3\ln\left(\tfrac{1}{3} z^I N_{I\bar{J}}(z,\bar{z})\,\bar{z}^J\right) \tag{2.10}$$

One has thus proven that the N=1 supergravity-chiral matter couplings give rise to a Kählerian scalar manifold[10]. In fact all Kählerian manifolds can occur, because (2.10) is invertible. I.e. for all "Kähler potentials G" there is a function

$$N(X,\bar{X}) = 3X^0\bar{X}^0 \exp\tfrac{1}{3}\mathcal{G}\left(\frac{X^A}{X^0}, \frac{\bar{X}^A}{\bar{X}^0}\right) \tag{2.11}$$

In the fermionic part a similar procedure has to be worked out. An S-gauge condition eliminates Ω^0 in function of the Ω^A. For the pure supergravity case (n=0) we write this schematically in fig. 2

Fig. 2: N=1 supergravity (old minimal) from superconformal

Weyl multiplet $e_\mu^{\ a}, \psi_\mu, A_\mu$

+

Compensating chiral multiplet X, Ω, h

⇓

Poincaré supergravity: $e_\mu^{\ a}, \psi_\mu, A_a, h$

3. - N=2 multiplets.

N=2 implies that in our super(conformal) group we will introduce apart from the Weyl group, 2 supersymmetries Q^i (i=1,2) and we will need also 2 special supersymmetries S^i. To obtain a supergroup one needs now also

$$\begin{aligned} &U(1) \otimes SU(2) && \text{for } d=4 \\ &SU(2) && \text{for } d=6 \end{aligned} \tag{3.1}$$

The Weyl multiplet contains now apart from gauge fields also "matter" (=non-gauge) fields. The multiplets consist of [2),4)]

$$
\begin{array}{lll}
\text{Weyl multiplet} & & \\
d=4 & \left(e_\mu{}^a, \psi_\mu{}^i, V_\mu{}^i{}_j, A_\mu, T_{ab}, \chi^i, D \right) & 24+24 \\
d=6 & \left(e_\mu{}^a, \psi_\mu{}^i, V_\mu{}^i{}_j, T^-_{abc}, \chi^i, D \right) & 40+40
\end{array}
\tag{3.2}
$$

After the graviton and gravitini, one finds in each case the gauge fields according to (3.1). Then there is an antisymmetric tensor, which in 6 dimensions satisfies an antiselfduality condition, a doublet spinor and a real scalar. The construction of these multiplets in non-trivial, but we refer to refs. 2),4) and 5). Having obtained this gauge multiplet one can consider other representations of the N=2 superconformal algebras.

Vector multiplets[4,12)] consist of the fields

$$
\begin{array}{lll}
\text{Vector multiplet} & d=4 & \left(X, \Omega^i, W_\mu, Y^{ij} \right) \\
 & & \qquad\qquad 8+8 \\
 & d=6 & \left(\Omega^i, W_\mu, Y^{ij} \right)
\end{array}
\tag{3.3}
$$

They contain in each case a doublet spinor, the real gauge vector, and a SU(2) triplet real (auxiliary) field. In 4 dimensions there is moreover a complex scalar, which can be understood as the fifth and sixth component of the W_μ of d=6. The gauge fields can gauge an an arbitrary group G commuting with the superconformal group. In this case all the fields in (3.3) are in the adjoint of G, and I give them an extra index I. In the supersymmetry algebra occurs a "covariant general coördinate transformation"

$$
\left[\delta_Q(\epsilon_1), \delta_Q(\epsilon_2) \right] = \delta_{CGCT}\left(\tfrac{1}{2} \bar{\epsilon}_2 \gamma_\mu \epsilon_1 \right) + \dots \tag{3.4}
$$

which includes, after gauging a group G, also a term

$$
\delta_{CGCT}(\xi^\mu) = \dots + \delta_G\left(-\xi^\mu W_\mu{}^I \right) \tag{3.5}
$$

According to my previous remark that the W_μ of d=6, contains the X of d=4, we expect therefore also a new term in d=4. Indeed we find

$$\underset{\text{in } d=4}{[\delta_Q(\epsilon_1), \delta_Q(\epsilon_2)]} = \dots + \delta_G(-4 X^I \bar{\epsilon}_{1i} \epsilon_{2j} \epsilon^{ij} + h.c.) \tag{3.6}$$

So, if X^I acquires a non-zero v.e.v. the corresponding generator is a central charge.

The scalar multiplet[1)13)] contains only spin 0 and ½ fields

$$\text{Scalar multiplet} \quad (A^i_\alpha, \zeta_\alpha) \qquad \alpha = 1, \dots, 2r \tag{3.7}$$

There are 2r Majorana spinors, and 4r fields, which satisfy a reality condition

$$\overline{(A^i_\alpha)} = \rho^{\alpha\beta} \epsilon_{ij} A^j_\beta \tag{3.8}$$

where $\rho^{\alpha\beta}$ is a matrix satisfying $\rho\bar{\rho} = -1$, which by redefinitions can be brought in the form

$$\rho = \begin{pmatrix} 0 & 1 & & & \\ -1 & 0 & & & \\ & & 0 & 1 & \\ & & -1 & 0 & \\ & & & & \ddots \end{pmatrix} \tag{3.9}$$

This implies in fact that the A^i_α are r quaternions. The index α can label also a representation of the gauge group G

$$\delta_G A^i_\alpha = \Lambda^I t_{I\alpha}{}^\beta A^i_\beta \tag{3.10}$$

This transformation should respect (3.8), therefore the part of G which acts non-trivially on the scalar multiplet should be subgroup of the general linear transformation group on r quaternions $Gl(r, Q)$.

Counting the number of real degrees of freedom one has at this point $4r$ bosonic and $8r$ fermionic components. So one can expect that the algebra is not closed on (3.7). In fact one has closure iff

$$\Gamma_\alpha \equiv \not{D} \zeta_\alpha + \ldots = 0$$
$$\mathcal{C}^i_\alpha \equiv D^a D_a A^i_\alpha + \ldots = 0 \qquad (3.11)$$

So these equations have the form of field equations and as we will show in section 4 one can indeed obtain an action such that (3.11) are field equations. Then the scalar multiplet is an on-shell multiplet. That is indeed the full situation in d=6. But in d=4 scalar multiplets exist off-shell if the algebra contains central charge. I will now describe how this is achieved in the tensor calculus.

Assuming that a gauge group acts on (3.7), the covariant derivatives in (3.11) should contain resp. $-\gamma^\mu W^I_\mu t_{I\alpha}{}^\beta \zeta_\beta$ and $W^I_\mu W^{\mu J}(t_I t_J A^i)_\alpha$. This being the case also in d=6 we expect also here X-dependent terms in (3.11) in d=4. And indeed we can for d=4 identify further terms in (3.11)

$$\Gamma_\alpha \equiv \not{D} \zeta_\alpha + 2g X^I (t_I \zeta)_\alpha + \ldots = 0$$
$$\mathcal{C}^i_\alpha \equiv D^a D_a A^i_\alpha + 2g^2 X^I \bar{X}^J \left(\{t_I, t_J\} A^i\right)_\alpha + \ldots = 0 \qquad (3.12)$$

(I mention here only relevant terms and omit unimportant factors. The exact formulae should be found in ref. 1). By choosing now some special transformation with corresponding $\langle X^I \rangle \neq 0$, (3.12) will become "conventional constraints". Let me introduce an infinite number of 2 r component ($\alpha = 1,\ldots,2r$) scalar multiplets. The first transforms in the second one under the action of the transformation "z", the second transforms in the third,...

$$\begin{pmatrix} A^i_\alpha \\ \zeta_\alpha \end{pmatrix} \xrightarrow{z} \begin{pmatrix} A^{i\,(z)}_\alpha \\ \zeta^{(z)}_\alpha \end{pmatrix} \xrightarrow{z} \begin{pmatrix} A^{i\,(zz)}_\alpha \\ \zeta^{(zz)}_\alpha \end{pmatrix} \xrightarrow{z} \begin{pmatrix} A^{i\,(zzz)}_\alpha \\ \zeta^{(zzz)}_\alpha \end{pmatrix} \rightarrow \qquad (3.13)$$

So at this point we have an infinite number of fields, which we can interpret as a system with an extra bosonic degree of freedom[14] as in the "harmonic superspace"[15]. We suppose then that $\langle X^z \rangle \neq 0$ such that (3.12) for the multiplet (A, ζ) determines $\zeta^{(z)}$ and $A^{(zz)}$. Applied on ($A^{(z)}$, $\zeta^{(z)}$) it gives $\zeta^{(zz)}$ and $A^{(zzz)}$, ... Finally only $A_\alpha^{i(z)}$ remains as extra (auxiliary) field, but the constraints are solved and one has an off-shell $8\,r + 8\,z$ component multiplet. As we have supposed $\langle X^z \rangle \neq 0$ it follows from (3.6) that Z is a "central charge transformation".

So in conclusion we have an on-shell scalar multiplet in d=6 and we can extend it to an off-shell multiplet in d=4 by introducing a gauged central charge.

<u>The linear multiplet</u> consist of fields

$$\text{Linear multiplet } d=4 \;(L^{ij}, \varphi^i, E_a, G) \qquad d=6 \;(L^{ij}, \varphi^i, E_a) \tag{3.14}$$

L^{ij} is a real triplet, φ^i a doublet spinor, E_a is a vector which satisfies a constraint

$$D^a E_a + \dots = 0 \tag{3.15}$$

G is the complex scalar which can be understood from the dimensional reduction of E_a from 6 to 4 dimensions.

In principle the linear multiplet can transform in a representation of the gauge group. However, if it is gauge inert then the constraint (3.15) can be solved for an antysimmetric tensor

$$\begin{aligned} E_a &= \tfrac{i}{2}\, e^{-1} e_{a\mu}\, \varepsilon^{\mu\nu\rho\sigma} \mathcal{D}_\nu E_{\rho\sigma} & d=4 \\ &= \tfrac{i}{24}\, e^{-1} e_{a\mu}\, \varepsilon^{\mu\nu\rho\sigma\lambda\tau} \mathcal{D}_\nu E_{\rho\sigma\lambda\tau} & d=6 \end{aligned} \tag{3.16}$$

<u>The nonlinear multiplet</u> is introduced in d=4 and d=6 in references 4) and 2). We will skip it here.

In d=6 there is still the <u>tensor multiplet</u>. Our first trial started from independent fields (σ, ψ^i, $B_{\mu\nu}$) where σ is a real scalar and $B_{\mu\nu}$ a gauge antisymmetric tensor. The transformations on these fields close only if one imposes constraints

$$\Gamma^i \equiv \hat{\not{\mathcal{D}}}\psi^i - \frac{1}{6}\sigma\chi^i - \frac{1}{12}\gamma\cdot T^-\psi^i = 0$$
$$\mathcal{C} \equiv \hat{\mathcal{D}}^a\hat{\mathcal{D}}_a\sigma - \frac{1}{6}\sigma D + \frac{1}{3}F^{+abc}T^-_{abc} + \frac{7}{6}\bar{\chi}\psi = 0$$
$$\mathcal{F}^-_{\mu\nu\rho} \equiv F^-_{\mu\nu\rho}(B) - 2\sigma T^-_{\mu\nu\rho} = 0 \qquad (3.17)$$

where $F^-(B)$ is the antiselfdual part of the B-curvature. We intended to obtain an action of the form

$$e^{-1}\mathcal{L}_T = \sigma\mathcal{C} - 4\bar{\psi}\Gamma - \frac{1}{6}F^+_{\mu\nu\rho}\mathcal{F}^{-\mu\nu\rho} + \bar{\psi}_\mu\gamma^\mu\sigma\Gamma \qquad (3.18)$$

The action is invariant at the linearized level, but turned out to be non invariant in order ψ_μ

$$\delta\mathcal{L}_T = \bar{\epsilon}\gamma^a\psi^\mu F^-_{abc}F^-_\mu{}^{bc} \qquad (3.19)$$

This system would describe the minimal coupling of tensor multiplets with supergravity. If it would work, one could couple an arbitrary number of tensor multiplets to d=6 supergravity. However, we will be able to couple only one tensor multiplet and this one is also necessary to obtain d=6 supergravity. This is due to difficulties with Lorentz-invariant couplings of (anti) selfdual tensors[16]. These occur in tensor multiplets and in the pure supergravity multiplet. In ref. 17) one coupled one tensor multiplet to supergravity. In that case there is one self dual and one antiselfdual tensor from which a Lorentz invariant action can be obtained. In the group manifold approach one constructed a pure d=6 supergravity theory[18]. One obtained also constraints analogous to our eqs. (3.17) for the tensor multiplets. The first two are field equations so they pose no problem for an on-shell description. The last one is a problem in ordinary field theory. In the group manifold approach it is a field equation for a field in a Q-direction. Therefore there is a consistent set of field equations transforming among themselves. The Lagrangian is not invariant however, again because of terms similar to (3.19).

We now turn to a different approach to the tensor multiplet which leads to a consistent action for one tensor multiplet coupled to supergravity and other matter multiplets (scalar, vector, linear). One starts as in superspace from a multiplet containing

tensor multiplet $(\sigma, \psi^i, F^+_{abc})$ (3.20)

where the last field is just a self dual antisymmetric tensor. The constraints for closure are then the first two of (3.17) and a new one

$$\Gamma^i = \mathcal{C} = 0$$
$$G_{ab} = D^c(F^+_{abc} - 2\sigma T^-_{abc}) - \hat{\hat{R}}_{ab}(Q)\psi - \tfrac{1}{6}\bar{\chi}\gamma_{ab}\psi \qquad (3.21)$$

But now we will solve the constraints as we will take the domain $\sigma \neq 0$. For the first two this defines

$$\chi = \sigma^{-1}\left(6\hat{\not{D}}\psi - \tfrac{1}{2}\gamma\cdot T^-\psi\right)$$
$$D = \sigma^{-1}\left(6\hat{D}^a\hat{D}_a\sigma + 2F^+_{abc}T^{-abc} + 7\bar{\chi}\psi\right) \qquad (3.22)$$

The last one can be interpreted as a Bianchi identity

$$F^+_{abc} + 2\sigma T^-_{abc} = 3\partial_{[a}B_{bc]} + 3\bar{\psi}_{[a}\gamma_{bc]}\psi + \tfrac{3}{2}\bar{\psi}_{[a}\gamma_b\psi_{c]}\sigma \qquad (3.23)$$

So this means that the fields χ^i, D and T^-_{abc} of the Weyl multiplet are not independent fields any more. They have been replaced by ψ^i, σ, and $B_{\mu\nu}$, which keeps the number of components the same. So in this way we have obtained a second Weyl multiplet

d=6 second version of Weyl multiplet
$(e_\mu{}^a, \psi_\mu{}^i, V_\mu{}^i{}_j, \sigma, \psi^i, B_{\mu\nu})$ (3.24)

This has been explained in more detail in the contribution of E. Bergshoeff to this conference. We still observe that in this second version of the Weyl multiplet the group is in fact enlarged by the gauge transformation of the antisymmetric tensor. This transformation occurs also in the $\{Q,Q\}$ anticommutator

$$[\delta(\epsilon_1), \delta(\epsilon_2)] = \ldots \quad + \delta_B\left(\tfrac{1}{2}\bar{\epsilon}_2\gamma_\nu\epsilon_1\sigma\right) \qquad (3.25)$$

which remains in the rigid symmetry limit as we imposed $\Gamma \neq 0$.

Before starting the section on actions we want to repeat which multiplets are on-shell and off-shell. In fact if multiplets are off-shell, one can consider different actions which include this multiplet. If a multiplet is on-shell, its field equation is essentially determined, so there is only one action. E.g. vector multiplets should better be off-shell because we want to be able to obtain its kinetic action (Yang-Mills action) but they should also appear in scalar multiplet actions if these transform in a gauge group. Obviously the Weyl multiplet appears in all actions and is off-shell. So is also the linear multiplet. The scalar multiplet is on-shell but can in d=4 be made off-shell by introducing central charges. We will however see in the next section that this off-shell extension did not allow to contruct more general actions. In our approach to the d=6 tensor multiplet we solved the constraints such that this system is also off-shell.

4. - Actions in d=4 N=2.

Vector multiplets in d=4 are essentially chiral superfields ($\bar{D}X = 0$) which satisfy an extra constraint ($D^{ij} X = \varepsilon^{ik}\varepsilon^{jl}\bar{D}_{kl}\bar{X}$). So we can construct a holomorphic function of X which defines a new chiral superfield from which an action can be obtained by a local superconformal generalization of

$$\int d\theta \; F(X^I) \tag{4.1}$$

To be able to define this action superconformal invariant, the Weyl Weyl weight of F should be 2, while X has weight 1, so

$$F(\lambda X^I) = \lambda^2 F(X^I) \tag{4.2}$$

By defining

$$N_{IJ} = \frac{1}{2} Re \left(\frac{\partial}{\partial X^I} \frac{\partial}{\partial X^J} F(X) \right) \tag{4.3}$$

we can write some relevant terms of the action as[1)]

$$e^{-1}\mathcal{L}_F = N_{IJ}\,\mathcal{D}_\mu X^I \mathcal{D}^\mu \bar{X}^J + (X^I N_{IJ}\bar{X}^J)(-\tfrac{1}{6}R + D)$$

$$+\frac{1}{8}N_{IJ}F^I_{\mu\nu}F^{\mu\nu J} - \frac{1}{64}(F_{IJ}-\bar{F}_{IJ})e^{-1}\varepsilon^{\mu\nu\rho\sigma}F^I_{\mu\nu}F^J_{\rho\sigma} + \dots \qquad (4.4)$$

One can notice that if F_{IJ} are imaginary constants, $N_{IJ} = 0$ and (4.4) reduces to a total divergence. The superspace or tensor calculus construction used a function F which is invariant under the gauge transformation

$$\delta_g X^I = \Lambda^K X^J f^I_{JK} \qquad (4.5)$$

However because of the previous remark, one could consider

$$\delta_g F(X) = ig\,\Lambda^I C_{I,JK} X^J X^K \qquad (4.6)$$

where the coefficients $C_{I,JK}$ are real. For space-time dependent Λ^I the variation of $\mathcal{L}_F$ is then not yet a total divergence, but

$$\delta_g \mathcal{L}_F = -\frac{i}{16}g\,\varepsilon^{\mu\nu\rho\sigma}\Lambda^I C_{I,JK} F^J_{\mu\nu}F^K_{\rho\sigma} \qquad (4.7)$$

We can add a term to the previous action to reobtain invariance as well w.r.t. gauge as to supersymmetry transformations.

$$\mathcal{L}_{F,add} = -\frac{i}{6}g\,\varepsilon^{\mu\nu\rho\sigma}C_{I,JK}W^I_\mu W^J_\nu\left(\partial_\rho W^K_\sigma - \frac{3}{16}g f^K_{LM}W^L_\rho W^M_\sigma\right) \qquad (4.8)$$

For the <u>scalar multiplets</u> we can also use some tensor calculus, but after elimination of the field $A^{(z)}$, by its field equation, we find only one type of action

$$e^{-1}\mathcal{L}_S = \left(A^\alpha_i \mathcal{C}^i_\beta + 2\bar{\zeta}^\alpha\Gamma_\beta + \bar{\Psi}^i_\mu\gamma^\mu A^\alpha_i\Gamma_\beta\right)d_\alpha{}^\beta \qquad (4.9)$$

$\mathcal{C}$ and Γ are the field equations introduced in (3.12) where the Z-transformation of (3.13) is not any more taken into account, so they are not identical to zero. The coefficients $d_\alpha{}^\beta$ are constants which form an hermitian matrix satisfying the quaternionic condition

$$\bar{d} = \rho^{+} d \rho \tag{4.10}$$

By field equations it is equivalent to one of the form

$$d = \begin{pmatrix} 1\!\!1_{2p} & 0 \\ 0 & -1\!\!1_{2q} \end{pmatrix} \tag{4.11}$$

where ρ is in the form (3.9). The gauge transformations allowed in (4.9) should now also respect d, which restricts them to

$$U(p, q, Q) \cong USp(2p, 2q) \tag{4.12}$$

After integrations by parts, ... (4.9) can be seen to have the bosonic content

$$e^{-1}\mathcal{L}_{S,B} = \Big[-\mathcal{D}_\mu A^i_\beta \mathcal{D}^\mu A^\alpha_i + \tfrac{1}{6} R A^i_\beta A^\alpha_i + \tfrac{1}{2} D A^i_\beta A^\alpha_i + 4g^2 A^i_\beta (t_I t_J)^\alpha{}_\gamma A^\gamma_i \bar{X}^I X^J + g A^i_\beta t_I{}^\alpha{}_\gamma A^\gamma_k Y^{Ijk} \varepsilon_{ij} \Big] d^\beta_\alpha \tag{4.13}$$

So the only freedom in the scalar multiplet action consists of the signature (p, q) in (4.11), and in the gauge group which must be a subgroup of (4.12). If one has introduced a gauge group with corresponding $\langle X^I \rangle \neq 0$ then the fields acquire a mass and this transformation is a central charge (see (3.6)). At this point one would argue that the signature should be completely + (q = 0) for positivity of the kinetic energy. I will in a moment explain when you need-signs.

Let me <u>combine</u> now the previous results (4.4) and (4.13). The terms which are important for this discussion are

$$\begin{aligned} \mathcal{L} = &-\tfrac{1}{2} e R \left(\tfrac{1}{3} X^I N_{IJ} \bar{X}^J - \tfrac{1}{3} A^\alpha_i d^\beta_\alpha A^i_\beta \right) \\ &+ e D \left(X^I N_{IJ} \bar{X}^J + \tfrac{1}{2} A^\alpha_i d^\beta_\alpha A^i_\beta \right) \\ &+ e N_{IJ} \mathcal{D}_\mu X^I \mathcal{D}^\mu \bar{X}^J - e d^\beta_\alpha \mathcal{D}_\mu A^\alpha_i \mathcal{D}^\mu A^i_\beta + \dots \end{aligned} \tag{4.14}$$

It is important to realize that X^I transforms under U(1) and is an SU(2) singlet, while A_i^α is U(1) inert, but is an SU(2) doublet

$$\mathcal{D}_\mu X^I = (\partial_\mu + i A_\mu) X^I \quad ; \quad \mathcal{D}_\mu A_i^\alpha = \partial_\mu A_i^\alpha - V_{\mu i}{}^j A_j^\alpha \tag{4.15}$$

Then I impose, as in section 2, a D-gauge which puts the part multiplying $-\frac{1}{2} eR$ equal to one, and use the field equation for the field D of the Weyl multiplet. This gives

$$X^I N_{IJ} \bar{X}^J = 1 \quad ; \quad A_i^\alpha d_\alpha{}^\beta A_\beta{}^i = -2 . \tag{4.16}$$

These equations define the scalars of one vector and one scalar multiplets. In fact (4.16) defines only their moduli, but their phases are then fixed by a gauge choice of respectively U(1) and SU(2). As in (2.6) I can define their independent coordinates

$$z^I = \frac{X^I}{X^0} \qquad ; \quad B_a{}^\alpha = A^{-1}{}^i{}_a A_i^\alpha \tag{4.17}$$

where a runs over 1,2 and X^0 and A_a^i are the scalars for which (4.16) is solved. Using the A_μ and $V_\mu{}^i{}_j$ field equations

$$\left(X^I \overleftrightarrow{\mathcal{D}}_\mu \bar{X}^J\right) N_{IJ} = 0 \quad ; \quad \left(A_i^\alpha \overleftrightarrow{\mathcal{D}}_\mu A_\beta^j\right) d_\alpha^\beta = 0 \tag{4.18}$$

one obtains then

$$\mathcal{L} = -\frac{1}{2} eR + e \mathcal{G}_{A\bar{B}} \partial_\mu z^A \partial^\mu \bar{z}^B + 2e \Delta_\alpha^\beta \partial_\mu B_i^\alpha \partial^\mu B_\beta^i \tag{4.19}$$

as kinetic terms for the graviton and the scalars. As in (2.9), (2.10) $\mathcal{G}_{A\bar{B}}$ defines a Kähler metric on the manifold of the Z-fields. Δ_α^β defines a quaternionic metric

$$\begin{aligned} \Delta_\alpha^\beta &= d_\alpha^\beta C^{-1} - 2 d_\alpha^\gamma B_\gamma^a B_a^\delta d_\delta^\beta C^{-2} \\ C &= B_\alpha^a d_\beta^\alpha B_a^\beta \end{aligned} \tag{4.20}$$

So the scalars of the scalar multiplets define a quaternionic manifold[19]. Before I give some more results on both manifolds, let me repeat what has been used to obtain pure N=2 supergravity ($n=0 \rightarrow I=0$; $r=1 \rightarrow \alpha=1,2$). Above I wrote that X^0 and A_i^α are eliminated by dilatational gauge and field equation for D. Their fermionic partners Ω^0 and ζ_α are likewise eliminated by an S-gauge and the χ^i field equations. In the vector multiplet (where Y_{ij}^0 is also an auxiliary field) remains the vector which is identified with the physical vector in N=2 supergravity. Its presence is a sign that we need this multiplet to be able to obtain N=2 supergravity. In schematic form we obtained the theory[4)] as in fig. 3.*

fig. 3: N=2, d=4

Weyl multiplet (e_μ^a, ψ_μ^i, $V_\mu{}^i{}_j$, A_μ, T_{ab}, χ^i, D)

+

Compensating vector (X^0, Ω^0, W_μ^0, Y_{ij}^0)

+

Compensating scalar (A_i^α, ζ^α, $A_i^{\alpha(z)}$)

$\Downarrow$

Poincaré supergravity (e_μ^a, ψ_μ^i, $V_a{}^i{}_j$, A_a, T_{ab}, χ^i, D, W_μ^0, Y_{ij}^0, u, $A_i^{\alpha(z)}$

One can obtain gauged N=2 supergravity theories in this formalism if the scalar compensating multiplet transforms under a gauge group. In this way one can obtain either SU(2) or U(1) gauged supergravity's. In the latter case the U(1) can be gauged[4,20)] by the vector compensating multiplet, or a physical vector multiplet or a combination of both. In this way one reobtains[20)] the different versions of gauged N=4 supergravity's[21)]. Also the compensating vector multiplet can be part of a non-compact gauge multiplet as e.g. for SO(2,1)[22)] or SU(3,1)[23)].

Consider in more detail the scalar multiplet couplings. As mentionned above one scalar multiplet does not consist of physical fields, but of auxiliary and gauge degrees of freedom. The second equation

----

* One real field out of X^0, A_i^α, denoted as u is auxiliary.

in (4.16) implies that in (4.11) there should be at least 1 minussign. The corresponding fields are then the non-physical ones. The isometry group is then USp (2p,2) and the scalars span the coset

$$\mathbb{H}P(p) = \frac{USp(2p,2)}{USp(2p)\otimes USp(2)} \tag{4.21}$$

In ref. 24 an extension of this mechanism is given. A U(1) group acting on the scalar multiplets is gauged by a vector multiplet for which no kinetic terms are introduced. The field equations of its auxiliary fields Y_{ij} imply then constraints on the scalar manifold and it is shown that one scalar multiplet is eliminated. Also it turns out that one has to start now with $d_\alpha{}^\beta$ (4.11) with q=2. The obtained manifolds are

$$X(p) = \frac{SU(p,2)}{SU(p)\otimes SU(2)\otimes U(1)} \tag{4.22}$$

In fact we suppose that this mechanism can be extended to construct all quaternionic manifolds listed in ref. 19 or 25.

$$\begin{array}{lll} \mathbb{H}P(n) \quad (\dim_H = n) & X(n) \quad (\dim n) & Y(n) = \dfrac{SO(n,4)}{SO(n)\otimes SO(4)} \quad (\dim n) \\ \dfrac{G_2}{SU(2)\otimes SU(2)} \quad (\dim 2) & \dfrac{F_4}{USp(6)\otimes USp(2)} \quad (\dim 7) & \dfrac{E_6}{SU(6)\otimes SU(2)} \quad (\dim 10) \\ \dfrac{E_7}{SO(12)\otimes SU(2)} \quad (\dim 16) & \dfrac{E_8}{E_7\otimes SU(2)} \quad (\dim 28) & \end{array} \tag{4.23}$$

The Kähler manifold spanned by the scalars of the vectormultiplet is not the most general one as it is the case in N=1. In N=1 we started with arbitrary functions of X and $\bar{X}$ while here one starts with a holomorphic function F(X). E.g. all the symmetric spaces which are Kählerian are the noncompact ones

$$
\begin{array}{lll}
I_{p,q} = \dfrac{SU(p,q)}{SU(p)\otimes SU(q)\otimes U(1)} & & \dim_{\mathbb{C}} = pq \\
II_m = \dfrac{SO^*(2m)}{U(m)} & & \frac{1}{2}m(m-1) \\
III_m = \dfrac{Sp(m)}{U(m)} & & \frac{1}{2}m(m+1) \\
IV_m = \dfrac{SO(m,2)}{SO(m)\otimes SO(2)} & & m \\
V = \dfrac{E_6}{SO(10)\otimes U(1)} & & 16 \\
VI = \dfrac{E_7}{E_6\otimes U(1)} & & 27
\end{array}
\tag{4.24}
$$

and their compact analogues. In this list of Cartan are the irreducible ones. One can moreover have reducible ones. In ref. 26) we showed that the N=2 vectormultiplets allow only noncompact manifolds, and only the following ones

$$
\begin{array}{l}
I_{1,m} \quad (\dim m) \\
III_3 \ (\dim 6)\ ;\ I_{3,3}\ (\dim 9)\ ;\ II_6\ (\dim 15)\ ;\ VI\ (\dim 27) \\
\text{Reducible}:\ I_{1,1}\otimes IV_n \quad (\dim n+1)
\end{array}
\tag{4.25}
$$

The first one is called minimal coupling because it is obtained by the function F

$$
F(X) = X^0X^0 - X^AX^A \tag{4.26}
$$

which is the straightforward choice as F should satisfy (4.2). The other ones allow gaugings such that supersymmetry is broken and the potential is zero. They were studied first in the context of N=2 d=5 supergravity and in their connection to the Jordan algebras [27]. In 4 dimensions they are studied thoroughly in ref. 28).

The actions for gauge vectors

$$\mathcal{L}_{YM} = -\frac{1}{4} e\, F_{\mu\nu}\, g^{\mu\rho} g^{\nu\sigma} F_{\rho\sigma} \tag{4.27}$$

are conformal invariant essentially because e has Weyl weight (-dimension), so -4 here. $g^{\mu\nu}$ has Weyl weight 2, and curvatures should be of zero Weyl weight. If we would go to 6 dimensions this Lagrangian is not Weyl invariant. Also if we consider 2-index antisymmetric tensors, so 3-index field strengths, we need one more $g^{\mu\nu}$, so the action is not Weyl invariant. Therefore to construct conformal actions for antisymmetric tensor multiplets one needs to introduce scalars to compensate for the non-dilatational invariance. For the linear multiplets in N=1 it was shown that the scalar present in the multiplet itself could be used for this[29], and this was called the "improved tensor multiplet". The same was done for N=2[30] using the scalar

$$L = \left(L^{ij} L_{ij}\right)^{1/2} \tag{4.28}$$

The action has the form

$$e^{-1}\mathcal{L}_L = -\frac{1}{2}\mathcal{D}_\mu L^{ij} \mathcal{D}^\mu L_{ij} L^{-1} + L\left(\frac{1}{3}R + D\right) + L^{-1}\left(E^\mu E_\mu + G\bar{G}\right)$$
$$-\frac{i}{2}\varepsilon^{\mu\nu\rho\sigma}\partial_\mu L_{ij} L^{jk} \partial_\nu L_{k\ell}\, \varepsilon^{i\ell} E_{\rho\sigma} L^{-3} + \ldots \tag{4.29}$$

The last mentionned term can not be written in terms of the $E_{\mu\nu}$ field strength which makes it more difficult to prove the equivalence to scalar multiplet couplings as it was done in N=1[31]. Without gravity this equivalence is proven in 30). This action can also be used as an alternative to the scalar multiplet as compensating multiplet. Different compensating multiplets give rise to Poincaré supergravity's with a different set of auxiliary fields. E.g. at the end of section 2, I gave in fig. 2 the construction of the "old minimal" formulation of N=1 supergravity[32]. If the compensating chiral multiplet is replaced by the linear multiplet (L , φ , $a_{\mu\nu}$), one obtains the "new minimal" set of auxiliary fields[33]. In that case the U(1) invariance of the conformal group would still remain in the final formulation. In N=2 the first formulation used a "nonlinear multiplet" as compensator[4], which we will not review here. The second formulation[4] was the one schematically given in fig. 3. The third one[30] uses the action (4.29), or rather its negative such that the dilatational gauge and the field equation for D give L=1. But by imposing e.g.

$$L^{ij} = \frac{1}{\sqrt{2}} \delta^{ij} \tag{4.30}$$

we break SU(2) only to its SO(2) subgroup. So the corresponding auxiliary field formulation has a remaining abelian invariance.

5. - Actions in d=6 N=2.

The scalar multiplet action in d=4 did not rely on its off-shell formulation. The on-shell formulation of the scalar multiplet is completely the same in d=4 and d=6, so we can also repeat everything involved in the construction of the action. Of course in the part dependent on the gauging there are no X-dependent terms. So one can not obtain in d=6 a massive scalar multiplet, and this is again related to the impossibility of central charges in d=6. The bosonic part of the scalar multiplet action is then[2]

$$e^{-1}\mathcal{L}_{S,B} = -\mathcal{D}_\mu A_i^\alpha \mathcal{D}^\mu A_\beta^i d_\alpha^\beta + A_i^\alpha d_\alpha^\beta A_\beta^i \left(\frac{1}{5}R + \frac{1}{15}D\right) \tag{5.1}$$

The field equation for D would make this action trivial. In d=4 the vector multiplet action was added to avoid this. However, in d=6 the vector multiplet does not contain scalars. Also, the argument in d=4 that a vector multiplet is needed because the physical d=4 N=2 supergravity representation contains a spin 1, is not any more valid. However as explained before the six dimensional theory needs a tensor multiplet. And this gives also the solution to the problem of the field equation for D. As (3.22) expressed D in terms of fields of the tensor multiplet, one should not vary w.r.t. D independently. In fact, this procedure is the one which is used also in the first formulation of d=4 N=2, but here this is also used for χ^i and T^-_{abc} of the Weyl multiplet and does not augment the number of independent field components. After making the conformal box in (3.22) explicit, the bosonic part of D is

$$\frac{4}{15}\sigma D = \mathcal{D}^a \mathcal{D}_a \sigma + \frac{1}{5}\sigma R + \frac{1}{3} F^+_{abc} T^{-abc} \tag{5.2}$$

Using this in (5.1) gives

$$e^{-1}\mathcal{L}_{S,B} = -\mathcal{D}_\mu A_i^\alpha \mathcal{D}^\mu A_\beta^i d^\beta_\alpha + \frac{1}{4} A_i^\alpha d^\beta_\alpha A_\beta^i R$$
$$+ \frac{1}{4} A_i^\alpha d^\beta_\alpha A_\beta^i \sigma^{-1} \mathcal{D}^a \mathcal{D}_a \sigma + \frac{1}{12} A_i^\alpha d^\beta_\alpha A_\beta^i F^+_{abc} T^{-abc} \tag{5.3}$$

Imposing then the D-gauge

$$A_i^\alpha d^\beta_\alpha A_\beta^i = -2 \tag{5.4}$$

eliminating the $V_\mu{}^i{}_j$, using (3.23) and a parametrization as (4.17) we get

$$e^{-1}\mathcal{L}_{S,B} = -\partial_\mu B^i_{\cdot\alpha} \partial^\mu B_i^\beta \Delta^\alpha_\beta - \frac{1}{2} R$$
$$-\frac{1}{2}\sigma^{-2}\partial_\mu\sigma\partial^\mu\sigma - \frac{3}{4}\sigma^{-2}\left(\partial_{[\mu} B_{\nu\rho]}\right)^2 \tag{5.5}$$

The scalars span again a quaternionic manifold and the constructions mentionned for d=4 could also be applicable in d=6. With 1 scalar multiplet we obtain Poincaré supergravity coupled to a tensor multiplet. Schematically the result is shown in fig. 4, which looks slightly different from figs. 1,2,3.

An important difference with the previous cases is that I have used here an on-shell multiplet (the scalar multiplet). So at the end I do not get on off-shell formulation of the supergravity theory

Fig. 4). N=2, d=6 Poincaré supergravity, with tensor multiplet from conformal methods.

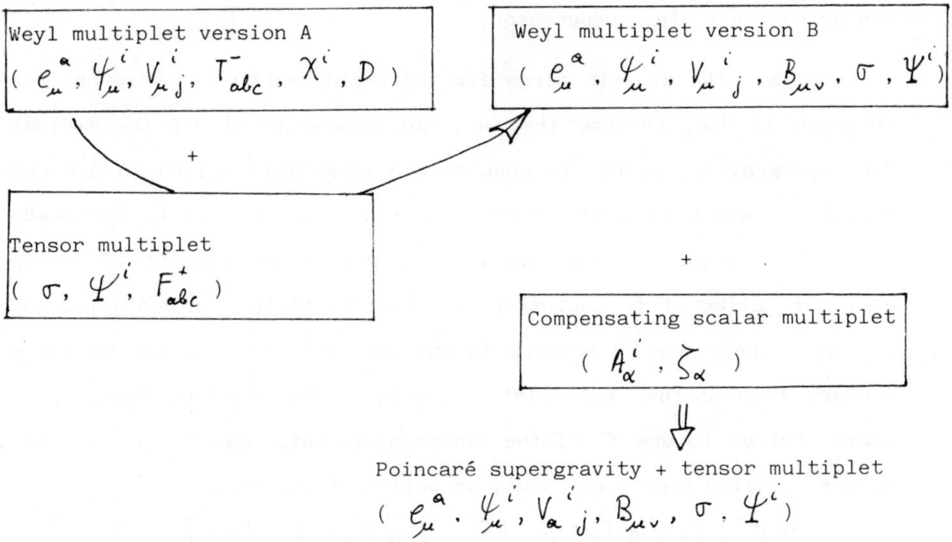

(although I mentionned in fig. 4 the auxiliary fields $V_{a\,j}^{\,i}$). Considering the alternatives of d=4, we see no use of the nonlinear multiplet. In d=4 the action was provided by the vector multiplet, and the nonlinear multiplet defined the field D. Here the compensator still has to provide the action and D is already defined by the tensor multiplet. The linear multiplet is however a better alternative. The action in d=6 [2] is analogous to (4.29), and we can perform the same manipulations as with (5.1). But as we use here an off-shell multiplet the result is also an off-shell

representation of the Poincaré theory. As with (4.30), SU(2) is broken to SO(2), so this formulation is also of the "new minimal" - type. One of the three vectors in $V_{a\,j}^{\,i}$ is still a gauge vector, and the final result in fig. 4 contains then also the auxiliary gauge antisymmetric tensor $E_{\mu\nu\rho\sigma}$. The advantage of the previous (on-shell) representation with the scalar compensating multiplet is that the symmetries of the quaternionic manifold are more transparent. Gauge groups such that the scalars in compensating and physical multiplets transform in one irreducible representation might only be possible using the scalar compensator.

So far, the kinetic terms for the vector multiplets were not yet obtained in d=6, because they are not needed to obtain the minimal d=6 supergravity theory. To construct a Yang-Mills action in d=6 one encounters again the difficulties discussed after (4.27). One needs scalars to compensate for the lack of Weyl invariance of (4.27) in d=6. But unlike of the solution of linear multipleto in d=4 N=1,2 and d=6 N=2, there are no scalars in the vector multiplet. So one needs scalars from another multiplet. In view of the previous results, it seems obvious to use σ of the tensor multiplet. Indeed, one can construct[2], using tensor calculus, an action of the form

$$e^{-1}\mathcal{L}_V = \mathrm{Tr}\Big\{\sigma\big[-\tfrac{1}{4}F_{\mu\nu}F^{\mu\nu} - 2\bar{\Omega}\not{D}\Omega + Y^{ij}Y_{ij}\big] + \frac{i}{16}e^{-1}\varepsilon^{\mu\nu\rho\sigma\lambda\tau}B_{\mu\nu}F_{\rho\sigma}F_{\lambda\tau} + \dots\Big\} \tag{5.6}$$

By introducing the tensor multiplet one has obtained also a coupling between the antisymmetric tensor and the Yang-Mills fields. This action looks very similar to the Yang-Mills-Einstein action of d=10[34] and especially with its conformal formulation[35]. It is clear from our construction that $B_{\mu\nu}$ is introduced independent of the gauge group, and is thus gauge inert. In this way it is similar to the 6-index gauge field in d=10 which occurs also in the Yang-Mills--Einstein action in the form

$$\mathrm{Tr}\big[\varepsilon^{\mu_1\dots\mu_{10}}B_{\mu_1\dots\mu_6}F_{\mu_7\mu_8}F_{\mu_9\mu_{10}}\big] \tag{5.7}$$

In 10-dimensions there exist a duality transformation to a 2-index gauge field $A_{\mu\nu}$ which does then transform under the gauge group. Here such a duality transformation transforms $B_{\mu\nu}$ in another antisymmetric tensor $A_{\mu\nu}$ [36]. One has then to rewrite (5.6) in function of the field strength $B_{\mu\nu\rho} = 3\partial_{[\mu} B_{\nu\rho]}$. The last term is

$$-\frac{i}{12}\varepsilon^{\mu\nu\rho\sigma\lambda\tau} B_{\mu\nu\rho}\left(W_\sigma^I \partial_\lambda W_\tau^I + \frac{1}{3} f_{IJK} W_\sigma^I W_\lambda^J W_\tau^K\right) \tag{5.8}$$

To perform the duality transformation one adds a term

$$\frac{i}{12}\varepsilon^{\mu\nu\rho\sigma\lambda\tau} B_{\mu\nu\rho}\partial_\sigma A_{\lambda\tau} \tag{5.9}$$

to the action. $A_{\lambda\tau}$ is a Lagrange multiplier, such that $B_{\mu\nu\rho}$ is the curl of the 2-index $B_{\mu\nu}$ by its field equation. But if $B_{\mu\nu\rho}$ is not by definition a curl, then (5.8) is not Yang-Mills invariant, and (5.9) has to restore the non-invariance by defining a Yang-Mills transformation for $A_{\mu\nu}$

$$\delta_{YM} A_{\mu\nu} = \Lambda^I \partial_{[\mu} W_{\nu]}^I \tag{5.10}$$

If one eliminates $B_{\mu\nu\rho}$ from the action, one obtains a term

$$-\frac{3}{2}\left(\partial_{[\mu} A_{\nu\rho]} - W_{[\mu}^I \partial_\nu W_{\rho]}^I - \frac{1}{3} f_{IJK} W_\mu^I W_\nu^J W_\rho^K\right) \tag{5.11}$$

It is in this form that the theory has been obtained in ref. 17).

6. - Conclusion.

Kähler manifolds occurring in N=1 and N=2 d=4 matter couplings have been investigated in great detail. However so far we do not know much about the quaternionic manifolds, the corresponding potentials, symmetry breakings, ... The study locks similar in d=4 or d=6 N=2 theories.

The treatment of antisymmetric tensors in d=6 is new and one could wonder whether this could help to understand the structure of d=10 theories. A Weyl multiplet has been constructed in d=10 [35]. But it is still an open question whether that can be used to extend the linearised Poincaré auxiliary field formulation of d=10[37] to the non--linear level or to obtain other off-shell representations.

Acknowledgements.

I thank the organizers Prof. T. Regge, R. d'Auria and P. Fré for the hospitality and the I.S.I. (Piemonte, Italia) for the financial support.

References.

1) - B. de Wit, P. Lauwers and A. Van Proeyen, Nucl. Phys. B255 (1985) 569.

2) - E. Bergshoeff, E. Sezgin and A. Van Proeyen, preprint Trieste IC/85/112, Nucl. Phys. B., to be published.

3) - M. Kaku, P. Townsend and P. van Nieuwenhuizen, Phys. Rev. D17 (1978) 3179.

4) - B. de Wit, J.W. van Holten and A. Van Proeyen, Nucl. Phys. B167 (1980) 186 (E: B172 (1980) 543); B184 (1981) 77 (E: B222 (1983) 516).

5) - B. de Wit, in "Supersymmetry and Supergravity '82", eds. S. Ferrara, J.G. Taylor and P. van Nieuwenhuizen (World Scientific Publishing Co. 1982) p. 85.

A. Van Proeyen in "Supersymmetry and Supergravity 1983", ed. B. Milewski (World Scientific Publishing Co. 1983)

6) - E. Cremmer, B. Julia, J. Scherk, S. Ferrara, L. Girardello and P.van Nieuwenhuizen, Phys. Lett. 79B (1978) 231, Nucl. Phys. B147 (1979) 105.

A. Van Proeyen, Nucl. Phys. B162 (1980) 376.

E. Cremmer, S. Ferrara, L. Girardello and A. Van Proeyen, Phys. Lett. 116B (1982) 231, Nucl. Phys. B212 (1983) 413.

T. Kugo and S. Uehara, Nucl. Phys. B222 (1983) 125.

7) - J. Wess and B. Zumino, Nucl. Phys. B70 (1974) 39.

S. Ferrara, Nucl. Phys. B77 (1974) 73.

8) - S. Ferrara, M. Kaku, P.K. Townsend and P. van Nieuwenhuizen, Nucl. Phys. B129 (1977) 125.

9) - S. Ferrara and A. Van Proeyen, Phys. Lett. 138B (1984) 77.

10) - J. Bagger and E. Witten, Phys. Lett. 115B (1982) 202; 118B (1982) 103.

J. Bagger, Nucl. Phys. B211 (1983) 302.

11) - L. Alvarez-Gaumé and D.Z. Freedman, in "unification of the Fundamental Particle Interactions", eds. S. Ferrara, J. Ellis and P. van Nieuwenhuizen (Plenum Press, New York, 1980).

12) - M. de Roo, J.W. van Holten, B. de Wit and A. Van Proeyen, Nucl. Phys. B173 (1980) 175.

13) - B. de Wit, J.W. van Holten and A. Van Proeyen, Phys. Lett. 95B (1980) 51.

14) - M.F. Sohnius, Nucl. Phys. B138 (1978) 109.

15) - A. Galperin, E. Ivanov, S. Kalitzin, V. Ogievetsky and E. Sokatchev, Class. Quantum Grav. 1 (1984) 447.

16) - N. Marcus and J.H. Schwarz, Phys. Lett. 115B (1982) 111.

17) - H. Nishino and E. Sezgin, Phys. Lett. 144B (1984) 187.

18) - R. D'Auria, P. Fré and T. Regge, Phys. Lett. 128B (1983) 44.

19) - J. Bagger and E. Witten, Nucl. Phys. B222 (1983) 1.

20) - B. de Wit and A. Van Proeyen, Nucl. Phys. B245 (1984) 89.

21) - E. Cremmer, S. Ferrara and J. Scherk, unpublished.

A. Das, M. Fischler and M. Roček, Phys. Rev. D16 (1977) 3427.

D.Z. Freedman and J.H. Schwarz, Nucl. Phys. B137 (1978) 333.

S.J. Gates Jr. and B. Zwielbach, Phys. Lett. 123B (1983) 200.

22) - B. de Wit, P.G. Lauwers, R. Philippe and A. Van Proeyen, Phys. Lett. 135B (1984) 295.

23) - M. Günaydin, G. Sierra and P.K. Townsend, Caltech preprint CALT-68-1123 (1984).

24) - P. Breitenlohner and M. Sohnius, Nucl. Phys. B187 (1981) 409.

25) - J.A. Wolf, J. Math. Mech. 14 (1965) 1033.

D.V. Alekseevshi, Math. USSR Izv. 9 (1975) 297.

26) - E. Cremmer and A. Van Proeyen, Class. Quantum Grav. 2 (1985) 445.

27) - M. Günaydin, G. Sierra and P.K. Townsend, Phys. Lett. 133B (1983) 72; Nucl. Phys. B242 (1984) 244; Phys. Lett. 144B (1984) 41.

28) - E. Cremmer, C. Kounnas, A. Van Proeyen, J.P. Derendinger, S. Ferrara, B. de Wit and L. Girardello, Nucl. Phys. B250 (1985) 385.

29) - B. de Wit, and M. Roček, Phys. Lett. 109B (1982) 439.

30) - B. de Wit, R. Philippe and A. Van Proeyen, Nucl. Phys. B219 (1983) 143.

31) - W. Siegel, Phys. Lett. 85B (1979) 333.

U. Lindström and M. Roček, Stocholm preprint n. 4, October 1982.

S. Ferrara, L. Girardello, T. Kugo and A. Van Proeyen, Nucl. Phys. B223 (83) 191.

32) - S. Ferrara and P. van Nieuwenhuizen, Phys. Lett. 74B (1978) 333.

K.S. Stelle and P.C. West, Phys. Lett. 74B (1978) 330.

33) - M.F. Sohnius and P.C. West, Phys. Lett. 105B (1981) 353.

34) - A. H. Chamseddine, Phys. Rev. D24 (1981) 3065.

E. Bergshoeff, M. de Roo, B. de Wit and P. van Nieuwenhuizen Nucl. Phys. B195 (1982) 97.

G. Chapline and N. Manton, Phys. Lett. 120B (1983) 105.

35) - E. Bergshoeff, M. de Roo and B. de Wit, Nucl. Phys. B217 (1983) 489.

36) - J. Gates and N. Nishino, Phys. Lett. 157B (1985) 157.

37) - P. Howe, H. Nicolai and A. Van Proeyen, Phys. Lett. 112B (1982) 446.

SUPERCONFORMAL INVARIANCE AND THE TENSOR MULTIPLET IN SIX DIMENSIONS

E. Bergshoeff
International Centre for Theoretical Physics, Trieste, Italy.

ABSTRACT

We present the complete structure of N = 2 conformal supergravity in six dimensions. The relation with the graded algebra Osp(6,2/1) is discussed. It is shown that there exist two different formulations of N = 2, d = 6 conformal supergravity. The field content of these two formulations consists of the same superconformal gauge fields but differ in the type of matter fields.

Some special properties of the N = 2, d = 6 tensor multiplet are discussed. It is shown that the on-shell constraints of the tensor multiplet coupled to conformal supergravity are algebraic equations for the matter fields of conformal supergravity which can be solved. By using these algebraic equations we find that the two formulations of N = 2, d = 6 conformal supergravity are equivalent.

1. INTRODUCTION

$N = 2$, $d = 6$ Poincare supergravity coupled to matter [1] is known to have some very interesting properties. First of all, it has been found ([2,3]) that for a special choice of matter multiplets and gauge group due to the Green-Schwartz mechanism [4] the theory is free of both gravitational and Yang-Mills gauge anomalies. A particular interesting anomaly cancellation occurs when the gauge group is chosen to be $E_6 \times E_7 \times U(1)$ [2]. Other anomaly free combinations have been obtained [3] by considering the dimensional reduction of the ten-dimensional anomaly free theory. Another interesting property of $N = 2$, $d = 6$ Poincare supergravity coupled to matter is that the theory compactifies on the product of four dimensional Minkowski space-time and a 2-sphere, yielding chiral fermions in $d = 4$ [5].

Clearly, it is desirable to construct complete actions for the coupling of matter multiplets to $N = 2$, $d = 6$ Poincare supergravity. It is well known that the superconformal tensor calculus ([6,7]) is a very convenient tool to construct general matter couplings. This has been used to construct general matter couplings to $N = 1$ [8], $N = 2$ [7,9] and $N = 4$ [10] Poincare supergravity in four dimensions. For reviews explaining details of the superconformal method we refer to Ref. [11]. Here, we will only review the first step in the superconformal program in six dimensions, i.e. the construction of the $N = 2$, $d = 6$ Weyl multiplet. For a discussion of the remaining steps in this program we refer to the contribution of A. Van Proeyen to this conference.

In Sec. 2 we discuss $N = 2$, $d = 6$ conformal supergravity as the gauge theory of the graded algebra Osp (6,2/1). We show that there exist two different formulations of the $N = 2$, $d = 6$ Weyl multiplet. The explicit transformations for one of these formulations are given. In Sec.3 some special properties of the $N = 2$, $d = 6$ tensor multiplet are discussed, which are used in Sec. 4 to construct the transformation rules of the second formulation of the $N = 2$, $d = 6$ Weyl multiplet. We present our conclusions in Sec. 5.

2. THE GAUGE THEORY OF THE GRADED ALGEBRA Osp (6,2/1)

As a first step in the construction of the $N = 2$, $d = 6$ Weyl multiplet we consider the superconformal algebra in six dimensions which is Osp(6,2/1). The $d = 6$ conformal algebra consists of translations P_a, rotations M_{ab}, dilatations D and special conformal transformations K_a. The supersymmetric extension of the algebra contains additional supersymmetries Q^i_α $(i = 1,2)$, "special" supersymmetries S^i_α and SU(2) rotations $U_{ij} = U_{ji}$. The spinor generators Q^i_α and S^i_α are Majorana-Weyl. For our notations and conventions we refer to [12]. In Appendix B of this reference one also finds the non zero commutators between the different generators of Osp(6,2/1).

To each generator T_A of the superconformal algebra we assign a gauge field h^A_μ in the following way:

$$\begin{aligned} h_\mu^A T_A = {} & e_\mu^a P_a + \omega_\mu^{ab} M_{ab} + b_\mu D + f_\mu^a K_a \\ & + \bar{\psi}_\mu^i Q_i + \bar{\phi}_\mu^i S_i + V_\mu^{ij} U_{ij} \quad . \end{aligned} \tag{2.1}$$

Using the structure constants of the superconformal algebra

$$[T_A, T_B\} \equiv T_A T_B \mp T_B T_A = f_{AB}{}^C T_C \tag{2.2}$$

and the basic rules

$$\begin{aligned} \delta h_\mu^A &= \partial_\mu \varepsilon^A + \varepsilon^C h_\mu^B f_{BC}{}^A \\ R_{\mu\nu}^A &= 2\partial_{[\mu} h_{\nu]}^A + h_\nu^C h_\mu^B f_{BC}{}^A \end{aligned} \tag{2.3}$$

one can immediately determine the gauge transformations of the superconformal gauge fields and calculate the curvature tensors $R^A_{\mu\nu}$. The results which are quite lengthy are given in [12].

The superconformal gauge fields defined in (2.1) describe <u>155 + 80</u> (bosonic + fermionic) off-shell field degrees of freedom. The bosonic degrees of freedom are described by the gauge fields e^a_μ (30), ω^{ab}_μ (75), V^{ij}_μ (15), f^a_μ (30) and b_μ (5). The fermionic degrees of freedom are described by the gauge fields ψ^i_μ (40) and ϕ^i_μ (40). The 30 degrees of freedom (d.o.f.) described by the sechsbein decompose into massive representations of the $d = 6$ Poincare group as a massive spin 2 representation (14 d.o.f.) plus lower spin representations. <u>By definition</u> the d = 6 Weyl multiplet is the smallest irreducible multiplet that contains the massive spin 2 representation.

To achieve a maximal irreducibility of the superconformal gauge field configuration we impose a maximal set of so-called conventional constraints [13] on the superconformal curvatures. For this purpose the following set of constraints suffices:

$$R_{\mu\nu}^a(P) = 0 \;\; (90) \;, \quad R_{\mu\nu}^{ab}(M) e_b^\nu = 0 \;\; (36) \;, \quad \gamma^\mu R_{\mu\nu}^i(Q) = 0 \;\; (48) \tag{2.4}$$

We have indicated the number of constraints between brackets. The set of constraints (2.4) enables one to derive algebraic expressions for the gauge fields ω_μ^{ab}, f_μ^a and ϕ_μ^i in terms of e_μ^a, b_μ, V_μ^{ij} and ψ_μ^i.

The constraints (2.4) eliminate <u>126 + 48</u> (bosonic + fermionic) field degrees of freedom. Hence after imposing these constraints we are left with <u>(14 + 15) + 32</u> d.o.f. The bosonic degrees of freedom are described by the independent gauge fields e_μ^a (14) and V_μ^{ij} (15), while the fermionic degrees of freedom are described by ψ_μ^i (32). The gauge field b_μ has no d.o.f. since under special conformal transformations K with parameter Λ_a^K it transforms as $\delta_K b_\mu = \Lambda_\mu^k$. It is well known from four dimensions that at this stage one must add matter fields to the superconformal gauge fields in such a way that the d.o.f. of gauge fields plus matter fields form a N = 2 massive spin-2 representation of the d = 6 super-Poincare algebra. As an example we have indicated in the table how this is achieved for the 24 + 24 N = 2, d = 4 Weyl multiplet [7].

<u>Table</u> - Massive spin-2 representation of the d = 4 and d = 6 super-Poincare algebra.

spin s	N = 2, d = 4	gauge fields	matter fields
2 (5)	1	1 e_μ^a	
3/2 (4)	4	4 ψ_μ^i	
1 (3)	6	4 $V_{\mu j}^i$, A_μ	2 T_{ab}^-
1/2 (2)	4		4 χ^i
0 (1)	1		1 D

spin s	N = 2, d = 6	gauge fields	matter fields
2 (14)	1	1 e_μ^a	
3/2 (8)	4	4 ψ_μ^i	
1 (5)	5	3 V_μ^{ij}	2 T_{abc}^- $B_{\mu\nu}$
1/2 (2)	4		4 χ^i or ψ^i
0 (1)	1		1 D σ

The numbers between brackets in the first column represent the dimensionality of each spin. Those in the second column denote the number of spins contained in a massive spin-2 representation. The numbers in the third and fourth column indicate the massive spin states, which are described by the super conformal gauge fields and the added matter fields respectively.

One can easily verify that in four dimensions the choice of matter fields is unique. This situation changes in six dimensions, where there are two possible sets of matter fields. As we have indicated in the table one set consists of an antisymmetric tensor T^-_{abc} of negative duality, an SU(2)-Majorana-Weyl spinor χ^i of negative chirality and a real scalar D. The other set contains an antisymmetric tensor gauge field $B_{\mu\nu}$, an SU(2) -Majorana-Weyl spinor ψ^i of positive chirality and a real scalar σ. The two sets of matter fields correspond to two different formulations of the 40 + 40 N = 2, d = 6 Weyl multiplet, which are from now on denoted by I and II respectively:

$$\text{formulation I:} \quad e_\mu^a, \psi_\mu^i, V_\mu^{ij}, T^-_{abc}, \chi^i, D$$

$$\text{formulation II:} \quad e_\mu^a, \psi_\mu^i, V_\mu^{ij}, B_{\mu\nu}, \psi^i, \sigma \tag{2.5}$$

Starting from the linear transformation rules of the superconformal gauge fields and the matter fields introduced above we can now construct the full non linear transformations of the N = 2, d = 6 Weyl multiplet by applying an iterative procedure which is described in detail in [12]. For the moment we apply this procedure only to formulation I. The result is given by

<u>Formulation I</u>

$$\delta e_\mu^a = \tfrac{1}{2}\bar{\varepsilon}\gamma^a\psi_\mu$$

$$\delta\psi_\mu^i = \mathcal{D}_\mu\varepsilon^i + \tfrac{1}{24}\gamma\cdot T^-\gamma_\mu\varepsilon^i + \gamma_\mu\eta^i$$

$$\delta V_\mu^{ij} = -4\bar{\varepsilon}^{(i}\phi_\mu^{j)} - \tfrac{1}{3}\bar{\varepsilon}^{(i}\gamma_\mu\chi^{j)} - 4\bar{\eta}^{(i}\psi_\mu^{j)}$$

$$\delta T^-_{abc} = -\tfrac{1}{32}\bar{\varepsilon}\gamma^{de}\gamma_{abc}\hat{R}_{de}(Q) - \tfrac{7}{96}\bar{\varepsilon}\gamma_{abc}\chi$$

$$\delta\chi^i = \tfrac{1}{8}\gamma\cdot T^-\overleftarrow{\hat{\not{\mathcal{D}}}}\varepsilon^i + \tfrac{3}{16}\gamma\cdot\hat{R}^{ij}(V)\varepsilon_j + \tfrac{1}{4}D\varepsilon^i + \tfrac{1}{2}\gamma\cdot T^-\eta^i$$

$$\delta D = \bar{\varepsilon}\hat{\not{\mathcal{D}}}\chi - 2\bar{\eta}\chi \tag{2.6}$$

where the positive chiral ε^i and the negative chiral η^i are the parameters of a Q and S transformation respectively. For the definition of the hatted curvatures and the supercovariant derivatives see [12]. In the next section we will now first discuss some special properties of the N = 2, d = 6 tensor multiplet, which will be needed in section 4 to derive the full nonlinear transformation rules of formulation II of the N = 2, d = 6 Weyl multiplet from those of formulation I.

3. THE N = 2, d = 6 TENSOR MULTIPLET

The components of the N = 2, d = 6 tensor multiplet consist of an antisymmetric tensor gauge field $B_{\mu\nu}$, an SU(2)-Majorana Weyl spinor ψ^i of negative chirality and a real scalar σ. In flat superspace the supersymmetry transformations of these components are given by:

$$\begin{aligned} \delta B_{\mu\nu} &= -\bar{\varepsilon}\gamma_{\mu\nu}\psi \\ \delta\psi^i &= \tfrac{1}{48}\gamma\cdot F^+(B)\varepsilon^i + \tfrac{1}{4}\not\partial\sigma\varepsilon^i \\ \delta\sigma &= \bar{\varepsilon}\psi \end{aligned} \tag{3.1}$$

where $F^+(B)$ is the self dual part of the B-curvature. The commutator of two supersymmetry transformations on these fields gives a translation provided that one imposes the following constraints:

$$\mathcal{F}^-_{abc} = F^-_{abc}(B) = 0 \quad , \Gamma^i = \not\partial\psi^i = 0 \quad , \mathcal{C} = \Box\sigma = 0 \tag{3.2}$$

In other words, the tensor multiplet is an on-shell multiplet.

When coupled to formulation I of conformal supergravity the full nonlinear transformation rules of the tensor multiplet can be derived by a standard iterative procedure which is described in [12]. The result is given by

$$\begin{aligned} \delta B_{\mu\nu} &= -\bar{\varepsilon}\gamma_{\mu\nu}\psi - \sigma\bar{\varepsilon}\gamma_{[\mu}\psi_{\nu]} \\ \delta\psi^i &= \tfrac{1}{48}\gamma\cdot\hat{F}^+(B)\varepsilon^i + \tfrac{1}{4}\hat{\not{D}}\sigma\varepsilon^i - \sigma\eta^i \\ \delta\sigma &= \bar{\varepsilon}\psi \end{aligned} \tag{3.3}$$

The explicit form of the supercovariant derivatives is given in [12]. The superconformal algebra does close on ($B_{\mu\nu}$, ψ^i, σ) provided one replaces the linearized constraints (3.2) by the following nonlinear ones:

$$\mathcal{F}^-_{abc} = \hat{F}^-_{abc}(B) - 2\sigma T^-_{abc} = 0$$

$$T^i = \hat{\not{D}}\psi^i - \frac{1}{6}\sigma\chi^i - \frac{1}{12}\gamma\cdot T^-\psi^i = 0$$

$$C = \left(\hat{D}^a\hat{D}_a - \frac{1}{6}D\right)\sigma + \frac{1}{3}F^+(B)\cdot T^- + \frac{7}{6}\bar{\chi}\psi = 0 \qquad (3.4)$$

It is at this point that a surprising thing happens. From (3.4) we see that the on-shell constraints of the nonlinear tensor multiplet are algebraic equations from which one can solve for the matter fields (T^-_{abc}, χ^i, σ) of formulation I of the N = 2, d = 6 Weyl multiplet in terms of the superconformal gauge fields (e^a_μ, ψ^i_μ, V^{ij}_μ) and the component fields ($B_{\mu\nu}$, ψ^i, σ) of the tensor multiplet. One immediately verifies that the explicit solution is given by:

$$\begin{aligned} T^-_{abc} &= \frac{1}{2}\sigma^{-1}\hat{F}^-_{abc}(B) \\ \chi^i &= 6\sigma^{-1}\left(\hat{\not{D}}\psi^i - \frac{1}{24}\sigma^{-1}\gamma\cdot\hat{F}^-(B)\psi^i\right) \\ D &= 6\sigma^{-1}\Big(\hat{D}^a\hat{D}_a\sigma + \frac{1}{6}\sigma^{-1}\hat{F}^+(B)\cdot\hat{F}^-(B) \\ &\qquad + 7\sigma^{-1}\bar{\psi}\hat{\not{D}}\psi - \frac{1}{24}\sigma^{-2}\bar{\psi}\gamma\cdot\hat{F}^-(B)\psi\Big) \end{aligned} \qquad (3.5)$$

The solution (3.5) can now be used to replace in formulation I of conformal supergravity the set of matter fields (T^-_{abc}, χ^i, σ) by the set ($B_{\mu\nu}$, ψ^i, σ) corresponding to formulation II of the Weyl multiplet. Solution (3.5) will be used in the next section to derive the full nonlinear transformation rules of formulation II of the Weyl multiplet from those of formulation I.

Before closing this section we wish to make some remarks about the $N = 2$, $d = 6$ tensor multiplet. Since the on-shell constraints (3.4) of the tensor multiplet can be explicitly solved (see (3.5)) one can view the Weyl multiplet (e_μ^a, ψ_μ^i, V_μ^{ij}, $B_{\mu\nu}$, ψ^i, σ) as an off-shell multiplet containing the component fields ($B_{\mu\nu}$, ψ^i, σ) of the tensor multiplet. Thus the superconformal algebra does close on the component fields $B_{\mu\nu}$, ψ^i and σ of the Weyl multiplet without using any constraint equations. It is known that one cannot write down a Lorentz-invariant supersymmetric action for the on-shell tensor multiplet in ordinary field theory [14]. (Such an action does exist however in the group manifold approach [15]). Starting with

$$\mathcal{L}\ \text{(on-shell tensor)} \propto \sigma \Box \sigma + \bar{\psi} \not{\partial} \psi + F^{\mu\nu\rho}(B) F_{\mu\nu\rho}(B) \tag{3.6}$$

the iterative Noether procedure fails in cancelling terms proportional to $(F^-(B))^2 \psi_\mu$, i.e.

$$\delta \mathcal{L} \propto \bar{\epsilon} \gamma^{(\mu} \psi^{\nu)} F^{-\ \lambda\rho}_{\mu}(B) F^-_{\nu\lambda\rho}(B) \tag{3.7}$$

Similarly we do not expect that one can write down an action for the conformal multiplet containing $B_{\mu\nu}, \psi^i, \sigma$ of the form

$$\mathcal{L}\ \text{(conformal)} \propto R + \bar{\psi}_\mu \gamma^{\mu\nu\rho} \psi_{\nu\rho} + F^{\mu\nu\rho}(B) F_{\mu\nu\rho}(B) \tag{3.8}$$

Here we have imposed the D and S gauge choices $\sigma = 1$ and $\psi = 0$. (see (4.2) for the transformation rules). It would be interesting to see whether one can write down a higher derivative action for the conformal multiplet of the form:

$$\mathcal{L}\ \text{(conformal)} \propto R_{\mu\nu}{}^{ab} R_{\mu\nu}{}^{ab} + \bar{\psi}_{\mu\nu} \not{\partial} \psi_{\mu\nu} + F^{\mu\nu\rho}(B) \Box F_{\mu\nu\rho}(B) \tag{3.9}$$

4. FORMULATION II OF $N = 2$, $d = 6$ CONFORMAL SUPERGRAVITY

After the preliminaries given in Sec. 3 it is now straightforward to derive the transformation rules of formulation II of $d = 6$ conformal supergravity from those of formulation I (see (2.6)). After substituting the solution (3.5) for the matter fields T^-_{abc}, χ^i and D into the transformation rules (2.6) of formulation I we obtain the following result:

$$\delta e_\mu^a = \frac{1}{2}\bar{\varepsilon}\gamma^a\psi_\mu$$

$$\delta\psi_\mu^i = \mathcal{D}_\mu\varepsilon^i + \frac{1}{48}\sigma^{-1}\gamma\cdot\hat{F}(B)\gamma_\mu\varepsilon^i + \gamma_\mu\eta^i$$

$$\delta V_\mu^{ij} = -4\bar{\varepsilon}^{(i}\phi_\mu^{j)} - 2\sigma^{-1}\bar{\varepsilon}^{(i}\gamma_\mu\hat{\not{\mathcal{D}}}\psi^{j)} + \frac{1}{12}\sigma^{-2}\bar{\varepsilon}^{(i}\gamma_\mu\gamma\cdot\hat{F}(B)\psi^{j)} - 4\bar{\eta}^{(i}\psi_\mu^{j)}$$

$$\delta B_{\mu\nu} = -\sigma\bar{\varepsilon}\gamma_{[\mu}\psi_{\nu]} - \bar{\varepsilon}\gamma_{\mu\nu}\psi$$

$$\delta\psi^i = \frac{1}{48}\gamma\cdot\hat{F}(B)\varepsilon^i + \frac{1}{4}\hat{\not{\mathcal{D}}}\sigma\varepsilon^i - \sigma\eta^i$$

$$\delta\sigma = \bar{\varepsilon}\psi \tag{4.1}$$

Restrictions on the Weyl multiplet (4.1) should lead to the on-shell Poincare theory. To show this we consider the linearized transformation rules and impose the K, D and S gauge choices $b_\mu = 0$, $\sigma = 1$ and $\psi^i = 0$. These gauge choices require compensating transformations, which lead to the following transformation rules:

$$\delta e_\mu^a = \frac{1}{2}\bar{\varepsilon}\gamma^a\psi_\mu$$

$$\delta\psi_\mu^i = \mathcal{D}_\mu\varepsilon^i + \frac{1}{8}\gamma^{ab}F_{ab\mu}(B)\varepsilon^i$$

$$\delta V_\mu^{ij} = \bar{\varepsilon}^{(i}R_\mu^{j)}$$

$$\delta B_{\mu\nu} = -\bar{\varepsilon}\gamma_{[\mu}\psi_{\nu]} \tag{4.2}$$

where $R^i_\mu = \gamma^\lambda(\partial_\lambda\psi^i_\mu - \partial_\mu\psi^i_\lambda)$ is the gravitino field equation. From (4.2) one can now obtain the on-shell Poincare theory by imposing the constraint

$$F^+_{abc}(B) = 0 \qquad \text{on shell-Poincare} \qquad (4.3)$$

One can easily verify that supersymmetry variations of (4.3) lead to the gravitino field equation $R^i_\mu = 0$, the Einstein equation $R_\mu{}^a(e) = 0$ and the equation $V^{ij}_\mu = 0$ (V^{ij}_μ is an auxiliary field).

As an alternative to (4.3) one can also impose the following weaker condition which reduces the conformal multiplet to the on-shell Poincare + tensor multiplet:

$$\partial^a F_{abc}(B) = 0 \qquad \text{on-shell Poincare + tensor} \qquad (4.4)$$

Supersymmetry variations of (4.4) now lead to the following conditions on e^a_μ, ψ^i_μ and V^{ij}_μ:

$$R(e) = 0 \quad , \quad \partial_{[\mu} R_{\nu]}{}^a(e) = 0$$

$$\gamma^{\mu\nu}\partial_\mu\psi^i_\nu = 0 \quad , \quad \not\partial\,\partial_{[\mu}\psi^i_{\nu]} = 0$$

$$\partial_{[\mu}V^{ij}_{\nu]} = 0 \qquad (4.5)$$

The conditions (4.5) have been encountered before in the context of the d = 10 Weyl multiplet [16]. A careful analysis of these conditions (see Appendix B of Ref. [16]) shows that the sechsbein describes a massless spin 2 and spin 0 state while the gravitino describes a massless spin 3/2 and spin 1/2 state (V^{ij}_μ does not describe any massless state). Hence e^a_μ, ψ^i_μ and $B_{\mu\nu}$ describe exactly the massless states of the on-shell Poincare plus tensor multiplet.

Finally, we note that a third possibility is to impose a constraint, which reduces the conformal multiplet to an on-shell tensor multiplet in flat spacetime:

$$F^-_{abc}(B) = 0 \qquad \text{on-shell tensor} \qquad (4.6)$$

Supersymmetry variations of (4.6) with constant parameter ε lead to $e_\mu^a \sim \delta_\mu^a \sigma$ and $\psi_\mu \sim \gamma_\mu \psi$ with $V_\mu^{ij} = \Box\sigma = \not{\partial}\psi = 0$.

We note the remarkable resemblance of the multiplet (4.1) to the $N = 1$, $d = 10$ Weyl multiplet (e_μ^a, ψ_μ, $B_{\mu_1 \ldots \mu_6}$) [16], the only difference being that in $d = 6$ we have an additional vector field V_μ^{ij} (this is so because the $d = 6$ superconformal algebra has a nonzero automorphism group) and in $d = 10$ we have a 6-index antisymmetric tensor field $B_{\mu_1 \ldots \mu_6}$ instead of a 2-index one. It appears that also the Yang-Mills coupled to conformal supergravity systems in $d = 6$ and $d = 10$ are very similar (c.p. Eq.(4.21) of [12] to Eq. (5.27) of [16]). An important difference between the $d = 6$ and $d = 10$ Weyl multiplets however is that in $d = 10$ the fields ψ^i and σ satisfy certain differential constraints [16]. Also the relation between the $d = 10$ Weyl multiplet and the $d = 10$ superconformal algebra is not well understood. Our hope is that the more simpler conformal supergravity theory in $d = 6$ will learn us something more about the structure of conformal supergravity in $d = 10$.

5. CONCLUSIONS

We have investigated the structure of $N = 2$ conformal supergravity in six dimensions. There exist two different formulations of the theory. These two formulations can be shown to be equivalent by using some special properties of the tensor multiplet in six dimensions.

Once the conformal supergravity theory is constructed it can be used to study general matter couplings to $N = 2$ Poincare supergravity in six dimensions. For this we refer to the contribution of Van Proeyen to this conference. Another application is the construction of off-shell formulations of $N = 2$, $d = 6$ Poincare supergravity plus a tensor multiplet. For instance, by using an $8 + 8$ linear multiplet as compensating multiplet one obtains the following $48 + 48$ off-shell multiplet [12]:

$$e_\mu^a,\ \psi_\mu^i,\ B_{\mu\nu},\ \psi^i,\ \sigma,\ V_a^{ij},\ E_{\mu\nu\rho\sigma} \qquad \text{off-shell Poincare + tensor} \tag{5.1}$$

which has only two auxiliary tensor gauge fields (V_a^{ij} and $E_{\mu\nu\rho\sigma}$).

Finally, it is clear that there exist a stiking resemblance between local conformal supersymmetry in six and ten dimensions. Therefore our hope is that the study of the more simpler conformally supersymmetric theories in six dimensions may lead to a better understanding of local conformal supersymmetry in ten dimensions.

This work is based on the paper by Bergshoeff, Sezgin and Van Proeyen [12].

REFERENCES

[1] H. Nishino and E. Sezgin, Phys. Lett. 144B, 187 (1984).

[2] S. Randjbar-Daemi, Abdus Salam, E. Sezgin and J. Strathdee, Phys. Lett. 151B, 351 (1985);
Abdus Salam and E. Sezgin, to appear in Physica Scripta (1985)

[3] M.B. Green, J.H. Schwarz and P.C. West, Nucl. Phys. B254, 327 (1985).

[4] M.B. Green and J.H. Schwarz, Phys. Lett. 149B, 117 (1984).

[5] Abdus Salam and E. Sezgin, Phys. Lett. 147B, 47 (1984).

[6] M. Kaku, P. Townsend and P. van Nieuwenhuizen, Phys. Rev. D17 (1978) 3179.

[7] B. de Wit, J.W. van Holten and A. Van Proeyen, Nucl. Phys. B167 (1980) 186 (E: B172 (1980) 543); B184 (1981) 77 (E: B222 (1983) 516).

[8] E. Cremmer, B. Julia, J. Scherk, S. Ferrara, L. Girardello and P. van Nieuwenhuizen, Nucl. Phys. B147 (1979) 105;
E. Cremmer, S. Ferrara, L. Girardello and A.Van Proeyen, Nucl. Phys. B212, (1983) 413;
T. Kuga and S. Uehara, Nucl. Phys. B222 (1983) 125;
S. Ferrara, L. Girardello, T. Kugo and A. Van Proeyen, Nucl. Phys. B223, (1983) 191.

[9] B. de Wit, P.G. Lauwers and A. Van Proeyen, Nucl. Phys. B255 (1985) 569.

[10] M. de Roo, Nucl. Phys. B255 (1985) 515; M. de Roo, Phys. Lett. 156B (1985), 331; M. de Roo and P. Wagemans, Groningen preprint (1985).

[11] B. de Wit, in Supersymmetry and Supergravity '82, Eds. S. Ferrara, J.G. Taylor and P. van Nieuwenhuizen (World Scientific Publishing Co., 1982) p. 85;
A. Van Proeyen in Supersymmetry and Supergravity 1983, Ed. B. Milewsky, (World Scientific Publishing Co., 1983).

[12] E. Bergshoeff, E. Sezgin and A. Van Proeyen, Nucl. Phys. B264 (1986) 653.

[13] S.J. Gates and W. Siegel, Nucl. Phys. B163, 519 (1984);
S.J. Gates, K.S. Stelle and P.C. West, Nucl. Phys. B169, 347 (1980).

[14] N. Marcus and J.H. Schwarz, Phys. Lett. 315B, 111 (1982).

[15] R. d'Auria, P. Fré and T. Regge, Phys. Lett. 128B, 44 (1983).

[16] E. Bergshoeff, M. de Roo and B. de Wit, Nucl. Phys. B217 (1983) 489.

MATTER COUPLED N = 3 SUPERGRAVITY [**]

L. Castellani [*], R. D' Auria [o*] and P. Fre' [+]

[*] Istituto Nazionale di Fisica Nucleare, Sez. di Torino

[o] Istituto di Fisica Teorica, Universita' di Torino

[+] C.E.R.N. - Geneva

ABSTRACT

The complete N = 3 matter coupling to supergravity is obtained in a geometrical framework. This coupling always exists if the 3n complex scalars of the n vector multiplets are co-ordinates of the Kähler-Grassmannian manifold SU(3,n)/SU(3) × SU(n) × U(1). Subgroups of SO(3,n) SU(3,n) of dimension 3+n can be gauged and give rise to a non-trivial scalar potential. The techniques used in this work allow for the calculation of scalar potentials of extended supergravities in any dimension without explicit construction of the Lagrangian. This opens the possibility of discussing patterns of partial supersymmetry and gauge symmetry breaking on a purely group-theoretical ground.

[**] This contribution is based on a recent paper by L. Castellani, A. Ceresole, R. D' Auria, S. Ferrara, P. Fre' and E. Maina, to be published in Nucl. Phys.B (see ref. 22)

1)- Introduction

Whether supersymmetry is implemented in Nature is far from being an answered question. In any case, supersymmetry must be spontaneously broken since physical particles do not combine into supermultiplets of equal mass. Accordingly, all phenomenological applications of supergravity [1)] have been based on the so-called superHiggs phenomenon which takes place when matter multiplets involving scalar fields are coupled to supergravity [2)].

In particular it appears that the low energy world corresponding to scales between 100 GeV and say 1 TeV should be described in terms of just one supersymmetry (N=1 theory) [3)]. This is mainly due to the fact that the only supersymmetric theory compatible with a chiral gauge theory has precisely one supercharge.

There is, however, a general consensus that the N=1 description is just an effective approximation of a more fundamental one, since it does not provide a finite theory of gravity and since it involves too many arbitrary functions and representation choices. Specifically the N=1 matter coupled supergravity involves, besides the choice of a gauge group G and of the G-representations to which the chiral multiplets (z^α, λ^α) are assigned, also the arbitrary selection of a Kähler manifold with Kähler potential $\mathcal{G}$ (z , z^*) determining the σ-model of the scalar fields. Furthermore the kinetic term of the vector fields V_μ^i contains an arbitrary analytic symmetric function f_{ij} (z), and one can introduce an optional "potential" term determined by $\mathcal{G}$.

In general, supersymmetric theories are more restricted if formulated in higher dimensions or with a larger number of supercharges. It would then be desirable to obtain the effective N=1 model through the spontaneous breakdown of a larger theory. In the case N $>$1 one is led to consider the question of partial supersymmetry breaking via the superHiggs mechanism, while for higher dimensional supergravities one deals with the issue of spontaneous compactification [4)]: here the number of Killing spinors admitted by the manifold of the compactified dimensions [5)] equals the number of unbroken supersymmetries of the 4-dimensional theory.

Higher-dimensional supergravities could, in turn, be a low energy limit of superstring theory [6], and might first compactify to a 4-dimensional theory with $N>1$ and then break to $N=1$ via a suitable extremum of its scalar field potential.

Therefore, any attempt to link supersymmetric phenomenology to more fundamental theories requires a firm control on the properties of matter coupled extended supergravities in four dimensions.

Specifically one needs the explicit form of the scalar field potential and the fermion supersymmetry transformation rules in order to study the problem of partial supersymmetry breaking.

The most relevant models have N=2,3,4 since, starting with N=5, up to N=8 one cannot couple supergravity to gauge multiplets and thus obtain theories with the phenomenological $SU_3 \otimes SU_2 \otimes U_1$.

The complete N=2 theories have been constructed [7] using the tensor calculus approach as in the N=1 case. The N=4 matter coupling to supergravity has been recentely constructed with the Noether procedure [8]. In this paper we couple N=3 matter to supergravity using an entirely geometrical method, the group manifold approach.

Let us review some of the features of the N=2 and N=4 couplings.

i) For N=2 there are two kinds of scalar manifolds because there are two independent multiplets containing scalar fields, the vector and the hypermultiplets. Scalar fields belonging to vector multiplets are associated to a restricted class of Kähler manifolds, whose Kähler potential can be computed in terms of a scalar analytic function $f(z^i)$ [7]. On the other hand, hypermultiplet scalars are co-ordinates of quaternionic manifolds which in general do not have a Kähler structure.

In contrast to the N=1 case a scalar potential is possible in N=2 supergravity only if some group is gauged. This feature still hold in all $N>2$ extended supergravities.

ii) In N=4 matter coupling the scalar manifold seems uniquely determined to be [9] $(SO(n,6) / SO(n)\otimes SO(6)) \otimes (SU(1,1) / U(1))$ where n is the number of vector multiplets. This manifold is non Kählerian, a feature shared by the scalar manifolds for N=2 hypermultiplets and N=8 supergravity Non - Kählerian manifolds are associated to scalar fields satisfying a reality condition due to the self-conjugation property of the multiplet.

For N=3 the matter multiplet is not self-conjugate and the scalar manifold turns out to be Kähler. As in the N=4 case there is a natural candidate, the homogeneous Kähler manifold $SU(3,n) / SU(3) \otimes SU(n) \otimes U(1)$. As explained later, the lack of arbitrariness in the coupling of N=3 and N=4 matter to supergravity may be traced back to the absence of auxiliary fields in these theories. Indeed arbitrary interactions are already absent in the global version of the theory, independently of the coupling to supergravity.

We notice that both N=3 and N=1 supergravities couple to Kählerian scalars. This gives reasonable hope for a partial breaking of N=3 to N=1. Another remarkable fact is that an N=3 theory in four dimensions can be obtained from a compactification of D=11 supergravity [10]. This is also true for N=1 and N=2, but not for N=4 [11].

The method we use to construct the N=3 matter coupled supergravity is the so called Group-Manifold Approach (4a) which turns out to be particulary economical and efficient for theories involving non linear σ - models. Given the scalar manifold G/H, one can introduce the vielbein and connections of G/H, together with the supervielbein (V^a, ψ_A) of the N-extended superspace and solve the Bianchi identities of the complete system under the assumption of rheonomy (4a)[20]. This yields the complete set of supersymmetry transformation rules, closing the on-shell algebra from which the potential can already be computed, as shown in [12] and [13]. Indeed the extra terms in the fermion transformation rules, due to the gauging of a suitable subgroup of the isometry group of the scalar manifold, are all one needs to work out the potential and therefore to study the possible supersymmetry and gauge symmetry breakings of these theories. We emphasize that these extra terms are obtained, in our approach, from pure group-theory without any reference to the Lagrangian.

Actually the scalar potential of all $N > 2$ extended supergravities depends only on the gauge structure constants and the coset space structure of the scalar manifold.

Our results are organized as follows:

- In section 2 we discuss the N=3 vector multiplet, the absence of auxiliary fields and of self-interactions in the global supersymmetry limit. Turning supergravity on we expect the scalars to become the co-ordinates of a coset manifold G/H, where G is the group of duality rotations on the vector field strengths. We find the unique answer: $G = SU(3,n)$ and $H = SU(3) \otimes SU(n) \otimes U(1)$, n being the number of vector multiplets involved.
- In section 3 we introduce the $SU(3,n) / SU(3) \otimes SU(n) \otimes U(1)$ formalism and we solve the Bianchi identities both with and without gauging of the σ- model. In particular we show that the possible gauge groups are either $SO(3) \times K_n$ or $SO(3,1) \times K_{n-3}$, K_m being any compact group of dimension m.
- In section 4 we calculate the scalar field potential.

The complete Lagrangian for N=3 gauged Supergravity is given in the Table. For its detailed derivation, we refer to 22).

2) - The N=3 vector multiplet and the G / H structure of the supergravity coupling

We begin by fixing our conventions for the indices:

a, b, c, d	= Lorentz flat indices: take 4 values
A, B, C	= SU(3) indices: take 3 values and enumerate the gravitinos
i, j, k, l	= SU(n) indices, n being the number of vector multiplets
Λ, Σ, Π	= SU(3,n) indices will take the values A, B, C plus the values i, j, k, l.

No curved indices are needed since we shall use the language of forms throughout the whole paper. In order to discuss the N=3 supersymmetric vector multiplet, we need the flat coframe$(\omega^{ab}, V^a, \psi_A)$ describing N=3 superPoincaré algebra in the dual language of 1-forms and Maurer-Cartan equations. Let V^a be the vierbein, ω^{ab} the spin connection, A_A a triplet of real 1-forms associated to the 3 central charges and ψ_A the 3 chiral gravitinos:

$$\psi_A = \gamma_5 \psi_A \qquad ; \quad \bar{\psi}^A = \psi_A^{\dagger} \gamma_0 \tag{2.1a}$$

$$\psi^A = -\gamma_5 \psi^A \;\; ; \;\; \psi^A = C(\bar{\psi}^A)^T \;\; ; \;\; \bar{\psi}^A = (\psi^A)^{\dagger}\gamma_0 \tag{2.1b}$$

The N=3 superPoincaré algebra reads:

$$R^a \underset{def}{=} \mathcal{D}V^a - i\,\bar{\psi}^A \gamma^a \psi_A = 0 \tag{2.2a}$$

$$F_A \underset{def}{=} dA_A - \frac{1}{2}\epsilon_{ABC}\bar{\psi}^B \wedge \psi^C - \frac{1}{2}\epsilon^{ABC}\bar{\psi}_B \wedge \psi_C = 0 \tag{2.2b}$$

$$\rho_A \underset{def}{=} \mathcal{D}\psi_A = 0 \tag{2.2c}$$

$$R^{ab} \underset{def}{=} d\omega^{ab} - \omega^{ac} \wedge \omega_c^{\;\;b} = 0 \tag{2.2d}$$

where the Lorentz covariant derivatives are (*) :

$$\mathcal{D}V^a = dV^a - \omega^{ab} \wedge V_b \;\; ; \; \mathcal{D}\psi_A = d\psi_A - \frac{1}{4}\omega^{ab} \wedge \gamma_{ab}\psi_A \tag{2.3}$$

This algebra admits an SO(3) automorphism group which can be extended to SU(3) only at the price of introducing, besides the "electric" potential A_A, a "magnetic" one B_A defined by the following equation:

$$\tilde{F}_A = dB_A - \frac{i}{2}\epsilon_{ABC}\bar{\psi}^B \wedge \psi^C + \frac{i}{2}\epsilon^{ABC}\bar{\psi}_B \wedge \psi_C \tag{2.4}$$

(*) The conventions for gamma matrices are given in ref. 22)

The dual Lie Algebra formed by equations (2.2) and (2.4) is now invariant under the following SU(3) infinitesimal transformations:

$$\delta_{SU_3} \psi_A = \Omega_{AB} \psi_B + i \mathcal{S}_{AB} \psi_B \tag{2.5a}$$

$$\delta_{SU_3} A_A = \Omega_{AB} A_B - \mathcal{S}_{AB} B_B$$

$$\delta_{SU_3} B_A = \Omega_{AB} B_B + \mathcal{S}_{AB} A_B \tag{2.5b}$$

where $\Omega_{AB} = -\Omega_{BA}$ are the parameters of the SO(3) subgroup and $\mathcal{S}_{AB} = \mathcal{S}_{BA}$ those of SU_3 / SO_3.

This is different from what happens in the N=2 case where the analogue of eq. (2.2b)

$$F = dA - \frac{1}{2} \epsilon^{AB} \bar{\psi}_A \wedge \psi_B - \frac{1}{2} \epsilon_{AB} \bar{\psi}^A \wedge \psi^B = 0 \tag{2.6}$$

is already invariant under the SU_2 -transformation:

$$\delta_{SU_2} \psi_A = \Omega_{AB} \psi_B + i \mathcal{S}_{AB} \psi_B \qquad \delta_{SU_2} A = 0 \tag{2.7}$$

without the need to introduce a magnetic potential B. This may explain why N=3 multiplets do not admit auxiliary fields and hence do not allow for an arbitrary choice of the σ-model structure contrary to what happens for N=1 and N=2. Indeed, for N=3 there are no hypermultiplets and the only available matter multiplets are the vector ones. Each of them has the following field content

$$\left[(1),\ 3(\tfrac{1}{2}) \oplus 3(\tfrac{1}{2}),\ 3(0^+) \oplus 3(0^-) \right]$$

which corresponds to a 1-form A, a 0-form left handed triplet of spinors

$$\gamma_5 \lambda_A = \lambda_A \tag{2.8}$$

a right handed singlet spinor

$$\gamma_5 \lambda = - \lambda \tag{2.9}$$

and a complex 0-form triplet of scalars z^A ($z_A = (z^A)^*$).

Consider now n of these multiplets labelled by an index i. Following the procedure already utilized for the N=1 [14], N=2 [15] and N=4 [16] vector multiplets, the most general rheonomic parametrization of the exterior derivatives is:

$$dz_i{}^A = z_i{}^A{}_{|a} V^a + \bar{\lambda}_{iB} \psi_C \epsilon^{ABC} + \bar{\lambda}_i \psi^A \tag{2.10a}$$

$$\begin{aligned} F_i = dA_i = F_i{}^{ab} V_a \wedge V_b + b_1 \bar{\lambda}_{iA} \gamma^a \psi^A \wedge V_a - \\ - b_1^* \bar{\lambda}^{Ai} \gamma^a \psi_A \wedge V_a + \frac{1}{2} z_i{}^A \bar{\psi}^B \wedge \psi^C \epsilon_{ABC} + \frac{1}{2} z^i_A \bar{\psi}_B \wedge \psi_C \epsilon^{ABC} \end{aligned} \tag{2.10b}$$

$$\begin{aligned} \mathcal{D}\lambda_{iA} = \mathcal{D}_a \lambda_{iA} V^a + b_2 z_i{}^B{}_{|a} \gamma^a \psi^C \epsilon_{ABC} + \\ + b_3 G_i{}^{ab} \gamma_{ab} \psi_A + \mathcal{M}_{iA}{}^{\cdot\cdot B} \psi_B \end{aligned} \tag{2.10c}$$

$$\mathcal{D}\lambda_i = \mathcal{D}_a \lambda_i V^a + b_4 z_i{}^A{}_{|a} \gamma^a \psi_A + \mathcal{N}_{iA} \psi^A \tag{2.10d}$$

where b_1 , b_2 , b_3 , b_4 are coefficients to be determined from Bianchi identities, $G_i{}^{ab}$ is defined by:

$$G_i{}^{ab} = \frac{1}{2} \left(F_i{}^{ab} - \frac{i}{2} \epsilon^{abcd} F_i{}^{cd} \right) \tag{2.11}$$

and $\mathcal{M}_{iA}{}^{\cdot\cdot B}$, $\mathcal{N}_{iA}$ are functions of the scalars which we can call the "auxiliary fields" (i.e., the value taken on shell by the auxiliary fields).

Implementing the Bianchi identities:

$$d^2 z_i^{\ A} = dF_i = \mathcal{D}^2 \lambda_{iA} = \mathcal{D}^2 \lambda_i = 0 \tag{2.12}$$

in the flat background described by eqs. (2.2), we find:

$$b_1 = \frac{i}{2} \qquad b_2 = -i \qquad b_3 = 1 \qquad b_4 = i \tag{2.13}$$

and

$$\mathcal{M}_{iA}^{\ \ \cdot\cdot B} = \mathcal{N}_{iA} = 0 \tag{2.14}$$

This last result follows from:

$$\mathcal{M}_{iB}^{\ \ \cdot\cdot R}\, \overline{\Psi}_C \Psi_R\, \epsilon^{ABC} = 0 \tag{2.15a}$$

$$\overline{\Psi}^A \Psi^B \mathcal{N}_{iB} = 0 \tag{2.15b}$$

These equations, because of the integrability condition

$$d^2 z_i^{\ A} = 0 \tag{2.16}$$

only have the trivial solution (2.14). The analogous equation one obtains for the N=2 multiplet[15)]

$$\epsilon^{BC}\, \overline{\Psi}_C \Psi_R\, \mathcal{M}_{iB}^{\ \ \cdot\cdot R} = 0 \tag{2.17}$$

admits instead the general solution

$$\mathcal{M}_{iB}^{\ \ \cdot\cdot R} = SU_2 \text{ Lie Algebra element} \tag{2.18}$$

The difference is due to the fact that SU_2 is an automorphism group for equation (2.6) while SU(3) is not an automorphism group for eq. (2.2b).

In other words, the N=2 scalars are SU_2 singlets while in the second case of N=3 they transform under SU_3. We stress that the transformation properties of the scalars are connected to those of the central charges, since the scalar fields show up as coefficients of the $\bar{\psi}_A \wedge \psi_B$ terms in the expansion of the vector field strenght Fi (see eq. (2.10b).

The conclusion, therefore, is that in the global supersymmetry limit the N=3 vector multiplet has no arbitrary self-interaction and gives rise only to a trivial σ-model (=flat manifold). Hence the non-linear σ- model structure is entirely determined by the coupling to supergravity and must correspond to a unique choice of the scalar manifold $\mathcal{M}$. Let us determine $\mathcal{M}$. First of all, the connection of $\mathcal{M}$ must act linearly on the SU(3) index of ψ_A; therefore the holonomy group $\mathcal{H}(\mathcal{M})$ must be of the form

$$\mathcal{H}(\mathcal{M}) = SU(3) \otimes \mathcal{H}' \tag{2.19}$$

where $\mathcal{H}'$ is some other group acting on the tangent space indices.

Secondly, if we assume that $\mathcal{M}$ is a homogeneous space we have

$$\mathcal{M} = \mathcal{G}/\mathcal{H}(\mathcal{M}) \tag{2.20}$$

The isometry group $\mathcal{G}$ is the group of duality rotations of the vector field strenghts; according to ref.[17] it satisfies the following constraints:

i) it is a subgroup of Sp[2x(3+n)], the maximal duality group;

ii) it has a 3+n complex representation or equivalently a 2x(3+n) real representation $\mathcal{D}$ (3+n is the total number of vector fields of supergravity $\oplus$ matter theory).

Finally, $\mathcal{G}/\mathcal{H}$ must have 3n complex dimensions corresponding to the 3n complex scalars.

One may easily verify that these requirements are satisfied, independently of n, by the choice

$$\mathcal{G} = SU(3, m) \tag{2.21a}$$

$$\mathcal{H} = SU(3) \times SU(m) \times U(1) \tag{2.21b}$$

Let us now discuss the possible gaugings of the vector fields. The basic idea is that, irrespectively of the specific theory, when we deal with a σ- model, the symmetries we can gauge are the isometries of the scalar manifold. Hence in our case the gauge group G must be a subgroup of $\mathcal{G}$ = SU(3,n). In supergravity, however, we have the further restriction that the isometry group $\mathcal{G}$ of the scalar manifold is also the duality group of the vector field strengths. This means that the available gauge fields, which are by definition in the adjoint representation of G , are also assigned, together with their duals, to an irreducible representation of $\mathcal{G}$. In our case this is the fundamental $\underline{3+n}$ of SU(3,n).

Let

$$D(T)_{\Lambda}{}^{\Sigma}$$

be the $(\underline{3+n}) \times (\underline{3+n})$ matrix representation of the generator $T \in$ SU(3,n) and let

$$J_{\Lambda\Sigma} = J^{\Lambda\Sigma} = \begin{pmatrix} \delta_{AB} & 0 \\ 0 & -\delta_{ij} \end{pmatrix} \tag{2.22}$$

be the SU(3,n) - invariant metric; then we have:

$$D^{\dagger}(T) J + J D(T) = 0 \tag{2.23}$$

Decomposing $D(T)$ into its real and imaginary parts:

$$D(T) = X(T) + i Y(T) \tag{2.24}$$

eq. (2.23) becomes

$$X^{T}(T) J + J X(T) = 0 \tag{2.25}$$

$$-Y^{T}(T) J + J Y(T) = 0 \tag{2.26}$$

The elements $T \in SU(3,n)$ for which $Y(T) = 0$ form, because of equation (2.25), the subgroup $SO(3,n) \subset SU(3,n)$. On the other hand the vector fields are arranged into a <u>3+n</u> of SU(3,n) by writing

$$H_\Sigma = A_\Sigma + i B_\Sigma \tag{2.27}$$

$$\delta H_\Sigma = D(T)_\Sigma{}^\Pi H_\Pi \tag{2.28}$$

where A_Σ is the "electric" potential (the physical field) and B_Σ is the "magnetic" potential defined from A_Σ via a duality transformation. The symmetry under duality is broken by the gauging since we can gauge only the electric potential which is the real part of the SU(3,n) vector <u>3+n</u>. This means that we can gauge only those generators of SU(3,n) transforming A_Σ into itself. Hence G must be a subgroup of SO(3,n):

$$G \subset SO(3,n) \tag{2.29}$$

Now the 3+n complex representation D must decompose into adj $\oplus$ adj when $\mathcal{G}$ is restricted to the gauge group $G \subset SO(3,n)$

$$D \xrightarrow[G]{} \text{adj} \oplus \text{adj} \tag{2.30}$$

This condition selects as possible gauge groups G either

$$G = SO(3) \times K_n \tag{2.31}$$

or

$$G = SO(3,1) \times K_{n-3} \tag{2.32}$$

(K_n meaning compact group of dimension n) canonically embedded in the SO(3,n) subgroup of SU(3,n). Indeed the fundamental $(\underline{3+n})^{\text{complex}}$ of SU(3,n) decomposes under SO(3,n) into two $(\underline{3+n})^{\text{real}}$

$$(\underline{3+n})^{\text{complex}} \xrightarrow[SO(3,n)]{} (3+n)^{\text{real}} \oplus (3+n)^{\text{real}} \tag{2.33}$$

and each of the two $\underline{(3+n)}^{real}$ can be identified with the adjoint of G given by either (2.31) or (2.32). The last statement follows from the fact that the adjoint matrices of an n-dimensional compact Lie Algebra are antisymmetric, and therefore also elements of the SO(n) Lie Algebra. Furthermore the action of the subgroup G $\subset$ SO(3,n) $\subset$ SU(3,n) on A_Σ must be the same whether we consider it as an element of the adjoint representation of G or as a vector of SO(3,n): if

$$T_\Omega \in G \subset SO(3,m) \subset SU(3,m) \tag{2.34}$$

is a generator of G , we must have

$$X(T_\Omega)_\Lambda{}^\Sigma = f_{\Omega\Lambda}{}^\Sigma \tag{2.35}$$

where $f_{\Omega\Lambda}{}^\Sigma$ are the G structure constants

$$[T_\Omega, T_\Lambda] = f_{\Omega\Lambda}{}^\Sigma T_\Sigma \tag{2.36}$$

Comparing eq. (2.36) with eq. (2.23) we see that $J_{\Lambda\Sigma}$ is the Killing metric of G and that $f_{\Lambda\Sigma\Gamma}$, defined by

$$f_{\Lambda\Sigma\Gamma} = f_{\Lambda\Sigma}{}^{\Gamma'} J_{\Gamma\Gamma'} \tag{2.37}$$

is completely antisymmetric in its three indices. This last equation will be essential to prove the supersymmetry of the Lagrangian.

3 - SU(3,n) / SU(3) × SU(n) × U(1) formalism and the solution of Bianchi identities.

We now introduce the $\mathcal{G}/\mathcal{H}$ coset representatives in terms of which we construct the full theory: our procedure is similar to that applied in ref. [8]) to the N=4 coupling; the essential difference is that we consider the composite vielbein and connections as forms over superspace rather than ordinary space.

Let $L_\Lambda{}^\Sigma(z)$ be the SU(3,n) element which represents the equivalence class (i.e. the point of the $\mathcal{G}/\mathcal{H}$ manifold) identified by the co-ordinates z (i.e. the scalar fields $z_\Lambda{}^i$)[18].

By definition L is an SU(3,n) matrix and hence satisfies the condition

$$L^\dagger J L = J \tag{3.1}$$

or in matrix notation:

$$(L^{-1})_\Lambda{}^\Sigma = J_{\Lambda\Pi} J^{\Sigma\Delta} (L_\Delta{}^\Pi)^* \tag{3.2}$$

Here $J_{\Lambda\Pi}$ is the SU(3,n) - invariant metric defined in (2.22).

L can now be used to write down the vielbein P and the connection Q of the gauged σ-model.

We first define

$$\Omega_\Lambda{}^\Pi = (L^{-1})_\Lambda{}^\Sigma \, dL_\Sigma{}^\Pi + g(L^{-1})_\Lambda{}^\Sigma f_\Sigma{}^{\Omega\Gamma} A_\Omega L_\Gamma{}^\Pi \tag{3.3}$$

where g is the gauge coupling constant and we see that when g=0, Ω reduces to the ordinary left-invariant 1-form over the coset manifold. When $g\neq 0$, Ω no longer satisfies the standard Cartan-Maurer equations and develops a non-vanishing SU(3,n) curvature. Indeed we find

$$d\Omega_\Lambda{}^\Pi + \Omega_\Lambda{}^\Gamma \wedge \Omega_\Gamma{}^\Pi = g(L^{-1})_\Lambda{}^\Omega f_\Omega{}^{\Sigma\Gamma} \mathcal{F}_\Sigma L_\Gamma{}^\Pi \tag{3.4}$$

where

$$\mathcal{F}_\Sigma \equiv dA_\Sigma + f_\Sigma{}^{\Omega\Delta} A_\Omega \wedge A_\Delta \tag{3.5}$$

is the field strength of the gauge group G .

Decomposing Ω into SU(3) $\otimes$ SU(n) $\otimes$ U(1) tensors we obtain both the scalar vielbein P and the composite connections Q.

We set

$$\Omega_i{}^A = P_i{}^A \quad ; \quad \Omega_A{}^i = P_A{}^i = (P_i{}^A)^* \tag{3.6}$$

$$\Omega_A{}^B = \Phi_A{}^B - m\,\delta_A^B\,\Phi^{\otimes} \tag{3.7}$$

$$\Omega_i{}^j = \Phi_i{}^j + 3\,\delta_i^j\,\Phi^{\otimes} \tag{3.8}$$

where $\Phi_A{}^B$ $(\Phi_A{}^A = 0)$ and $\Phi_i{}^j$ $(\Phi_i{}^i = 0)$ are the SU(3) and SU(n) connections respectively, while $Q^{\otimes}$ is the U(1) connection.

The geometric structure of the matter coupled N=3 supergravity theory is now obtained by merging together the coframe of the pure supergravity sector (ω^{ab}, V^a, ψ_A) with the coframe of the σ-model ($P_A{}^i, \Phi_A{}^B, \Phi_i{}^j, \Phi^{\otimes}$) via the definition of a new set of curvatures. Let us recall eqs. (2.2), valid in the global limit, and replace them with

$$R^a \equiv \mathcal{D}V^a - i\,\bar\psi^A \gamma^a \psi_A \tag{3.9a}$$

$$\rho_A \equiv \nabla\psi_A = \mathcal{D}\psi_A + \Phi_A{}^B{}_\wedge\psi_B + p\,\Phi^{\otimes}{}_\wedge\psi_A \tag{3.9b}$$

$$F_\Lambda \equiv \mathcal{F}_\Lambda - \frac{1}{2} L_\Lambda{}^A \epsilon_{ABC}\bar\psi^B{}_\wedge\psi^C - \frac{1}{2} L^{-1}{}_A{}^{\pi} J_{\Lambda\pi}\epsilon^{ABC}\bar\psi_B\psi_C \tag{3.9c}$$

$$R^{ab} \equiv d\omega^{ab} - \omega^{ac}{}_\wedge\omega_c{}^b \tag{3.9d}$$

where p is the U(1) charge of the gravitino which we shall shortly determine. Equation (3.9c) encompasses in a single structure both the vectors pertaining to supergravity and those pertaining to the matter multiplets. We now supplement the list of curvatures with the SU(3) × SU(n) × U(1) covariant derivatives of the spin ½ fermions. These are the λ_{iA} and λ_i spinors already introduced in eqs. (2.10) plus the spin ½ field $\chi_{\otimes}$ of the supergravity multiplet, which is a singlet under SU(3) and SU(n):

$$\gamma_5\chi_{\otimes} = \chi_{\otimes} \qquad \bar\chi^{\otimes} = \chi^{\dagger}_{\otimes}\gamma_0 \tag{3.10a}$$

$$\chi^{\otimes} = C(\bar{\chi}^{\otimes})^T \qquad \bar{\chi}_{\otimes} = \chi^{\otimes\dagger}\gamma_0 \tag{3.10b}$$

Hence we write:

$$\nabla\chi_{\otimes} \equiv \mathcal{D}\chi_{\otimes} + q\,\phi^{\otimes}\chi_{\otimes} \tag{3.11a}$$

$$\nabla\lambda_i \equiv \mathcal{D}\lambda_i + \phi_i{}^j\lambda_j + r\,\phi^{\otimes}\lambda_i \tag{3.11b}$$

$$\nabla\lambda_{iA} \equiv \mathcal{D}\lambda_{iA} + \phi_i{}^j\lambda_{jA} + \phi_A{}^B\lambda_{iB} + s\,\phi^{\otimes}\lambda_{iA} \tag{3.11c}$$

where the U(1) charges q, r, s will be determined by consistency with the charge p of the gravitino.

From equations (3.11) we see that the spinors λ_i, λ_{iA} have been assigned to the fundamental representation of SU(n).

From equations (3.9) and (3.11), by taking a second covariant derivative we obtain the Bianchi identities:

$$\mathcal{D}R^a + R^{ab}\wedge V_b + i\bar{\rho}^A\gamma^a\psi_A - i\bar{\psi}^A\gamma^a\rho_A = 0 \tag{3.12a}$$

$$\nabla\rho_A + \frac{1}{4}R^{ab}\wedge\gamma_{ab}\psi_A - R_A{}^B\wedge\psi_B - p\,R^{\otimes}\wedge\psi_A = 0 \tag{3.12b}$$

$$\nabla F_\Lambda + \frac{1}{2}L_\Lambda{}^i P_i{}^A\wedge\epsilon_{ABC}\bar{\psi}^B\wedge\psi^C - \frac{1}{2}P_A{}^i(L^{-1})_i{}^\Pi J_{\Lambda\Pi}\epsilon^{ABC}\bar{\psi}_B\psi_C - L_\Lambda{}^A\epsilon_{ABC}\bar{\psi}^B\rho^C - (L^{-1})_A{}^\Pi J_{\Lambda\Pi}\epsilon^{ABC}\bar{\psi}_B\rho_C = 0 \tag{3.12c}$$

$$\mathcal{D}R^{ab} = 0 \tag{3.12d}$$

$$\nabla^2 \chi_{\otimes} = -\frac{1}{4} R^{ab} \gamma_{ab} \chi_{\otimes} + q R^{\otimes} \chi_{\otimes} \tag{3.12e}$$

$$\nabla^2 \lambda_{iA} = -\frac{1}{4} R^{ab} \gamma_{ab} \lambda_{iA} + R_i{}^j \lambda_{jA} + R_A{}^B \lambda_{iB} + s R^{\otimes} \lambda_{iA} \tag{3.12f}$$

$$\nabla^2 \lambda_i = -\frac{1}{4} R^{ab} \gamma_{ab} \lambda_i + R_i{}^j \lambda_j + r R^{\otimes} \lambda_i \tag{3.12g}$$

where the curvatures of the composite SU(3)×SU(n)×U(1) connections have been introduced. They are defined by

$$R_A{}^B \equiv d\Phi_A{}^B + \Phi_A{}^M \wedge \Phi_M{}^B \tag{3.13}$$

$$R_i{}^j \equiv d\Phi_i{}^j + \Phi_i{}^m \wedge \Phi_m{}^j \tag{3.14}$$

$$R^{\otimes} \equiv d\Phi^{\otimes} \tag{3.15}$$

These curvatures are completely determined by the structure of the coset manifold, specifically by eq. (3.4).

We now introduce "boosted structure constants" which are functions of the scalar fields:

$$C_\Lambda{}^{\Pi\Gamma} = C_\Lambda{}^{\Pi\Gamma}(z) = (L^{-1})_\Lambda{}^{\Lambda'} L_{\Pi'}{}^{\Pi} L_{\Gamma'}{}^{\Gamma} f_{\Lambda'}{}^{\Pi'\Gamma'} \tag{3.16a}$$

$$C^\Lambda{}_{\Pi\Gamma} = C^\Lambda{}_{\Pi\Gamma}(z) = L_{\Lambda'}{}^{\Lambda} (L^{-1})_{\Pi}{}^{\Pi'} (L^{-1})_{\Gamma}{}^{\Gamma'} f_{\Pi'\Gamma'}{}^{\Lambda'} \tag{3.16b}$$

$$C_\Lambda{}^{\Pi\Gamma} = J_{\Lambda\Lambda'} J^{\Pi\Pi'} J^{\Gamma\Gamma'} (C^{\Lambda'}{}_{\Pi'\Gamma'})^* \tag{3.16c}$$

and from eq. (3.4) we obtain

$$R_A{}^B = g\{(L^{-1})_A{}^\Lambda L_\Sigma{}^B - \frac{1}{3}\delta_A{}^B (L^{-1})_M{}^\Lambda L_\Sigma{}^M\} f_\Lambda{}^{\cdot\Omega\Sigma} F_\Omega$$

$$+ \frac{1}{2} g\left(C_A{}^{\cdot PB} - \frac{1}{3}\delta_A{}^B C_M{}^{\cdot PM}\right)\epsilon_{PRS}\bar{\Psi}^R \wedge \Psi^S +$$

$$+ \frac{1}{2} g\left(C_{AP}{}^{\cdot\cdot B} - \frac{1}{3}\delta_A{}^B C_{MP}{}^{\cdot\cdot M}\right)\epsilon^{PRS}\bar{\Psi}_R \wedge \Psi_S -$$

$$- \left(P_A{}^m \wedge P_m{}^B - \frac{1}{3}\delta_A{}^B P_M{}^m \wedge P_m{}^M\right) \tag{3.17}$$

$$R_i{}^j = g\{(L^{-1})_i{}^\Lambda L_\Sigma{}^j - \frac{1}{m}\delta_i{}^j (L^{-1})_m{}^\Lambda L_\Sigma{}^m\} f_\Lambda{}^{\cdot\Omega\Sigma} F_\Omega$$

$$+ \frac{1}{2} g\{C_i{}^{\cdot Pj} - \frac{1}{m}\delta_i{}^j C_m{}^{\cdot Pm}\}\epsilon_{PRS}\bar{\Psi}^R \wedge \Psi^S$$

$$+ \frac{1}{2} g\{C_{iP}{}^{\cdot\cdot j} - \frac{1}{m}\delta_i{}^j C_{mP}{}^{\cdot\cdot m}\}\epsilon^{PRS}\bar{\Psi}_R \wedge \Psi_S$$

$$- \left(P_i{}^M \wedge P_M{}^j - \frac{1}{m}\delta_i{}^j P_m{}^M \wedge P_M{}^m\right) \tag{3.18}$$

$$R^{\otimes} = -\frac{g}{3m}(L^{-1})_M{}^{\cdot\Lambda} L_\Sigma{}^M f_\Lambda{}^{\cdot\Omega\Sigma} F_\Omega +$$

$$- \frac{g}{6m} C_M{}^{\cdot AM}\epsilon_{ABC}\bar{\Psi}^B \wedge \Psi^C - \frac{g}{6m} C_{MA}{}^{\cdot\cdot M}\epsilon^{ABC}\bar{\Psi}_B \wedge \Psi_C$$

$$+ \frac{1}{3m} P_M{}^m \wedge P_m{}^M . \tag{3.19}$$

where we have added and subtracted the spinor terms necessary to complete the definition of F_Λ . If we set $g = 0$, eqs. (3.17), (3.18) and (3.19) reduce to the ordinary Maurer-Cartan equations of SU(3,n). Furthermore from eq. (3.4) we also obtain

$$\nabla P_i{}^A = dP_i{}^A + Q_i{}^j P_j{}^A + P_i{}^B Q_B{}^{\cdot A} + (3+m) Q^{\circ} P_i{}^A \qquad (3.20)$$

which we can call the "internal torsion equation" since it states that the covariant derivative of the composite vielbein $P_i{}^A$ vanishes. Eq. (3.20) fixes the U(1) weight of $P_i{}^A$ from which we deduce all the others. At this point, we write down the parametrization of all the curvatures, or exterior derivatives, and compute the coefficients by implementation of the Bianchi identities. We stress that our base space is superspace, for which an independent basis is provided by (V^a, ψ_A) : the vielbein $P_i{}^A$ is not an independent form, being given in terms of the scalar field differentials. Hence similarly to dz^i_A in the global case, $P_i{}^A$ must be expanded along the complete basis (V^a, ψ_A).
In full analogy to eq. (2.10a) we write

$$P_i{}^A = P_i{}^A{}_{|a} V^a + \bar{\lambda}_{iB} \psi_C \epsilon^{ABC} + \bar{\lambda}_i \psi^A \qquad (3.21)$$

identifying the spinor fields λ_{iB}, λ_i with the spinor derivatives of the scalar fields.

By comparison with (3.20) and (3.9c), eq. (3.21) fixes all the U(1) weights of the spinor fields except q:

$$p = \frac{m}{2} \quad ; \quad r = 3\left(1 + \frac{m}{2}\right) \quad ; \quad s = 3 + \frac{m}{2} \qquad (3.22)$$

The weight q is fixed by writing the most general rheonomic parametrization of the F_Λ -curvature:

$$F_\Lambda = F_\Lambda{}^{ab} V_a V_b + b_1 L_\Lambda{}^i \bar{\lambda}_{iA} \gamma^a \psi^A \wedge V_a$$
$$+ b_1^* (L^{-1})_i{}^\pi J_{\Lambda\pi} \bar{\lambda}^{iA} \gamma^a \psi_A V_a +$$
$$+ i L_\Lambda{}^A \bar{\chi}^\otimes \gamma^a \psi_A \wedge V_a + \qquad (3.23)$$
$$+ i (L^{-1})_A{}^\pi J_{\Lambda\pi} \bar{\chi}_\otimes \gamma^a \psi^A{}_\wedge V_a$$

and is the analogue of eq. (2.10b). It implies

$$q = \frac{3}{2} m \qquad (3.24)$$

In order to write the rheonomic parametrization of the fermionic curvatures ρ_A, $\nabla\chi_\otimes$, $\nabla\lambda_{iA}$, $\nabla\lambda_i$, it is convenient to use the following notation. Given the space-time components $F_\Lambda{}^{ab}$ of F_Λ, we introduce new complex tensors $G_A{}^{ab}$ and $G_i{}^{ab}$ which have a definite duality:

$$\epsilon^{abcd} G_A{}^{cd} = -2i\, G_A{}^{ab} \quad ; \quad G^A{}_{ab} = (G_A{}^{ab})^* \qquad (3.25a)$$

$$\epsilon^{abcd} G_i{}^{cd} = +2i\, G_i{}^{ab} \quad ; \quad G^i{}_{ab} = (G_i{}^{ab})^* \qquad (3.25b)$$

and are related to $F_\Lambda{}^{ab}$ by the following equation:

$$F_\Lambda{}^{ab} = L_\Lambda{}^A (G_A{}^{ab} + a_1 \bar{\lambda}^i \gamma^{ab} \lambda_{Ai})$$
$$+ L_\Lambda{}^i (G_i{}^{ab} + a_2 \bar{\chi}^\otimes \gamma^{ab} \lambda_i)$$
$$+ (L^{-1})_A{}^\pi J_{\Lambda\pi} (G^A{}_{ab} + a_1^* \bar{\lambda}_i \gamma^{ab} \lambda^{Ai}) \qquad (3.26)$$
$$- (L^{-1})_i{}^\pi J_{\Lambda\pi} (G^i{}_{ab} + a_2^* \bar{\chi}_\otimes \gamma^{ab} \lambda^i)$$

involving two new coefficients to be determined by Bianchi identities. Eq. (3.26) is invertible and one can solve it for G_A^{ab} or G_i^{ab} in terms of F_Λ^{ab}, $\tilde{F}_\Lambda^{ab}$ and the spinor currents. We do not need this explicit inversion: we just need to know that it exists. With the above notation we can write the remaining parametrizations as:

$$\begin{aligned}
\rho_A = {} & \rho_{A|ab} V^a \wedge V^b + a_3 G^B_{ab} \gamma^a \psi^C V^b \epsilon_{ABC} \\
& + a_4 \psi_A \wedge \bar{\chi}^{\otimes} \gamma^a \chi_{\otimes} V_a + a_5 \gamma_{ab} \psi_A \bar{\chi}^{\otimes} \gamma^a \chi_{\otimes} V^b \\
& + a_6 \gamma_a \psi^B V^a \bar{\lambda}^{Ci} \lambda_i \epsilon_{ABC} + a_7 \gamma_a \psi^B V_b \bar{\lambda}^{Ci} \gamma^{ab} \lambda_i \epsilon_{ABC} \\
& + a_8 \psi_A \bar{\lambda}^i \gamma^a \lambda_i V_a + a_9 \gamma_{ab} \psi_A \bar{\lambda}^i \gamma^a \lambda_i V^b \\
& + a_{10} \psi_A \bar{\lambda}^{iB} \gamma^a \lambda_{iB} V_a + a_{11} \gamma_{ab} \psi_A \bar{\lambda}^{iB} \gamma^a \lambda_{iB} V^b \\
& + a_{12} \psi_B \bar{\lambda}^{iB} \gamma^a \lambda_{iA} V_a \\
& + a_{13} \gamma_{ab} \psi_B \bar{\lambda}^{iB} \gamma^a \lambda_{Ai} V^b \\
& + a_{14} \chi_{\otimes} \bar{\psi}^B \psi^C \epsilon_{ABC} \\
& + g\, a_{15} C_A{}^{PQ} \gamma_m \psi^R V^m \epsilon_{PQR} \\
& + g\, a_{16} C_M{}^{MC} \gamma_m \psi^C V^m \epsilon_{ABC}
\end{aligned} \tag{3.27}$$

$$\begin{aligned}
\nabla \chi_{\otimes} = {} & \nabla_a \chi_{\otimes} V^a + a_{17} G^A_{ab} \gamma^{ab} \psi_A \\
& + a_{18} \psi_A \bar{\lambda}^{Ai} \lambda_i + g\, a_{19} C_M{}^{MB} \psi_B
\end{aligned} \tag{3.28}$$

$$\nabla\lambda_{iA} = \nabla_a\lambda_{iA} V^a + b_2 P_i{}^B{}_{|a}\gamma^a\psi^C\epsilon_{ABC} + b_3 G_i{}^{ab}\gamma_{ab}\psi_A + a_{20}\psi_A\bar{\lambda}_i\chi^{\otimes} + g a_{21} C_{iA}{}^B\psi_B + g a_{22} C_{iM}{}^M\psi_A \tag{3.29}$$

$$\nabla\lambda_i = \nabla_a\lambda_i V^a + b_4 P_i{}^A{}_{|a}\gamma^a\psi_A + a_{23}\psi^A\bar{\lambda}_{iA}\chi_{\otimes} + g a_{24} C_i{}^{AB}\psi^C\epsilon_{ABC} \tag{3.30}$$

where we have distinguished between the b-coefficients already existing in the global limit (see eqs. (2.10)) and the a-coefficients introduced by the supergravity coupling; moreover we have explicitly written the gauge coupling constant in front of the terms involving the C - functions since they disappear at g=0. Inserting eqs. (3.27) - (3.30) into the Bianchi identities (3.12) fixes all the coefficients. The result is given in eqs. (3.38). Comparing with eq. (2.13), we see that the coefficients existing before the coupling to supergravity remain unchanged. On the other hand, comparing eqs. (2.10c) and (2.10d) with (3.29) and (3.30), we see that the shifts of the fermion transformation rules, forbidden at the global level by the absence of "auxiliary fields", show up in the local theory when $g \neq 0$. Indeed we can identify the functions $\mathcal{M}_{iA}{}^B$ and $\mathcal{N}_{iA}$:

$$\mathcal{M}_{iA}{}^B = g\left(a_{21} C_{iA}{}^B + a_{22}\,\delta^B_A C_{iM}{}^M\right) \tag{3.31a}$$

$$\mathcal{N}_{iA} = g\, a_{24} C_i{}^{PQ}\epsilon_{APQ} \tag{3.31b}$$

and we see that the supersymmetry transformations of λ_{iA} and λ_i in the presence of gauging differ from the non-gauged ones only by the addition of the terms (3.33). We find:

$$(\delta_\epsilon \lambda_{iA})\ \text{gauged} = (\delta_\epsilon \lambda_{iA})\ \text{ungauged} + \mathcal{M}_{iA}{}^{B}\,\epsilon_B \qquad (3.32)$$

$$(\delta_\epsilon \lambda_{i})\ \text{gauged} = (\delta_\epsilon \lambda_{i})\ \text{ungauged} + \mathcal{N}_{iA}\,\epsilon^A \qquad (3.33)$$

This follows from the usual identification of the supersymmetry transformation with the Lie-derivative in the fermionic directions and the use of eqs. (3.29) and (3.30).
With a similar argument we deduce:

$$(\delta_\epsilon \psi_A)\ \text{gauged} = (\delta_\epsilon \psi_A)\ \text{ungauged} + S_{AB}\gamma_m \epsilon^B V^m \qquad (3.34)$$

$$(\delta_\epsilon \chi_\otimes)\ \text{gauged} = (\delta_\epsilon \chi_\otimes)\ \text{ungauged} + \mathcal{U}^A \epsilon_A \qquad (3.35)$$

where

$$S_{AB} = g\left(a_{15} C_A{}^{PQ}\epsilon_{BPQ} + a_{16} C_M{}^{MC}\epsilon_{ABC}\right) \qquad (3.36)$$

$$\mathcal{U}^A = g\, a_{19} C_M{}^{MA} \qquad (3.37)$$

In this way, we have determined all the transformation rules from the pure algebraic structure of the theory. Using a general theorem [11)] we are now able to calculate the potential without reference to the Lagrangian. For completeness, however, in the next section we derive the complete action.

Here we give the a- and b- coefficients appearing in the fermionic curvatures (3.27) - (3.30):

$$
\begin{array}{llll}
b_1 = \frac{i}{2} & b_2 = -i & b_3 = 1 & b_4 = i \\
a_1 = -\frac{1}{8} & a_2 = -\frac{1}{4} & a_3 = 2i & a_4 = i \\
a_5 = \frac{i}{2} & a_6 = 0 & a_7 = \frac{i}{4} & a_8 = 0 \\
a_9 = \frac{i}{8} & a_{10} = 0 & a_{11} = -\frac{i}{8} & a_{12} = \frac{i}{4} \\
a_{13} = \frac{i}{4} & a_{14} = -1 & a_{15} = -\frac{i}{4} & a_{16} = -\frac{i}{4} \\
a_{17} = \frac{1}{2} & a_{18} = \frac{1}{4} & a_{19} = -\frac{1}{4} & a_{20} = 1 \\
a_{21} = -1 & a_{22} = \frac{1}{2} & a_{23} = -2 & a_{24} = -\frac{1}{2}
\end{array}
\tag{3.38}
$$

4 - The scalar potential

Let us consider the term $L_{(\text{potential})}$ in the Table:

$$
\mathcal{L}_{(\text{Potential})} = -g^2 W(z)\, \epsilon_{abcd} V^a \wedge V^b \wedge V^c \wedge V^d \tag{4.4}
$$

where $W(z)$ is the scalar potential. Supersymmetry invariance of the action requires a cancellation between the contributions of $\mathcal{L}_{(\text{Potential})}$ and those of $\mathcal{L}_{(2\ \text{Fermi})}$ (cfr. Table) . This occurs if and only if the following equation holds true:

$$
-g^2 W(z)\, \delta_A^B = 2\, S_{AM} S^{MB} - \frac{2}{3} \mathcal{U}_A \mathcal{U}^B - \frac{1}{6} \mathcal{N}_{iA} \mathcal{N}^{iB} - \frac{1}{6} \mathcal{M}^{iM}{}_A \mathcal{M}_{iM}{}^B \tag{4.5}
$$

Eq. (4.5) requires that the traceless part of the r.h.s. cancel identically. This non-trivial condition provides an excellent check on all the previous computations. We refer to [22] for more details. Eq. (4.5) is a general result, essentially model-independent, which links the potential to the fermion shifts [12].

The potential can now be easily computed taking the trace of eq.(4.5). From the Jacoby identities for the group G and the definition of C functions one obtains:

$$|C_i|^2 = |C_P|^2 + |C_A{}^{iP}|^2 - |C_i{}^{PQ}|^2 \tag{4.6}$$

Using eq. (4.6) we arrive at the simple result:

$$12\, g^2 W(z) = \frac{1}{2}|C_{iA}{}^B|^2 + \frac{1}{2}|C_i{}^{PQ}|^2 - \left(|C_A{}^{PQ}|^2 - |C_P|^2\right) \tag{4.7}$$

The last term in the bracket is the contribution from the gravitino mass matrix $|S_{AB}|^2 = \frac{1}{8}\left(|C_A{}^{PQ}|^2 - |C_P|^2\right)$.

Note that in the absence of matter multiplets $W = -1/2$ and we recover N=3 anti-De Sitter supergravity [21].

Conclusions

In this paper we have constructed the complete N=3 matter coupled supergravity for an arbitrary number of vector multiplets gauging one of the following two classes of groups G:

G = SO(3) x (compact gauge group)

G = SO(1,3) x (compact gauge group)

For the study of the potential and of possible partial supersymmetry breakings, we refer to L. Girardello's contribution in these Proceedings (see also ref. 23))

References

1) See, for instance:

P. Fayet, 21st Intern. Conf. on High Energy Phys. (Paris 1982), eds. P. Petiau and M. Porneuf, C3/ page 673;

S. Ferrara, Intern. Europhysics Conf. on High Energy Physics, (Brighton 1983), eds. J. Guy and C. Costain, (Rutherford Appleton Lab., Chilton, Didcot U.K.), page 522;

J. Ellis, Proc. Int. Symp. on Lepton and Photon Interactions at High Energies, (Ithaca 1983) eds. D.G. Cassel and D.L. Kreinick, (F.R. Newman Lab. of Nucl. Studies, Cornell Univ., Ithaca, N.Y.) page 439;

R. Barbieri, ibid., page 479;

H.P. Nilles, Phys. rep. 110 C (1984), 1;

H.E. Haber and G.L. Kane, Phys. Rep. 117 C (1985) 75;

D. Nanopoulos, 22nd Int. Conf. on High Energy Physics, Leipzig (1984) page 36.

2) D.V. Volkon and V.A. Soroka, Sov. Phys. JETP Lett. 18 (1973) 312;

S. Deser and B. Zumino, Phys. Rev. Lett. 38 (1977) 1433;

E. Cremmer et al., Nucl. Phys. B147 (1979) 105; Phys. Lett. 79B (1978) 231.

3) E. Cremmer et al., Nucl. Phys. B212 (1983) 413;

J. Bagger and E. Witten, Phys. Lett. 115B (1982) 202;

J. Bagger, Nucl. Phys. B211 (1983) 302.

4) P.G.O. Freund and M.A. Rubin, Phys. Lett. 97B (1980) 233;

Moreover, for reviews of spontaneous compactifications of D=11 supergravity see:

a) L. Castellani, R. D'Auria and P. Fré in "Supergravity and Supersymmetry '83" edited by B. Milewski, World Scientific Publishing Company, Singapore;

b) M.J. Duff in "Supersymmetry and Supergravity '84", proceedings of Trieste School on Supergravity and Supersymmetry 1984, (eds. B. de Wit, P. Fayet and P. van Nieuwenhuizen), World Scientific Publishing Company, page 212;

c) P. van Nieuwenhuizen, ibid. page 239;

d) P. Fré, ibid. page 324;

e) H. Nicolai, ibid. page 368.

5) M.J. Duff, Nucl. Phys. B219 (1983) 389.

6) M.B. Green and J.H. Schwarz, CALT-68-1182, CALT-68-1194
P. Candelas et al., Santa Barbara preprint 1984,
"Vacuum Configurations for Superstrings".
For an introduction to Superstrings, see:
J.H. Schwarz, Phys. Reports 89 (1982) 223;
M.B. Green, Surveys in High Energy, Physics 3 (1983) 127.

7) B. de Wit, P.G. Lauwers, R. Philippe, Su S.Q. and A. Van Proeyen
Phys. Lett. 134B (1984) 37;
B. de Wit and A. Van Proeyen, Nucl. Phys. B245 (1984) 89;
E. Cremmer et al., Nucl. Phys. B250 (1985) 385;
B. de Wit et al., Bonn preprint (1984) HE 84-37.

8) M. de Roo, Groningen preprints (1985): "Matter Coupling in N=4 Supergravity" and "Gauged N=4 Matter Coupling";
E. Bergshoeff, I.G. Koh and E. Sezgin, Trieste preprint IC/85/14 (1985);
M. Awada and P.K. Townsend, Cambridge preprint (1985);
"N=4 Maxwell-Einstein Supergravity in five dimensions and its SU(2) gauging".

9) J.P. Derendinger and S. Ferrara, Trieste conference (As in 4b,c,d,e) page 159.

10) L. Castellani and L.J. Romans, Nucl. Phys. B238 (1984) 683.

11) L. Castellani, L.J. Romans and N.P. Warner, Nucl. Phys. B241 (1984) 429.

12) S. Cecotti, L. Girardello and M. Porrati, LPTENS 84-10 (1984);
see also Phys. Lett. 151B (1985) 367.

13) L. Castellani et al., CERN preprint TH 4179 (1985).

14) P. Fré, Lett. al Nuovo Cimento 30 (1981) 507.

15) P. Fré, Nucl. Phys. B187 (1981) 376.

16) R. D'Auria - P. Fré and A.J. da Silva, Nucl. Phys. B196 (1982) 205.

17) M.K. Gaillard and B. Zumino, Nucl. Phys. B193 (1981) 221.

18) On the formalism of coset representatives, see:
R. Gilmore, "Lie Groups, Lie Algebras and some of their applications" (John Wiley, New York, 1974);
A. Salam and J. Strathdee, Ann. of Phys. 141 (1982) 316;
R. D'Auria and P. Fré, Ann. of Phys. 157 (1984) 1.

19) R. D'Auria, P. Fré, P.K. Townsend and P. van Nieuwenhuizen, Ann. of Phys. 155 (1984) 423.

20) P. van Nieuwenhuizen, Proceedings of Trieste School on Supergravity Supersymmetry - CUP (J.G. Taylor, S. Ferrara and P. van Nieuwenhuizen, editors) (1985).

21) D.Z. Freedman and A. Das, Nucl. Phys. B120 (1977) 221;
P. Fré, Nucl. Phys. B186 (1981) 44.

22) L. Castellani, A. Ceresole, S. Ferrara, R. D' Auria, P. Fre' and E. Maina, CERN preprint TH 4180/85.

23) S. Ferrara, P. Fre' and L. Girardello, CERN preprint TH 4239/85

TABLE : THE LAGRANGIAN OF MATTER COUPLED N = 3 SUPERGRAVITY

The Lagrangian is split as follows:

$$\mathcal{L} = \mathcal{L}_{(\text{ungauged})} + \Delta\mathcal{L}_{(\text{gauging})}$$

where $\mathcal{L}_{(\text{ungauged})}$ is the Lagrangian at g=0, and is further decomposed as:

$$\mathcal{L}_{(\text{ungauged})} = \mathcal{L}_{\text{KIN}} + \mathcal{L}_{\text{PAULI}} + \mathcal{L}_{\text{TORSION}} + \mathcal{L}_{\text{4FERMI}} + \mathcal{L}^{(4V)}_{\text{4FERMI}}$$

The full Lagrangian is obtained by introducing G-covariant derivatives inside the definition of $P_A{}^i$ and of the Q-connections and by adding $\Delta\mathcal{L}_{(\text{gauging})}$, which we write as:

$$\Delta\mathcal{L}_{(\text{gauging})} = \mathcal{L}_{\text{POTENTIAL}} + \mathcal{L}_{\text{2-FERMI}} + \mathcal{L}^{\text{4-V}}_{\text{4-FERMI}}$$

The action is in first-order formalism: hence $\overset{(+)}{F}{}_\Lambda{}^{ab}$, $\overset{(-)}{F}{}_\Lambda{}^{ab}$ and $P^i_{A|a}$ must be varied as indepentent dynamical variables. SU(3) x SU(n) invariance is manifest due to the presence of scalar fields providing the metrics W^{AB}, W^{ij} (for their explicit expressions see ref. 22)). In the limit where the scalar fields are zero, however, $W^{AB} \to \delta^{AB}$, $W^{ij} \to \delta^{ij}$ and global invariance under SU(3) x SU(n) is lost.

$$\mathcal{L}_{(\text{ungauged})}$$

$$\mathcal{L}_{KIN} = \epsilon_{abcd} R^{ab}{}_\wedge V^c{}_\wedge V^d - 4(\bar\psi^A \gamma_a \rho_A + \bar\rho^A \gamma_a \psi_A) V^a$$

$$- \frac{i}{3}\Big[4(\bar\chi^{\otimes}\gamma_a \nabla\chi_{\otimes} - \nabla\bar\chi^{\otimes}\gamma_a\chi_{\otimes}) + (\bar\lambda^i\gamma_a\nabla\lambda_i - \nabla\bar\lambda^i\gamma_a\lambda_i)$$

$$+ (\bar\lambda^{Ai}\gamma_a\nabla\lambda_{Ai} - \nabla\bar\lambda^{Ai}\gamma_a\lambda_{Ai})\Big] V_b{}_\wedge V_c{}_\wedge V_d\, \epsilon^{abcd} +$$

$$+ \frac{1}{3}\Big[P^i{}_{A|a}(P_i{}^A - \bar\lambda_{iB}\psi_C\epsilon^{ABC} - \bar\lambda_i\psi^A) +$$

$$+ P_i{}^A{}_{|a}(P^i{}_A - \bar\lambda^{iB}\psi^C\epsilon_{ABC} - \bar\lambda^i\psi_A)\Big] V_b{}_\wedge V_c{}_\wedge V_d\,\epsilon^{abcd}$$

$$+ \frac{1}{3}\Big(\mathcal{N}^{\Lambda\Sigma}\overset{(+)}{F}{}_\Lambda^{ab}\overset{(+)}{F}{}_\Sigma^{ab} + \bar{\mathcal{N}}^{\Lambda\Sigma}\overset{(-)}{F}{}_\Lambda^{ab}\overset{(-)}{F}{}_\Sigma^{ab} - \frac{1}{2}P^i{}_{A|a}P_i{}^A{}_{|a}\Big)\cdot$$

$$\cdot\,\epsilon_{c_1\ldots c_4} V^{c_1}{}_\wedge\ldots{}_\wedge V^{c_4} - 8i\Big(\mathcal{N}^{\Lambda\Sigma}\overset{(+)}{F}{}_\Lambda^{ab} - \bar{\mathcal{N}}^{\Lambda\Sigma}\overset{(-)}{F}{}_\Lambda^{ab}\Big)\cdot$$

$$\cdot\Big(F_\Sigma - iL_\Sigma{}^A\bar\chi^{\otimes}\gamma_c\psi_A{}_\wedge V^c - \frac{i}{2}L_\Sigma{}^i\bar\lambda_{iA}\gamma^c\psi^A{}_\wedge V^c -$$

$$- i(L_\Sigma{}^A)^*\bar\chi_{\otimes}\gamma_c\psi^A V^c - \frac{i}{2}(L_\Sigma{}^i)^*\bar\lambda_{iA}\gamma_c\psi^A{}_\wedge V^c\Big)_\wedge V_a V_b$$

$$\mathcal{L}_{(PAULI)} = F_\Lambda\Big\{4i\big(\bar{\mathcal{N}}_{\Lambda\Sigma}L_\Sigma{}^A\epsilon_{ABC}\bar\psi^B{}_\wedge\psi^C - \mathcal{N}_{\Lambda\Sigma}(L_\Sigma{}^A)^*\epsilon^{ABC}\cdot$$

$$\cdot\,\bar\psi_B{}_\wedge\psi_C\big) + 4\big(\mathcal{N}_{\Lambda\Sigma}L_\Sigma{}^i\bar\lambda_{iA}\gamma^a\psi^A - \bar{\mathcal{N}}_{\Lambda\Sigma}(L_\Sigma{}^i)^*\bar\lambda^{iA}\gamma^a\psi_A\big)V^a$$

$$- 8\big(\mathcal{N}_{\Lambda\Sigma}L_\Sigma{}^A\bar\chi^{\otimes}\gamma^a\psi_A - \bar{\mathcal{N}}_{\Lambda\Sigma}(L_\Sigma{}^A)^*\bar\chi_{\otimes}\gamma^a\psi^A\big)V_a$$

$$+ 2\big(\bar{\mathcal{N}}_{\Lambda\Sigma}L_\Sigma{}^A\bar\lambda^i\gamma_{ab}\lambda_{Ai} - \mathcal{N}_{\Lambda\Sigma}(L_\Sigma{}^A)^*\bar\lambda_i\gamma_{ab}\lambda^{Ai}\big)_\wedge V_a{}_\wedge V_b$$

$$+ 4i\big(\mathcal{N}_{\Lambda\Sigma}L_\Sigma{}^i\bar\chi^{\otimes}\gamma_{ab}\lambda_i - \bar{\mathcal{N}}_{\Lambda\Sigma}(L_\Sigma{}^i)^*\bar\chi_{\otimes}\gamma_{ab}\lambda^i\big)V^a{}_\wedge V^b$$

$$- 2i\big(P_i{}^A\bar\lambda^{iB}\gamma_{ab}\psi^C\epsilon_{ABC} - P_A{}^i\bar\lambda_{iB}\gamma_{ab}\psi_C\epsilon^{ABC}\big)_\wedge V^a{}_\wedge V^b$$

$$+ 2i\big(P_i{}^A\bar\lambda^i\gamma_{ab}\psi_A - P_A{}^i\bar\lambda_i\gamma_{ab}\psi^A\big)_\wedge V^a{}_\wedge V^b\Big\}$$

$$\mathcal{L}_{(\text{TORSION})} = 4 R_a \wedge V^a \wedge \left[\bar\chi^{\otimes} \gamma_b \chi_{\otimes} - \bar\lambda^i \gamma_b \lambda_i + \bar\lambda^{Ai} \gamma_b \lambda_{Ai} \right] \wedge V^b$$

$$\mathcal{L}_{(\text{4FERMI})} = i \left[\bar\psi^A \wedge \psi^B \wedge \bar\psi^P \wedge \psi^Q \epsilon_{ABC} \epsilon_{PQR} W^{CR} - \bar\psi_A \wedge \psi_B \wedge \bar\psi_P \wedge \psi_Q \cdot \right.$$

$$\cdot\, \epsilon^{ABC} \epsilon^{PQR} W_{CR} \quad + 2i \left(\bar\lambda^{iA} \gamma_a \lambda_{iB} \bar\psi_A \wedge \gamma_b \psi^B - \right.$$

$$- \bar\lambda^{iA} \gamma_a \lambda_{iA} \bar\psi_B \gamma_b \psi^B + \bar\lambda^i \gamma_a \lambda_i \bar\psi_A \wedge \gamma^b \psi^A \big) \wedge V_a \wedge V_b$$

$$- \frac{i}{2} \left(\bar\psi_A \wedge \psi_B \bar\lambda^{Ai} \gamma_{ab} \lambda^{Bj} W_{ij} - \bar\psi^A \wedge \psi^B \bar\lambda_{Ai} \gamma_{ab} \lambda_{Bj} W^{ij} \right) \wedge V^a \wedge V^b$$

$$- \frac{i}{2} \left(\bar\psi_A \wedge \gamma_{ab} \psi_B \bar\lambda^{Ai} \lambda^{Bj} W_{ij} - \bar\psi^A \wedge \gamma^{ab} \psi^B \bar\lambda_{Ai} \lambda_{Bj} W^{ij} \right) \wedge V^a \wedge V^b$$

$$- \frac{i}{3} \left(\bar\lambda^i \lambda_{iA} \bar\chi_{\otimes} \gamma_a \psi^A + \bar\lambda_i \lambda^{iA} \bar\chi^{\otimes} \gamma_a \psi_A \right) V^b V^c V^d \epsilon_{abcd}$$

$$+ \left(\bar\lambda_i \gamma_{ab} \lambda^{iA} \bar\chi^{\otimes} \gamma_c \psi_A - \bar\lambda^i \gamma_{ab} \lambda_{iA} \bar\chi_{\otimes} \gamma_c \psi^A \right) V^a V^b V^c$$

$$- i \left(\bar\lambda^i \gamma_a \lambda^{jA} \bar\chi_{\otimes} \psi_A W_{ij} + \bar\lambda_i \gamma_a \lambda_{ja} \bar\chi^{\otimes} \psi^A W^{ij} \right) V_b V_c V_d \epsilon^{abcd}$$

$$- \left(\bar\lambda^i \gamma_a \lambda^{jA} \bar\chi_{\otimes} \gamma_{bc} \psi_A W_{ij} + \bar\lambda_i \gamma_a \lambda_{jA} \bar\chi^{\otimes} \gamma_{bc} \psi^A W^{ij} \right) V^a V^b V^c$$

$$- i \left(\bar\psi_A \gamma_{ab} \psi_B \bar\chi^{\otimes} \chi^{\otimes} W^{AB} - \bar\psi^A \gamma_{ab} \psi^B \bar\chi_{\otimes} \chi_{\otimes} W_{AB} \right) V^a V^b$$

$$\mathcal{L}^{(4V)}_{(\text{4FERMI})} = \frac{1}{96} \epsilon_{abcd} V^a V^b V^c V^d \Big\{ - 4 \bar\lambda_i \chi^{\otimes} \bar\lambda^i \chi_{\otimes} +$$

$$+ \left(4 \bar\chi^{\otimes} \gamma^m \chi_{\otimes} + \bar\lambda^i \gamma^m \lambda_i - \bar\lambda^{Ai} \gamma^m \lambda_{Ai} \right)^2 +$$

$$+ \bar\lambda_{Ai} \gamma^{ab} \lambda^i \left(3 W^{AB} \bar\lambda^i \gamma^{ab} \lambda_{Bi} - \bar\lambda_i \gamma^{ab} \lambda^{Ai} \right) \Big\} \quad + \text{h.c.}$$

$\Delta\mathcal{L}_{\text{(gauging)}}$

$$\mathcal{L}_{\text{POTENTIAL}} = -g^2\, W(z)\, \epsilon_{abcd}\, V^a \wedge V^b \wedge V^c \wedge V^d$$

$$\begin{aligned}
\mathcal{L}_{\text{2-FERMI}} = &+ \frac{2i}{3} \Big[2\bar{\chi}^{\otimes} \gamma^a \psi_A \mathcal{U}^A + 2\bar{\chi}_{\otimes} \gamma^a \psi^A \mathcal{U}_A \\
&+ \bar{\lambda}^i \gamma^a \psi^A \mathcal{N}_{iA} + \bar{\lambda}_i \gamma^a \psi_A \mathcal{N}^{iA} + \bar{\lambda}_{Ai} \gamma^a \psi^B \mathcal{M}^{iA}{}_B \\
&+ \bar{\lambda}^{Ai} \gamma^a \psi_B \mathcal{M}_{iA}{}^B \Big] V^b V^c V^d \epsilon_{abcd} \\
&- 4 \left(S_{AB} \bar{\psi}^A \gamma^{ab} \psi^B - S^{AB} \bar{\psi}_A \gamma_{ab} \psi_B \right) V_a V_b
\end{aligned}$$

$$\begin{aligned}
\mathcal{L}^{\text{4-V}}_{\text{2-FERMI}} = &-\frac{1}{6} \Big\{ - C^i \bar{\chi}^{\otimes} \lambda_i - C_i{}^{PQ} \bar{\chi}^{\otimes} \lambda^{iR} \epsilon_{PQR} + \\
&+ \frac{1}{2} C^A \bar{\lambda}^i \lambda_{Ai} + C_i{}^{Aj} \bar{\lambda}^i \lambda_{Aj} + \frac{1}{2} C_A{}^{ij} \lambda_{Bi} \lambda_{Cj} \epsilon^{ABC} \\
&+ C_i \bar{\chi}_{\otimes} \lambda^i + C^i{}_{PQ} \bar{\chi}_{\otimes} \lambda_{iR} \epsilon^{PQR} \\
&+ \frac{1}{2} C_A \bar{\lambda}_i \lambda^{Ai} + C^i{}_{Aj} \bar{\lambda}_i \lambda^{Aj} \\
&+ \frac{1}{2} C^A{}_{ij} \bar{\lambda}^{Bi} \lambda^{Cj} \epsilon_{ABC} \Big\} \epsilon_{abcd} V^a V^b V^c V^d
\end{aligned}$$

EXTENDED SUPERGRAVITY THEORIES: FUNCTIONAL IDENTITIES, PARTIAL BREAKING OF SUPERSYMMETRY AND EXCEPTIONAL MODELS

L. Girardello

Department of Physics

University of California, Los Angeles, CA 90024

and

Dipartimento di Fisica and INFN, Milano, Italy

1. INTRODUCTION

N = 1 supergravity is by now viewed as a likely intermediate step in any scheme of unification of the fundamental interactions. Its lack of renormalizability and the inherent indeterminacy of the matter couplings, demand a consistent, quantum framework in which the arbitrariness is removed.

Tentative guiding principles are the extended supergravity theories and the more recent, promising, superstring avenue. Here we shall discuss some aspects of extended supergravities in the light of the above point of view.

Extended supergravity theories can be obtained via spontaneous compactification (reduction) of super Kaluza-Klein theories in higher dimensions[1] or can be formulated directly in 4-D. We will consider this last formulation here. For $N \geq 3$, there is no superconformal tensor calculus as for $N = 1,2,$[2] but progress in the construction of $N \geq 3$ theories in 4-D as well as a better understanding of their structure comes from the Ward-like identities for local supersymmetry.[3] The main one relates the transformation properties of the spin 3/2-and spin-1/2 fields to the scalar potential, so the derivation of the bosonic part of the Lagrangian is greatly symplified as the recent construction of the $N = 3$ theory[4] based on the group manifold approach, has shown. These identities will be presented in Section 1. As a matter of fact more general Ward identities, involving spin-1 fields, can be derived.

Besides their intrinsic interest, these functional identities also provide a powerful tool in the analysis of the possible mechanisms and patterns of spontaneous breaking of local supersymmetry.[3,5,6] The search for spontaneous partial breaking can be reformulated in a much more effective way.[6]

In this general framework we expect the physically relevant N = 1 theories as resulting from a partial breaking of a higher N theory in Minkowski space.

It turns out that this is very unlikely to happen, as long as semi-simple compact groups are gauged. It is instead very easy to have breaking to N = 1 in an anti-de Sitter background. In general also the higher-D theories lead to partial breaking in 4-D anti-de Sitter space. Examples are known with partial breaking in flat space. They result respectively from a generalized dimensional reduction[7] of the N = 1 theory in 11-D or from the spontaneous compactification of the 10-D low energy theory[8] from superstring theory[9] and of a 6-D Maxwell-Einstein supergravity model.[10] This last model has, however, a U(1) anomaly; its removal leads to N = 0 or N = 2.[11]

Our discussion is organized as follows: In Section 2 we present some Ward identities briefly and discuss their meanings and consequences. In Section 3 we indicate a systematic way of building models of extended supergravity with assigned patterns of spontaneous breaking of local supersymmetry. In the last section we propose a method of bypassing the difficulty of partial breaking in flat space with an example of N = 2 theory.

2. WARD IDENTITIES FOR LOCAL SUPERSYMMETRY

Consider a general N-extended, locally supersymmetric theory. The relevant structure of the Lagrangian density is

$$\mathcal{L} = \mathcal{L}_B + \mathcal{L}_F^{(2)} + \mathcal{L}_R \tag{1}$$

where

$$\mathcal{L}_B = -\frac{1}{2}eR - eV(\phi)$$

$$\mathcal{L}_F^{(2)} = \frac{i}{2}e\bar{\psi}_\mu{}^A \gamma^{\mu\nu\rho} D_\nu{}^{AB} \psi_\rho{}^B + \frac{e}{2}\chi^i Z_i{}^j(\phi)\, \gamma^\mu \nabla_\mu \chi_j$$

$$+ \psi_\mu{}^A J_A{}^\mu - \frac{1}{2}e\chi^i M_{ij}(\phi)\, \chi^j + \ldots \tag{2}$$

and $\mathcal{L}_R$ contains all the remaining terms which are not important for our considerations. The fields $\psi_\mu{}^A$ are the spin-3/2 fields (gravitini) $A = 1,2,\ldots,N$, χ^i the spin-1/2 and ϕ_α the spin-0 fields. Furthermore,

$$D_\mu{}^{AB} = \nabla_\mu \delta^{AB} + i\gamma_\mu S^{AB}(\phi) \tag{3}$$

and the symmetric matrix S^{AB}, depending on the spin-0 fields, is the gravitino mass matrix. The Noether supercurrents have the explicit form:

$$J_\mu{}^A = \gamma^\mu\, \Sigma_A{}^i(\phi)\, \chi_i + \ldots\ . \tag{4}$$

The relevant transformation properties of the fermions (fermionic shifts) are:

$$\delta\psi_\mu{}^A = \nabla_\mu \varepsilon^A + i\, S^{AB}(\phi)\, \gamma_\mu \varepsilon^B + \ldots \tag{5}$$

$$\delta\chi^i = \Sigma_A{}^i(\phi)\, \varepsilon^A + \ldots \tag{6}$$

and for the scalars:

$$\delta\phi_\alpha = \bar{\chi}^i \varepsilon^A L_{\alpha A_i}(\phi) \ . \tag{7}$$

The relation between the expressions on the right-hand side of (4) and (6) as well as the fact that the matrix S^{AB} in (5) is the gravitino mass matrix, are well known consequences of local supersymmetry. The functions S^{AB} and $\Sigma_A{}^i$ will be called fermionic shifts. We can now perform a supersymmetry transformation on the action for the Lagrangian of Eqs. (1) and (2), followed by a functional derivation with respect to $\psi_\mu{}^A$ on a homogeneous bosonic background or, equivalently, setting all spin-1/2, 1, 3/2 fields to zero as well as all derivatives:

$$\frac{\delta}{\delta\psi_\mu{}^A} \int d^4x\, \delta_\varepsilon \mathcal{L} = 0 \ . \tag{8}$$

Standard manipulations lead to the Ward-like identities

$$-6S_{AC}(\phi)\, S^{CB}(\phi) + \Sigma_A{}^i(\phi)\, Z_i{}^j\, \Sigma_j{}^B(\phi) = -2V(\phi)\, \delta_A{}^B \tag{9}$$

valid for all values of the spin-0 fields ϕ_α. If there is no spin-1/2 ghost, $Z_j{}^i(\phi)$ is a positive definite matrix. This is the main functional identity of local supersymmetry. It can also be derived,[3] as a consequence of an algebra similar to the one of global supersymmetry even in the case of spontaneous breaking of local supersymmetry.

We'll discuss now some consequences of relation (9).

--The knowledge of the fermionic shifts (5) and (6) determines completely the scalar potential of the theory. This can be viewed as the generalization of the global case where the potential is a quadratic form in the auxiliary fields (of N = 1), which are indeed fermionic (spin-1/2) shifts. The so-called T-identities of Ref. 12 are an example of saturation of the functional identity (9).

--The necessary and sufficient condition for broken supersymmetry is that at least one of the spin-1/2 shifts, $\Sigma_A{}^i$, be non-zero at an extremum of the potential. Since the spin-3/2 and spin-1/2 contributions have opposite sign in Eq. (9), the potential at the extrema can be null or negative (Minkowski or anti-de Sitter background). In such backgrounds and for unitary theories, i.e. without ghosts, the invariance of the spin-3/2 fields automatically implies invariance of the spin-1/2 fields so, as an alternative, one has to consider only the shifts of the spin-3/2 fields. By exploiting this result, we can show[13] that a general N = 2 theory formulated according to the standard tensor calculus[2] cannot break to N = 1 in a Minkowski background. It amounts to showing that $S^{AB}S_{BC}$ cannot be a 2 × 2 matrix of rank 1.

--Partial or sequential spontaneous breaking of local supersymmetry at different scales, i.e. with different gravitino masses, is not forbidden as it is in the global extended case.

One can also obtain a functional identity[4,5] which involves the first derivative of the potential and hence of direct bearing on the spontaneous breaking problem. This identity can be derived, under the same conditions set for the derivation of Eq. (8), by considering

$$\frac{\delta}{\delta\chi^i} \int d^4x\, \delta_\varepsilon \mathcal{L} = 0 \tag{10}$$

which reads

$$M_{ij}(\phi)\, \Sigma_A{}^i(\phi) + 4Z_{ij}(\phi)\, \Sigma_j{}^A(\phi)\, S_{BA}(\phi) = - \frac{\partial V}{\partial \phi_\alpha} L_{\alpha A_i}(\phi)\ . \tag{11}$$

The presence of at least one unbroken supersymmetry to which there corresponds a Killing spinor η^A (i.e. such that $\delta\psi_\mu{}^A = D_\mu \eta^A = 0$ and $\delta\chi^i = \Sigma_A{}^i \eta^A = 0\ \forall i$), ensures that the potential remains stationary (Minkowski or adS bckg.) as one can easily see by contracting previous Ward identities contracting with η^A.

From positive energy arguments[3] or from these Ward identities,[6] one can derive the following stronger result: the existence of at least one non-trivial solution η^A to the algebraic equations

$$\Sigma_A{}^i(\phi)\, \eta^A = 0 \tag{12}$$

is sufficient to ensure at least one unbroken supersymmetry and, moreover, the values of the scalar fields ϕ for which such a solution exists identify a stationary point of the potential. The surviving supersymmetry guarantees the global stability of the stationary point. In other words the condition (12) selects a stationary supersymmetric manifold, and we can make a systematic analysis of the supersymmetry breaking patterns without looking at the potential at all.

3. GENERAL CRIITERIA FOR THE DETERMINATION OF MODELS WITH ASSIGNED PATTERNS OF SUPERSYMMETRY BREAKING

We are concerned with the spontaneous breaking of local supersymmetry via spin-0 condensates, so the final objects of interests are the fermionic shifts of Eqs. (5) and (6). These are, in turn, for $N \geq 3$, of purely group theoretical origin.[12,4,14]

If we want to investigate the possible existence of N-extended supergravity models with breaking of N' charges, $1 \leq N' < N$, we can start by assuming this "final" structure with N-N' unbroken supersymmetries as a germ for the construction of the model itself. In other words, by exploiting the final argument of the previous section, we assign a breaking pattern and see if a model can be built to reproduce that pattern.

In order to pursue and to illustrate this program, we recall here the essentials of the structure of the extended supergravity theories. For $N \geq 3$ there are no more arbitrary functions of the scalars, as there are for N = 2.[15] and the N = 1[16] cases (one and two arbitrary functions respectively). The scalar fields are coordinates of a complex diffeentiable manifold $\mathcal{M}$, a coset space G/H. The only liberty regards the number of matter vector multiplets (for N = 3,4) and the group to be gauged. The spin-0 fields are contained in the supergravity multiplets ($N \geq 4$) and in the vector multiplets ($N \geq 2$), whereas for N = 1,2 there are also scalar or hypermultiplets. The noncompact group G, the isometry group of $\mathcal{M}$, is the group of duality rotations of all the vector field strengths and it leaves the equations of motion invariant in the absence of gauging. G must have a rep. D of $2[\tilde{N}+n]$ real or $\tilde{N}+n$, complex dimensions $\tilde{N}$ being the number of vector fields from supergravity and n from matter multiplets. H is the isotropy group of $\mathcal{M}$ and it has the form SU(N) × H', where H' = 1 or U(1) for N ≤ 4. The invariance group G' of $\mathcal{L}$ is in general smaller than G and it is the subgroup of the duality rotations that does not mix the electric and the magnetic fields.

A subgroup K of G' ⊂ G can be gauged and the symmetry G is broken at least down to K. The available gauge fields, which, together with their dual, are assigned to the rep. D of G, have to be in the adjoint rep. of K, hence the requirement $D_K \to \text{adj } K \oplus \text{adj } K$. These considerations apply for the gauging of compact groups of the form SO(N) × K_n, K_n being any semisimple compact group of dim n. Noncompact gauging are also possible[14,17] since G' can contain some noncompact group such taht G' → adj K. For N > 4, K_n is absent and only subgroups of SO(N) can be gauged (or its contractions.[14] In coset space the fermionic shifts in (4) and (5) are appropriate contractions of the boosted structure constants $C_\Lambda{}^{\Sigma\Delta}$ defined by:

$$C_\Lambda{}^{\Sigma\Delta}(\phi) = (L^{-1}(\phi))_\Lambda{}^{\Lambda'} L_{\Sigma'}{}^{\Sigma}(\phi)\, L_{\Delta'}{}^{\Delta}(\phi)\, f_{\Lambda'}{}^{\Sigma'\Delta'} \qquad (13)$$

where $f_{\Lambda'}{}^{\Sigma'\Delta'}$ are the structure constants of K_n and the matrix $L_{\Lambda'}{}^{\Lambda}$ is a coset representative of G on G/H and depends on the scalar fields. Of course, in the absence of gauging, the shifts vanish and there is no potential.

We can now indicate a criterion for analyzing the possible patterns of the spontaneous breaking, i.e. of the super Higgs and Higgs effects, by a concrete example for the N = 3 case,[6] where G = SU(3,n) and H = SU(3) ⊗ U(n).[4]

A question we want to ask is if there exist N = 3 models with one unbroken charge, without any request, for the moment, on the surviving gauge group. Our criterion consists in requiring surfaces in the general scalar manifold $M = SU(3,n)/SU(3) \otimes U(n)$ with special symmetry properties, compactible with one unbroken supersymmetry and, for example, the two massive gravitini having the same mass. This is, of course, one of the possible super Higgs effects: we are asking for one breaking scale, whereas there might well be cases with two scales: i.e. two different masses for the gravitini associated with the two broken charges, not to mention the possibility of a total breaking with 3 scales. These situations are more difficult to analyze but the criterion generalizes in a straightforward way.

So we can decide to ask for a breaking to N = 1 with the two massive gravitini transforming as a doublet of the subgroup $SU(2)_1$ of SU(3), the automorphism group of the charges. The surviving charge is the SU(2) singlet. The scalar fields $Z_A{}^i$, where A is the SU(3) index, A = 1,2,3, and i the adj K index, are the complex coordinates of the manifold $\mathcal{M}$. The stationary submanifold $\mathcal{M}' \subset \mathcal{M}$, which breaks N = 3 to N = 1 must have an isotropy group $H' \subset H = SU(3) \times U(n)$, i.e., it must be composed of those points whose coordinates are singlets under H'.

The interesting case here is when H' is taken to be the diagonal subgroup $SU(2)_1 \otimes SU(2)_2$, where $SU(2)_2$ is a subgroup of the gauge group K (still to be determined). The group K is then required to contain, in its adj. rep., doublets under its subgroup $SU(2)_2$, i.e.:

$$\text{adj } K \underset{SU(2)\times N_k}{\rightarrow} (3,1) \oplus (1, \text{adj } N_k) \oplus (2,D') \qquad (14)$$

where D' is a pseudo real rep. of N_k, the normalizer of $SU(2)_2$ in K. A very specific pattern of local supersymmetry breaking has been so assigned.

It remains to be seen which gauge groups K are compatible with the above request. The candidate models have been classified:[6] the hunting ground for that specific breaking pattern has been severely restricted. For each model (defined by a group K within the list so constructed) we can compute the shifts of the spin-1/2 in terms of the boosted structure constants of K.

We now impose that these shifts are zero along one of the three supersymmetries. In other words, we assume a nontrivial solution η^A, say in the direction A = 3, for the equations (12), which read then as a set of algebraic equations $\Sigma_3{}^i(Z) = 0$ for the scalar fields. Their solution spans the submanifold M', which, in virtue of the arguments of Section 2 is <u>automatically</u> a stationary and stable surface for the model under consideration. The evaluation of the gravitino masses for the submanifold solution M' leads finally to the determination of the nature of the space-time background: if one of the masses is zero, the background is Minkowski, otherwise adS.

We have explicitly studied two cases: for $K = SO(\nu+4)$ and $K = E_6$;[6] both lead to AdS backg. The equations obtained admit solutions only if the ratio between the SO(3) and the K gauge coupling constants has a certain value, which, together with the cosmological term is the same for both cases. All possible patterns, for any $N \geq 3$ extended model, can be discussed in this way.

3. A METHOD TO DERIVE PARTIAL BREAKING IN FLAT SPACE

As we already said, partial breaking in flat space is unlikely with gauging of semisimple compact groups, whereas the adS situation is always possible. In particular we know that the general N = 2 supergravity models as are known from the superconformal tensor calculus[2] (STC) cannot have only one unbroken charge in flat space.[3,13] On the other hand the set of theories which can be constructed according to the STC is an "open" set and there might be theories which can be obtained from them only via a limiting procedure. A possible strategy, then, consists in starting from a theory which already presents a partial super Higgs phase in adS space and in trying to "flatten" the model. If some parameter is still free, it might be possible, by taking suitable singular limits, to recover a theory with a partial super Higgs phase in Minkowski space. More precisely, there might exist a singular field redefinition under which the original fields diverge at the regular values of the new ones, while the Lagrangian and the supersymmetry transformations remain finite in terms of the new fields.

This may not be the only way of "flattening" a theory: one can contract, from the very beginning, the gauge group in such a way to evade the general arguments which seem to make partial super Higgs impossible in flat space. In other words, one might avoid solving the theory in adS, as a starting point. The above strategy has been applied[18] to a N = 2 model,[19] originally formulated according to the SCT rules.[2,14] The model is based on two vector multiplets (a compensator and a matter one), with minimal coupling, and one physical hypermultiplet A_i^{α} ($i,\alpha = 1,2$) and the compensating hypermultiplet. The compensating vector multiplet is used to gauge an SO(2) internal symmetry of the compensating hypermultiplet (with coupling constant g'_o) whereas the matter vector multiplet gauges an SO(2) subgroup of the physical hypermultiplet (gauge constant g_o).

In Ref. 19, it was shown that the model has stationary solutions

$$|A|^2 = A_\alpha{}^i A_i{}^\alpha = 2t^2/1-t^2 \;;\quad |x|^2 = t^2/1-t^2 \tag{15}$$

with cosmological term

$$\Lambda = -6g_o{}^2 t^2/1-t^2 \;;\quad t = \frac{g_o'}{g_o}\,, \qquad t^2 < 1\,. \tag{16}$$

The classical potential does not fix the direction and the phases of the scalar v.e.v. and so the gravitino mass matrix turns out to be sliding with the mass ratio:

$$R = \frac{|m_1|^2}{|m_2|^2} = \frac{1 + t^4 + 2t^2 \cos(\phi+\theta)}{1 + t^4 + 2t^2 \cos(\phi-\theta)}\,. \tag{17}$$

For the choice $\phi = \theta = \pi/2 \pmod \pi$, one supersymmetry is broken in adS space; the two gravitini have masses

$$m_1 = \sqrt{-\Lambda/3} \quad;\quad m_2 = \sqrt{2}\, g_o t \quad (1+t^2)/(1-t^2)^{3/2}\,. \tag{18}$$

From Eqs. (16) and (18), it is obvious that in the limit $t^2 \to 1$ with $g = g_o(1-t^2)^{-3/2}$ fixed, the cosmological constant Λ and m_1 vanish, whereas m_2 has a finite limit. All fields of spin-0, 1/2, 1 can then be redefined via rescalings which are singular for $t^2 \to 1$, leading to new covariant and supercovariant derivatives. The new 2-parameter supersymmetry transformations for these new fields are finite for $t^2 \to 1$ and so is their Lagrangian, albeit in a subtle way. The model so obtained can be interpreted as a N = 2 theory with partial breaking to N = 1 in Minkowski space.

We discuss now the main features of this limit theory.

The scalar potential vanishes identically; in particular the physical scalar live in the same SU(1,1)/U(1) Kaehler manifold of the standard N = 1 flat potential.[20] One of the two gravitini is massless and the other one has a sliding mass. The two physical spin-1 fields have the same sliding mass as well. The Lagrangian is invariant with respect to a parity operation which, on scalars, is just complex conjugation. The parity assignment of the two spin-1 fields are opposite. This is in agreement with group theory considerations:[21] the N = 1 massive spin-3/2 supermultiplet must contain, besides one spin-3/2 and one spin-1/2, a vector and a pseudovector. This also shows that a partial super Higgs effect has to be accompanied by a Higgs effect which requires, in turn, the presence of one N = 2 vector multiplet and one hypermultiplet, at least. This new parity operation is not connected with the original one of the STC, for which all fields of spin-1 are true vectors.

This construction can be generalized to the case of N = 2 flat models with any given ratio R of the gravitino masses, by expanding the fields around a different solution of the same model. In this way we can have simultaneously a flat potential and hierarchial spontaneous breaking of supersymmetry. Since many models with N > 2 admit partial breaking in adS space, more general flat models with partial breaking can be constructed by this method. It seems, in any case, that only Abelian gauge interactions are allowed in the relevant singular limit, apart from spectator non-Abelian gauge interactions which do not play any role in the above supersymmetry breaking mechanism. Thus, at the moment, we are not able to generate chiral fermions which remains the main problem for $N > 2$.

The $N = 3$ models discussed in Section 3 do not lend themselves to this procedure since the ratio of the 2-coupling constants is fixed. This result is obviously not affected by the presence of Abelian factors in the gauge group K. More general situations are under investigation.

REFERENCES

1. See, for example, P. Fré, Proceedings of the Trieste Spring School, 1984 (World Scientific, 1984); M. J. Duff, B. E. W. Nillson and C. N. Pope, Phys. Reports (to appear).

2. For a review, see B. de Wit, in Supergravity 1981, eds. S. Ferrara and S. G. Taylor, 1874, Cambridge University Press; A. Van Proeyen, Proceedings of the 1983 Karpacz School of Supergravity and Supersymmetry, ed. B. Milewski (World Scientific, 1984).

3. S. Cecotti, L. Girardello, M. Porrati, Ecole Normale Superieure, LPTENS 84-10 Paris, to appear in Nuclear Physics B; and CERN TH 4256/85 to appear in Proceedings of the Ninth Johns Hopkins Workshop, Florence 1985 (World Scientific 1985).

4. L. Castellani, A. Ceresole, R. D'Auria, P. Fré, S. Ferrara, E. Marino, Phys. Lett. 161B (1985) and CERN-TH 4180/85, to appear in Nuclear Physics B.

5. S. Ferrara, L. Maiani, CERN-TH 3121/85, to appear in the Proceedings of the Vth Silarg Symposium (World Scientific 1986).

6. S. Ferrara, P. Fré, L. Girardello, CERN-TH 4239/85, to appear in Nuclear Physics B.

7. J. Scherk and J. H. Schwarz, Phys. Lett. 82B, 60 (1979); Nucl. Phys. B153, 61 (1979); E. Cremmer, J. Scherk, J. H. Schwarz, Phys. Lett. 84B, 83 (1979).

8. P. Candelas, C. T. Horowitz, A. Strominger, and E. Witten, Nucl. Phys. B258, 46 (1985).

9. M. B. Green, J. H. Schwarz, Phys. Lett. 149B, 117 (1984).

10. A. Salam, E. Sezgin, Phys. Lett. 147B, 47 (1984).

11. For a discussion of this point, see the second paper of Ref. 3.

12. B. de Wit and H. Nicolai, Phys. Lett. 108B, 285 (1981) and Nucl. Phys. B208, 323 (1982).

13. S. Cecotti, L. Girardello, M. Porrati, Phys. Lett. 145B, 61 (1984).

14. M. de Roo, Phys. Lett. 156B, 331 (1985); Nucl. Phys. B255, 515 (1985); E. Bergshoeff, J. G. Koh, and E. Szegin, Phys. Lett. 155B, 71 (1985).

15. B. de Wit, P. G. Lauwers, A. Van Proeyen, Nucl. Phys. B225, 569 (1985).

16. E. Cremmer, S. Ferrara, L. Girardello, A. Van Proeyen, Phys. Lett. 116B, 231 (1982); and Nucl. Phys. B212, 413 (1983).

17. C. M. Hull, Phys. Lett. 148B, 297 (1984) and references therein.

18. S. Cecotti, L. Girardello, M. Porrati, CERN-TH-4269/85, to appear in Physics Letters B.

19. S. Cecotti, L. Girardello, M. Porrati, Phys. Lett. 151B, 367 (1985).

20. E. Cremmer, S. Ferrara, C. Kounnas, and D. V. Nanopoulos, Phys. Lett. 133B, 61 (1983); J. Ellis, C. Kounnas, and D. V. Nanopoulos, Nucl. Phys. B241, 406 (1984); S. Ferrara, A. Van Proeyen, Phys. Lett. 138B, 77 (1984).

21. S. Ferrara, P. van Nieuwenhuizen, Phys. Lett. 127B, 70 (1983).

Part Three: Phenomenology and Supersymmetry

PHENOMENOLOGY FROM SUPERSTRINGS*)

L.E. Ibañez

CERN - Geneva

A B S T R A C T

I discuss several aspects of recent attempts to make contact between the superstring and low energy phenomenology. This includes a discussion of the truncation procedure for going from ten to four dimensions. I also consider the origin of the scale invariances which appear in the low energy N = 1 d = 4 effective supergravity Lagrangian. Low energy phenomenological models are discussed and it is argued that a gauge group of the form $SU(3)\times SU(2)_L\times U(1)_Y\times U(1)_Z$ is singled out in order to allow for a solution of the ν-mass problem. The extra $U(1)_Z$ generator is uniquely fixed by the condition $Y_Z(\nu_R) = 0$. The radiative breaking of the residual gauge symmetries is described. The phenomenological requirement $M'_{Z_0} \gg M_{Z_0}$ can only be obtained for particular values of the low energy soft mass terms.

---

*) Much of the original work presented has been done in collaboration with J.P. Derendinger and H.P. Nilles.

1. - INTRODUCTION

In the last few years much effort has been dedicated to the construction of phenomenologically consistent models with low energy supersymmetry[1]. The final aim of these models is to use the no-renormalization properties of supersymmetry in order to understand the smallness of the weak scale M_W compared to the Planck mass M_P.

People soon realized that by far the most economical and aesthetical way to keep low energy supersymmetry consistent with experiment is to couple the SUSY standard model to N = 1 supergravity[2]. In these "low energy supergravity models" (LESM), supersymmetry is broken in a SU(3)×SU(2)×U(1) (3×2×1 in compact notation) singlet sector ("hidden sector")[2]. The reason for doing this is that in the Lagrangian of the "observable" world one then gets SUSY breaking soft terms which are universal for all particles (at least at the GUT scale). This universality is desired in order that the SUSY-GYM mechanism does not fail .A remarkable fact of this class of models is that the $SU(2)_L \times U(1)_Y$ symmetry is broken as a radiative effect of supersymmetry breaking in a natural way[3)-7)]. Thus one does not have to add ad hoc (e.g., negative mass^2) terms in the Higgs potential to induce the breaking. Supersymmetry breaking + renormalization group equations (which drive m_H^2 negative for large ranges of the parameters) imply the breaking of the SU(2)×U(1) symmetry.

Unfortunately, N = 1 supergravity theories are not very much constrained. The particle content is not fixed and the Lagrangian is specified by two arbitrary functions[8], the Kähler potential:

$$G(z,z^*) = K(z,z^*) + \log|W(z)|^2 \qquad (1)$$

and the chiral gauge function $f_{ab}(z)$. W(z) is the superpotential and z denotes all the scalars in the theory. From phenomenological reasons, we know some of the terms which must be included in W(z) (i.e., the ones corresponding to the usual Yukawa couplings). Otherwise the form of the Kähler potential and $f_{ab}(z)$ is unknown. Fortunately, practically any reasonable form of G and f (which includes SUSY breaking in the hidden sector) will do in order to get universal SUSY breaking soft terms and radiative SU(2)×U(1) breaking[6),7)]. Then one parametrizes our ignorance of G, f and the structure of the hidden sector by assuming general SUSY breaking soft terms[2),9)]. These include (i) gaugino masses M; (ii) scalar masses m; (iii) trilinear scalar couplings mAW(z)+h.c: (iv) bilinear soft Higgs mass^2 Bμm. All these are assumed to be consistent with the

GUT theory at the M_x scale. The term (iv) is required in order to get a consistent minimum of the Higgs potential (at least in the minimal case) and requires that the superpotential contains a mass term

$$W_H = \mu H\bar{H} \tag{2}$$

where one must have[7] $\mu \sim m$. This is necessary to avoid a dangerous massless axion. Thus, as a whole, the minimal low energy supergravity extension of the standard model contains five free parameters: M, m, μ, A, B. Wide ranges of these parameters allow for radiative $SU(2)\times U(1)$ breaking. There are also strong constraints amongst the masses of s-particles since there are 64 new particles and their masses are determined essentially by those five parameters[7].

This scheme is simple and elegant but is certainly only an effective theory valid at energies $\lesssim M_P$. We know that N = 1 supergravity is a non-renormalizable theory and hence makes no sense for energies $\gg M_P$. Furthermore, there is no real unification of chiral, gauge and gravitational interactions but a mere coexistence of these three types of interations. Thus we lack an understanding of things like the origin of Yukawa couplings and family replication.

The only known candidates for a renormalizable (finite) theory of gravity are ten-dimensional superstring theories[10]. There are five of them known at present: type I, type II-a and type II-b, and the heterotic strings[11] with gauge groups $E_8\times E_8$ and SO(32). Only type-I and the heterotics contain ten-dimensional gauge bosons and hence it seems that only these can give rise to low energy chiral fermions after compactification to four dimensions. These three theories are also gauge and gravitational anomaly free [type-I string only for the SO(32) gauge group]. This important property has only recently been realized[12] and is the origin of recent interest in string theories.

The low energy limit of type-I and the heterotic strings is some form of ten-dimensional supergravity[13] coupled to super-Yang-Mills with gauge group $E_8\times E_8$ or SO(32). The fact that we have elementary gauge bosons present allows for the possibility of obtaining chiral fermions upon compactification of the extra six dimensions on a non-trivial manifold K_6. This compactification may or may not leave an unbroken supersymmetry in four dimensions. It is easy to obtain reasonable compactifications on simple (coset) manifolds which do not preserve supersymmetry. If one wants to leave an unbroken supersymmetry in D = 4 it seems that one has to compactify the ten-dimensional Lagrangian on a very restricted[14] class of six-dimensional manifolds: K_6 must be a Ricci-flat Kähler manifold with an SU(3) holonomy group ("Calabi-Yau" manifolds). Consistency does also require [14] that the gauge fields with indices in extra dimensions A^m (m=5-10) acquire vacuum configurations equal to the connection in K_6, $\langle A^m \rangle = \omega^m$. This identification breaks the gauge symmetry down to $E_6 \times E'_8$ or SO(26). The latter is not phenomenologically interesting since it has a vector-like content under SU(3)×SU(2)×U(1). On the other hand, the E_6 group obtained from the $E_8 \times E_8$ heterotic string is a very sensible grand unification group. Is there any preference from the superstring to compactifiy on Calabi-Yau manifolds so that one supersymmetry remains unbroken? There is not any convincing answer to this question at present but it seems that Ricci-flatness[15] of the manifold is required on general grounds if you want to maintain the conformal invariance of the string (finite two-dimensional string σ model). This also requires the identification $A^m = \omega^m$ in the case of the heterotic string. Thus there may be already an indication that the heterotic string prefers to leave an unbroken supersymmetry in four dimensions although this is probably premature.

After the compactification process described above, one is left with an $E_6 \times E'_8$ gauge theory coupled to N = 1 supergravity. But <u>this is precisely the kind of structure we were hoping for</u> in the low energy supergravity models! The chiral multiplet content is a number of E_6 families ($\chi/2$ of them χ = Euler characteristic of K_6) plus some extra gauge singlets (no E_8 chiral matter). Thus we have an "observable" E_6 world and a "hidden sector" of singlets and E_8 gauge particles which only communicate with the observable world through gravitational interactions. It is certainly remarkable that this structure which seems to be required by low energy phenomenology appears for free in the low energy limit of strings.

In principle, once we know what is the "correct" K_6 manifold we could calculate all the relevant low energy couplings. Also the Kähler potentials $G(z,z^*)$ and $f(z)$ would be calculable and thus all the low energy parameters M, m, A, B. In practice Calabi-Yau are very complicated manifolds with unknown metric and hence the above-mentioned computations are not possible at present. Also, as we will discuss below, it is still not clear how the supersymmetry breaking in the hidden sector takes place as well as how the breaking of the E_6 gauge theory to the standard model is induced. Still it is interesting to try and guess how Calabi-Yau compactification would look like. We try to extract some information about these questions in the following sections.

FROM TEN TO FOUR DIMENSIONS

It is well known that the massless sector of type-I and the heterotic strings corresponds to that of ten-dimensional supergravity coupled to super-Yang-Mills. The field theory limit of these strings is obtained in an expansion on (Energy/$\alpha^{-\frac{1}{2}}$), it is not really a zero-slope limit. The complete (even tree level) field theory limit Lagrangians of these string theories have not been obtained yet. However, they are known to contain the standard supergravity Lagrangian[13)] plus some additional terms. These additional terms include the modifications[12)] due to the presence of a Lorentz Chern-Simons term ω_3^L as well as[11),14),16)] R^2 terms and several others. The bosonic sector of the standard supergravity Lagrangian[13)] is

$$E^{-1}\mathcal{L} = -\frac{1}{2}R + \frac{9}{16}\left(\frac{\partial_M\varphi}{\varphi}\right)^2 - \frac{1}{4}\varphi^{-3/4}F_{MN}^{a}F_{a}^{MN} + \frac{3}{4}\varphi^{-3/2}H_{MNP}H^{MNP} \quad (3)$$

where we have reabsorbed g in the definition of ϕ. This describes the interactions of the graviton and gauge fields with a scalar particle ϕ and a two-index antisymmetric tensor B_{MN} with field strength[12)]

$$H_{MNP} = \partial_{[M}B_{NP]} - \left(\omega^G_{MNP} - \omega^L_{MNP}\right) \quad (4)$$

The Yang-Mills Chern-Simons term is

$$\omega^G_{MNP} = \mathrm{Tr}\left(A_{[M}F_{NP]} - \frac{2}{3}A_{[M}A_{N}A_{P]}\right) \quad (5)$$

and the Lorentz Chern-Simons term

$$\omega^L_{MNP} = \mathrm{tr}\left(\omega_{[M}R_{NP]} - \frac{2}{3}\omega_{[M}\omega_{N}\omega_{P]}\right) \quad (6)$$

is only present in the low energy limit of the string[12)]. Here R_{NP} is the curvature tensor and ω_m is the Lorentz connection, the trace taken place over tangent space indices. Further terms are known to exist in the low energy limit of the heterotic string[11),14),16)] including higher derivative terms like:

$$\frac{1}{4}\varphi^{-3/4}\,\mathrm{tr}(R_{MN}R^{MN}) \quad (7)$$

and possibly[17)] R^4, H^4 as well as other terms (all these from tree-level string scattering amplitudes). One important point to remark is that the addition of the Lorentz Chern-Simons term ω^L inside H_{MNP} as well as (7) breaks the supersymmetry invariance of the standard supergravity Lagrangian and still is not known if a supersymmetrization exists or the whole superstring is required to restore supersymmetry.

Another important point to notice is that the Lagrangian (even after the addition of the extra terms) presents a classical scale invariance[18] under which it gets an overall (classically irrelevant) factor:

$$\left.\begin{array}{l} g_{MN} \longrightarrow \lambda\, g_{MN} \\ \varphi \longrightarrow \lambda^{-4/3}\, \varphi \end{array}\right\} \longrightarrow E^{-1}\mathcal{L} \rightarrow \lambda^{-1}\mathcal{L}E^{-1} \tag{8}$$

(one also has to scale appropriately the fermion fields). Of course, this scale invariance is broken by loop effects since then the overall factor is relevant. Thus, e.g., the one-loop terms relevant for the cancellation of anomalies like[12]

$$\varepsilon^{MNPQRSTUVW} B_{MN}\, \mathrm{tr}(F_{PQ}F_{RS})\, \mathrm{tr}(F_{TU}F_{VW}) \tag{9}$$

do violate explicitly this scale invariance. The classical scale invariance can be traced back to the fact that both vertex operators and inverse propagators in the string scale in the same way under dilatations[18]. Then all tree-level graphs scale in the same way.

To find out the precise form of the low energy four-dimensional field theory obtained after compactification we would have to expand the ten-dimensional fields around the background manifold K_6 and integrate out the heavy modes. Since the metric of K_6 is unknown at the moment we cannot do this. However, one can try to guess how compactification on a Calabi-Yau manifold could look like by making some truncation[18] on the ten-dimension field theory which preserves supersymmetry. One assumes that the tangent space is explicitly broken down to SO(1,3)×SO(6) through an ansatz

$$g_{MN} = \begin{pmatrix} \exp(-3\sigma)\, g_{\mu\nu} & 0 \\ 0 & \exp(\sigma)\, g_{mn} \end{pmatrix} \quad \begin{array}{l} \mu,\nu = 1\text{-}4 \\ m,n = 5\text{-}10 \end{array} \tag{10}$$

σ is a real scalar field and g_{mn} is the internal metric tensor with unit determinant. One thus has a zehnbein determinant $E = \exp(-3\sigma)e$, e being the $D = 4$ vierbein determinant. The ansatz in (10) is chosen so that one gets the canonical Einstein Lagrangian. One then assumes that the light fields are invariant under translations on x^m, $m = 5\text{-}10$. This would give rise to an $N = 4$, $d = 4$ theory, where the four SUSY generators would sit on a $\underline{4}$ of SU(4)~SO(6). To get rid of the three unwanted supersymmetries one splits $\underline{4} = \underline{3}+\underline{1}$ and requires invariance under this SU(3) for the light fields. For the fields with gauge quantum numbers, owing to the identification $\langle A_m^a \rangle = \omega_m^a$ $[a \in SU(3)]$, one requires invariance under diagonal $SU(3)+SU(3)_G$, where $SU(3)_G$ is the subgroup in E_8 which commutes with E_6. Under these conditions, one trivially gets that the light fields are

a) $E_6 \times E'_8$ gauge fields and gauginos;
b) one E_6 family ($\underline{27}$) of scalars from A_m^a and SUSY partners χ^a;
c) the two "dilaton" fields ϕ and σ;
d) the two-parity partners of the above: a pseudoscalar η from the ansatz $B_{mn} = \varepsilon_{mn}\eta$, m,n = 5-10; a pseudoscalar θ from the duality transformation $H_{\mu\nu\rho} \sim \varepsilon_{\mu\nu\rho\sigma}\partial^\sigma\theta$, $\mu\nu\rho\zeta$ = 1-4;
e) fermionic partners of the scalars, which come from ψ_m (m=5-10) and λ;
f) graviton and gravitino.

It is easy to obtain[18] a truncated version of the standard supergravity Lagrangian by setting $g_{mn} = \delta_{mn}$ in (10). However, from the truncated bosonic Lagrangian one gets from Eq. (3), it is not easy at all to recognize the corresponding Kähler potential G(z,z*). It is more appropriate to obtain G from the fermionic bilinear sector[19] since we know that in the D = 4, N = 1 supergravity Lagrangian[8] the gravitino mass term is given by

$$-e\, e^{G/2}\, \bar{\tilde{\psi}}_{\mu_L} \gamma^{\mu\nu} \gamma^5 \psi_{\nu_L} \tag{11}$$

so that G will be easily determined by looking at the truncation of the D = 10 bilinears:

$$E^{-1}\mathcal{L}_{3/2} = -\frac{i}{2}\bar{\psi}_M \Gamma^{MNP} D_N \psi_P - \frac{i\sqrt{2}}{16}\varphi^{-3/4} H_{MNP}\left(\bar{\psi}_Q \Gamma^{QMNPR}\psi_R\right) + \cdots \tag{12}$$

The Weyl-Majorana-Rarita-Schwinger ψ_M spinor has the SO(3,1)×SO(6) spinorial structure $\psi_M = (\psi_{iM_L}, \tilde{\psi}^i_{M_R})$, i = 1-4 (SO(6)spinor). After a shift and a rescaling

$$\psi'_{\mu i_L} = e^{3/4\,\alpha}\left(\psi_{\mu i_L} + \frac{i}{2}\gamma_\mu \gamma^5 (\Gamma^m)_{ij}\, \tilde{\psi}^j_{m_R}\right) \tag{13}$$

in order to get a minimal gravitino kinetic term, one gets[19] a gravitino mass term from Eq. (12):

$$-e\left(e^{-3\sigma}\varphi^{-3/4}\right)\bar{\tilde{\Psi}}_{\mu_L}\gamma^{\mu\nu}\gamma^5(H.\Gamma)\Psi_\nu \tag{14}$$

where $H.\Gamma \sim \varepsilon^{ijk}H_{ijk}$, ijk being SU(3) holomorphic indices. From Eqs. (4) and (5) one observes that in the present truncation[20),18)]

$$(H.\Gamma) \sim tr(A_i A_j A_k)\varepsilon^{ijk} \sim tr(C_x C_x C_x) \tag{15}$$

where the C_x's are the zero mode scalars transforming as $\underline{27}$'s. Comparing Eqs. (11) and (14) one obtains for the truncated Kähler potential[18)]

$$G = \mathrm{Log}\left(\frac{1}{16}e^{-6\sigma}\varphi^{-3/2}\right) + \mathrm{Log}|W|^2 \tag{16}$$

where $W = 8\sqrt{2}\,C_x C_y C_z d^{xyz}$ is the truncated superpotential (d is the symmetric E_6 tensor). We thus have that <u>the superpotential is related to the gauge Chern-Simons symbol</u>[20),18)]. The dependence of G on ϕ and σ can be easily traced back[19)] to an overall factor $e^{-3\sigma}$ from the D = 10 vielbein times the $\phi^{-3/4}$ factor present in Eq. (12). The other rescaling of fields compensates after normalization of the gravitino kinetic terms. Notice that there is no explicit dependence of G on the pseudoscalar partners of σ and ϕ, η and θ. This signals the presence of two "Peccei-Quinn" symmetries in the Lagrangian, imaginary translations of η and θ. Examining the bosonic kinetic terms is easy to find the combinations of ϕ and σ which are parity partners of η and θ. They are[18)]

$$S = \varphi^{-3/4}e^{3\sigma} + i\theta \tag{17a}$$

$$T = e^{\sigma}\varphi^{3/4} + |C_x|^2 + i\eta \tag{17b}$$

In terms of these one can rewrite Eq. (16):

$$G = -\mathrm{Log}(S+S^*) - 3\,\mathrm{Log}(T+T^* - 2|C_x|^2) + \mathrm{Log}|W|^2 \tag{18}$$

Concerning the gauge kinetic function one obtains after truncation:

$$-\frac{E}{4}\varphi^{-3/4}F^a_{MN}F_a^{MN} \longrightarrow -\frac{e}{4}\left(e^{-3\sigma}\varphi^{-3/4}\right)\left(e^{6\sigma}\right)F^a_{\mu\nu}F_a^{\mu\nu} \tag{19}$$

so that $f_{ab} = \delta_{ab} S$. The coupling of θ to $F\tilde{F}$ is obtained after the duality transformation[18),19)]. The T-dependent sector of G is similar to the one in the so-called "no-scale models"[21)].

It is interesting at this point to discuss what is the origin of the structure of G and to what extent it is a consequence of string properties or the truncation procedure[22)]. The S dependence of (18) may be easily traced back to the scale invariance[18)] of the string tree amplitudes we discussed above. The classical invariance of Eq. (8) when seen from the truncated D = 4 Lagrangian amounts to

$$\left.\begin{array}{l} S \rightarrow \lambda S \\ T \rightarrow T \\ g_{\mu\nu} \rightarrow \lambda g_{\mu\nu} \end{array}\right\} \quad e^{-1}\mathcal{L} \rightarrow \lambda^{-1} e^{-1}\mathcal{L} \tag{20}$$

Since the scalar potential implies no space-time contractions it has necessarily to be of the general form:

$$V(S,T,C_x) = \frac{1}{S}\, v(C_x,T) \tag{21}$$

with $V(C_x,T)$ scale invariant. Recalling the general N = 1 supergravity formula[8)] for the scalar potential (proportional to e^G) one obtains that the term $-\log(s+s^*)$ should be present in G[22)]. Since the symmetry in (20) is only classical, there is in principle no reason to expect that kind of S-behaviour for the complete low energy effective Kähler potential. It is just a consequence of the classical scale invariance of tree string amplitudes.

The T-behaviour of G is obviously related to the compactification process[22)]. It can be intuitively understood as follows. In the truncation procedure, whenever there is a contraction of <u>internal</u> (m = 5-10) world indices one gets [see Eq. (10] an $e^{-\sigma}$ (in fact a T^{-1}) factor. This just reminds us that e^{σ} is a measure of the size of the compact manifold K_6. The matter scalar fields C_x come from the A_m's and hence whenever you have a C_x field you expect a $T^{-\frac{1}{2}}$ factor from the internal contraction. Thus you expect a scale invariance under[22)]

$$\begin{array}{l} T \rightarrow (\lambda')^2\, T \\ C_x \rightarrow \lambda' C_x \\ S \rightarrow S \end{array} \tag{22}$$

In particular, the $|W|^2$ term in the truncated Lagrangian can only appear in the scaled form

$$\frac{|W|^2}{(\mathrm{Re}T)^3} \tag{23}$$

in order to preserve the scale invariance in Eq. (22). This leads us to the structure in Eq. (18). In particular the factor 3 in front of the T-dependent sector of G is just counting the number of internal contractions[22] one has in $|\omega_{ijk}|^2$. Since the superpotential transforms as the SU(3)-invariant three-form ε_{ijk}, the factor 3 in Eq. (18) is a reflection of the SU(3) holonomy. This second scale invariance would be broken if in the truncated D = 4 Lagrangian there remain terms proportional to vacuum expectation values of heavy fields [e.g., proportional to $F^a_{mn} \wedge F^a_{pq}$ ($a \in SU(3)$), or ω_m, R^2, etc.]. Interestingly enough, no such terms are present from the truncation of the tree level D = 10 supergravity Lagrangian (at least from the terms known at the moment to exist). Scale breaking terms would be obtained from the following terms with world indices in internal dimensions

$$-\frac{1}{2}R \; ; \; \frac{1}{4}\varphi^{-3/4}(R^2 - F^2) \; ; \; \frac{3}{4}\varphi^{-3/2}(\omega_3^G - \omega_3^L)^2 \; ; \; \ldots \tag{24}$$

However, all these terms disappear for Ricci-flat manifolds if the identification $\langle A^a_m \rangle = \omega_m$ is made, as required in a supersymmetric truncation. It is, however, not likely that this feature remains in the complete low energy limit of the string. For example, R^4 terms would violate the invariance unless its vacuum expectation values were cancelled by some F^4 terms. Also, terms like that in (9) violate this invariance since a v.e.v. $\langle F^a_{[mn} F^a_{pq]} \rangle$ would give rise[23] to low energy terms[*),19),24)]

$$\varepsilon\,(\mathrm{Im}T)\left[(F^{\mu\nu}\tilde{F}_{\mu\nu})_{E_6} - (F^{\mu\nu}\tilde{F}_{\mu\nu})_{E_8'} - 5(R^{\mu\nu}\tilde{R}_{\mu\nu})\right] \;, \; \varepsilon \sim 10^{-5} \tag{25}$$

explicitly violating (22). It is reasonable to expect that similar scaling arguments can be already stated at the level of the string itself propagating on a non-trivial background[22].

From the truncated Kähler potential one easily obtains the scalar potential[18] (real parts of S and T are understood)

---

*) These terms by themselves are non-supersymmetric. A supersymmetrization would require additional one-loop terms like, e.g., $(F^2)^2$ which leads to couplings like $\mathrm{ReT}(F^2)$ upon truncation.

$$V = \frac{S^{-1}T^{-3}}{16}\left[|W|^2 + \frac{T}{3}|W_c'|^2\right] + \frac{9}{2}T^{-2}S^{-1}(C_x^* \lambda^a C_x)^2 \tag{26}$$

where one can immediately recognize the scale invariances discussed above. As it stands, (26) is minimized for $C_x = 0$ and/or $S,T \to \infty$ (free, uncompactified theory) in a supersymmetric minimum. For $C_x \simeq 0$ (physically interesting case) there is a classical degeneracy with respect to any value of S and T. Eventually, one expects that this degeneracy must be lifted (although it turns out not to be so easy) and one has to normalize all kinetic terms to canonical form. This can be done for the observable fields through the redefinitions[19)]

$$\begin{aligned} \chi', C_x' &= \sqrt{3}\, T^{-1/2} C_x, \chi \\ A_\mu' &= S^{1/2} A_\mu \\ g &= \mathrm{Re}\, S^{-1/2} \end{aligned} \tag{27}$$

which lead to the following rescaled potential

$$V_r = |W_r|^2 + \left|\frac{\partial W}{\partial C_x}\right|^2 + \frac{g^2}{2}(C_x^* \lambda' C_x)^2 \tag{28}$$

with $W_r = 1/3\sqrt{8/3}\, g d_{xyz} C^x C^y C^z$ (the C's have canonical kinetic terms). Notice that in the limit $M_P \to \infty$ (28) is just a globally supersymmetric potential. Notice also that if there were a constant term c inside W_r (corresponding to a v.e.v. $H_{ijk} = c\,\varepsilon_{ijk}$)[20),25),26)] supersymmetry would be broken since

$$m_{3/2} = e^{\langle G\rangle/2} = \frac{\langle W\rangle}{4S^{1/2}T^{3/2}} \neq 0 \tag{29}$$

In this case one would get low energy soft SUSY breaking parameters[19),20)]:

$$A = 1 \quad ; \quad m = 0 \quad ; \quad M = F_S f^{\prime S} = e^{\langle G\rangle/2} = m_{3/2} \tag{30}$$

Unfortunately one would have a cosmological constant $V_0 = m_{3/2}^2\, M_P^2$. Also since the masses of the C's vanish but the trilinear scalar couplings do not ($A \neq 0$), the observable fields C would acquire unwanted large v.e.v.'s. Furthermore, since in fact the fields S and T are undetermined, Eq. (29) and $V_0 = m_{3/2}^2\, M_P^2$ lead to S and/or $T \to \infty$ (non-interacting and/or uncompactified theory). Thus a simple naive truncation of the theory plus the simplest implementation of residual supersymmetry breaking (a H_{ijk} v.e.v.) seems to lead to problems.

There is another possible supersymmetry breaking mechanism already built in the $E_8 \times E_8$ heterotic string. The "hidden" E_8 (or some subgroup) gauge interactions may give rise to gaugino condensation[20),25)]. In the presence of a non-trivial gauge kinetic function f_{ab} (as is here the case) this condensation gives rise to supersymmetry breaking[27)]. If $\langle \chi_8 \chi_8 \rangle = \Lambda^3$ the scale of supersymmetry breaking[27),28)] will be $M_{ss}^2 \sim \Lambda^3/M_P$ and the gravitino will gain a mass $m_{3/2} \sim \Lambda^3/M_P^2$ through the super-Higgs effect from terms such as

$$\mathrm{Tr}\,(\chi_8 \Gamma_{mnp} \chi_8)\, \psi_\mu \Gamma^\mu \Gamma^{mnp} \lambda \tag{31}$$

in the D = 10 supergravity Lagrangian.

The "hidden" gauge group could be E_8 or may be some subgroup[19),29)] obtained from it through symmetry breaking by non-contractible Wilson loops[14)]. In fact if you assume $g_6(M_x) = g_8(M_x)$ (as seems to be required[30)] by general tree level string arguments), an unbroken E_8 would lead to gaugino condensates with $\Lambda \sim M_P$, as one can easily see from renormalization group arguments[19)]. In order to get a relatively light gravitino (say, $m_{3/2} \lesssim 10^6$ GeV) the hidden E_8 must be broken upon compactification down to a group with small rank non-Abelian factors[19)] like SU(2), SU(3) or SO(5). Otherwise the hidden gauge coupling evolves too fast and the condensation scale appears very close to the compactification scale. This could affect the compactification process. It could well be that higher-loop string effects allowed for a situation with $g_8(M_x) < g_6(M_x)$ in which case E_8 could be broken to larger groups and still we could have a relatively light gravitino. This we want if we are interested in relating somehow the gravitino mass to the weak scale.

The presence of gaugino condensation modifies the scalar potential of N = 1 supergravity in the following form[27)]:

$$V = (G''^{-1})_i^j \left[e^{G/2} G'_j + \frac{f'_j}{4} \langle \chi\chi \rangle \right] \left[e^{G/2} G'^i + \frac{f'^i}{4} \langle \chi\chi \rangle \right] - 3\, e^G \tag{32}$$

where i,j = S,T,C_x. Recalling that f = S one gets[25)], instead of (26),

$$V = \frac{S^{-1} T^{-3}}{16} \left| W - (S+S^*) 4 S^{1/2} T^{3/2} \langle \chi\chi \rangle \right|^2 + (\text{rest as in } (26)) \tag{33}$$

where we must remember that $\langle\chi\chi\rangle$ is a "running" condensate since the value of g (or $\langle S\rangle$) has not yet been determined. Notice that if there is a constant in the superpotential W, the value of S will adjust itself[25] to cancel it. This leads to an interesting situation with a zero cosmological constant but broken supersymmetry (recall $G_T' \neq 0$). However, although supersymmetry is broken in the hidden sector, the observable sector remains supersymmetric (at the tree level) since $M = F_s f'^s = F_s = 0$ and also $A = 0$. This is not necessarily problematic since one expects that radiative corrections[31)-33)] will eventually transmit the breaking of supersymmetry to the observable sector. Unfortunately, radiative corrections also induce[32] a negative cosmological constant $V_0 \sim -m_{3/2}^2 M_P^2$. This is a disaster because the value of $m_{3/2} \sim s^{-1/2}T^{-3/2}$ is still undetermined (T is still not fixed) and then it adjusts itself to be $m_{3/2} \sim M_P$. This has three problems: (i) if $m_{3/2} \sim M_P$ then $\Lambda \sim M_P$ and there are strong interactions on the compactification scale, sheding doubts on the whole present naive approach; (ii) if $m_{3/2} \sim M_P$ it is difficult to understand how one can protect the scalars from acquiring very large masses once SUSY is broken; (iii) there is a large cosmological constant.

It seems that naïve application of the simple truncation discussed above leads to problems. This is perhaps not surprising since the procedure we followed was probably too naïve. Instead of expanding our ten-dimensional fields around a realistic manifold and integrating out the heavy fields we just truncated the theory and set $g_{mn} = \delta_{mn}$. This truncation misses, e.g., the extra zero modes of the Calabi-Yau metric g_{mn}. For example, for some manifolds with $b_{1,1} = 1$ there is one of these extra zero modes for each light <u>27</u> generation. This sort of "generation structure" in the metric zero modes of this type of manifolds is intriguing[19]. The truncation is also not realistic in that one does not get several <u>27</u> generations. Also, we only truncated the known parts of the ten-dimensional field theory obtained from the string. Thus, for example, if there are further terms one could obtain additional contributions to the truncated Lagrangian

$$\begin{aligned} F^4 &\longrightarrow \frac{|W_{,c}|^4}{(\mathrm{Re}\,T)^4} \\ H^4 &\longrightarrow \frac{|W|^4}{(\mathrm{Re}\,T)^6} \\ &\vdots \end{aligned} \tag{34}$$

etc. They could change the conclusions we obtained above. The inclusion of this type of terms is equivalent to stating that we should consider a higher derivatrive version of the D = 4, N = 1 supergravity Lagrangian [see the last paper in Ref. 17)]. A further point to remark is that maybe one cannot ignore the effect of the excited string states. This is probably the case[34] since Planck mass and string scale should be of the same order of magnitude. Of course, string effects could completely modify the situation we have discussed. Whatever the case may be, it seems that the phenomenological problem one obtains after truncation is that the degeneracy signalled by the dilatons S and T still remains. It seems to me that a successful phenomenology would require a mechanism to fix the values of S and T upon compactification. However, it must be such that $m_{3/2} \ll M_P$ (say $m_{3/2} \lesssim 10^6$ GeV), otherwise one does not understand how one could protect the masses of scalars from quadratic divergences. Dynamical supersymmetry breaking through a condensate could explain the hierarchy $m_{3/2} \ll M_P$ if $g_8(M_x) < g_6(M_x)$ or if E_8 is broken upon compactification to a subgroup with small non-Abelian factors.

LOW ENERGY SUPERGRAVITY AND SUPERSTRING MODELS

Compactification on a "Calabi-Yau" manifold of the $E_8 \times E_8$ ten-dimensional Lagrangian leads to an E_6 SUSY-GUT coupled to N = 1 supergravity. To obtain a realistic theory one has to (i) break the E_6 symmetry down to the standard model and (ii) give masses to the unwanted states in the fundamental $\underline{27}$'s of E_6. The first task can be achieved if the gauge fields in extra dimensions get trapped in the non-simply connected structure of the Calabi-Yau manifold[30),35)] (Wilson loop or Hosotani mechanism). This needs in general to be supplemented by further symmetry breaking induced by the light non-singlet chiral multiplets of the theory[37),38)]. The second task is more complicated and leads to some phenomenological problems in superstring models.

The low energy non-singlet chiral fields are given by[30)]

$$n_g\,(\underline{27}) + b_{1,1}\,(\underline{27}_H + \overline{\underline{27}}_H) \qquad (35)$$

where n_g = 3 or 4 (hopefully). To fix the notation let us show the SU(5) content as well as a tentative quark-lepton assignment of the states inside a $\underline{27}$

$$27_i = (10 + \bar{5} + 1)_i + (5 + \bar{5})_i + 1_i$$

$$\downarrow$$

$$(U_R, Q_L, E_R)_i \quad (L, d_R)_i \quad \nu_{R_i} \quad (D, \bar{D}; H, \bar{H})_i \quad N_i \qquad (36)$$

where i = 1-3(4) families. To indicate that a certain particle is associated with the light objects from $\underline{27}+\overline{\underline{27}}$ we will add an H sub-index. The most general superpotential[30] allowed by the existence of an original E_6 symmetry (before possible Wilson-loop breaking) is

$$\begin{aligned} W = \sum_{\text{families}} & h_L HLE_R + h_d HQd_R + h_u \bar{H} Q U_R + \\ & + h_\nu \bar{H} L \nu_R + h'_\nu \bar{D} d_R \nu_R + \\ & + \lambda_2 N H \bar{H} + \lambda_3 N D \bar{D} + \\ & + \lambda_B Q_L Q_L D + \lambda'_B U_R d_R \bar{D} + \lambda''_B L Q_L D \end{aligned} \qquad (37)$$

where, if there are extra light fields from $\underline{27}+\overline{\underline{27}}$, the corresponding extra couplings may also be present.

The main problem with E_6 [39] is that it is too big. An E_6 family contains the usual stuff plus twelve unobserved particles: D, $\bar{D}$, H, $\bar{H}$, ν_R, N. If these particles remain light [or some couplings in (37) are not forbidden] several phenomenological disasters may happen. The states D and $\bar{D}$ mediate too fast proton decay unless $m_{D,\bar{D}} \gtrsim 10^{10}$ GeV. If the right-handed neutrinos ν_R are light ($m_{\nu_R} \lesssim$ 1-10 TeV) then the left-handed neutrinos get induced masses which are incompatible with present phenomenology.

The neutrino mass problem was solved in the old left-right symmetric GUTs by assuming the existence of large ν_R Majorana masses. This is a natural explanation to understand the smallness of ν_L masses: the ν_R's are singlets under the standard symmetries and hence they are allowed to gain large masses. This large mass was given by an explicit Yukawa coupling to the appropriate Higgs field [an SO(10) $\underline{126}$, an E_6 $\underline{351}$] or through some radiative corrections[40]. The latter does not work in the supersymmetric case[41] since the radiative corrections are hierarchically suppressed. The former does not work in superstring models since we only have adjoint + fundamental E_6 fields to do the breaking. One possibility[38] is to assume the existence of some symmetry setting $h_\nu = 0$ (i.e., forbidding Dirac ν-masses). I find that this explanation is a step back in our understanding of the ν-mass problem. The traditional explanation of the massiveness of ν_R is quite appealing and should not be abandoned so soon. Moreover, if $h_\nu = 0 = h'_\nu$ the ν_R's would only interact very weakly and would lead to cosmological problems[19]. The situation would be solved if there were non-renormalizable terms in the effective superpotential coming from the higher string modes. Thus a term of the form[19]

$$O(1/M_s)\, \underline{27}\times\underline{27}\times\overline{\underline{27}}_H\times\overline{\underline{27}}_H \longrightarrow \nu_R\times\nu_R\times\frac{\langle\bar{N}_R\rangle^2}{M_s} \tag{38}$$

where one of the fields inside $\overline{\underline{27}}_H$ $(\bar{N}_R)$ has the quantum numbers of a ν_R and a $\langle\bar{N}_R\rangle \sim v$, would give ν_R masses of the order of v^2/M_S. I find this possibility more appealing than just setting $h_\nu = 0$ in Eq. (36). A second possibility is present[19] if there are renormalizable couplings of the form $\overline{27}_H\times\underline{27}\times\underline{1}$ to some E_6 singlets which combine with the ν_R's to form superheavy Dirac particles if a large $\langle\overline{27}_H\rangle = \langle\bar{N}_R\rangle$ is present. Another mechanism[19] one should consider in the case in which compactification and string scales were really overlapping, is that excited states themselves could acquire large v.e.v.'s giving direct masses to the ν_R's. In the heterotic string there are excited states with the right quantum numbers (351's of E_6) to give such masses. Whatever be the case, it seems that a solution to the ν-mass problem would prefer[42] that the <u>E_6 generators under which ν_R is not a singlet should not remain at low energies</u>. This has strong implications for low energy physics, as we will see later on.

The fast proton-decay problem seems hard to solve unless there is some symmetry[30] forbidding some couplings in Eq. (37) (i.e., $\lambda_B = \lambda'_B = 0$). A discrete symmetry under which $D,\bar{D} \to (D,\bar{D})$ would forbid the undesired terms. This fast proton-decay problem is not new for superstring models, it is similar to the doublet-triplet splitting problem ("second hierarchy problem") of usual SU(5) models. Unlike the ν-mass problem, this one never had a tremendously appealing solution so that assuming the existence of a discrete symmetry is not a step backwards in our understanding, we stay where we were. The alternative to forbid these couplings is to envisage a mechanism[37),38),31),19] to give to all the $D,\bar{D}$ fields a large mass. But if we want to have both large masses for $D,\bar{D}$ and the ν_R's, only standard gauge interactions $[SU(3)\times SU(2)\times U(1)]$ would be left at low energies. This would lead to a model really analogous to the ones of usual low energy supergravity[7)] so that we will not consider this case any longer here. If, on the other hand, we just insist on heavy ν_R's the extra Z'_0 remaining at low energies necessarily has a vanishing coupling to ν_R so that it is uniquely fixed[42)] to be the one coupling to the Y_Z generator in the Table.

	Q_L	U^C_L	e^C_L	d^C_L	L	H	$\bar{D}$	$\bar{H}$	D	N_R	N	
$Y(\times\sqrt{3/5})$	$\frac{1}{6}$	$-\frac{2}{3}$	1	$\frac{1}{3}$	$-\frac{1}{2}$	$-\frac{1}{2}$	$\frac{1}{3}$	$\frac{1}{2}$	$-\frac{1}{3}$	0	0	unbroken
$Y_Z(\times\frac{1}{2}\sqrt{10})$	-1	-1	-1	-2	-2	3	3	2	2	0	-5	unbroken
$Y_B(\times\frac{1}{4}\sqrt{6})$	-2	-2	-2	4	4	-2	-2	4	4	-8	-2	broken

- Table I -

Y_Z is a linear combination of the two U(1)'s inside E_6 which commute with SU(5). Although we know that the generator Y_Z must remain unbroken down to low energies, one could think of possible "non-Abelian" extensions[37),38)] of our weak group $SU(2)_L\times U(1)_Y\times U(1)_Z$. However, it is well known from renormalization group studies in usual SUSY-GUT theories[44),37),38)] that those usually lead to bad predictions for $\sin\theta_W$ and M_x. Let us forget this fact for the moment and let us consider a classification of possible low energy superstring models. According to the possible low energy chiral superfield content, one has[19)]

A) Minimal model with just three 27's and no extra Higgses;

B-I) Model with three 27's and light Weinberg-Salam (W-S) doublets from $27+\overline{27}$;

B-II) Model with three 27's and light $N_H+\bar{N}_H$ from $27+\overline{27}$;

B-III) Model with three 27's and light $N_R+\bar{N}_R$ from $27+\overline{27}$;

and we assume that the low energy gauge group is of the form Standard × (U(1) or $U(1)^2$ or SU(2)×U(1)). All these possibilities are discussed in some detail in Ref. 19) where it is also shown how the residual gauge symmetries may be broken radiatively. At first sight the model B-I seems to be the most attractive[30], since one can get rid of all the dangerous (proton-decay mediating) $D,\bar{D}$ fields when the fields $(N_H+\bar{N}_H)$ (which are necessarily light along with the W-S doublets) acquire a large v.e.v. This large v.e.v., for N_H and $\bar{N}_H$ are induced in a natural way by radiative corrections[31),19)], in the same way that the W-S doublets get their v.e.v.'s in low energy supergravity models[3)-7)]. Model B-I has, however, several problems. First, the prediction for $\sin^2\theta_W$ and M_x is rather bad[44),38)] ($\sin^2\theta_W \simeq 0.29$, $M_x \simeq 10^{24}$ GeV). Second, this model necessarily has an $SU(2)_R$ gauge symmetry. This implies that the couplings $h_\nu = h_L$ in Eq. (37) so that if one forbids Dirac ν-masses, one also forbids*) charged lepton masses![19)]. Thus there is a ν-mass problem. Although a B-I model could exist avoiding all these problems, it will probably be quite contrived.

The class of models which seems to accommodate better the existence of a light $U(1)_Z$ with the quantum numbers of the Table is the B-III models. In these one can assume E_6 is broken down to $SU(3)\times SU(2)_L\times U(1)^3$ at the compactification scale through some Wilson loop configuration[30),35)-37),19)] like, e.g.,

$$U_g = \mathbb{1}_C \times \begin{pmatrix} \alpha & & \\ & \alpha & \\ & & \alpha^{-2} \end{pmatrix}_L \times \begin{pmatrix} \alpha^{-2} & & \\ & \beta & \\ & & \gamma \end{pmatrix}_R \tag{39}$$

where $\alpha,\beta,\gamma \in Z^n$ and $\beta \neq \alpha^{-2}$ [to avoid an $SU(2)_R$ symmetry]. This configuration would leave invariant an $\bar{N}_R$ field from the $\overline{27}$ which would be light along with his opposite chirality partner N_R. If these $(N_R+\bar{N}_R)$ fields get a v.e.v. the unbroken gauge group would be, precisely,

$$SU(3)\times SU(2)_L \times U(1)_Y \times U(1)_Z \tag{40}$$

It is, however, not clear how to give a v.e.v. to the N_R, $\bar{N}_R$ fields and avoid at the same time giving a large mass to some fields we want to remain light[19)]. Anyway the B-III models have the right low-energy content to do what we want. In order to give also large masses to the ν_R's one has to assume the existence of couplings like those of Eq. (38). Still, it could turn out that the ν_R's get

---

*) This problem may be avoided if there are extra Higgs doublets and the fields H and $\bar{H}$ with non-vanishing v.e.v.'s are not $SU(2)_R$ partners. However, I find this situation rather contrived.

their large masses from direct coupling to higher string states, as we discussed above, or through some other presently unknown mechanism. In this case, models A and B-II could be equally interesting and could lead to the low energy group in Eq. (40). Particularly attractive is the minimal case A which does not assume a very special gauge configuration leading to any remnant Higgs, there is a perfect Higgs-quark-lepton symmetry.

We have now to consider how the symmetry (40) is broken down to $SU(3)_C \times U(1)_{em}$. The low energy chiral content will be now that of three E_6 families except for the ν_R's which we assume have gained large masses, since ν_R-masses are not forbidden by any gauge symmetry $[Y_Z(\nu_R) = Y(\nu_R) = 0]$. The low energy effective superpotential will be of the form

$$W = \sum_{\text{families}} h_L H L E_R + h_d H Q d_R + h_u \bar{H} Q U_R + + \lambda_2 N H \bar{H} + \lambda_3 N D \bar{D} \tag{41}$$

where we have also assumed there is some symmetry forbidding baryon number violating couplings. Once supersymmetry is broken in the "hidden sector" we will have soft SUSY breaking terms as discussed in the introduction. Since, as we discussed in the previous paragraph, the breaking of the residual supersymmetry is still not completely understood, we will assume for the moment general soft couplings M, m and A. There are three sets of N_i, H_i, $\bar{H}_i$ (i=1-3) fields. Let us assume that only one set has large enough Yukawa couplings λ_2, λ_3 and h_t. The relevant scalar potential for the neutral directions will then be[42)]

$$\begin{aligned} V(H,\bar{H},N) = {} & \frac{g_z^2}{2}\left(2|\bar{H}|^2+3|H|^2-5|N|^2\right)^2 + \frac{1}{8}(g^2+g'^2)(|H|^2-|\bar{H}|^2)^2 + \\ & + \lambda_2^2 |H\bar{H}|^2 + \lambda_2^2 |N|^2 (|H|^2+|\bar{H}|^2) + \\ & + m_N^2 |N|^2 + m_H^2 |H|^2 + m_{\bar{H}}^2 |\bar{H}|^2 + m(\lambda_2 A_2 N H \bar{H} + h.c.) \end{aligned} \tag{42}$$

where g_z is the $U(1)_z$ gauge coupling constant. We want to find minima with $N \gg v, \bar{v}$ ($\langle H \rangle \equiv v$, $\langle \bar{H} \rangle \equiv \bar{v}$). This is because we have three sets of W-S Higgses which will induce flavour changing[*)] neutral currents (FCNC) unless they are heavy enough[19)]. They will only be very heavy if m (and hence N) is large enough

*) Recall that a diagonalization of quark mass matrices only implies diagonalization of Higgs couplings if there is only one set of Higgses H+$\bar{H}$.

(~1 TeV). Of course one also wants that $N \gg \nu,\bar{\nu}$ because of neutral current experimental constraints[45),46)] but these are much milder than the FCNC constraints. A situation in which $N \gg \nu,\bar{\nu}$ requires that m_N^2 gets negative at low energies faster than $m_{\bar{H}}^2$ and m_H^2. Interestingly enough this situation is indicated by the renormalization group equations[19),42)]

$$\frac{dm_N^2}{dt} = \frac{\alpha_z}{\pi}(25)M_{\tilde{z}}^2 - 3\left(\frac{\lambda_3}{4\pi}\right)^2\left(m_{\bar{D}}^2 + m_D^2 + m_N^2 + A_3^2 m^2\right) - 2\left(\frac{\lambda_2}{4\pi}\right)^2\left(m_{\bar{H}}^2 + m_H^2 + m_N^2 + A_2^2 m^2\right) \tag{43a}$$

$$\frac{dm_{\bar{H}}^2}{dt} = \frac{3\alpha_2}{4\pi}M_2^2 + \frac{\alpha_1}{4\pi}M_1^2 + \frac{4\alpha_z}{\pi}M_{\tilde{z}}^2 - 3\left(\frac{h_t}{4\pi}\right)^2\left(m_Q^2 + m_U^2 + m_{\bar{H}}^2 + A_t^2 m^2\right) - \left(\frac{\lambda_2}{4\pi}\right)^2\left(m_H^2 + m_{\bar{H}}^2 + m_N^2 + A_2^2 m^2\right) \tag{43b}$$

$$\frac{dm_H^2}{dt} = \frac{3\alpha_2}{4\pi}M_2^2 + \frac{\alpha_1}{4\pi}M_1^2 + \frac{9\alpha_z}{\pi}M_{\tilde{z}}^2 - \left(\frac{\lambda_2}{4\pi}\right)^2\left(m_H^2 + m_{\bar{H}}^2 + m_N^2 + A_2^2 m^2\right) \tag{43c}$$

which give $m_N^2 \ll m^2_{\bar{H},H}$ for $\lambda_3 \gtrsim h_t$. The existence of the λ_3 Yukawa coupling is important, otherwise it is difficult to avoid $N \sim \nu,\bar{\nu}$. One can easily integrate analytically[42)] the renormalization group equations in the approximation in which one neglects the effects of λ_2 in the renormalization group equations (but not in the scalar potential, of course) and check numerically that this situation is possible. However, for general soft terms M, m, A, it is difficult to obtain $N \gg \nu,\bar{\nu}$ (even if $m_N^2 \ll m^2_{\bar{H},H}$) without doing some kind of fine-tuning[*)]. This is because of the first term ($\frac{1}{2}D_z^2$) in Eq. (42). It tends to give v.e.v.'s $\nu,\bar{\nu} \sim N$ unless we compensate appropriately that negative contribution to the $H,\bar{H}$-mass^2 with positive contributions from $(m^2_{H,\bar{H}}+\lambda^2N^2)|H|^2+|\bar{H}|^2$. This requires some fine-tuning of independent terms particularly m_H^2 and $m^2_{\bar{H}}$[42)]. We want to remark that this is not a specific property of the $U(1)_z$ considered but will be the case

---

*) This fact has also been independently pointed out by the authors of Ref. 46).

for any other extra U(1)'. However, one must recall here that the usual procedure of "cutting" tghe evolution of the soft parameters at a given scale $Q \sim m$ and then using the parameters at that scale to minimize the potential is just an approximation to the problem of calculating the complete effective scalar potential. In particular, the soft parameters (e.g., m_H^2, $m_{\bar{H}}^2$) themselves depend on the fields (upon the Coleman-Weinberg renormalization prescription $Q = \phi$) and in principle one should include this dependence when minimizing the potential. Then the argumentation leading to the conclusion that a fine-tuning is required does not necessarily apply.

There is an interesting exception[42),47)] to this fine-tuning problem. If $A_2 = 0$ in Eq. (42) one can immediately check that there is a minimum with:

$$\nu = \bar{\nu} = 0 \quad ; \quad N^2 = \frac{-m_N^2}{25 g_z^2} \tag{44}$$

which is obtained if the renormalization group equations run m_N^2 towards negative values [as indicated by Eqs. (42)]. This is similar to what happens[6)] in low energy supergravity models when B = 0: one gets a minimum with $\nu = 0$ and $\bar{\nu}^2 = (-4m_{\bar{H}}^2)/(g^2+g'^2)$. Unfortunately, A_2 cannot be strictly zero because (also as in the old low energy supergravity models) then there would be a massless (harmful) axion [this is why we needed the coupling in Eq. (2)]. Of course, if there is some symmetry forbidding FCNC transitions one does not need to make M'_{Z_0} much bigger than M_Z and no real fine-tuning problem exists.

This scheme for understanding the low energy symmetry breaking has other interesting phenomenological features which will be discussed in a forthcoming paper[42)]. An interesting property of the $U(1)_z$ charge of the Table is that all quark-lepton fields have the same sign as N for the $U(1)_z$ change. This means that they all get positive D-term contributions to their scalar masses. This is desirable to avoid unwanted colour- and charge-breaking minima which may appear since the D^2 contributions are potentially very large ($\sim g_z^2 N^2$). This property does not appear in other U(1) charges . Another general property of these models is the fact that $M'_{Z_0} \gg M_{Z_0}$. This requires that all squarks and sleptons be quite heavy since they have several large contributions to their masses: from the soft masses $\sim m$ and from the large D-terms $\sim g_Z N$. One also expects relatively heavy gauginos [in order to avoid too light "charginos" with mass $\sim (M_W^2)/(\mu) \sim (M_W^2)/(\lambda_2 N)$].

Let us conclude by stating that although we are far from understanding the detailed connection between the superstring itself and the low energy phenomenology, several general properties are already encouraging. In particular, this scheme seems to fit very nicely with the low energy supergravity models developped in the last few years. Still we would like to understand better the compactificaiton process (maybe on the string itself?) as well as the breaking of the residual supersymmetry.

I thank my collaborators J.P. Derendinger and H.P. Nilles for many useful discussions. I also thank the organizers of the First Torino Meeting for their kind invitation.

REFERENCES

1) For a review, see H.P. Nilles - Physics Reports 110 (1984) 1.

2) L.E. Ibañez - Phys.Lett. 118B (1982) 73;
P. Nath, R. Arnowitt and A.H. Chamseddine - Phys.Rev.Lett. 49 (1982) 970;
R. Barbieri, S. Ferrara and C.A. Savoy - Phys.Lett. 119B (1982) 343.

3) L.E. Ibañez and G.G. Ross - Phys.Lett. 110B (1982) 215.

4) L. Alvarez-Gaumé, L. Claudson and M. Wise - Nucl.Phys. B207 (1982) 16;
B. Ovrut and C. Nappi - Phys.Lett. 113B (1982) 65;
M. Dine and W. Fischler - Phys.Lett. 110B (1982) 227;
J. Ellis, L.E. Ibañez and G.G. Ross - Phys.Lett. 113B (1982) 227.

5) K. Inoue et al. - Progr.Theor.Phys. 68 (1982) 927.

6) L.E. Ibañez - Nucl.Phys. B218 (1983) 514;
L.E. Ibañez and C. Lopez - Phys.Lett. 126B (1983) 54;
L. Alvarez-Gaumé, J. Polchinski and M. Wise - Nucl.Phys. B221 (1983) 495;
J. Ellis, J. Hagelin, D.V. Nanopoulos and K. Tamvakis - Phys.Lett. 125B (1983) 275.

7) L.E. Ibañez and C. Lopez - Nucl.Phys. B233 (1984) 511;
S. Jones and G.G. Ross - Phys.Lett. 155B (1984) 69;
C. Kounnas, A. Lahanas, D.V. Nanopoulos and M. Quiros - Nucl.Phys. B236 (1984) 438;
L.E. Ibañez, C. Lopez and C. Muñoz - Nucl.Phys. B256 (1985) 218.

8) E. Cremmer, S. Ferrara, L. Girardello and A. Van Proeyen - Nucl.Phys. B212 (1983) 413.

9) H.P. Nilles, M. Srednicki and D. Wyler - Phys.Lett. 120B (1982) 346.

10) For reviews, see:
J. Schwarz - Physics Reports 89 (1982) 223;
M. Green - Surveys in High Energy Phys. 3 (1982) 127;
L. Brink - CERN Preprint TH. 4006 (1984).

11) D. Gross, J. Harvey, E. Martinec and R. Rohm - Phys.Rev.Lett. 54 (1985) 502; Nucl.Phys. B256 (1985) 253; Princeton Preprint (1985).

12) M.B. Green and J. Schwarz - Phys.Lett. 149B (1984) 117; 151B (1985) 21.

13) A.H. Chamseddine - Nucl.Phys. B185 (1981) 403;
F. Bergshoeff, M. de Roo, B. de Wit and P. van Nieuwenhuizen - Nucl.Phys. B195 (1982) 97;
G. Chapline and N. Manton - Phys.Lett. 120B (1983) 105.

14) P. Candelas, G. Horowitz, A. Strominger and E. Witten - Nucl.Phys, in press.

15) P. Candelas et al. - "Superstring Phenomenology", Proceedings of the Argonne Symposium for Anomalies, Geometry and Topology, published by World Scientific Co., Singapore (1985), and references therein.

16) C. Callan, D. Friedan, E. Martinec and M. Perry - Princeton Preprint (1985).

17) L. Romans and N. Warner - CALT-68-1291 (1985);
R. Nepomechie - Washington (Seattle) Preprint 40048-20, P5 (1985);
S. Cecotti, S. Ferrara, L. Girardello and M. Porrati - CERN Preprint TH. 4253 (1985).

18) E. Witten - Phys.Lett. 155B (1985) 151.

19) J.P. Derendinger, L.E. Ibañez and H.P. Nilles - CERN Preprint TH. 4228 (1985), to appear in Nucl.Phys.B.

20) J.P. Derendinger, L.E. Ibañez and H.P. Nilles - Phys.Lett. 155B (1985) 65.

21) E. Cremmer, S. Ferrara, C. Kounnas and K. Tamvakis - Phys.Lett. 133B (1983) 61;
J. Ellis, A. Lahanas, D.V. Nanopoulos and K. Tamvakis - Phys.Lett. 1345B (1984) 429;
J. Ellis, C. Kounnas and D.V. Nanopoulos - Nucl.Phys. B241 (1984) 406; Nucl.Phys. B247 (1984) 373.

22) L.E. Ibañez and H.P. Nilles - unpublished.

23) E. Witten - Phys.Lett. 153B (1985) 243.

24) K. Choi and J.E. Kim - Seoul Preprint SNUHE 85/10 (1985).

25) M. Dine, R. Rohm, N. Seiberg and E. Witten - Phys.Lett. 156B (1985) 55.

26) R. Nepomechie, Y. Wu and A. Zee - Washington Preprint 40048-02 P5 (1985).

27) S. Ferrara, L. Girardello and H.P. Nilles - Phys.Lett. 125B (1983) 457.

28) H.P. Nilles - Phys.Lett. 115B (1982) 193; Nucl.Phys. B217 (1983) 366.

29) E. Cohen, J. Ellis, C. Gomez and D.V. Nanopoulos - CERN Preprint TH. 4159 (1985).

30) E. Witten - "Symmetry Breaking Patterns in Superstring Models", Princeton Preprint (1985).

31) M. Mangano - "Low Energy Aspects of Superstring Theories", Princeton Preprint (1985).

32) J. Breit, B. Ovrut and G. Segrè - Univ. of Pennsylvania Preprint (1985);
P. Binetruy and M.K. Gaillard - LBL Preprint (1985).

33) E. Cohen, J. Ellis, K. Enqvist and D.V. Nanopoulos - CERN Preprint TH. 4195 (1985).

34) M. Dine and N. Seiberg - "Is the Superstring Weakly Coupled?", "Couplings and Scales in...", Princeton Preprints (1985);
V. Kaplunovski - "Mass Scales of the String Unification", Princeton Preprint (1985).

35) J. Breit, B. Ovrut and G. Segrè - "E_6 Symmetry Breaking in the Superstring Theory", Univ. of Pennsylvania Preprint (1985).

36) Y. Hosotani - Phys.Lett. 126B (1983) 309; 129B (1983) 193.

37) S. Cecotti, J.P. Derendinger, S. Ferrara, L. Girardello and M. Roncadelli - Phys.Lett. 156B (1985) 318.

38) M. Dine, V. Kaplunovsky, M. Mangano, C. Nappi and N. Seiberg - "Superstring Model Building", Princeton Preprint (1985).

39) Y. Achiman and B. Stech - Phys.Lett. 77B (1978) 389;
P. Ramond - Sanibel Symposium (1979); Caltech Report CALT-68-709 (1979);
R. Barbieri and D.V. Nanopoulos - Phys.Lett. 91B (1980) 369.

40) E. Witten - Phys.Lett. 91B (1980) 81.

41) L.E. Ibañez - Phys.Lett. 117B (1982) 403.

42) L.E. Ibañez, J. Mas and H.P. Nilles - in preparation.

43) H.P. Nilles, M. Srednicki and D. Wyler - Phys.Lett. 124B (1983) 337.

44) L.E. Ibañez - Phys.Lett. 114B (1982) 243; Phys.Lett. 126B (1983) 196.

45) S.M. Barr - Washinton Preprint 40048-20 P5 (1985).

46) M. Drees, N. Falck and M. Glück - Dortmund Preprint DO-TH 85/25 (1985).

47) This was apparently noticed also by the Dortmund group. E. Reya - private communication to H.P. Nilles.

48) E. Cohen, J. Ellis, K. Enqvist and D.V. Nanopoulos - CERN Preprint TH. 4222 (1985).

CALCULATING CONDENSATES IN SUPERSYMMETRIC GAUGE THEORIES

Daniele AMATI

Theoretical Physics Division
CERN
1211 Geneva 23, Switzerland

Supersymmetric theories are known to possess remarkable properties. Cancellations between bosonic and fermionic excitations generate mild ultra-violet behaviours so that some classical properties are not modified by quantum corrections (non-renormalization theorems). Supersymmetry allows us also to infer on vacuum properties: the index theorem provides necessary conditions for a spontaneous breaking of SUSY. More astonishingly, some vacuum expectation values (condensates) may be explicitly and exactly computed as was first noticed by the ITEP group (cf. bibliography) for the SUSY Yang-Mills theory.

In this talk I will explain why this happens and try to convey the flavour of how SUSY allows for non-perturbative computations that would appear hopeless in ordinary gauge theories.

I will limit the discussion to supersymmetric QCD, remaining very sketchy in defining notations, giving formulae and referençes. The theory will contain a gauge field supermultiplet

$$\left(\lambda_\alpha , F^{\mu\nu} \right)$$

in the adjoint representation of the SU(N) gauge group and M (flavour) chiral matter supermultiplets

$$\left(\varphi_i, \psi_{\alpha,i}\right), \quad \left(\hat{\varphi}^i, \hat{\psi}_\alpha^i\right) \qquad i = 1, \dots, M$$

in the N and $\bar{N}$ fundamental representation, respectively. Their conjugates

$$\left(\varphi^{*i}, \bar{\psi}_{\dot\alpha}^i\right), \quad \left(\hat{\varphi}^*_i, \bar{\hat{\psi}}_{\dot\alpha,i}\right)$$

are antichiral superfields in $\bar{N}$ and N, respectively. Masses appear explicitly in the action through terms containing m_i $(i = 1,\dots,M)$ and chiral matter superfields or m_i^* and antichiral superfields.

We will assume that the vacuum preserves both supersymmetry and gauge invariance. This will lead to consistent results which thus represent vacuum properties (condensates) of a supersymmetric confining theory. We shall then briefly discuss the possibility of a complementary Higgs picture.

The first crucial property that allows non-perturbative computations is that supersymmetry gives information on the dependence of some Green functions on space-time and on the masses. Indeed, let us define

$$G^{p,q}_{(x_1,\dots,x_{p+q})} = \langle \hat{\varphi}^{i_1}\varphi_{j_1}(x_1)\dots\hat{\varphi}^{i_p}\varphi_{j_p}(x_p)\,\lambda\lambda(x_{p+1})\dots\lambda\lambda(x_{p+q})\rangle \qquad (1)$$

where $\lambda\lambda(x)$ and $\tilde{\phi}^i\phi_j(x)$ are local operators with saturated spin and colour indices so they represent Lorentz scalar gauge invariants. They are also the lowest component of a chiral supermultiplet - which we will generally denote by $O_\ell(x_\ell)$ - due to the fact that they are products of lowest components.

Supersymmetry then implies that $G^{p,q}(x_1,\dots,x_{p+q})$ is

a) independent on all space-time variables x_ℓ, $\ell = 1,\dots,p+q$;
b) independent on all m_i^*;
c) dependent on m_i as

$$G^{p,q}_{\cdot} = \text{const.}\left(\prod_{k=1}^{M} m_k\right)^{\frac{p+q}{N}} \prod_{\ell=1}^{p}\left(m_{i_\ell}\, m_{j_\ell}\right)^{-1/2} \qquad (2)$$

It is easy to understand the reason for these properties. Let us call χ_ℓ the second component of the chiral superfield whose lowest component is O_ℓ. Then

$$\{\bar{Q}^{\dot\alpha}, O_\ell\} = 0 \;, \quad \{\bar{Q}^{\dot\alpha}, \chi_\ell^{\alpha}\} = \sqrt{2}\,(\bar\sigma_\mu)^{\dot\alpha\alpha}\frac{\partial O_\ell}{\partial x_\mu}$$

the generators $\bar{Q}$ acting as the lowering operator in a chiral superfield. Therefore

$$(\bar\sigma_\mu)^{\dot\alpha\alpha}\frac{\partial}{\partial x_{\ell\mu}} G^{p,q}(x_1,\dots x_{p+q}) = \frac{1}{\sqrt{2}}\langle O_1(x_1)\cdots\{\bar{Q}^{\dot\alpha}\chi_\ell^{\alpha}(x_\ell)\}\cdots O_{p+q}(x_{p+q})\rangle = 0$$
$$\ell = 1,2,\dots,p+q$$

because $\bar{Q}$ anticommuting with all the O_j may be brought to the beginning or to the end of the operator chain where it annihilates the vacuum. This proves property a). Property b) may be shown in an analogous way. $\partial G/\partial m_i^*$ implies an insertion of an m_i^* Lagrangian term which, as said before, is the last component of an antichiral superfield thus given by an anticommutator of $\bar{Q}$ with the second component of that superfield ($\bar{Q}$ is a raising operator for antichiral supermultiplets). Again $\bar{Q}$ will overcome all O_ℓ on the left or the right of the insertion, thus annihilating the vacuum and producing property b). The m_i mass insertion being the last component of a chiral supermultiplet, will not lead to a vanishing result; its action turns out to be proportional to the sum of the charges of all operators O_ℓ in (1) under a non-anomalous axial global U(1) group which is broken by the mass m_i. By explicitly computing these charges, one obtains the property c).

Another important consequence of supersymmetry is the anomalous commutator

$$\left\{ \bar{Q}^{\dot{\alpha}}, \bar{\Psi}^{i}_{\dot{\alpha}} \varphi_i \right\} = m_i \, \hat{\varphi}^i \varphi_i + \frac{1}{32\pi^2} \lambda\lambda \tag{3}$$

where the λλ term in Eq. (3) is provided by quantum effects which, as for the Adler-Bardeen anqmaly, may be related to the need of a regularization procedure.

Properties a) to c) show already why SUSY gauge theories allow for exact computations unprecedented in conventional field theories. Indeed, if a Green function is known not to depend on a variable, one may compute it for limiting values where an approximation procedure becomes exact.

The constant value (x_i independent) of the Green functions (1) may, for instance, be computed in the short-distance limit $|x_i - x_j| \ll 1/\Lambda$ where Λ is the renormalization invariant scale of the theory. If in that region we may justify a strictly supersymmetric approximation implying corrections which are higher powers in an expansion in $|x_i - x_j|\Lambda$, then it must happen that:

i) within that approximation the Green functions of (1) should turn out to be constant; this is a check that the approximation preserves supersymmetry;

ii) the corrections to that approximation are strictly zero.

If, for instance, the quantum numbers of the operators in the right-hand side of (1) are such that unit topological charge configuration may contribute, then the single instanton approximation - as long as it preserves SUSY - will give the exact result. The usual arguments showing that the instanton dominates short distances imply that other configurations (such as instanton-anti-instanton excitations) would give corrections depressed by powers of $((x_i - x_j)\Lambda^2)^{\beta_1}$ where β_1 is the first coefficient of the β function which is positive if the theory is asymptotically free. On the other hand, for $N > M$, the Green functions

$$\mathcal{G}^{M,N-M}(x_1 \ldots x_N) = \langle \hat{\varphi}^{i_1} \varphi_{j_1}(x_1) \ldots \hat{\varphi}^{i_M} \varphi_{j_M}(x_M)\, \lambda\lambda(x_{M+1}) \ldots \lambda\lambda(x_N) \rangle \tag{4}$$

with $i_1, \ldots, i_M$ $(j_1, \ldots, j_M)$ all different so as to exhaust the M flavour indices, are not only x_i independent, due to a), but also m_i independent due to b) and c). Therefore if we are able to compute the instanton contribution in a large mass limit

$$m_\ell \gg \frac{1}{|x_i - x_j|} \gg \Lambda \qquad \text{all } \ell, \text{ all } i,j \tag{5}$$

we will find the exact constant value of the Green function of Eq. (4) for every configuration of the M different flavour indices i_ℓ and the M different j_ℓ.

An approximation of this kind is available. The fact that it leads to results which are actually computable (without too much pain!) is again a typical consequence of supersymmetric cancellations. As for the integration over gauge superfield fluctuations, the bosonic modes cancel the fermionic non-zero modes. The collective instanton co-ordinates (position and width of the instanton) turn out to be the SUSY partners of the fermionic gauge zero modes and their contributions are explicitly calculable. Likewise boson and fermion (large) mass degeneracy plays a basic rôle.

The final result - after performing the integrations over the collective co-ordinates - is finite and explicitly computable for any N and $M < N$ and, as expected, is independent on x_ℓ and m_i. In particular for $N = 2$ we find

$$M=1 \quad \langle \tilde{\varphi}^1 \varphi_1(x_1) \lambda\lambda(x_2) \rangle = \frac{2^8}{5} g^{-4} \pi^2 \Lambda^5$$

$$M=2 \begin{cases} \langle \tilde{\varphi}^1 \varphi_1(x_1)\, \tilde{\varphi}^2 \varphi_2(x_2) \rangle = \frac{16}{5} g^{-4} \Lambda^2 \\ \langle \tilde{\varphi}^1 \varphi_2(x_1)\, \tilde{\varphi}^2 \varphi_1(x_2) \rangle = 0 \end{cases} \tag{6}$$

This diagonal flavour structure is true for all M and N > M, i.e., the only non-vanishing flavour configuration is

$$G^{M,N-M} = \langle \tilde{\varphi}^1 \varphi_1(x_1)\, \tilde{\varphi}^2 \varphi_2(x_2) \cdots \tilde{\varphi}^M \varphi_M(x_M) | \lambda(x_{M+1}) \cdots \lambda\lambda(x_N) \rangle \tag{7}$$

$G^{M,N-M}$ being a constant provides thus an evaluation of the Green function also for large distances, i.e., for

$$|x_i - x_j| \gg \frac{1}{m_\ell}, \frac{1}{\Lambda} \tag{8}$$

Clustering therefore implies that the constant we obtained is an evaluation of

$$G^{M,N-M} = \left(\langle \lambda\lambda \rangle\right)^{N-M} \prod_{i=1}^{M} \langle \tilde{\varphi}^i \varphi_i \rangle$$

$$\langle \tilde{\varphi}^i \varphi_j \rangle = 0 \quad \text{for } i \neq j \tag{9}$$

Therefore for every M and M < N we have obtained a specific combination of the condensates $\langle\lambda\lambda\rangle$ and $\langle\tilde{\phi}^i \phi_i\rangle$. In order to evaluate them separately it is now sufficient to use Eq. (3) whose vacuum expectation value implies

$$\frac{1}{16\pi^2} \langle \lambda\lambda \rangle = m_i \langle \tilde{\varphi}^i \varphi_i \rangle \tag{10}$$

For the N = 2 case explicitly given in (6) we find, for instance,

$$\langle\lambda\lambda\rangle_{\substack{N=2\\M=1}} = g^{-2}64\pi^2\sqrt{\frac{m\Lambda^5}{5}} \qquad \langle\hat{\varphi}\varphi\rangle_{\substack{N=2\\M=1}} = g^{-2}4\sqrt{\frac{\Lambda^5}{5m}}$$

$$\langle\lambda\lambda\rangle_{N=M=2} = 2^6\pi^2 g^{-2}\Lambda^2\sqrt{\frac{m_1 m_2}{5}} \quad , \quad \langle\hat{\varphi}^1\varphi_2\rangle_{N=M=2} = 0$$

$$\langle\hat{\varphi}^1\varphi_1\rangle_{N=M=2} = 4g^{-2}\Lambda^2\sqrt{\frac{m_2}{5m_1}} \quad , \quad \langle\hat{\varphi}^2\varphi_2\rangle_{N=M=2} = 4g^{-2}\Lambda^2\sqrt{\frac{m_1}{5m_2}}$$

(11)

The dependence of the values obtained for Green functions or condensates on m_i, g and Λ are very general and, as we shall now discuss, have a direct physical meaning. Indeed, for all colours and flavours, these dependences are predictable so that the actual instanton computation explained above is needed only to obtain an overall purely numerical coefficient for every value of N and M. It is sufficient to illustrate the origin of the m_i, Λ and g factors in a specific Green function. Let us do it by analyzing the result

$$\langle\lambda\lambda(x_1)\ldots\lambda\lambda(x_N)\rangle_{N,M} = \text{number}\cdot g^{-2N}\Lambda^{3N-M}\prod_{i=1}^{M} m_i \tag{12}$$

The $\prod_{i=1}^{M} m_i$ originates from the determinant over the M fermion matter fields. The 't Hooft determinant of the quantum fluctuations of the gauge field around the instanton gives rise to a factor μ^{β_1} where μ is the subtraction point mass scale. The instanton action provides a factor $e^{-8\pi^2/g^2}$ which, when written together with μ^{β_1}, gives

$$\left[\mu\exp-\frac{8\pi^2}{\beta_1 g^2}\right]^{\beta_1} \equiv \Lambda^{\beta_1} \tag{13}$$

If we recall that $\beta_1 = 3N-M$, we recognize that the factor $\Lambda^{\beta_1} \prod_{i=1}^{M} m_i$ we have already obtained has dimension 3N which is the dimension of the left-hand side of Eq. (12). Therefore, all the remaining factors should depend only on g.

It is easy to realize that the power of g in (12) is dictated by renormalization group invariance. Indeed, calling

$$m_{inv} = m \exp - \int^{g} \frac{\gamma(g)}{\beta(g)} dg \qquad \Lambda_{inv} = \mu \exp - \int^{g} \frac{dg}{\beta(g)} \tag{14}$$

with γ the anomalous dimension of m, it is easy to check that

$$g^{-2N} \left(\prod_{i=1}^{M} m_i \right) \mu^{3N-M} e^{-\frac{8\pi^2}{g^2}} = \Lambda_{inv}^{3N-M} \prod_{i=1}^{M} m_{i\,inv.} \tag{15}$$

provided

$$\beta(g) = - \frac{g^3}{16\pi^2} \frac{3N-M+M\gamma(g)}{1 - Ng^2/8\pi^2} \tag{16}$$

Equation (16) represents a relation between β and γ, in agreement with the first few known coefficients, i.e.

$$\beta(g) = - \frac{\beta_1 g^3}{16\pi^2} + \frac{\beta_2 g^5}{(16\pi^2)^2} + O(g^7)$$

$$\beta_1 = 3N-M \quad , \quad \beta_2 = -6N^2 + 4MN - 2M/N$$

$$\gamma(g) = - \frac{1}{8\pi^2} \frac{N^2-1}{N} g^2 + O(g^4) \tag{17}$$

Let me remark that the functional dependence on m_i and Λ of Eq. (12) is just the one needed in order to satisfy the decoupling theorem. The Green function implies only gaugino fields and therefore the matter fields appear only as quantum fluctuations. The decoupling would imply that if the mass of a matter field, let us say m_M, becomes much larger than Λ

$$m_M \gg \Lambda > m_i \qquad i = 1, \ldots, M-1$$

then the Mth flavour should cease to be excited and the Green function of Eq. (12) should become the one calculated for the M-1 flavour case. Or

$$\langle\lambda\lambda\rangle_{N,M} = \Lambda^{3-M/N} \prod_{i=1}^{M} m_i^{1/N} \xrightarrow[m_M \gg \Lambda]{} \langle\lambda\lambda\rangle_{N,M-1} = \tilde{\Lambda}^{3-M/N} \prod_{i=1}^{M-1} m_i^{1/N} \tag{18}$$

where $\tilde{\Lambda}$ is the scale of the M-1 theory. This scale is determined from the condition that the two theories should coincide at the (decoupling) transition among them. This means that if g are the running coupling constants,

$$g_M(q) \sim g_{M-1}(q) \quad \text{for} \quad q \sim m_M \tag{19}$$

implying

$$(3N-M) \log \frac{m_M}{\Lambda} = (3N-M+1) \log \frac{m_M}{\tilde{\Lambda}}$$

thus

$$\tilde{\Lambda} = \Lambda \left(\frac{m_M}{\Lambda}\right)^{\frac{1}{3N-M+1}} \quad \therefore \quad \Lambda < \tilde{\Lambda} < m_M \tag{20}$$

It is easy to see that Eq. (20) implies indeed the correct decoupling of Eq. (18).

It is possible to get an independent check of the reliability of the calculations done by computing in a completely different way another Green function besides the m independent $G^{M,N-M}$ of Eq. (4). This is equivalent to the derivation of the anomaly relation from a non-perturbative dynamical computation. Indeed, it is possible to directly compute the Green function of Eq. (12) by considering its derivative with respect to all masses m_i. This will imply the insertion of mass operators which are bilinear in either boson or fermion matter fields. Here it is possible to show that a miracle similar to that of SUSY Yang-Mills theory occurs. This is that matter bosonic modes and fermionic non-zero modes cancel exactly so that the only matter fields entering the calculation are the M fermionic zero modes that may be exactly computed and give, of course, an m_i independent result. A trivial m_i integration leads again to the expression (12) where the numerical coefficient, when computed (as for the N = 2, M = 1,2 case discussed above), coincides exactly with those obtained through the use of the anomaly condition (10).

That this coincidence, and thus the evaluation of the anomaly, is far from trivial may be recognized by the fact that the matter zero modes which provide the evaluation of the Green function of Eq. (12) do not give a correct evaluation of the mass independent Green function of Eq. (4). Naively, one could had hoped that a mass independent result could be obtained by a formal development in m_i whose zero-order term is described by zero modes. But the hope is jeopardized by the fact that the coefficients of a formal mass expansion are plagued with mass singularities. We thus find the curious fact that the small mass limit of a mass independent Green function differs - numerically and in the flavour structure - from its value for the strictly massless theory!

This is well illustrated by the N = M = 2 case where

	$m_i \neq 0$	$m_i \equiv 0$
$\langle\tilde{\phi}^1\phi_1(x)\tilde{\phi}^2\phi_2(y)\rangle$	C	5/12 C
$\langle\tilde{\phi}^1\phi_2(x)\tilde{\phi}^2\phi_1(y)\rangle$	0	-5/12 C
$\langle\det\tilde{\phi}(x).\det\phi(y)\rangle$	0	5/6 C

$C = 16/5\ g^{-4}\Lambda^4$

$\det\phi$ being the determinant of ϕ in colour and flavour indices. This extends to all N = M where for $m_i \to 0$ squark condensates are finite and differ from those of the massless theory. Gluino condensates vanish for both cases as seen from property c) or from the anomaly relation (10). These same relations show that for $M < N$ when $m_i \to 0$, $\langle \tilde{\phi}^i \phi_i \rangle \to \infty$ thus revealing runaway vacua.

The massless case is peculiar in another sense: it possesses a flat potential in terms of the classical value for the corresponding field ϕ. Therefore it could suggest a $\langle \phi \rangle = 0$ Higgs picture which we have excluded in the preceding approach by requiring a gauge invariant vacuum. An argument sometimes advanced in advocating this alternative is that a very large (actually infinite) value for $\langle \tilde{\phi}\phi \rangle$ is hard to justify in terms of pure quantum fluctuation around $\langle \phi \rangle = 0$ as happens in a confining phase for $m \to 0$. For $M < N$ both methods imply ill-defined vacua (infinite vacuum expectation value). For $M = N$, the confining picture gives definite values of $\langle \tilde{\phi}\phi \rangle$ while the Higgs one allows SUSY vacua with arbitrary $\langle \phi \rangle$ and $\langle \tilde{\phi} \rangle$ thus with arbitrary $\langle \tilde{\phi}\phi \rangle$. It could be that the theory allows many inequivalent vacua. This is perhaps suggested by an effective Lagrangian analysis which, for the massless $M = N$ case, suggests a sort of complementarity between the Higgs and the confining pictures. It has even been proposed that there may be a complementarity between the confining and the Higgs picture for the massive case, even if in this case there are no flat directions in the superpotential, so that the only classical minimum is at $\phi = \tilde{\phi} = 0$.

The techniques used to compute condensates in SQCD have been applied to evaluate behaviours of other Green functions as well as for other supersymmetric matter fields. In some cases, evidence has been found for a spontaneous breaking of supersymmetry.

It is remarkable how the constraints and cancellations implied by supersymmetry permit computations that would be unthinkable in conventional theory thus allowing theoretical predictions that extend far beyond the perturbative framework.

BIBLIOGRAPHY

Super Yang-Mills

V.A. Novikov, M.A. Shifman, A.I. Vainshtein, M.B. Voloshin and V.I. Zacharov - Nucl.Phys. B229 (1983) 394;

V.A. Novikov, M.A. Shifman, A.I. Vainshtein and V.I. Zakharov - Nucl.Phys. B229 (1983) 381; ibid. 407.

Super QCD

D. Amati, Y. Meurice, G.C. Rossi and G. Veneziano - CERN Preprint TH.4201 (1985); Nucl.Phys., in press;

D. Amati, G.C. Rossi and G. Veneziano - Nucl.Phys. B249 (1985) 1;

G.C. Rossi and G. Veneziano - Phys.Lett. 138B (1984) 195;

M.G. Schmidt - Phys.Lett. 141B (1984) 236.

Super SU(5)

Y. Meurice and G. Veneziano - Phys.Lett. 141B (1984) 69.

Anomaly Relation

T.E. Clark, O. Piguet and K. Sibold - Nucl.Phys. 159B (1979) 1;

S.J. Gates, N. Grisaru, M. Rocek and W. Siegel - Superspace (Benjamin-Cummings, Reading, MA, 1983), Chapter 6;

K. Konishi - Phys.Lett. 135B (1984) 439.

Effective Lagrangians

A.C. Davis, M. Dine and N. Seiberg - Phys.Lett. 125B (1983) 487;

J.-M. Gérard and H.P. Nilles - Phys.Lett. 129 (1983) 243;

A. Masiero, R. Pettorino, M. Roncadelli and G. Veneziano - CERN Preprint TH. 4166 (1985);

H.P. Nilles - Phys.Lett. 129B (1983) 103;

M. Peskin - SLAC-PUB 3061 (1983);

T.R. Taylor, G. Veneziano and S. Yankielowicz - Nucl.Phys. B218 (1983) 493;

G. Veneziano - Phys.Lett. 124B (1983) 357;

G. Veneziano and S. Yankielowicz - Phys.Lett. 113B (1982) 321.

Higgs Picture

I.A. Affleck, M. Dine and N. Seiberg - Phys.Rev.Lett. 51 (1983) 1026; Nucl.Phys. B241 (1984) 493; Phys.Rev.Lett. 52 (1984) 1677;

J. Fuchs and M.G. Schmidt - Preprint HD-THEP-85-10 (1985);

V.A. Novikov, M.A. Shifman, A.I. Vainshtein and V.I. Zakharov - ITEP Preprint 31 (1985).

SUPERCOMPOSITENESS

G. Veneziano
CERN, Geneva, Switzerland

1. - OUTLINE

The outline of this seminar will be as follows. I shall start by motivating the study of composite models and, more specifically, that of supersymmetric composite models (supercompositeness) within our present theoretical ideas. Next, I shall recall some properties of supersymmetric (SUSY) gauge theories (in particular of SQCD) that make them a priori good candidates for preon dynamics, referring to D. Amati's[1] talk for much of the details. Finally, I shall describe a one-family model which comes close to being realistic. It will be instructive to describe the model at three successive levels of approximation: i) the massless, supersymmetric limit, ii) the R-symmetric limit with SUSY breaking; iii) the full model, including weak gauging and R-symmetry breaking. We shall see that, despite the many good features it possesses, the model is ruled out its prediction of some light unobserved new fermions [of mass O(1-10 GeV)]. We shall finally discuss some possible remedies, which however will make the model unappealing, and draw some conclusions.

2. - WHY (SUPER)COMPOSITENESS

Several things point in the direction of a possible composite structure for quarks and leptons[2].

i) Previous history: over the centuries we have encountered several successive layers in physics: molecules, atoms, nuclei, nucleons, quarks. Is this possibly the end of it?

ii) The generation puzzle (why ≥ 3 families?) and the large number of free parameters in the standard model call for some explanation: within present ideas these can be based on some horizontal symmetry and/or on compositeness.

iii) The quantum numbers of quarks and leptons within a family are strictly connected. Possible explanations of this fact call for grand unification ideas (a simple group under which quarks and leptons form an irreducible representation) or for some common constituents to quarks and leptons (compositeness), or both.

iv) On the more theoretical side, we have the fine-tuning problem for the elementary Higgs particles of the standard model. Resolving it calls for either composite Higgses or for supersymmetry or both, in which case composite fermions are also there.

A question comes then naturally to mind: if we take the grand unification road, is that mutually exclusive with the compositeness road? The answer is quite obviously no. Again (theoretical physics) history has shown us the happy coexistence in grand unified schemes of perturbative gauge interactions $[SU(2)_L \times U(1)_Y]$ and of strong, binding forces $[SU(3)_c]$. Since it is more popular today to replace GUTs by superstrings, let us ask directly the question:

Q: if superstrings are the theory of everything do we need quark-lepton compositeness?

The answer is not easy, but my guess is that, unless we are very lucky, we do still need new composite models. The argument runs as follows: superstrings give <u>elementary</u> (though stringy) gauge bosons, gravitinos, gravitons and this is naturally so. However, if there would be no composite structure in quarks and leptons, strings would also contain the individual number of each family in their light spectrum. That would amount to say that the phenomenon of compositeness that occurs repeatedly over six orders of magnitude of length scales (10^{-7} to 10^{-13} cm) is limited to that range and occurs no more in the next 20 orders (up to 10^{-33} cm), i.e., up to Planck's length!

Actually, one could argue that there is something intrinsic which distinguishes the successive layers: this is the dimensionless number r.m (size times mass). This number runs from something larger than 10^5 for molecules and atoms to $A^{4/3}$ for nuclei of mass number A to O(1) for the hadrons. Quarks and leptons would be the first bound systems for which $r.m \ll 1$, say, $< O(10^{-6})$. Could this be a "proof" that quarks and leptons <u>must</u> be elementary? The naive argument for a positive answer is that for a bound system: $r.m > r.\Delta p \gtrsim r.1/r = 1$. However in the last few decades we learned, through counter-examples, that this naive (and essentially non-relativistic) argument does not need to be valid. There are (at least) two known exceptions:

1) for bosons: $r.m = O(\ll 1)$ if it is a (pseudo) Nambu-Goldstone boson;
2) for fermions: $r.m. = O(\ll 1)$ when a (slightly broken) chiral symmetry prevents a fermionic mass term.

Both cases 1) and 2) come out naturally within the context of non-Abelian gauge theories and amusing interrelations between 1) and 2) further occur if such gauge theories are supersymmetric as we shall see.

Let us think again about superstrings. These lead to (almost) unique and very large gauge groups G like SO(32) and $E_8 \times E_8{}'$, which get broken down in the compactification process, leaving still large unbroken gauge symmetries, and presumably, N = 1 supersymmetry. The latter will break down [perhaps by E'_8 gluino condensation[3]] at some scale M_{SSB}. Now, if there are unbroken subgroups of G whose confinement scale $\Lambda \gg M_{SSB}$, then one has to face, at scales $O(\Lambda)$, the strong coupling regime of a SUSY gauge theory with "small" soft breaking terms (i.e., small compared to Λ). This is the kind of theories we shall consider.

We shall argue that the spectrum of such theories can be very different from the one we are accustomed to in QCD (which represents the opposite limit, $\Lambda_{QCD} \ll M_{SSB}$) and that it can contain light composite fermions bearing resemblance to quarks and leptons. We name the (almost) SUSY gauge interaction supercolour and the game supercompositeness, arguing that it avoids some of the difficulties of the standard (i.e., non-SUSY) compositeness game. It is also much more predictive (thanks to SUSY) and, unfortunately, this will make the specific model we shall consider unrealistic.

3. - SQCD WITH $N_c = N_f = N$

I shall recall here some peculiar properties of supersymmetry QCD (SQCD) in the case of an identical number of colours and flavours ($N_c = N_f = N$), since those properties are crucially employed in the composite model I shall present afterwards. I refer to Amati's talk[1] and to the literature for details. The gauge group is $SU(N_c)_{SC}$ with scale Λ_{SC} (SC for supercolour).

In the massless, SUSY limit SQCD with $N_c = N_f = N$ has a bosonic global symmetry:

$$G_F = SU(N) \times SU(N) \times U(1)_V \times U(1)_X \tag{1}$$

where $SU(N) \times SU(N) \times U(1)_V$ is the usual QCD global symmetry [after removing the anomalous $U(1)_A$ classical symmetry] and $U(1)_X$, a non-anomalous R-symmetry, acts as follows:

$$U(1)_X : \lambda \to e^{-i\beta}\lambda \; ; \; (\varphi,\tilde{\varphi}) \to (\varphi,\tilde{\varphi}) \; ; \; (\psi,\tilde{\psi}) \to e^{i\beta}(\psi,\tilde{\psi}) \qquad (2)$$

We denote by $W = (\lambda, F_{\mu\nu})$ the gauge superfield (containing gaugino and gauge fields) belonging to the adjoint of SU(N), and by $\Phi = (\phi_i, \psi_i)$, $\tilde{\Phi} = (\tilde{\phi}_j, \tilde{\psi}_j)$ the chiral matter superfields ($i,j = 1,\dots,N_f$, colour index suppressed) belonging to the N and $\bar{N}$ representations, respectively. The crucial $U(1)_X$ symmetry can be broken in two ways:

i) Explicitly, e.g., by a supersymmetric mass terms, or by a (SUSY breaking) gluino mass. It is not broken by a purely bosonic mass term $|\phi^2|$, $\phi\tilde{\phi}$.

ii) Spontaneously, e.g., by condensates such as $\langle\lambda\lambda\rangle$ as $\langle\psi\bar{\psi}\rangle$. It is not broken by the scalar condensate $\langle\phi\tilde{\phi}\rangle$.

It can be easily shown that, in the massless limit, both $\langle\lambda\lambda\rangle$ and $\langle\psi\tilde{\psi}\rangle$ also break SUSY. It thus follows that SUSY protects the axial $U(1)_X$ symmetry and thus protects the mass of any fermion which has $U(1)_X$ charge.

On the other hand, $\langle\phi\tilde{\phi}\rangle$ is allowed by SUSY and can be shown to occur[1]. Clearly, if

$$\langle \varphi_i \tilde{\varphi}_j \rangle \equiv \langle \varphi_{\alpha,i} \tilde{\varphi}^{\alpha}_j \rangle = v_{ij} \qquad (\alpha = 1 \dots N_c)$$

G_F is broken to a subgroup $H = H' \otimes U(1)_X$, the vacuum stability group. It is known that this results in Nambu-Goldstone bosons living in G/H. Here, however, because of SUSY, NG-bosons drag along fermions and more bosons. Thus, in SUSY theories, some fermionic masses are protected by the NG phenomenon[4].

Furthermore, since the NG-bosons carry no $U(1)_X$ charge (like $\phi\tilde{\phi}$), their fermionic left-handed partners have one unit of $U(1)_X$ charge (R-symmetry!) Thus some fermionic masses are also protected by a chiral symmetry [double protection mechanism[5]]. The two questions, on which the chances of success of supercompositeness depend, are now:

1) Can we add SUSY breaking terms so that (most) bosons pick up a large mass while (most) fermions remain massless?

2) If yes, can we identify the massless fermions with quarks and leptons?

We shall see how to answer these questions with a specific attempt.

4. - A SPECIFIC ONE-FAMILY MODEL

i) The SUSY massless limit

For the moment we leave N arbitrary (N > 2) which could be useful for describing families. The non-perturbative properties of SQCD are neatly summarized in the effective Lagrangian[6)]

$$L_{eff} = L_{kin} + L_{an.}$$

$$L_{kin} = G_F\text{-invariant } D\text{-terms}$$

$$L_{an} = \int d^2\theta \; S \log\det T/\Lambda^2 + h.c.$$

$$S \equiv g^2/32\pi^2 \cdot W^2 \quad ; \quad T_{ij} \equiv \Phi_{\alpha,i}\, \tilde{\Phi}^{\alpha}_{j}$$

Minimizing the potential, one finds that the vacuum manifold (which respects SUSY) has a huge degeneracy corresponding not just to G_F/H but to its complexification, as expected on general bases[7)]. More specifically one finds[1)]:

$$\langle S \rangle = \langle \lambda\lambda \rangle = 0 \quad , \quad \langle T_{ij} \rangle \neq 0 \; ; \langle \det T_{ij} \rangle = c\Lambda^{2N}$$

with c a calculable constant. Except for S and det T all the superfields remain strictly massless at this stage.

ii) Breaking SUSY while preserving $U(1)_X$

I shall now argue that the answer to the first question at the end of Section 3 is positive. In order to see that, we add to our fundamental Lagrangian purely bosonic mass terms of the type:

$$L_{SSB} = -\sum_i \mu_i^2 |\phi_i|^2 - \tilde{\mu}_i^2 |\tilde{\phi}_i|^2$$

We can mimic such terms in L_{eff} by[8)]

$$L^{eff}_{SSB} = - \sum_{ij} \bar{\mu}_{ij}^2 \, T_{ij} T_{ij}^* \Big|_{\theta=\bar{\theta}=0}$$

These terms break G_F [without breaking $U(1)_X$] and, as a result, remove the huge vacuum degeneracy of the massless theory. The vacuum aligns with the perturbation [the same happens if one uses a SUSY mass term[1)]] and, by minimizing the new potential, one finds:

$$\langle T_{ij} \rangle = c\,\delta_{ij}\,\bar{\mu}_{ii}^{-1} \; ; \quad \prod_1^{\tilde{N}} \langle T_{ii} \rangle = \Lambda^{2N}$$

thus

$$c = \Lambda^2 \prod_{i=1}^{\tilde{N}} (\bar{\mu}_{ii})^{1/N}$$

Unlike in the case of a SUSY mass term, however, we still find

$$\langle \lambda\lambda \rangle = \langle \psi\tilde{\psi} \rangle = 0$$

The outcome is that most bosons get mass $O(\bar{\mu})$ (N bosons become pseudo NG bosons), while all fermions remain massless at this level (and satisfy 't Hooft's anomaly matching conditions)[9)]

Before proceeding, let us make the model more specific[10)]. For one family we take $N = 6$. We represent the bosons and the fermions in T_{ij} as 6×6 matrices:

$$T_{ij}\Big|_{\theta=0} = \begin{array}{c} \begin{array}{ccc} 1,2,3 & 4 & 5,6 \end{array} \\ \left(\begin{array}{c|c|c} \text{octect} & lq & sq. \\ \hline \text{leptoquarks} & & se. \\ \hline \text{squarks} & \text{sleptons} & \text{Higgs} \end{array}\right) \end{array} ; \quad T_{ij}^{\theta} = \left(\begin{array}{c|c|c} \text{octect} & lq & \text{quarks} \\ \hline \text{leptoquarks} & & \text{leptons} \\ \hline \text{quarks} & \text{leptons} & \text{sHiggs} \end{array}\right)$$

We have written here some physical assignments to the various entries anticipating the properties of our bound states under the low energy group, that we are now going to discuss.

iii) Weak gauging and R-symmetry breaking

We are now considering the effects due to the "weak" gauge interactions which we define to be those of the standard $SU(3)_C \times SU(2)_L \times U(1)_Y$ model (colour is in the weak coupling regime at Λ_{SC}). At the same time, we shall break the R-symmetry, e.g., through an $SU(3)_C$-gluino mass term $m_{\tilde{g}}$. A better way to describe the weak-gauging is to start from the beginning from an $SU(6)_{SC} \times SU(3)_C \times SU(2)_L \times U(1)_Y$ gauge symmetry and give the representation of all the basic superfields under this group. This is done in Table 1, which also indicates the transformation properties of the composite fields, justifying the names given to them previously. Of course, our assignments avoid all gauge anomalies in the underlying theory. Also, the SUSY breaking masses have to respect all the gauge symmetries, but this still leaves considerable freedom in the choice of the $\bar{\mu}$'s so that a rather asymmetric vacuum can be achieved.

One important remark[10] is necessary at this point: if we break the R-symmetry by $m_{\tilde{g}}$, radiative corrections induce a mass m_λ (λ is the SC-gaugino) of order of $\alpha_c m_{\tilde{g}}$ as seen from the diagram

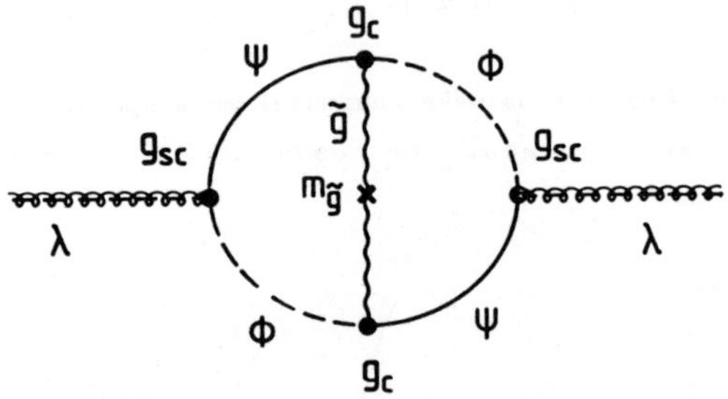

In order not to commit the fine-tuning crime, we shall add a term $m_\lambda \bar{\lambda}\lambda \sim m_\lambda S_{\theta=0}$ to our effective Lagrangian [with $m_\lambda = 0\ (\alpha_c m_{\tilde{g}})$]. We can now look at the vacuum and the low-lying spectrum of our model.

L.H. S.field \ Transfor. under	$SU(6)_{SC}$	$SU(3)_C$	$SU(2)_L$	$U(1)_Y$	$Q = T_3+Y/2$	
$\Phi_\alpha^{1,2,3}$	6	$\bar{3}$	1	-1/3	-1/6	
Φ_α^4	6	1	1	1	1/2	
$\Phi_\alpha^{5,6}$	6	1	2	0	±1/2	
$\tilde{\Phi}^\alpha_{1,2,3}$	$\bar{6}$	3	1	1/3	1/6	
$\tilde{\Phi}^\alpha_{4,5}$ 2 x	$\bar{6}$	1	1	-1	-1/2	
$\tilde{\Phi}^\alpha_6$	$\bar{6}$	1	1	1	1/2	
$\binom{u}{d} = \Phi_\alpha^{5,6}\tilde{\Phi}^\alpha_{1,2,3}$	1	3	2	1/3	2/3 -1/3	}
$\binom{\nu}{e} = \Phi_\alpha^{5,6}\tilde{\Phi}^\alpha_4$	1	1	2	-1	0 -1	}
$\bar{u} = \Phi_\alpha^{1,2,3}\tilde{\Phi}^\alpha_5$	1	$\bar{3}$	1	-4/3	-2/3	} 1 family (with ν_R)
$\bar{d} = \Phi_\alpha^{1\ 2\ 3}\tilde{\Phi}^\alpha_6$	1	$\bar{3}$	1	2/3	1/3	}
$\nu^c = \Phi_\alpha^4\ \tilde{\Phi}^\alpha_5$	1	1	1	0	0	}
$e^+ = \Phi_\alpha^4\ \tilde{\Phi}^\alpha_6$	1	1	1	2	+1	}
$\bar{\ell} q = \Phi_\alpha^4\ \Phi^\alpha_{1,2,3}$	1	3	1	4/3	2/3	B=1/3, L=-1
$\ell\bar{q} = \Phi_\alpha^{1,2,3}\tilde{\Phi}^\alpha_4$	1	$\bar{3}$	1	-4/3	-2/3	B=-1/3, L=1
Oct $= \Phi_\alpha^{1,2,3}\tilde{\Phi}^\alpha_{1,2,3}$	1	8	1	0	0	B = L = 0

<u>Table 1</u>

The $SU(6)_{SC}$, $SU(3)_C$, $SU(2)_L$, $U(1)_Y$ and $U(1)_{em}$ quantum numbers of the preonic superfields $(\Phi,\tilde{\Phi})$ and of the composite superfields $(\Phi\tilde{\Phi})$.

The vacuum is practically unchanged[10)] except that now

$$\langle \lambda\lambda \rangle \sim m_\lambda \cdot \min \langle T_{ii} \rangle^2 = O\left(\alpha_c m_{\tilde{g}} \Lambda^2 \cdot \bar{\mu}_L^2 / \bar{\mu}_H^2\right)$$

where $\bar{\mu}_L (\bar{\mu}_H)$ are the smallest (largest) $\bar{\mu}$ parameters. The vacuum is easily seen to respect $SU(3)_C$, as a consequence of the vector-like nature of our preons under $SU(3)_C$, and to break $SU(2)_L \times U(1)_Y$ to $U(1)_Q$, as a consequence of the chiral nature of our preons under the latter group. The Fermi scale gets set at[10)]

$$\frac{m_W}{g_2} \simeq 300\ \text{GeV} = \Lambda_{SC} \left(\frac{\bar{\mu}_{1,4}}{\bar{\mu}_{5,6}}\right)^{2/3}$$

Since $\Lambda_{SC} > 1$ TeV, we clearly need $\bar{\mu}_1$ [$= \bar{\mu}_2 = \bar{\mu}_3$ for $SU(3)_C$], $\bar{\mu}_4 \ll \bar{\mu}_{5,6}$. In other words, we have to choose rather asymmetric values for the $\bar{\mu}$'s in order to tilt the vacuum at an asymmetric point where

$$\langle T_{55} \rangle, \langle T_{66} \rangle \ll \langle T_{11} \rangle, \langle T_{44} \rangle$$

Let us now turn to the spectrum. Bosonic masses are not appreciably shifted (we take $m_\lambda \ll \bar{\mu}$) but now, finally, fermions get masses.

There are tree level masses, coming from the extra term $m_\lambda S_{\theta=0}$ and radiative masses coming from the diagram.

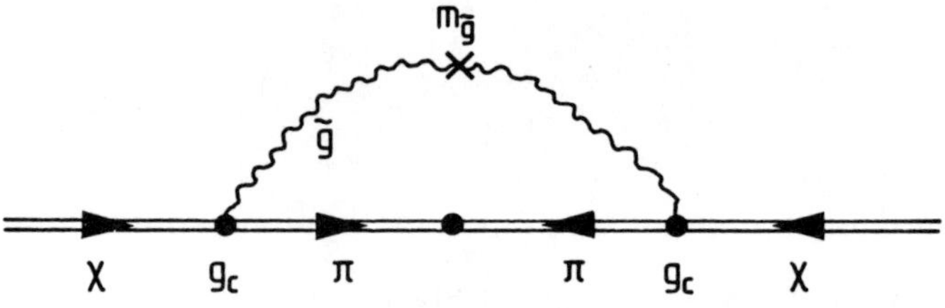

The latter gives

$$m_{u,d} = \alpha_c m_{\tilde{g}} \bar{\mu}_1 / \bar{\mu}_{5,6}$$

$$m_e = O(\alpha m_{\tilde{\gamma}}) \quad ; \quad m_\nu \sim 0$$

$$m_{\ell q} \sim \alpha_c m_{\tilde{g}} < m_{oct}$$

which are in a nice hierarchy. Unfortunately, adding to the radiative masses the tree level ones leads to a much more "democratic" mass matrix, i.e., to two conspicuous problems[10]:

a) $m_\nu \sim m_e$

b) $m_{\ell q} \sim \alpha_c m_{\tilde{g}} = 0$ (1 GeV)

For the neutrino a rather elegant solution can be found by adding one extra total singlet N in a way similar to the one employed in certain grand unified models. The physical neutrino becomes strictly massless and acquires a tiny component $(O(m_\ell/\Lambda_{SC}))$ of N. For the leptoquarks (ℓq), triplets of $SU(3)_C$ with L = 1, B = −1/3, Q = −2/3, the situation is more dramatic. We have only found an ugly solution involving spectators, whose only raison d'être, is that of ... eating up leptoquarks!

The full situation concerning the, alas unphysical, mass spectrum is summarized in Table 2.

CONCLUSIONS

i) In spite of several encouraging features (e.g., a confining theory with light composite fermions "predicted" rather than assumed) the model we have presented is ruled out (unless unnaturally modified) by lack of sufficient hierarchy in fermionic masses.

ii) There are hints that models with $N_c < N_f$ could have an easier life since VEVs tend to to be $\ll \Lambda_{SC}$ in this case. A possibility could be to add a family label ($N_f = 6 \to N_f = 10, 14$) without changing N_c. Or perhaps we should change the SC group and the preon numbers altogether (so to have a chiral model, for instance).

iii) Unfortunately guiding principles and quick ways of analyzing models are still badly insufficient.

Fermions	Level II Mass		Level III Mass	
Leptons	0		$\alpha_c\, m_{\tilde g}(\bar\mu_4/\bar\mu_{5,6})$	
Quarks	0		$\alpha_c\, m_{\tilde g}(\bar\mu/\bar\mu_{5,6})$	
Spin-½ leptoquarks	0		$\alpha_c\, m_{\tilde g}(\bar\mu\bar\mu_4/\bar\mu^2+\bar\mu_4^2)$	
Octet	0		$\alpha_c(C_A/C_F)m_{\tilde g}\log(\bar\mu/m_{\tilde g})$	
Higgsinos	0		eaten by $\tilde W$, $\tilde Z$	
χ_λ, χ_0	Λ_{SC}		Λ_{SC}	
χ_0', χ_0''	0		$\alpha_c m_{\tilde g}$, $\alpha_c m_{\tilde g}(\bar\mu/\bar\mu_{5,6})^2$	
X_F, $\tilde X_F$	0		$\alpha_c m_{\tilde g}\, f(\bar\mu_i/\bar\mu_j)$	
Gluino	-		$m_{\tilde g}$ (input)	
$\tilde W$, $\tilde Z$	-		$m_{W,Z} \pm \max(\alpha_c m_{\tilde g}, (\bar\mu_5^2-\bar\mu_6^2)/\bar\mu_{5,6}^2\, m_{W,Z})$	
Bosons	**Level II (mass)^2**		**Level III (mass)^2**	
Sleptons	$\bar\mu_{5,6}^2$		$\bar\mu_{5,6}^2$	
Squarks	$\bar\mu_{5,6}^2$		$\bar\mu_{5,6}^2$	
Scalar leptoquarks	$(\bar\mu\pm\bar\mu_4)^2$		$(\bar\mu\pm\bar\mu_4)^2$	
	scalar	**pseudoscalar**	**scalar**	**pseudoscalar**
Octet	$\bar\mu^2$	0	$\bar\mu^2$	$(\alpha_c m_{\tilde g}\log\bar\mu/m_{\tilde g})^2$
Higgses	$\bar\mu_{5,6}^2$	0	$\bar\mu_{5,6}^2+O(m_W^2)$	eaten by W,Z
π_λ, π_o	Λ_{SC}^2	Λ_{SC}^2	Λ_{SC}^2	Λ_{SC}^2
π_o', π_o''	$\bar\mu^2$, $\bar\mu_{5,6}^2$	0, 0	$\bar\mu^2$, $\bar\mu_{5,6}^2$	$\lesssim \Lambda_{QCD}^4/\Lambda_{SC}^2$; 0
$X + \tilde X$	$\bar\mu_i^2$	$\bar\mu_i^2$	$\bar\mu_i^2$	$\bar\mu_i^2$
$X - \tilde X$	$\bar\mu_i^2$	0	$\bar\mu_i^2$	0
W,Z.	-	-	$g_2^2\Lambda_{SC}^2\left(\dfrac{\bar\mu\bar\mu_4^{1/3}(\bar\mu_5^2+\bar\mu_6^2)}{(\bar\mu_5\bar\mu_6)^{5/3}}\right)$	$m_W/\cos\theta_W$

Table 2

Order of magnitude of the fermionic and bosonic masses at the levels II and III of the theory. See Sections 4 and 5 of the text for details.

iv) Maybe a completely new ingredient is missing in the game, e.g., we find no need for a repetitive family structure. Even then, the properties of strongly interacting SUSY gauge theories are likely to play an important rôle.

REFERENCES

1) D. Amati, these proceedings and references therein.

2) See, e.g., H. Harari, talk given at the St. Vincent meeting (1985); Weizmann Institute preprint (1985).

3) M. Dine, R. Rohm, N. Seiberg and E. Witten, Phys. Lett. 156B (1985) 55.

4) W. Buchmüller, R.D. Peccei and T. Yanagida, Phys. Lett. 124B (1983) 67.

5) R. Barbieri, A. Masiero and G. Veneziano, Phys. Lett. 128B (1983) 179.

6) T.R. Taylor, G. Veneziano and S. Yankielowicz, Nucl. Phys. B218 (1983) 493.

7) B.A. Ovrut and J. Wess, Phys. Rev. D24 (1982) 409; see also W. Lerche, Nucl. Phys. B238 (1984) 582.

8) A. Masiero and G. Veneziano, Nucl. Phys. B249 (1985) 593.

9) G. 't Hooft, Proc. of Cargèse Summer School (1979), eds. G. 't Hooft et al., (Plenum Press, New York and London, 1980).

10) A. Masiero, R. Pettorino, M. Roncadelli and G. Veneziano, Nucl. Phys. B261 (1985) 633.

11) H. Georgi and D.V. Nanopoulos, Nucl. Phys. B159 (1979) 166; see also D. Wyler and L. Wolfenstein, Nucl. Phys. B218 (1983) 205.

THE POSSIBLE STRONG INTERACTING SECTOR OF THE ELECTROWEAK THEORY*

R Gatto

Departement de Physique Theorique
Universite de Geneve
1211 Geneve 4, SWITZERLAND

*Partially supported by the Swiss National Science Foundation.

1. The present standard model of electroweak interactions is renormalizable and in principle contains no logical inconsistencies. There is also no evidence at this time for any experimental deviation from its predictions. The fact that the Higgs boson has not yet been seen is no evidence against the model. One simply did not expect to see the Higgs boson in all the experiments carried out so far, and one might only finally obtain some rough lower limit on its mass. For low mass Higgs, the Wilczek [1)] suggestion, of looking for quarkonium decay in $H+\gamma$, has attracted interest in view of the data presented by J. Lee Franzini [2)]. From the lowest order calculation, one expects for the ratio of $\Upsilon\to H\gamma$ to $\Upsilon\to\mu^+\mu^-$ someting of the order of three to two percent for sufficiently low Higgs mass, decreasing linearly with the Higgs-mass square to the end of phase space. Unfortunately the radiative QCD-corrections are large, and as such not quite reliably calculable. The existing calculations go unfortunately in the direction of substantially reducing the theoretical ratio [3)], such as to make comparison with present experiment essentially inconclusive.

2. In spite of the overall theoretical and experimental consistency of the standard model it is generally believed that its scalar sector has inherent pathologies. As a matter of fact, it is the effective use of the model that suggests either a complex dynamics or the convenience of modifications or extensions. A rough insight in the situation can be obtained by following the renormalization group evolution of a quartic self coupling λ

(as defined by a potential $V = \frac{1}{2}\lambda\varphi^4 + \ldots.$) which is given by $d\lambda/dt = (3/4\pi^2)\lambda^2$ with, say, $t = \ln(M/m_Z)$. In order to use the model as an effective theory the mass scale M cannot go beyond a limit such that $\lambda(M) \sim 1$. Within these limits one can roughly employ the semiclassical value $m_H = (250\text{ GeV})\lambda(m_H)^{\frac{1}{2}}$ for the Higgs mass. The cut-off mass M has then to satisfy in this sketchy model the very critical upper bound $M < m_H \exp\left[\frac{4\pi^2}{3}\left(\left(\frac{250}{m_H}\right)^2 - 1\right)\right]$. As we had said, there is in principle no need for a cut-off mass. Furthermore the real physical situation is more complex. Nevertheless, the upper bound, taken literally, allows M to extend to exceedingly high values, of order or larger than the Planck mass, only provided m_H stays below some limit of ~ 125 GeV. But already for $m_H \sim 240$ GeV the mass M comes down to values less than 1TeV, and, of course, one would have to cut-off at m_H itself if m_H were 250 GeV. When one goes back, however, from the effective utilization of the model to its deeper dynamical content, one encounters the central problem of the possible triviality [4] of φ^4 , and whether, and how, it extends to the Higgs sector of the standard model [5].

3. The suggestion of composite Higgs fields has been advanced in various forms and along different models (technicolor, composites, etc.). Another, not necessarily alternative, suggestion employs the non-renormalization theorem of supersymmetric theories. Such theorem provides for stabilization of the Higgs mass against radiative corrections. A common feature of these

suggestions is that they lead, at higher energies, usually in the 100 GeV - 1 TeV scale, to new physical phenomena. This is apparent in models where the scale of the spontaneous symmetry breaking of the electroweak theory is produced dynamically, as a condensate scale. At energies related to this scale some new interaction must become strong -- the exact relation depending on the particular model. For models using supersymmetry the reason is different. Since supersymmetry has anyway to be broken, on experimental grounds, one has to worry on how the exact compensations, between bosonic and fermionic loops, that keep stable the imput scalar mass in the fully supersymmetric limit, still remain approximately valid in the case of broken supersymmetry. In order not to heavily disturb such compensations, the masses of the bosons and of the related fermions in a supermultiplicity must not be too much split ones from the others. This condition on the fermion-boson mass splittings requires that new physics, that of the superpartners, be not so far away from the region of electroweak-breaking, so again below one TeV.

4. Leaving apart these proposals, the work by Veltman and by Lee, Quigg, and Thacker shows that partial-wave unitarity may be violated by the lowest order scattering graphs among longitudinal gauge-boson components and Higgs [6,7]. Such violation occurs at energies around 1 TeV. It concerns the perturbation theory amplitudes. Again, it does not imply internal inconsistency of the renormalizable electroweak theory. It is related to the fact that for large Higgs masses the longitudinal-scalar sector of the

standard model becomes strong-interacting.

The exact correspondence between Lagrangian parameters and the physically measured masses and couplings gets lost in the strong interacting regime occuring for large Higgs masses. In particular the mass parameter m_H , which is related in the initial potential to the values of its couplings, and which gives the Higgs mass after the classical symmetry breaking mechanism, now corresponds to the physical Higgs mass only as long as it is small. For larger m_H the theory moves into the strong interacting region and the parameter m_H ceases to be identifiable with the Higgs pole position. In such a situation a $J = 0$ pole remains, but it goes deeper in the complex plane for increasing m_H . The real part saturates at a bound of the order of 0.8 TeV [8,9].

5. For large m_H values, the entire longitudinal-scalar sector undergoes a strong interaction regime. One can advantageously use a result, by Cornwall, Levine, and Tiktopoulos, which allows one to relate the high energy amplitudes among longitudinally polarized gauge bosons to corresponding amplitudes among the respective would-be Goldstones [10].

Studies have been carried out of the possible measurements that would reveal deviations at large m_H [11]. A paper by Chanowitz and Gaillard contains a detailed discussion [12]. In a recent paper [13] we had derived an effective Lagrangian for the possible strong longitudinal-scalar sector in the large N-limit.

Large N here means substitution of the weak O(4) in the Higgs sector with O(N) for large N. Such an effective Lagrangian exhibits the J = 0 "Higgs remnant", as found by Einhorn [8]. In addition it gives all the multiple vertices between longitudinal bosons, dynamically generated in the strong regime. A different approach was developed in reference 14) for possible J = 1 poles. Vector poles are of great phenomenological interest. Such vector bound states, having the same quantum numbers of the γ, the W and the Z, would enlarge and modify, perhaps in an observable way, the standard mixing scheme. The standard phenomenology of W and Z might be slightly modified from the standard model behaviour, may be with observable deviations. Moreover, one could hope to produce such new bound states at the energies accessible at the Tevatron or in future machines, such as the SSC.

6. For the discussion of possible J = 1 poles an approach was used, in ref. 14), which goes back to work by Callan, Coleman, Wess, and Zumino [15] on non-linear realizations of symmetry. The approach exploits the techniques recently developed to treat hidden local symmetries of the non-linear σ-model [16]. Supergravity theories are an example in which hidden local symmetries were proposed, by Cremmer and Julia, to obtain a more convenient interpretation [18]. The hidden symmetry description has been exploited recently to describe vector mesons (ρ, ω, etc.) in strong interactions [19].

From work due to Appelquist and Bernard [20] one knows that, when one formally takes the limit $m_H \to \infty$ in the generating functional for the scalar sector of the standard electroweak model, one obtains the gauged non-linear σ-model. This model is formulated in the quotient space $SU(2)_L \otimes SU(2)_R / SU(2)_V$. The limit is only formal. The value of the isoscalar field is frozen to its classical value. The picture will however be changed by the occurrence of the quantum fluctuations.

The gauged non linear σ-model when translated into the unitary gauge gives massive gauge fields. By including loop effects one finds divergent expression, logarithmic at one-loop. Such logarithmic divergences correspond in the original model, which was renormalizable, to terms which depend logarithmically on the Higgs mass m_H (which corresponds no longer in general to the position of a physical scalar pole of the theory) and which therefore are infinite when m_H tends to infinity. So the Higgs mass indeed reappears as an effective cut-off when higher order terms are computed in a cut-off theory approach. When calculating one loop effects at large m_H, the choice of the gauge invariant potential appears to be inessential, as intuitively expected. It appears however that two-loop effects, in the same calculation, give terms which are sensitive to the explicit self-couplings in the scalar sector [22,23].

7. Quantum fluctuations are expected to become important for energies of some TeV. Due to this, only the description of bound

states below, say, 1 TeV, or even less, could be trusted in this approach. Formally one introduces explicit gauge boson degrees of freedom. They correspond to the gauge freedom of a local hidden $SU(2)_V$. The fields which correspond to such gauge bosons are, to be definite, auxiliary fields in the classical Lagrangian. This means that when they are eliminated, always at classical level, one re-obtains the original σ- model. In a particular example in two dimensions it has however been shown [17] that radiative effects have the effect of giving rise to a kinetic term for the new gauge bosons.

One is tempted to assume that this happens also in the real electroweak model. Then one would be in presence of new composite massive gauge bosons, corresponding to a local vector SU(2) group, and they are new dynamical degrees of freedom of the theory. Alternatively, it may happen that such an assumption, of a dynamically generated kinetic term, is not valid in the concrete case of the standard electroweak theory. In this case the model proposed in ref. (14) would not be deductible from the standard model. It would rather correspond to a new proposal, to be tested against experiment. In fact the possible significance of the formal limit $m_H \to \infty$ leaves much room for doubts. We have discussed the role of two-loop effects of the linear model [22].

8. Central to the discussion is, clearly, the question of the possible triviality[4] of φ^4 and whether it extends, and in which way, to the Higgs sector of the electroweak model [23]. Such ques-

tions go beyond perturbation theory and most probable the answer to them will not be readable from formal manipulations of the functional integral. The extreme possibility,of direct extension of the triviality to the actual electroweak scalar sector, would imply serious doubts on the whole mechanism of symmetry breaking, that however works so well experimentally. Excluding such a possibility one may have to deal with different possible scenarios.

One such scenario [24] is based on the assumption of a stable fixed point for the ratio of the quartic scalar coupling to the squared U(1)-coupling. Such stable point would prevent the theory from falling into the conjectured, but probable, triviality of the pure scalar model. In this way, through the use of the renormalization group equations one obtains a limit ~ 150 GeV for the Higgs mass [24].

Another possible scenario envisages that the triviality of the isolated scalar theory may turn out, in the real theory, into an asymptotic independence of the observables from the strength of the ϕ^4-coupling. Such a circumstance would imply increased calculability with respect to the perturbative approach. For instance, the Higgs mass may become calculable. A first attempt in this direction is due to Montvay [25], by carrying out a lattice calculation for the gauged SU(2) model. The results suggest a very weak or inexistent dependence on the quartic scalar coupling constant. The model is incomplete. No U(1) is included, and no fermion loops. Also finite size effects may bring substantial corrections.

If these incompleteness are ignored then one has a prediction, by Montvay, of m_H/m_W of the order of 6, that is a Higgs mass of $\sim$ 500 GeV. Quite evidently we are in the strong interacting regime.

9. As one sees, there are large uncertainties in this field, which reflect dynamical complexity of the theory. For such reasons we cannot affirm that the model developed in ref. (14) necessarily follows from the standard electroweak Lagrangian. If it is so, one should consider it as a new proposal, although having at least some formal connection to the $m_H \to \infty$ limit of the electroweak Lagrangain, at least in the classical treatment. In this sense the quantitative phenomenological predictions should be tested experimentally, and we shall especially discuss them in view of possible measurements in the now available energy regions.

The $SU(2)_L \otimes SU(2)_R$ invariance of the scalar sector of the standard model can be made evident by writing the usual complex scalar spinor (φ^o, φ^-) as a 2 x 2 matrix

$$M = \sqrt{2}\begin{bmatrix} \varphi^o & -\varphi_-^* \\ \varphi_- & \varphi_o^* \end{bmatrix} . \tag{1}$$

The standard scalar Lagrangian can then be written as

$$\mathcal{L} = \frac{1}{4}\mathrm{Tr}[(\partial_\mu M)^+(\partial^\mu M)] -$$

$$- \frac{1}{8f^2}m_H^2[\frac{1}{2}\mathrm{Tr}(M^+M) - f^2]^2 \tag{2}$$

and f is the usual vacuum expectation value, $f = 246$ GeV. Clearly the transformation $M \to V_L M V_R^+$, where V_L and V_R are matrices of $SU(2)_L$ and $SU(2)_R$ respectively, leaves $\mathcal{L}$ invariant. Turning on the $SU(2)_L \otimes U(1)$ gauge couplings, one obtains the usual Higgs mechanism and massive Z, W together with the massless photon. When one performs the formal $m_H \to \infty$ limit in the generating functional, M^+M is frozen to the value f^2, and one obtains the gauged non-linear σ-model with the Lagrangian

$$\mathcal{L} = \frac{1}{4} f^2 \mathrm{Tr}[(D^\mu U)^+ (D_\mu U)] \qquad (3)$$

where D^μ is the covariant $SU(2)_L \otimes U(1)$ derivative and $U = f^{-1}M$ is unitary, $U^+U = 1$.

For a description of the scalar sector evidencing the hidden local symmetry of the model one introduces local group-elements $L(x)$, $R(x)$ belonging to $SU(2)_L$ and $SU(2)_R$ respectively. Under global transformations of $SU(2)_L \otimes SU(2)_R$ these group elements are multiplied to their left by the corresponding global group transformation. In addition one can ask invariance under right-multiplication by a local group element of the unbroken $SU(2)_V$.

10. Before turning to the particular application it is perhaps simpler to summarize the general formulation. One starts by considering the local map $g(x)$ of the Minkowski space into a compact connected Lie group G. One defines the Maurer-Cartan form ω, which is globally invariant under left group multiplication

and antihermitean. One also has $F_{\mu\nu}(\omega) = 0$. For the connected subgroup H (to be called unbroken subgroup) one similarly has a local map $h(x)$. One decomposes $g(x)$ as $g(x) = \exp[iq(x)]h(x)$. The coordinates on the quotient space are $q_i(x) = \mathrm{Trace}(q(x)X_i)$, where X_i are the generators of G not belonging to $\mathrm{Lie}(H)$. They provide for the non-linear realization. Assume now right-multiplication invariance under the local H. Decomposing the form ω as $\omega = \omega_{/\!/} + \omega_{\perp}$, where parallel and orthogonal means with respect to H, one has, under the local right-multiplication: $\omega_{/\!/} \to h^{+}\omega_{/\!/}h + h^{+}\partial h$, whereas $\omega_{\perp} \to h^{+}\omega_{\perp}h$. Of course the coordinates $q_i(x)$ transform in general non-linearly under the restriction of the global G into the global H. When one specializes to a symmetric space, $\mathrm{Lie}[G]$ can be written, in standard base, as $[T_\mu, T_\nu] = if_{\mu\nu\lambda}T_\lambda$, $[T_\mu, X_i] = ig_{\mu ij}X_j$, $[X_i, X_j] = ig_{ij\mu}T_\mu$, where $T_\mu \in \mathrm{Lie}[H]$. There is a parity operation as automorphism, as well known. By gauge transforming $\mathrm{Tr}(\omega_{\perp}^2)$ with $h^{-1}(x)$ one reobtains the standard non-linear realization of Callan, Coleman, Wess and Zumino. One has $\omega_{\perp} = [\exp(-iq(x))\partial\exp(iq(x))]_{\perp}$, and, by use of Baker-Campbell-Hausdorf, $\omega_{\perp}$ becomes $\omega_{\perp} = i[\mathrm{adj}(iq(x))]^{-1}.\sinh[\mathrm{adj}(iq(x))]\partial q(x)$, where $\mathrm{adj}A$ is the adjoint of A in $\mathrm{Lie}[G]$. Then $f^2\mathrm{Trace}[\omega_{\perp}^2] = -f^2\mathrm{Trace}[(\partial q(x))^2 - \frac{1}{3}\partial q(x).[q(x),[q(x),\partial q(x)]] + \ldots]$ giving the standard results (curvent algebra, PCAC, etc). Now introduce the gauge field η for the local $H : \eta \to h^{+}\eta h + h^{+}\partial h$. Define $Dg(x) = g(x)(\overleftarrow{\partial} + \eta)$ (note the action to the left). One can then introduce $\zeta = \omega_{/\!/} - \eta$, and one has $\zeta \to h^{+}\zeta h$. This allows for a new term $f'^2\mathrm{Tr}[\zeta^2]$, where one has now f' in addition to f

(in QCD $f = f_\pi$). If η develops no kinetic term it remains an auxiliary field and can be algebraically eliminated reobtaining the standard theory of non-linear realization.

11. Let us now turn to our situation, where we had global transformations with respect to $SU(2)_L \otimes SU(2)_R$, and we had introduced local group elements $L(x)$ and $R(x)$ belonging respectively to $SU(2)_L$ and to $SU(2)_R$. The Maurer-Cartan form $\omega^\mu dx_\mu$, where $\omega^\mu = (\omega^\mu_L, \omega^\mu_R) = (L^+\partial^\mu L, R^+\partial^\mu R)$, can be decomposed into a component $\omega^\mu_{/\!/}$, parallel to the subgroup $SU(2)_V$, and a component $\omega^\mu_\perp$, orthogonal to $SU(2)_V$: $\omega^\mu = (\omega^\mu_{/\!/})_a T^a_V + (\omega^\mu_\perp)_a T^a_A$, where T^a_V and T^a_A are the vector and axial vector generators of $SU(2)_L \otimes SU(2)_R$. One has

$$(\omega^\mu_{/\!/})_a = \mathrm{Tr}[\tfrac{1}{2}\tau^a(L^+\partial^\mu L + R^+\partial^\mu R)]\ ,\ (\omega^\mu_\perp)_a = \mathrm{Tr}[\tfrac{1}{2}\tau^a(L^+\partial^\mu L - R^+\partial^\mu R)] \qquad (4)$$

The parallel Maurer-Cartan component transforms under the local $SU(2)_V$ invariance according to such local invariance, whereas the orthogonal component transforms as if such symmetry were only global (that is it develops no inhomogeneous term under the gauge transformation). In addition one introduces the $SU(2)_V$ gauge field: $\eta^\mu = (\eta^\mu)_a T^a_V$. One considers terms which are invariant under the whole set of global $SU(2)_L \otimes SU(2)_R$ transformations and local $SU(2)_V$ transformations, as defined. For constructing an effective Lagrangian one limits to terms containing at most two derivatives and which are linearly independent. The simplest term is $-f^2 Tr[\omega_\perp^2]$. If one writes $U = LR^+$ one can rewrite such term as

$$-f^2 Tr[\omega_\perp^2] = -\frac{1}{4}f^2 Tr[(L^+\partial^\mu L - R^+\partial^\mu R)^2] = \frac{1}{4}f^2 Tr[(\partial^\mu U)^+(\partial_\mu U)] \tag{5}$$

showing that one has only reexpressed the σ-model in terms of alternative degrees of freedom. However the availability of the field η^μ allows for a new term $-f'^2 Tr[(\omega_{/\!/} - \eta)^2]$, with a new independent vacuum value f'. On the other hand in absence of a kinetic term for η^μ the field equations derived by adding the two terms would simply lead back to the original non-linear σ-model. The main physical assumption is thus that η^μ becomes a dynamical field.

The gauging of $SU(2)_L \otimes U(1)$ only implies the substitution of the ordinary derivatives with covariant left- and right-derivatives, acting on the left or right group-elements respectively:

$$D_\mu^{(L)} L = \partial_\mu L + W_\mu^{(0)} L - L\eta_\mu \ , \ D_\mu^{(R)} R = \partial_\mu R + Y_\mu R - R\eta_\mu \tag{6}$$

where $Y_\mu = Y_\mu \cdot \frac{1}{2}\tau_3$, $W_\mu^{(0)} = \vec{W}_\mu^{(0)} \cdot \frac{1}{2}\tau$, $\eta_\mu = \vec{\eta}_\mu \cdot \frac{1}{2}\tau$, and we have added a superscript zero to W to allow us later for use of the simpler symbols for the physical field that will emerge after mixing. Notice the different positioning of for instance W_μ^o and η_μ in the covariant derivatives. The final Lagrangian will contain the term built up from the transverse Maurer-Cartan component, $-\frac{1}{4}f^2 Tr[(L^+D_\mu^{(L)} L - R^+D_\mu^{(R)} R)^2]$, corresponding to the field

$(\omega_{/\!/} - \eta)$ and which has the form $-\frac{1}{4}f'^2 \mathrm{Tr}[(L^+ D_\mu^{(L)} L + R^+ D_\mu^{(R)} R)^2]$, and the kinetic energies of the gauge bosons $W_\mu^{(0)}$, Y_μ, and also of η_μ.

The Higgs mechanism gives masses to all gauge bosons, except for the photon. All scalar degrees of freedom are absorbed. Formally one finds that one has to perform the following gauge transformation $(\Omega = RL^+)$:

$$\vec{W}^{(0)} = \Omega^+ \vec{\tilde{W}} \Omega + \Omega^+ \partial \Omega \; , \quad \vec{\eta} = R^+ \vec{\tilde{V}} R + R^+ \partial R \tag{7}$$

Finally one performs a rescaling of the fields according to $\tilde{W} \to g\tilde{W}$, $Y \to g'Y$, $2\tilde{V} \to g''\tilde{V}$ and after separate diagonalizations of the 2×2 charged and 3×3 neutral sectors one derives the physical vector boson states $W^\pm$, $V^\pm$ and A, Z^o, V^o with masses and mixing angles.

12. The masses and mixings are shown in Table 1, for both the charged sector $(W^\pm, V^\pm)$ and the neutral sector (photon, Z^o, V^o). As we have said, in absence of kinetic term for the gauge field of the hidden local symmetry, one can only recover the original gauged non-linear σ-model. The rescaling we have performed for the field $\tilde{V}_\mu$ allows however for a different way of looking at such a limit. When $g'' \to \infty$ the limit is again reobtained. Therefore $g'' \to \infty$ must lead back to the standard electroweak theory. This indeed happens quite evidently from the expressions for the

masses and mixings. The fermions, quarks and leptons, are those of the standard families, lefthanded fermions ψ_L and righthanded fermions ψ_R. Under the local $SU(2)_V$ they are assumed to be singlets. Their couplings are then uniquely determined

$$\bar{\psi}_L i\gamma^\mu(\partial_\mu + \vec{W}^{(0)}_\mu\frac{\vec{\tau}}{2} + \frac{1}{2}(B-L)Y_\mu)\psi_L + \bar{\psi}_R i\gamma^\mu(\partial_\mu + (\frac{\tau_3}{2} + \frac{1}{2}(B-L)Y_\mu)\psi_R$$

from which one obtains the couplings, as given in Table 2.

An overall fit of the electroweak data shows that even for relatively low values of g" the predictions of the standard electroweak model are essentially reproduced within the present experimental errors. Examples of predicted deviations, derived after fitting the low energy phenomenology, are shown in Table 3 (they are shown as possible examples). In fitting the present electroweak data the strategy is the following: one takes e, G_F, X as imputs to determine the constants g, g', and f of the model. One then fixes M_V and verifies that already for rather low values of g" one has c almost zero, ρ almost equal to 1, and M_W and M_Z not far from their standard model values. Better measurements of M_W and M_Z will provide sensitive tests. To avoid the uncertainties in $\sin^2\theta_W$ it is usual to consider the "$\sin^2\theta_W$-independent" relation

$$(1 - \frac{M_W^2}{M_Z^2})^{1/2} M_W = (\frac{\pi\alpha(M)}{\sqrt{2}G_F})^{1/2} = 38.65\text{GeV} \qquad (8)$$

of the standard model [26]. We find that the value 38.65GeV on the right-hand-side is increased in our model by a factor $\sim[1 + (g/g")^2 + \frac{1}{2}(g/g")^2(M_W^2/M_V^2$ and we again expect that, for renormalization at $M \sim M_W$, the bulk of the radiative corrections is already taken into account by the substitution of α of Josephson effect with $\alpha(M)$ in Eq. (1). This test is essentially independent of M_V and in Fig. 1 some predictions are reported for various g/g". High energy tests, whenever possible, would obviously be resolutive. Fig. 2 shows a predicted forward-backward asymmetry in $e^+e^- \to \mu^+\mu^-$ for $M_V = 250\text{GeV}$ and $g/g" = 0.22$ whereas Fig. 3 shows a predicted single V production in $p - \bar{p}$ collider, for the same parameters.

Acknowledgements

I would like to thank R. Casalbuoni, S. De Curtis, and D. Dominici for their active and useful collaboration and for the numerous discussions.

References

1) F. Wilczek, Phys. Rev. Lett. 39, 1304 (1977).

2) J. Lee Franzini, in "Physics in Collision 5" ed B. Aubert and L. Montanet, éditions Frontières, Gif-sur-Yvette, p. 145 (1985).

3) M. Vysotosky, Phys. Lett. 97B, 159 (1980).

4) See for instance the discussion in J. Fröhlich, in "Progress in Gauge Field Theory", Lectures at Cargese 1983, ed. by G. t Hooft et al., Plenum Press (1984).

5) M.A. Beg and R.C. Furlong, Phys. Rev. D31, 1370 (1985)
P. Hasenfratz (private communication)
J.E. Callaway and R. Petronzio, CERN TH 4270/85.

6) M. Veltman, Acta Physica Polonica B8, 475 (1977).

7) B.W. Lee, C. Quigg and H.B. Thacker, Phys. Rev. D16, 1519 (1979).

8) M.B. Einhorn, Nucl. Phys. 264B, 75 (1984).

9) R. Casalbuoni, D. Dominici, and R. Gatto, Phys. Lett. 147B, 419 (1984).

10) J.M. Cornwall, D.N. Levine, and G. Tiktopoulos, Phys. Rev. D11, 1145 (1974) see also C.G. Vayonakis, Lett. Nuovo Cimento 17, 383 (1976).

11) M.S. Chanowitz and M.K. Gaillard, Phys. Lett. 142B, 85 (1984;
M.K. Gaillard, in $p\bar{p}$ Options for the supercollider, p. 192 (1984;
R.N. Cahn and S. Dawson, Phys. Lett. 136B, 196 (1984);
Phys. Lett. 138B, 464 (1984);
M.K. Gaillard, in Proc. DPF Summer Study on the Design and

Utilization of the SSC, Eds R. Donaldson and J. Morfin (Fermilab 1985);
P.Q. Hung and H.P. Thacker, Phys. Rev. D32, 2866 (1985);
S. Dawson, Nucl. Phys. B 249, 42 (1985);
R.N. Cahn, Nucl. Phys. B 255, 341 (1985);
G.L. Kane, UM,TH 85-14, preprint University of Michigan;
M.J. Duncan, G.L. Kane, and W.W. Repko, UM,TH 85-18, preprint University of Michigan;
J. Lindfors, LAPP preprint TH 147 (1985).

12) M.S. Chanowitz and M.K. Gaillard, LBL, 19470, preprint, Berkeley (1985).

13) R. Casalbuoni, D. Dominici, and R. Gatto, Phys. Lett. 147B, 419 (1984).

14) R. Casalbuoni, S. De Curtis, D. Dominici, and R. Gatto, Phys. Lett. 155B, 95 (1985).

15) S. Coleman, J. Wess, and B. Zumino, Phys. Rev. 177, 2239 (1969).

16) C. Callan, S. Coleman, J. Wess, and B. Zumino, Phys. Rev. 177, 2247 (1969).

17) V. Golo and A.M. Perelemov, Phys. Lett. 79B, 112 (1978);
A. D'Adda, P. Di Vecchia, and M. Luscher, Nucl. Phys. B146, 73 (1978); ibid. B152, 125 (1979);
A.P. Balachandran, A. Stern, and G. Trahern, Phys. Rev. D19, 2416 (1979).

18) E. Cremmer and B. Julia, Phys. Letter 80B, 48, 1978; Nuclear Phys. B159, 141 (1979); J. Ellis, M.K. Gaillard, and B. Zumino, Phys. Lett. 94B, 343, (1980); Acta Physica Pol. B13, 253 (1982).

19) M. Bando, T. Kugo, S. Uehara, K. Yamawaki, T. Yanagida, Phys. Rev. Lett. 54, 1215 (1985); M. Bando, T. Kugo, K. Yamawaki, Nucl. Phys. B259, 493 (1985); M. Abud, G. Maiella, F. Nicodemi, R. Pettorino, K. Yoshida, Phys. Lett. 159B, 155 (1985).

20) T. Appelquist and C. Bernard, Phys. Rev. 22D, 200 (1980); and 23D, 425 (1981).

21) A. Longhitano, Phys. Rev. D22, 1166 (1980); Nucl. Phys. B188, 118 (1981).

22) J.J. van der Bij and M. Veltman, Nucl. Phys. B231, 205 (1984); J.J. van der Bij, Nucl. Phys. B161, 341 (1985).

23) G. Passarino, Phys. Lett. 156B, 231 (1985).

24) M.A. Beg, C. Panagistakopoulos, and A. Sirlin, Phys. Rev. Lett. 52, 883 (1984); J.E. Callaway, Nucl. Phys. B233, 189 (1984); R. Dashen and H. Neuberger, Phys. Rev. Lett. 50, 1897 (1983); K.S. Babu and E. Ma, Phys. Rev. D31, 2861 (1984); A. Bovier and D. Wyler, preprint ETH Zürich (1984); E. Ma, Phys. Rev. D31, 322 (1985).

25) I. Montvay, Desy preprint, 85-050 (1985).

26) See L. Maiani, in Proceedings of the International Conference on Unified Theories, Venice, March 1981; M. Consoli, L. Maiani, and S. Lo Presti, Nucl. Phys. B223, 474 (1983).

Table 1 : The physical vector boson states and their masses. Masses and mixing angles depend on the gauge couplings g, g' of $SU(2)_L \otimes U(1)$ and the new "strong" coupling g", on the vacuum expectation value f and on the new constant f'. We call α the ratio f'^2/f^2. The limit $g'' \to \infty$ gives back the coupling and masses of the standard electroweak model, as expected.

Charged sector : $(W^{\pm}, V^{\pm})$

$$W_i = \tilde{W}_i \cos\varphi - \tilde{V}_i \sin\varphi \quad (i = 1, 2) \qquad M_W^2 = \frac{f^2}{8}[c - (c^2 - 4\alpha g^2 g''^2)^{\frac{1}{2}}]$$

$$tg2\varphi = \frac{2\alpha g'' g}{g^2(1+\alpha) - g''^2\alpha} \qquad c = g^2(1+\alpha) + g''^2\alpha$$

$$V_i = \tilde{W}_i \sin\varphi + \tilde{V}_i \cos\varphi \qquad M_V^2 = \frac{f^2}{8}[c + (c^2 - 4\alpha g^2 g''^2)^{\frac{1}{2}}]$$

$c^2 - 4\alpha g^2 g''^2 \geq 0$ always $(M_V \geq M_W)$

<u>Neutral sector</u> : (A = photon field, Z^o, V^o)

$$A = A_2 \cos\psi + \tilde{V}_3 \sin\psi \qquad tg2\xi = \frac{2\alpha' G G''}{G^2(1+\alpha') - G''^2\alpha'}$$

$$Z^o = A_1 \cos\xi + A_2 \sin\xi \sin\psi - \tilde{V}_3 \sin\xi \cos\psi \qquad tg\psi = \frac{2gg'}{g''G}$$

$$V^o = A_1 \sin\xi - A_2 \cos\xi \sin\psi + \tilde{V}_3 \cos\xi \cos\psi \qquad \alpha' = \frac{(g^2-g'^2)^2}{G^4}$$

where

$$A_1 = (g\tilde{W}_3 - g'Y)/G \qquad G = (g^2 + g'^2)^{\frac{1}{2}}$$

$$A_2 = (g'\tilde{W}_3 + gY)/G \qquad G'' = \frac{g^2+g'^2}{g^2-g'^2}\left(\frac{4g^2g'^2}{g^2+g'^2} + g''^2\right)^{\frac{1}{2}}$$

$$M_A^2 = 0$$

$$M^2_{Z^o} = \frac{f^2}{8}[d - (d^2 - 4\alpha' G^2 G''^2)^{\frac{1}{2}}] \qquad d = G^2(1 + \alpha') + G''^2\alpha'$$

$$M^2_{V^o} = \frac{f^2}{8}[d + (d^2 - 4\alpha' G^2 G''^2)^{\frac{1}{2}}] \qquad d - 4\alpha' G^2 G''^2 \geq 0 \text{ always } (M_{V^o} \geq M_{Z^o})$$

Table 2 : Couplings of the physical vector bosons $W^{\pm}$, $V^{\pm}$ and A^{o}, Z^{o}, V^{o} to the fermions (for notations see Table 2)

Charged couplings :

$$(h_W W_\mu^{(i)} + h_V V_\mu^{(i)}).T_L^i \quad , \quad (i = 1,2);\ h_W = g\cos\varphi \ , \ h_V = g\sin\varphi$$

Neutral couplings :

$$eA_\mu Q_{em} + Z_\mu^o[\mathcal{A}T_L^3 + \mathcal{B}Q_{em}] + V_\mu^o[\mathcal{E}T_L^3 + \mathcal{D}Q_{em}]$$

$$e = \text{electric charge} = gg'G^{-1}\cos\psi$$

$$\mathcal{A} = G\cos\xi \qquad \mathcal{B} = -g'^2G^{-1}\cos\xi.(1 - \frac{g}{g'}\sin\psi\ tg\xi)$$

$$\mathcal{E} = G\sin\xi \qquad \mathcal{D} = -g'^2G^{-1}\sin\xi.(1 + \frac{g}{g'}\sin\psi\ ctg\xi)$$

Form of the effective low energy neutral current hamiltonian

$$H^{NC} = \sqrt{2}G_F\rho[(T_L^3 + XQ_{em})^2 + cQ_{em}^2]$$

(in standard model $\rho = 1$, $X = -\sin^2\theta_W$, $c = 0$)

<u>Table 3</u> : Examples of fits to the electroweak data. The parameters g, g', f are uniquely determined from the values of α, from Josephson effect, G_F from μ-decay, and X from the average values of ν and $\bar{\nu}$ deep-inelastic data. One then predicts the parameter c in the effective neutral current hamiltonian ($c = 0$ for the standard model), $\rho - 1$, and the relative deviations $\Delta M_W/M_W$ and $\Delta M_Z/M_Z$ from the standard model values.

g/g''	c	$\rho - 1$	$\Delta M_W/M_W$	$\Delta M_Z/M_Z$
0.16	0.0005	-0.0013	0.012	0.016
0.19	0.0007	-0.0020	0.017	0.023
0.22	0.0011	-0.0027	0.024	0.032

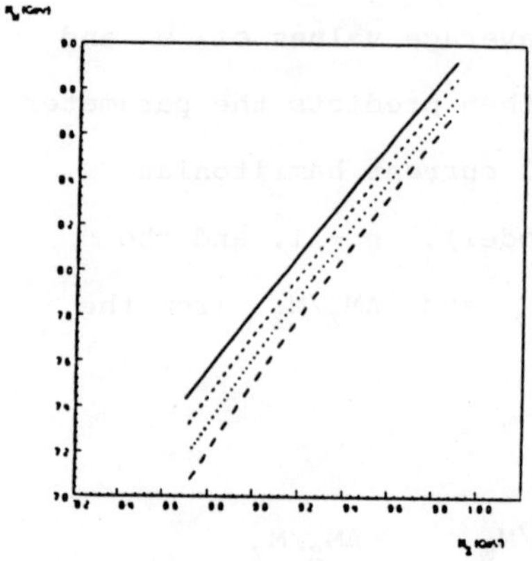

Fig. 1 M_W versus M_Z. Solid line is for the standard model; other lines are for the present model and different g/g": 0.16 (.-); 0.22 (...); 0.26 (.—).

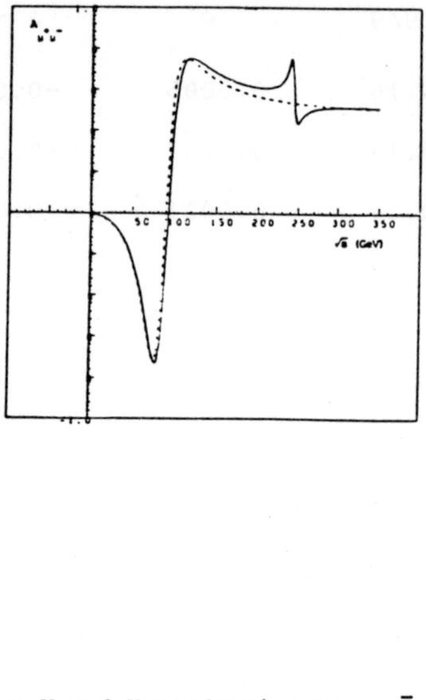

Fig. 2 $e^+e^- \to \mu^+\mu^-$ forward-backward symmetry for M_V = 250 GeV and g/g" = 0.22 (dotted line is for the standard model).

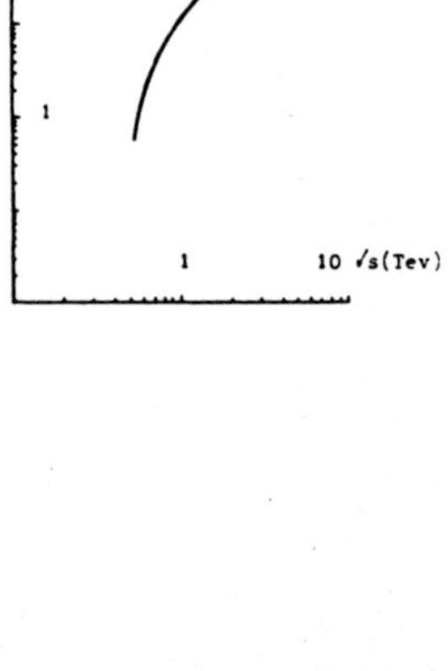

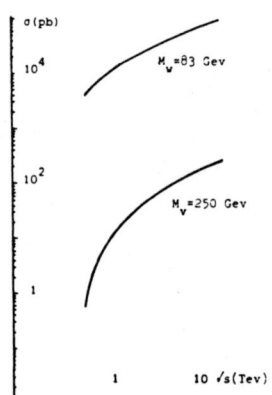

Fig. 3 Inclusive W and V production from $p\bar{p}$ for M_V = 250 GeV, g/g" = 0.22.

YUKAWA COUPLINGS FROM HIGHER DIMENSIONS

A.N. Schellekens

CERN - Geneva

A B S T R A C T

Several aspects of the problem of obtaining fermion masses from extra dimensional manifolds are discussed, and illustrated with an example of compactification of an $E_8 \times E_8$ theory on the manifold $SU(3)/U(1)^2$.

Attempts at getting phenomenology from extra dimensions have long been frustrated by the difficulty obtaining light particles in the right representations of the right gauge groups. In particular it was impossible to get chiral fermions from pure (super)gravity, unless one considered the seemingly unattractive possibility of adding Yang-Mills fields. All this has changed radically since the discovery of the $E_8 \times E_8$ heterotic string theory[1),2)]: the presence of an $E_8 \times E_8$ Yang-Mills group is essential for the consistency of the theory, and upon compactification to four dimensions one finds almost inevitably several generations of chiral fermions in the right representations of several possible unified gauge groups[3),4)].

Early investigations have shown that it is still quite difficult to understand exactly what happens between the Planck scale (M_{pl}) and the weak interaction scale (M_W)[5)]. But there is at least one feature of these theories that is not affected by these uncertainties, namely the appearance of several generations of fermions. It is a general property of Einstein-Yang-Mills theories with compactification due to identification of the gauge field and the spin connection that fermion generations appear with a multiplicity given by some topological invariant of the internal manifold. In compactifications of superstrings this is usually $\frac{1}{2}\chi$[3)] (where χ is the Euler-characteristic of the manifold), but in some cases it can be a more complicated invariant[4)]. Unfortunately the number of possible manifolds on which the superstring can be compactified is so large that nothing can be said about the number of generations, except that there is no reason why there should be just one.

A crucial test for any theory that claims to explain the repetition of generations is whether it can also explain the differences between them, namely the quark and lepton masses. In the $E_8 \times E_8$ superstring theory the simplest possibility is that the fermions get their masses from vacuum expectation values of a set of elementary Higgs bosons. If we restrict ourselves to the field theory limit of the superstring, the only candidate Higgs bosons are the scalar components of the dimensionally reduced ten-dimensional Yang-Mills fields.

The determination of the fermion masses consists of two parts: calculating the v.e.v.'s of the Higgs bosons, and calculating the Yukawa couplings of the Higgs bosons to the fermions. These are to a large extent independent problems. Of the two, the first is the most difficult and will require many speculative assumptions. It involves first of all an understanding of the origin of the weak scale. If supersymmetry plays a role, one has to understand supersymmetry breaking, and deal with a complicated potential involving squarks and sleptons. The renormalization group evolution of the parameters of the Higgs potential from M_{pl} to M_W is another serious problem, obviously sensitive to whatever happens in between. Finally, as will be discussed below, even determining the Higgs potential at M_{pl} is not as simple as one might think.

In most respects the situation is much better for Yukawa couplings, since one only has to consider three point vertices with a dimensionless coupling constant. The unknown physics between M_{pl} and M_W still enters into the problem via the renormalization group corrections, but at least those corrections are proportional to the Yukawa couplings themselves, which is not the case for Higgs self-interactions. For sufficiently small Yukawa couplings the dominant correction is due to flavour independent gluon interactions, which will not destroy the structure of the Yukawa coupling matrices. Therefore, many corrections will cancel in mass ratios. If the Yukawa couplings at M_{pl} are large, they may approach infra-red fixed points[6)], which would partly or totally obscure any information about Planck scale physics. This is an especially serious danger for four generation models.

These considerations lead to the following approach. The chances of understanding the Higgs system in sufficient detail seem remote, but at least all the uncertainties can be packaged neatly into a small number of parameters, the v.e.v's. For the moment we will treat these parameters as incalculable. The Yukawa couplings on the other hand can be calculated with sufficient reliability. To compare the result of such a calculation with the known masses one has to eliminate the free parameters describing the Higgs v.e.v.'s. If there were just one parameter one could simply compare mass ratios. In practice the analysis is usually not that simple. In this way one might learn about properties of the internal manifold directly from the known fermion masses, even without an understanding of the M_W/M_{pl} hierarchy, or of any physics between M_W and M_{pl}.

One of the problems one faces immediately is that in a model of this kind, quark and lepton masses tend to be of the same order of magnitude as M_W, since Yukawa and gauge coupling constants are related at M_{pl}. Renormalization group corrections are likely to make the situation worse rather than better, at least for quark masses. The calculation of Yukawa coupling constants involves a different overlap integral on the manifold than gauge couplings. The massless modes of the gauge field do not depend on the internal coordinates, whereas the scalar modes will in general depend on them. Therefore, it is possible that small numbers appear as a result of these integrals. If that is not the case, the problem will have to be solved by the Higgs system. This is only possible if one can find a Higgs boson that does not couple to some of the fermions, and which can therefore give a mass to the W-boson without doing too much damage in the fermion sector.

This latter solution is not very elegant, since it just shifts the problem, to the Higgs sector, but at least we can check whether the Yukawa couplings allow us to do that. On any manifold with some fixed metric it should be possible to obtain rigorous upper and lower bounds on quantities like $(\Sigma m_u^2)/M_W^2$, $(\Sigma m_d^2)/M_W^2$ or $(\Sigma m_u^2)/(\Sigma m_d^2)$, where the sums are over all generations. Whether these bounds are saturated depends on the Higgs system, but the bounds themselves are only properties of the manifold and the metric. This may already be sufficient to rule out many possible compactifications.

The calculation of the Yukawa coupling constants is in principle straightforward. One determines the massless fermion modes on the manifold, and one decides which scalar modes to consider as candidate Higgs bosons. In supersymmetric models the obvious modes to choose are the massless ones. In a more general context, where the solution to the hierarchy problem is left as an open question, one has to find another criterion to select the relevant modes. The Yukawa coupling constants are then determined by calculating overlap integrals of two fermion and one Higgs boson mode on the manifold. Furthermore, one needs the bilinear fermion and scalar terms in order to properly normalize and diagonalize the fields. In this process all higher modes are ignored, and one has to make sure that this is allowed. There are (at least) two potential dangers.

The first problem (emphasized recently by Wetterich[7)]) is that massless modes in one vacuum differ from those in another by higher mode contributions. Therefore, Yukawa couplings in one vacuum have nothing to do with those in another vacuum of the same theory. This is an important problem only when one considers two vacua that differ from each other by shifts of order M_{pl} (for example, vacua with or without symmetry breaking by the Hosotani mechanism[8)]). If fields are shifted by amounts of order $M \ll M_{pl}$, one only misses corrections of order (M/M_{pl}).

The second problem has to do with non-linear field redefinitions. These are known to play an important role in compactifications of D = 11 supergravity[9)]. It is not always obvious which parametrization of the higher dimensional fields is the most suitable for compactification. A well-known example is the

Kaluza-Klein ansatz for the metric, which can be written in several different forms, all leading to the same mass spectrum. Since different parametrizations are related to each other by non-linear transformations of the higher dimensional fields, one may expect to get different four-dimensional interaction terms. As a generic example we consider a higher dimensional ϕ^3 theory, compactified to three dimensions. After a harmonic expansion in modes ϕ_k (eigenfunctions of the internal space Laplacian) one may get a Lagrangian of the form

$$\mathcal{L} = \phi_0 \Box \phi_0 + \sum_{k=1}^{\infty} \phi_k (\Box - M_k^2) \phi_k + \sum_{k,l,m=0}^{\infty} \lambda^{klm} \phi_k \phi_l \phi_m$$

where ϕ_0 is a massless scalar. The quantity of interest is λ^{000}, a dimensionless coupling constant in three dimensions. A general non-linear reparametrization is obtained by replacing $\phi(x,y)$ by $f(\phi(x,y))$ in the higher dimensional theory, where f is some invertible function which s equal to the identity in the linear approximation. Such a reparametrization can always be written as a non-linear transformation of the modes ϕ_k:

$$\phi_k \rightarrow \phi_k' = \phi_k + \sum_{i,j} a_{ij}^{k} \phi_i \phi_j + \sum_{i,j,l} b_{ijl}^{k} \phi_i \phi_j \phi_l + \dots$$

One clearly gets a ϕ_0^3 term from this shift, but it is equal to

$$a_{00}^{0} \phi_0 \Box \phi_0^2$$

so that it has two extra momenta. The coefficient has a dimension $[\text{mass}]^{-2}$, and will be of order M_{pl}^{-2}, so that this term will have negligible phenomenological consequences. This is clearly generally true for cubic couplings of massless fields.

To demonstrate that this is not always an irrelevant point we can consider a four-point vertex in four dimensions. The same transformation we made before produces now the following additional ϕ_0^4 terms

$$(a_{00}^{0})^2 \phi_0^2 \Box \phi_0^2 + \sum_{k=1}^{\infty} (a_{00}^{k})^2 \phi_0^2 (\Box - M_k^2) \phi_{00}^2$$

$$+ 2 b_{000}^{0} \phi_0 \Box \phi_0^3 \quad + 3 \sum_{k=0}^{\infty} \lambda^{00k} a_{00}^{k} \phi_0^4$$

Now $a_{00}^{k} \sim 1/M_{pl}$ and $M_k \sim M_{pl}$, so that the second term produces contributions of order 1. There may be additional contributions from the fourth term, resulting from a shift of the trilinear interaction terms. Calculating the four-scalar vertex simply as an overlap integral of four massless modes would obviously be meaningless.

On the other hand, a physical quantity should not depend on field redefinitions. One can understand the origin of the problem by considering Green's functions for four massless bosons. At tree level there are not only contributions from four-point vertices, but also from exchanges of massive bosons, at least if there are vertices coupling two massless and one massive boson. In that case it is clearly incorrect to truncate the theory to the massless modes. If nonlinear redefinitions are made, the ϕ_0^4 terms change by amounts of order 1, while the coefficients of the $\phi_0^2\phi_k$ terms will get corrections of order M_{pl}. The correct ϕ_0^4 coupling is obtained if the four scalar Green's function of the truncated theory is equal to the one for the untruncated theory. Therefore, one has to define the modes in such a way that $\phi_0^2\phi_k$ terms are absent. This can always be done, but unfortunately it requires knowledge of not just the massless modes, but also the massive ones, unless the dangerous terms are absent due to some symmetry of the theory.

Several aspects of the problem of obtaining fermion masses from compactification can be illustrated by a model studied recently by Pilch and the author[10). We compactify the $E_8 \times E_8$ field theory limit of the heterotic string on the manifold $SU(3)/U(1)^2$. This manifold has Euler characteristic $\chi = 6$, so that there will be three generations. Because it is a coset manifold, explicit calculations can be made rather easily. For the same reason we consider only SU(3) invariant metrics. There is a two parameter family of such metrics (plus an irrelevant overall scale), and a one-parameter subset of Kähler metrics. The holonomy groups are thus SO(6) or SU(3) × U(1). We compactify by setting the gauge fields equal to the spin connection. Although we will not insist on consistency with the complete string theory (this is not relevant for our purpose), it is desirable to have consistency and anomaly cancellation at least in the field theory limit. To cancel all anomalies we can use the Green-Schwarz[1)] mechanism; a consistency condition is then that $\int dH$ vanishes on all closed four-dimensional submanifolds[11)]. For both holonomy groups this leads uniquely to an unbroken SO(10) Yang-Mills group in four dimensions. (For other Kähler manifolds, with a less restrictive charge quantization condition, there are additional possibilities[4)].) Furthermore, there is, of course, an SU(3) gauge symmetry due to the isometries of the manifold, and an unbroken E_8' symmetry, which we will ignore.

Topological arguments show that there are three generations and no mirror generations. They are singlets of the isometry group. The scalar sector of the theory is harder to discuss than the gauge boson and fermion sector, since their masses are not under control. It is, however, worth noticing that the (248) of E_8 contains scalars in precisely the correct SO(10) representations (45, 16 and 10) to produce the Higgs content of the minial SO(10) model[12)]. If we follow that suggestion, the (45) and the (16) would break SO(10) to SU(3) × SU(2) × U(1), while the (10) would be responsible for electroweak symmetry breaking. Since our manifold is simply connected, the mechanism of Ref. 8) cannot work. We will have to assume that at a scale M_{GUT}, a few orders of magnitude less than M_{pl}, an SO(10) breaking perturbation appears on the SO(10) - invariant background. Fortunately, the details of this are irrelevant, as long as the gauge symmetry is broken. (One possible mechanism might be a dynamical symmetry breaking by the SU(3) gauge group.) It is, however, important that $M_{GUT} \ll M_{pl}$, because we will calculate the Yukawa couplings in the SO(10) background, so that corrections of order M_{GUT}/M_{pl} are ignored.

The massless fermions can only couple to scalars that are singlets of SU(3) and (10)'s of SO(10). The complete harmonic expansion of all scalar components of the Yang-Mills fields contains just three complex fields with that property. We will regard all of them as possible Higgs bosons. We have now precisely the same field content as the model of[12)]. (The other scalar modes do not affect the fermion masses, but might still contribute to m_W. We will assume that this is not the case.)

Unlike these authors we do not have the liberty to choose our Yukawa couplings: they can be calculated. The details and results of that calculation can be found in Ref. 10). We obtain the following formulas for the up and down quark mass matrices and the W-mass:

$$m_u = \tfrac{1}{2} g \left(Z_i M_i + \zeta_i N_i \right)$$

$$m_d = \tfrac{1}{2} g \left(\bar{Z}_i M_i - \bar{\zeta}_i N_i \right)$$

$$m_W = \tfrac{1}{2} g \sqrt{Z_i \bar{Z}_i + \zeta_i \bar{\zeta}_i}$$

where M_i and N_i are real matrices given in (10) and Z_i and ζ_i (i=1...3) are six complex parameters describing the v.e.v.'s of the Higgs bosons.

This system can be analyzed as follows. First we calculate $(\sum m_u^2)/m_W^2$ and $(\sum m_d^2)/m_W^2$, and find that both quantities are bounded between ~1 and 4 for most choices of the metric. This eliminates immediately most of the parameter space, with the exception of a very narrow strip around a line on which the lower bound vanishes.*) This turns out to be precisely the line on which the metric is Kähler. We can now saturate the lower bound for $\sum m_d^2$ on this line and calculate all up quark masses. It should be emphasized that whether this bound is saturated is a question about the Higgs-potential, which we cannot answer. But the resulting up quark masses, as well as the bound itself, are purely a property of the internal manifold.

We find that the up quark masses are 0, 0 and $3\sqrt{2}\cos\phi\cos 2\phi\, m_W$, where ϕ is a parameter of the metric along the Kähler line. In the allowed range for ϕ the top quark mass varies between 0 and 96 GeV (the value 0 is obtained for a singular metric). Both m_u and m_c vanish in this limit, which is fine because m_c^2/m_W^2 is indeed a small number.

The renormalization group corrections to those results are difficult to estimate, since there are diagrams that mix all coupling matrices. Without a calculation we can at least show that the lower bound on $\sum m^2{}_d$ for Kähler metrics is not seriously affected. This bound can be understood as follows. For Kähler metrics there exists a <u>complex</u> linear combination ϕ_c of the six real scalars which decouples from all fermions. In other words, if we complexify the kinetic terms and Yukawa couplings, then the new Lagrangian has a symmetry $\phi_c \to -\phi_c$. This symmetry is completely protected if it is also respected by Higgs self-interactions. Otherwise, the corrections to the coupling of ϕ_c will most likely be of order g^5 (we assume that cubic interactions, which have a dimensional coupling constant which must be much less than M_{pl}, do not play a role. Quartic interactions have a coupling constant of order g^2). Therefore, it should still be possible to make $(\sum m_d^2)/m_W^2$ sufficiently small. The $O(g^5)$ corrections may, however, be distributed over the three down quarks in an incorrect way. Furthermore, one would expect corrections of order $g^3 m_W$ to m_u and m_c. Therefore, there may not be a problem for m_s and m_c, but m_u and m_d will almost certainly be too large.

---

*) One obtains the same conclusion by studying $(\Sigma m_d^2)/(\Sigma m_u^2)$, which eliminates the dependence on m_W.

This is all we can learn by focusing on the parameters of the manifold. One can fit the remaining parameters of the Higgs system and obtain the correct quark masses[10)] (although not the correct K.M. angles). Because small ratios like m_u/m_c are not explained by fitting these parameters, we do not learn much from this, except that there are no rigorous bounds preventing us from obtaining the right masses.

Perhaps this example does not quite make it as the "theory of everything". But there are many other compactifications, and some with better prospects of solving the hierarchy problem. Yukawa couplings provide the cleanest and most direct test of the idea that fermion generations are due to extra dimensions. Calculating them may not always be as easy as in our example (see however Ref. 13), but it might provide crucial information about the structure of the internal manifold.

ACKNOWLEDGEMENTS

Most of the work reported here was done in collaboration with K. Pilch at Stony Brook. I am grateful to H. Nicolai and B. de Wit for useful discussions about consistent truncations.

REFERENCES

1) M.B. Green and J.H. Schwarz, Phys. Lett. 149B (1984) 117.

2) D. Gross, J. Harvey, E. Martinec and R. Rohm, Nucl. Phys. B256 (1985) 253.

3) P. Candelas, G. Horowitz, A. Strominger and E. Witten, Nucl. Phys. B258 (1985) 46.

4) K. Pilch and A.N. Schellekens, Nucl. Phys. B259 (1985) 637.

5) M. Dine, V. Kaplunovsky, M. Mangano, C. Nappi and N. Seiberg, Princeton Preprint (1985);
J.D. Breit, B.A. Ovrut and G. Segré, Univ. of Pennsylvania Preprint (1985);
S. Cecotti, J.P. Derendinger, S. Ferrara, L. Girardello and M. Roncadelli, Phys. Lett. 156B (1985) 318;
J.P. Derendinger, L.E. Ibañez and H.P. Nilles, CERN-TH.4228/85;
E. Witten, Nucl. Phys. B258 (1985) 75;
E. Cohen, J. Ellis, C. Gomez and D.V. Nanopoulos, CERN-TH.4159/85;
E. Cohen, J. Ellis, K. Enqvist and D.V. Nanopoulos, CERN-TH.4195/85 and CERN-TH.4222/85.

6) B. Pendleton and G.G. Ross, Phys. Lett. 98B (1981) 291.

7) C. Wetterich, CERN-TH.4154/85.

8) Y. Hosotani, Phys. Lett. 126B (1983) 309; 129B (1983) 193.

9) B. de Wit and H. Nicolai, CERN-TH.4291 (1985), to appear in the proceedings of the Cambridge Supergravity Workshop, July 1985.

10) K. Pilch and A.N. Schellekens, "Do Quarks Know About Kähler Metrics", Stony Brook Preprint ITP-SB-85-50, to appear in Phys. Lett. B.

11) E. Witten, Phys. Lett. 149B (1984) 351.

12) H. Georgi and D.V. Nanopoulos, Nucl. Phys. B155 (1979) 52.

13) A. Strominger and E. Witten, Princeton Preprint;
A. Strominger, Preprints NSF-ITP-85-105 and NSF-ITP-85-109.

Part Four: Superstrings and Extra Dimensions

ALGEBRAS, LATTICES AND VERTEX OPERATORS

Peter Goddard

Department of Applied Mathematics and Theoretical Physics
University of Cambridge, Silver Street, Cambridge CB3 9EW
U.K.

1. Introduction

In this lecture I shall describe how the vertex operators of string theory associate an algebra with a (suitable) lattice. This will be done with a variant of the Frenkel-Kac[1] construction, following an approach developed in 1983 by Olive and myself[2]; a similar approach has also been given by Frenkel[3]. This construction provides concrete realizations of the exceptional Lie group E_8 and of the triality properties of SO(8), which underly the supersymmetry of superstring theories[4]. It yields naturally compact Lie algebras, affine Kac-Moody algebras, Clifford algebras (and their affinisations), hyperbolic Kac-Moody algebras, depending on the lattice chosen and the points on it which are considered. Further development[5] of the Kac-Frenkel construction yields the Griess algebra [6] which is used to prove the existence of the Monster group, the largest sporadic finite simple group.

The construction provides the algebras with a natural string theory interpretation[2]: they are symmetries of the physical states for a string whose momenta lie on the lattice, i.e. a string moving on the torus dual to the lattice.

2. Lattices

Let us begin by reviewing some properties of lattices. (See e.g. [7] or [2] for more details.) A lattice is a set of points in a vector space of the form

$$\Lambda = \{ \textstyle\sum n_i e_i : n_i \in \mathbb{Z} \} \tag{1}$$

where $\{e_i\}$ is a basis for the space, but not necessarily an orthonormal one. Such a lattice is said to be integral if

$$x \cdot y \in \mathbb{Z} \quad \text{for all } x, y \in \Lambda \tag{2}$$

i.e. $\Lambda \subset \Lambda^*$, the lattice dual to Λ,

$$\Lambda^* = \{ x : x \cdot y \in \mathbb{Z} \text{ for all } y \in \Lambda \} \tag{3}$$

A lattice is said to unimodular if it has one point per unit volume. It is easy to see that a lattice is self-dual if and only if it is both integral and unimodular. A lattice is called even if the squared lengths of all of its points are even integers.

Even self-dual lattices are not only particularly interesting to consider from a mathematical point of view; their study is also motivated by physical considerations. Self-dual lattices occur naturally in monopole theory if one seeks a duality between electricity and magnetism[8]. They also appear to be necessary for the consistency of certain string theories[9].

An even self-dual Euclidean lattice must have a dimension which is a multiple of 8. Thus the lowest dimensional such lattices are 8-dimensional. Up to isomorphism there is only one, the root lattice of E_8. In dimension 16 there are two inequivalent even self-dual lattices, the root lattice of $E_8 \times E_8$ and the weight lattice of $Spin(32)/\mathbb{Z}_2$. (For each of E_8 and $E_8 \times E_8$ the weight and root lattices coincide.) In dimension 24 there precisely 24 inequivalent such lattices, including one for which the minimum squared distance between points is 4, namely the Leech lattice. After this the number grows very rapidly.

In Minkowski space the situation is simpler. A Lorentzian even self-dual lattice must have a dimension of the form $8n+2$, where n is an integer, and it is essentially unique for each n. It is usually denoted $II^{8n+1,1}$ and consists of points of the form $\underset{\sim}{m}$ and $\underset{\sim}{m}+\underset{\sim}{\tfrac{1}{2}}$ where

$$\underset{\sim}{m} = (m_1, m_2, \ldots, m_{8n+1}; m_{8n+2}), m_i \in \mathbb{Z}, \sum m_i \in 2\mathbb{Z}, \quad (4)$$

and

$$\underset{\sim}{\tfrac{1}{2}} = (\tfrac{1}{2}, \tfrac{1}{2}, \ldots, \tfrac{1}{2}; \tfrac{1}{2}). \quad (5)$$

It is easy to check that the lattice consisting of such points is even and self-dual.

There is a relationship between Euclidean even self-dual lattices in dimension 8n and light-like vectors in $II^{8n+1,1}$. Given a primitive light-like vector k in the Lorentzian lattice(i.e. k is not a multiple of any other lattice vector except -k), we consider the points of this lattice modulo translations by k. In this way we obtain an even self-dual Euclidean lattice. It can be shown that all such lattices can be obtained in this way and that there is a one to one correspondence between their equivalence classes and the primitive light-like vectors k modulo automorphisms of the lattice $II^{8n+1,1}$. This construction is clearly related to the polarization states of a massless vector particle (photon) in the appropriate dimension.

3. String theory and Vertex Operators

A history of the string is described by a world sheet $x^\mu(\sigma,\tau)$, $0 \leq \sigma \leq \pi$, $-\infty < \tau < \infty$, where μ ranges over d values. The motion of the string is determined by the action

$$S = k \int [(\dot{x}.x')^2 - \dot{x}^2 x'^2]^{1/2} \, d\sigma d\tau \tag{6}$$

where

$$\dot{x} = \frac{\partial x}{\partial \tau}, \quad x' = \frac{\partial x}{\partial \sigma}, \quad \text{and} \quad k = T_0/c. \tag{7}$$

is a constant, with T_0 having the dimensions of tension. If the reparametrization invariance of S is used to impose the orthonormality conditions,

$$\dot{x}.x' = 0, \qquad \dot{x}^2 + x'^2 = 0, \tag{8}$$

the equations of motion and boundary conditions take the form

$$\ddot{x} = x'' \tag{9}$$

and

$$x' = 0 \quad at \quad \sigma = 0, \pi. \tag{10}$$

These have the general solution

$$\frac{1}{\kappa} x^\mu(\sigma,\tau) = q^\mu + p^\mu \tau + \sum_{n=1}^{\infty} (q_n^\mu \cos n\tau + \dot{q}_n^\mu \sin n\tau) \cos n\sigma \tag{11}$$

where the constant κ has been introduced to absorb the dimensions of length. It is convenient to introduce harmonic oscillator variables

$$\alpha_n^\mu = \frac{1}{2} \{ \dot{q}_n^\mu - in q_n^\mu \} \tag{12}$$

so that

$$\alpha_n^{\mu\dagger} = \alpha_{-n}^\mu, \tag{13}$$

where, classically, the dagger denotes complex conjugation, to become hermitian conjugation on quantisation. We can then rewrite eq.(11) as

$$\frac{1}{\kappa} x^\mu(\sigma,\tau) = q^\mu + p^\mu \tau + \frac{i}{2} \sum_{n\neq 0} \left\{ \frac{\alpha_n^\mu}{n} e^{-in(\tau+\sigma)} + \frac{\alpha_n^\mu}{n} e^{-in(\tau-\sigma)} \right\} \tag{14}$$

$$\equiv \frac{1}{2} Q^\mu(e^{i(\tau+\sigma)}) + \frac{1}{2} Q^\mu(e^{i(\tau-\sigma)}), \tag{15}$$

for

$$Q^\mu(z) = q^\mu - i p^\mu \log z + i \sum_{n\neq 0} \frac{\alpha_n^\mu}{n} z^{-n}. \tag{16}$$

In terms of the harmonic oscillator variables (12), the canonical commutation relations take the form

$$[\alpha_m^\mu, \alpha_n^\nu] = m g^{\mu\nu} \delta_{m,-n} \tag{17}$$

$$[q^\mu, p^\nu] = i g^{\mu\nu} \tag{18}$$

provided that

$$\kappa^2 k \pi = \hbar . \tag{19}$$

The creation and annihilation operators α_n^μ act in a Hilbert space which they generate from certain "vacuum" states $|\gamma\rangle$ which carry only momentum,

$$\alpha_n^\mu |\gamma\rangle = 0, \quad n>0; \qquad p^\mu|\gamma\rangle = \gamma^\mu|\gamma\rangle \tag{20}$$

and can be generated from a true vacuum $|0\rangle$ in the usual way,

$$|\gamma\rangle = e^{iq.\gamma}|0\rangle \tag{21}$$

The constraints (8) correspond, in terms of α_n^μ, to the vanishing of

$$L_n = \frac{1}{2} \sum_{m=-\infty}^{\infty} :\alpha_m . \alpha_{n-m}: \tag{22}$$

where the normal ordering has been introduced to ensure that

$$L_0 = \frac{1}{2} + \sum_{n=1}^{\infty} \alpha_n^+ . \alpha_n \tag{23}$$

is well-defined. So defined, these operators satisfy the Virasoro algebra,

$$[L_m, L_n] = (m-n) L_{m+n} + \frac{d}{12} m(m^2-1)\, \delta_{m,-n} \tag{24}$$

In this covariant treatment of the quantum theory, the classical constraints (8) are applied quantum mechanically, as conditions

$$L_n|\psi\rangle = 0, \qquad n>0, \tag{25a}$$

$$L_0|\psi\rangle = \lambda|\psi\rangle \tag{25b}$$

where λ is an arbitrary constant which can be regarded as an ambiguity associated with normal ordering. It can be shown that the space of physical states, defined by eqs.(25), is free of ghosts, i.e. negative norm states, resulting from the Lorentzian metric in eq.(17), if and only if[10-12]

<u>either</u> $d=26$ and $\lambda = 1$ (26)

<u>or</u> $1 \leq d \leq 25$ and $\lambda \leqslant 1$. (27)

However the spectrum of states for the free open string that we have been discussing is much neater if d = 26 and $\lambda = 1$ and it is only in this case that an apparently reasonably consistent theory of interacting strings exists. Even this theory possesses a tachyon, the ground state of the string described by unexcited states $|\gamma\rangle$, of the form of eq.(21), satisfying eqs.(25) with $\lambda = 1$ and so having

$$\gamma^2 = 2 . \tag{28}$$

With the metric we are using, such a momentum γ is space-like. To avoid this difficulty one needs to move to the d = 10 theory of Neveu, Schwarz and Ramond[13], which also invloves fermionic degrees of freedom and has a sector which is tachyon-free and supersymmetric in ten-dimensional Minkowski space, the "superstring" theory of Green and Schwarz[4].

Living with the tachyon, the interactions of this d = 26 bosonic string are defined perturbatively, starting from tree diagram Born terms. If we introduce the vertex operator

$$U(\gamma, z) = \, :\exp\{i\gamma . Q(z)\}: \tag{29}$$

$$\equiv \exp\{i\gamma . Q_<(z)\} \exp\{i\gamma . Q_0(z)\} \exp\{i\gamma . Q_<(z)\} \tag{30}$$

where

$$Q_>^\mu(z) = i \sum_{n>0} \frac{\alpha_n^\mu}{n} z^{-n}, \quad Q_<^\mu(z) = i \sum_{n<0} \frac{\alpha_n^\mu}{n} z^{-n} \tag{31}$$

and

$$Q_0^\mu(z) = q^\mu - ip^\mu \log z \tag{32}$$

so that

$$e^{i\gamma . Q_0(z)} = z^{\frac{1}{2}\gamma^2} e^{i\gamma . q} z^{\gamma . p}, \tag{33}$$

the Born term amplitude for the interaction of N ground state strings of momenta $\gamma_1, \ldots, \gamma_N$ (with $\gamma_i^2 = 2$, $\sum \gamma_i = 0$) is obtained by symmetrising

$$\int \langle 0 | U_{\gamma_1, \ldots, \gamma_N}(z_1, \ldots, z_N) | 0 \rangle \prod_{i=1}^{N} \frac{dz_i}{z_i} \Big/ d\Upsilon \tag{34}$$

where

$$U_{\gamma_1, \ldots \gamma_N}(z_1, \ldots, z_N) = U(\gamma_1, z_1) \ldots U(\gamma_N, z_N), \tag{35}$$

$$d\Upsilon = \frac{dz_a \, dz_b \, dz_c}{(z_a - z_b)(z_b - z_c)(z_c - z_a)}, \tag{36}$$

a, b, c being any fixed three of the indices 1, ... , N, and the integral is taken with the variables $z_1, \ldots, z_N$ varying over the real line, subject to the constraints that they maintain their order $z_1 > z_2 > \ldots > z_N$, and the points z_a, z_b, z_c remain fixed (as is implied by dividing by $d\gamma$). Independence of the choice of z_a, z_b, z_c follows from the invariance of the integrand of eq.(34) under the group of Mobius transformations

$$z \longmapsto (az+b)/(cz+d) \tag{37}$$

The way that one proves this Mobius invariance and, more generally, the fact that the only states that couple as intermediate states in such amplitudes are physical states in the sense of eqs.(25) is to exploit the relationship of the vertex (29) to the Virasoro algebra (24), which has the Mobius algebra as the three-dimensional subalgebra given by n = 1, 0, -1,

$$[L_n, U(\gamma, z)] = z^n\left(z\frac{d}{dz} + \frac{\gamma^2}{2}n\right)U(\gamma, z). \tag{38}$$

A particular consequence, if $\gamma^2 = 2$, is that then

$$[L_n, U(\gamma, z)] = z\frac{d}{dz}\left(z^n U(\gamma, z)\right). \tag{39}$$

It follows that

$$A_\gamma = \frac{1}{2\pi i}\oint \frac{dz}{z} U(\gamma, z) \tag{40}$$

where the integration contour encircles the origin once, is a physical state creation operator,

$$[L_n, A_\gamma] = 0, \tag{41}$$

i.e. it maps physical states into physical states, provided that it is well-defined.

Consider A_γ acting on a state with momentum λ. The only problem is with the factor $\exp\{i\gamma . Q_0(z)\}$,

$$\exp\{i\gamma . Q_0(z)\}|\lambda\rangle = z^{\frac{1}{2}\gamma^2 + \gamma . \lambda}|\lambda\rangle. \tag{42}$$

Since $\gamma^2 = 2$, this will be single valued if $\gamma . \lambda$ is an integer. The easiest way to achieve this is for both γ and λ to be members of the same integral lattice, Λ. We will have operators A_γ acting within the physical state space if we restrict the momenta or, at least, some of their components, to lie on an integral lattice. This is not such an unfamiliar restriction. If we consider a particle moving on a torus, formed by taking a d-dimensional space modulo some lattice Γ, its wave function $\exp\{i\gamma . x/\hbar\}$ will be single-valued on the torus provided that

$$\frac{\gamma . a}{2\pi\hbar} \in \mathbb{Z} \qquad \text{i.e.} \qquad \gamma \in 2\pi\hbar\Gamma^* = \Lambda, \text{ say.} \tag{43}$$

At this point we shall rather abruptly leave our discussion of string theory amplitudes because we have enough to motivate the introduction of vertex operators with momenta lying on an integral lattice. This motivation has necessarily been rather sketchy in order to avoid too big a digression.

4. The Algebra Associated with an Integral Lattice

We shall now explore further the string model formalism of section 3, taking the momenta to lie on a lattice Λ. For the purposes of exploring the mathematics of the construction, the could be taken to lie in a space with any signature we wish; it could be Lorentzian, Euclidean or there could even be certain directions which were totally null, i.e. the metric tensor might be singular. (More complicatedly it could be that only the projection of the momenta on a certain subspace lies on a discrete lattice, corresponding to a configuration space which is the product of Minkowski space and a torus; this is indeed the situation in the applications which are considered to be promising from the point of view of physical relevance.)

Specifically we consider the Hilbert space, $\mathcal{H}$, generated by the operators α_n^μ, $n \in \mathbb{Z}$, $1 \leq \mu \leq d$, satisfying the commutation relations

$$[\alpha_m^\mu, \alpha_n^\nu] = m g^{\mu\nu} \delta_{m,-n} \tag{44}$$

and the hermiticity conditions

$$\alpha_m^{\mu\dagger} = \alpha_{-m}^\mu \tag{45}$$

from states $|\gamma\rangle$, $\gamma \in \Lambda \subset \mathbb{R}^d$, obeying

$$p^\mu |\gamma\rangle = \gamma^\mu |\gamma\rangle \tag{46a}$$

$$\alpha_m^\mu |\gamma\rangle = 0, \quad m > 0, \tag{46b}$$

$$\langle \gamma' | \gamma \rangle = \delta_{\gamma'\gamma}, \tag{46c}$$

where $p^\mu \equiv \alpha_0^\mu$. In this context we cannot define the position operator q^μ but we can define $\exp\{i\gamma.q\}$ for $\gamma \in \Lambda$ by

$$e^{i\gamma.q} |\gamma'\rangle = |\gamma + \gamma'\rangle \tag{47}$$

and the condition it commutes with α_n^μ, $n \neq 0$. Then the whole of is generated from the vacuum vector $|0\rangle$ by α_n^μ and $\exp(i\gamma.q)$, $\gamma \in \Lambda$.

We wish to investigate the algebra generated by the operators

$$A_r = \frac{1}{2\pi i} \oint \frac{dz}{z} U(r,z), \tag{48}$$

for

$$r \in \Lambda_2 = \{r \in \Lambda : r^2 = 2\} \tag{49}$$

How does one calculate $[A_r, A_s]$? Consider

$$A_r A_s = \frac{1}{2\pi i}\oint \frac{dz}{z} U(r,z) \frac{1}{2\pi i}\oint \frac{d\zeta}{\zeta} U(s,\zeta). \tag{50}$$

To normal order the integrand in eq.(50) we use

$$\begin{aligned} &\exp\{ir.Q_>(z)\} \exp\{is.Q_<(\zeta)\} \\ &\quad = (1-\zeta/z)^{r.s} \exp\{is.Q_<(\zeta)\} \exp\{ir.Q_>(z)\}, \text{ if } |\zeta|<|z|, \end{aligned} \tag{51}$$

which follows from

$$[Q_>^\mu(z), Q_<^\nu(\zeta)] = -g^{\mu\nu}\log(1-\zeta/z), \quad \text{if } |\zeta|<|z|, \tag{52}$$

and

$$e^A e^B = e^{[A,B]} e^B e^A, \tag{53}$$

which holds if [A,B] is a c-number. It follows that

$$\begin{aligned} U(r,z)U(s,\zeta) &= \exp\{ir.Q_<(z) + is.Q_<(\zeta)\} \\ &\quad .(z-\zeta)^{r.s} z\zeta\, e^{i(r+s).q} z^{r.p} \zeta^{s.p} \\ &\quad .\exp\{ir.Q_>(z) + is.Q_>(\zeta)\}, \quad \text{for } |\zeta|<|z| \\ &\equiv U_{r,s}(z,\zeta) \end{aligned} \tag{54}$$

The right hand side of eq.(54) is a regular function of z, ζ and except for $z = 0$, $\zeta = 0$ and $z = \zeta$. If we calculate the product in the other order we see that

$$U(s,\zeta)U(r,z)$$

will be given by the same function of z and ζ apart from a factor of $(-1)^{r.s}$, which could be either +1 or -1. Thus what we can easily calculate is

$$A_r A_s - (-1)^{r.s} A_s A_r = \frac{1}{(2\pi i)^2}\left\{ \oint\frac{d\zeta}{\zeta}\oint_{|z|>|\zeta|}\frac{dz}{z} - \oint\frac{d\zeta}{\zeta}\oint_{|\zeta|>|z|}\frac{dz}{z} \right\} U_{r,s}(z,\zeta). \tag{55}$$

The z contours in these two integrals, illustrated in Figure 1, have a difference which is equivalent to a small loop about ζ. Thus

$$A_r A_s - (-1)^{r.s} A_s A_r = \frac{1}{(2\pi i)^2}\oint_0 \frac{d\zeta}{\zeta}\oint_\zeta \frac{dz}{z} U_{r,s}(z,\zeta), \tag{56}$$

where the z contour positively encircles ζ once and the ζ

contour positively encircles the origin.

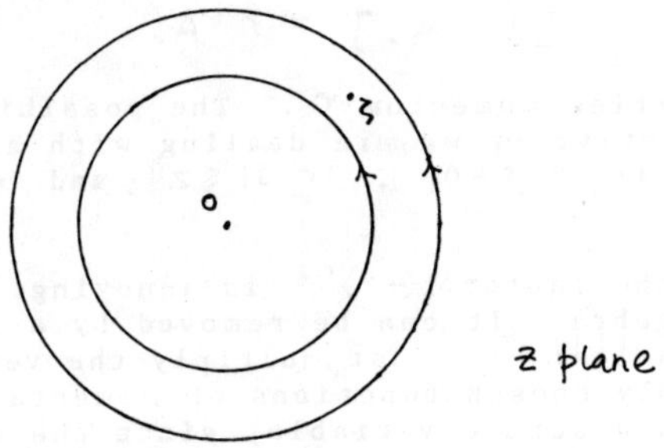

Figure 1

In certain circumstances (56) is easy to evaluate. Firstly, it vanishes if $r.s \geq 0$ because in this case $U_{r,s}(z,\zeta)$ is non-singular at $z=\zeta$. Secondly if $r.s=-1$, there is a simple pole at $z=\zeta$ and so

$$A_r A_s - (-1)^{r.s} A_s A_r = \frac{1}{2\pi i}\oint d\zeta\, e^{i(r+s).q}\, \zeta^{(r+s).p} \exp\{i(r+s).Q_<(\zeta)\} . \exp\{i(r+s).Q_>(\zeta)\},$$

$$= \frac{1}{2\pi i}\oint \frac{d\zeta}{\zeta} U(r+s,\zeta), \qquad (57)$$

$$= A_{r+s},$$

using eqs. (29)-(33) and since $(r+s)^2 = 2$ in this case. Thirdly, if $r.s=-2$ there is a double pole at $z=\zeta$ whose contribution is particularly easy to evaluate in the special case where $r=-s$. In this case it gives

$$A_r A_s - (-1)^{r.s} A_s A_r = \frac{1}{2\pi i}\oint \frac{d\zeta}{\zeta}\, r.P(\zeta), \qquad (58)$$

where

$$P^\mu(z) = iz\frac{dQ^\mu(z)}{dz}, \qquad (59a)$$

$$= \sum_{n=-\infty}^{\infty} \alpha_n^\mu z^{-n}. \qquad (59b)$$

Thus we have calculated the results:

$$A_r A_s - (-1)^{r.s} A_s A_r = 0, \quad \text{if } r.s \geq 0, \qquad (60a)$$

$$= A_{r+s}, \quad \text{if } r.s=-1, \qquad (60b)$$

$$A_r A_s - (-1)^{r.s} A_s A_r = r.p \quad \text{if } r = -s. \tag{60c}$$

Additionally we have that

$$[p^r, A_r] = r^r A_r \tag{60d}$$

because A_r carries momentum r. The possibilities listed in eqs. (60) are exhaustive if we are dealing with a Euclidean lattice because then, if $r, s \in \Lambda_2$, $|r.s| \leq 2$; and $r.s = -2$ if and only if $r = -s$.

Clearly the factor $(-1)^{r.s}$ is annoying. Without it we would have a Lie algebra. It can be removed by a trick introduced by Frenkel and Kac[1]: we just multiply the vertex operators A_r by certain suitably chosen functions of momenta (which are thus functions of a discrete variable, since the momenta lie on a lattice). Suppose we denote by Λ_R the sublattice of Λ generated by the points $r \in \Lambda_2$. For each $u \in \Lambda_R$, we introduce a function of momentum c_u with properties such that if $\hat{c}_u = \exp(iq.u)\, c_u$

$$\hat{c}_u \hat{c}_v = (-1)^{u.v} \hat{c}_v \hat{c}_u, \tag{61a}$$

$$\hat{c}_u \hat{c}_{-u} = 1, \tag{61b}$$

$$\hat{c}_u \hat{c}_v = \epsilon(u,v) \hat{c}_{u+v} \tag{61c}$$

where for each pair $u, v \in \Lambda_R$, $\epsilon(u,v)$ is either +1 or −1. If we can find such an object it will solve our problems because we can then set

$$E^r = A_r c_r, \qquad r \in \Lambda_2, \tag{62}$$

and we have in consequence

$$[E^r, E^s] = 0, \quad \text{if } r.s \geq 0, \tag{63a}$$

$$= \epsilon(r,s) E^{r+s}, \quad \text{if } r.s = -1, \tag{63b}$$

$$= r.p, \quad \text{if } r = -s, \tag{63c}$$

and

$$[p^r, E^r] = r^r E^r, \tag{63d}$$

$$[p^r, p^v] = 0. \tag{63e}$$

Further, provided that c_u is hermitian, p^r, E^r satisfy the hermiticity conditions

$$E^{r\dagger} = E^{-r}, \quad p^{r\dagger} = p^{-r} \tag{64}$$

Explanations of how to construct c_u are given in refs. [2] and [14].

In the Euclidean case, eqs.(63) define a closed finite dimensional Lie algebra g_Λ, which we can regard as the Lie algebra of some compact Lie group G_Λ associated with the integral lattice Λ. The momentum operators $p^r, 1 \leq r \leq d$, form a Cartan subalgebra for g_Λ and E^r are the step operators, so that

$$\text{rank } g_\Lambda = d, \tag{65a}$$

$$\dim g_\Lambda = d + |\Lambda_2| \tag{65b}$$

using $|\Lambda_2|$ for the number of points of length squared 2 on Λ; Λ_2 is the set of roots of g_Λ. (We have not completely established that there are no elements of g_Λ commuting with all the p^r which are not linear combinations of them. For this see ref. [2].) The algebra g_Λ will be non-abelian provided that Λ_2 is non-empty, and it will be semi-simple provided that $\dim \Lambda_R = \dim \Lambda$.

We can generalise what we have said here a little: it is not really necessary for the lattice Λ to be integral. We need only that the operators E^r are well-defined for $r \in \Lambda_2$ and this requires that

$$e^{ir.Q_0(z)}|\lambda\rangle = z^{1+\lambda.r}|\lambda+r\rangle \tag{66}$$

be single valued [cf. eq.(42)]. This happens if and only if $\lambda.r \in \mathbb{Z}$ whenever $r \in \Lambda_2$ and $\lambda \in \Lambda$. This condition can be equivalently written

$$\Lambda_R \subset \Lambda^* \tag{67}$$

which is a weaker condition than that we previously imposed, that the lattice be integral, i.e. $\Lambda \subset \Lambda^*$.

Let us consider some examples of Euclidean lattices. The simplest possibility is to take $\Lambda = \mathbb{Z}^d$, $d > 1$, the hypercubic lattice in d dimensions. There are $2d(d-1)$ points of squared length 2 on this lattice. These are the points for which only two of the coordinates are non-zero and these are each ± 1. Thus Λ_R consists of points of the form

$$(n_1, n_2, \ldots, n_d), \quad n_i \in \mathbb{Z}, \quad \sum n_i \in 2\mathbb{Z} \tag{68}$$

It is not difficult to check that Λ_2 is the root system of $so(2d)$ and, of course, the dimensions check

$$d + |\Lambda_2| = d(2d-1) = \dim so(2d) \tag{69}$$

If we exponentiate the generators we have constructed we get a specific group G_Λ with g_Λ as algebra, that is we get a specific global structure. The torus from which we have started is essentially the maximal torus, the maximal abelian subgroup of G_Λ. We can determine this global structure by looking at which eigenvalues of the generators p^r, $1 \leq r \leq d$, occur, and this is given exactly by the original lattice Λ. In other words Λ is the set of weights of G_Λ. The condition that λ be a weight for g_Λ is that

$$2\, r.\lambda / r^2 \in \mathbb{Z} \quad \text{for all roots } r \tag{70}$$

and this is precisely (67) because in our construction all the roots of g_Λ inevitably have $r^2 = 2$. In the case $\Lambda = \mathbb{Z}^d$, the points of length 1, which are just those with only one non-zero coordinate and that coordinate being ± 1, correspond to the weights of the vector representation of SO(2d). Since these weights generate additively all of $\mathbb{Z}^d$, we have

$$G_{\mathbb{Z}^d} = SO(2d) \tag{71}$$

The largest lattice permitted, if we are to have g_Λ equal to so(2d), is the dual of (68), namely the lattice consisting of the points $(n_1, n_2, \ldots, n_d)$ where

$$\underline{\text{either}} \quad n_i \in \mathbb{Z}, \; 1 \leq i \leq d, \qquad \underline{\text{or}} \quad n_i + \tfrac{1}{2} \in \mathbb{Z}, \; 1 \leq i \leq d. \tag{72}$$

This is the lattice of all weights of so(2d). Taking this lattice instead of $\mathbb{Z}^d$, we still have g_Λ = so(2d), provided that d $\neq$ 8, but we have G_Λ = Spin(2d), the simply connected covering group of SO(2d). On the other hand, the smallest possible choice is to take Λ to be just Λ_R, in which case we would have

$$G_\Lambda = SO(2d)/\mathbb{Z}_2, \text{ where } \mathbb{Z} = \{\pm 1\} \subset SO(2d).$$

If d is even, there are yet other possible choices for Λ. We could take Λ to be the lattice consisting of Λ_R together with the points

$$(n_1 + \tfrac{1}{2}, n_2 + \tfrac{1}{2}, \ldots, n_d + \tfrac{1}{2}), \; n_i \in \mathbb{Z}, \; \Sigma n_i \in 2\mathbb{Z}. \tag{73}$$

This lattice is only integral if d is a multiple of 4 and, in that case it is, like $\mathbb{Z}^d$, self-dual, i.e. $\Lambda = \Lambda^*$. But when d is a multiple of 8 it has the additional property, not possessed by $\mathbb{Z}^d$, of being even, like $II^{8n+1,1}$, which it resembles in construction. In general, this choice of Λ produces a group G_Λ, which is Spin(2d) divided by a different subgroup, not isomorphic to SO(2d). The exceptions occur when d = 4 or 8. If d = 8, the lattice has extra points of squared length 2, namely the 128 permitted points each of whose coordinates is $\pm\frac{1}{2}$. The corresponding operators E^r increase in number from 112 to 240 and in fact $G_\Lambda = E_8$. This is the essentially unique self-dual Euclidean lattice of dimension 8 mentioned in Section 2. [Of course, it is open to us not to add these extra points and stick with SO(16).] If d = 4, the resulting group G_Λ is actually isomorphic to SO(8). Actually, the condition in (73) that Σn_i be even could be replaced by the condition that Σn_i be odd; for any d this would produce an isomorphic G_Λ and for d = 4 it produces a third copy of SO(8). These three so(8) algebras are related by outer isomorphisms associated with the symmetry of the corresponding Dynkin diagram; see Fig. 2. This is known as the trialty property of SO(8).

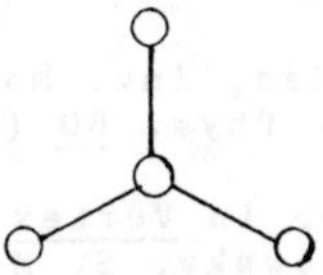

Figure 2

When d = 16 the lattice specified by adding the points of (73) to Λ_R , is again even and self dual. The only other possibility for such a lattice, with dimension 16, is the direct sum of two E_8 lattices. These even self dual lattices produce the groups $Spin(32)/\mathbb{Z}_2$ and $E_8 \times E_8$, respectively, which have excited a very great deal of interest recently[9].

Another exercise is to consider the possibilities in one dimension. There if Λ is to satisfy (67) and have points r with $r^2 = 2$, it must be either $\sqrt{2}\mathbb{Z}$ or $(1/\sqrt{2})\mathbb{Z}$. In the former case G_Λ = SO(3), in the latter SU(2). In two dimensions taking $\Lambda = \mathbb{Z}^2$ gives G_Λ = SO(4) = $SU(2) \times SU(2)/\mathbb{Z}_2$ whilst taking Λ = $\sqrt{2}\,\mathbb{Z}^2$ gives G_Λ = $SU(2) \times SU(2)$.

In general, we see that Λ is the weight lattice of some group with algebra g_Λ , a sublattice of the lattice of all weights of or, equivalently of the weight lattice of the simply connected group with algebra g_Λ. This Lie algebra is inevitably simply-laced, that is all the roots have the same squared length, 2. If g_Λ is simple, the only possibilities are that it is one of A_n = su(n+1), D_n = so(2n), E_6, E_7 and E_8; otherwise it is a direct sum of copies of these and u(1) factors. All simply-laced compact Lie groups can be obtained in this way.

5. Further Developments

In this lecture we have only been able to touch on some aspects of an extensive and fast growing body of knowledge about vertex operators, finite and infinite dimensional Lie algebras, and their applications in theoretical physics. Other topics include: points of squared length 1, their correspondence to Neveu-Schwarz and Ramond fermions, and the fermion-boson equivalence; points of squared length 4 and their relation to the Virasoro and Griess algebras[5,15]; vertex operators for non-simply-laced groups, interacting fermions and the Freundental magic square[16]; lattices in positive semi-definite spaces and untwisted affine Kac-Moody algebras; Lorentzian lattices and hyperbolic Kac-Moody algebras; and the interrelation of these ideas with the representation theory of Kac-Moody and Virasoro algebras. For a recent review of many of these topics see ref. [16].

References

[1] I.B. Frenkel and V.G. Kac, Inv. Math. 62 (1980) 23; G. Segal, Commun. Math. Phys. 80 (1981) 301.

[2] P. Goddard and D. Olive, in Vertex Operators in Mathematics and Physics, ed. J. Lepowsky, S. Mandelstam and I. Singer (Springer-Verlag, 1984), p.51.

[3] I.B. Frenkel, Am. Math. Soc. Lectures in Applied Mathematics 21 (1985) 325.

[4] F. Gliozzi, D. Olive and J. Scherk, Nucl. Phys. B122 (1977) 253; M. Green and J. Schwarz, Nucl. Phys. B181 (1981) 502.

[5] I.B. Frenkel, J. Lepowsky and A. Meurman, in Vertex Operators in Mathematics and Physics, ed. J. Lepowsky, S. Mandelstam and I. Singer (Springer-Verlag, 1984), p.231.

[6] R.L. Griess, Inv. Math. 69 (1982) 1.

[7] J.-P. Serre, A Course in Arithmetic (Springer-Verlag, 1973).

[8] D. Olive, in Monopoles in Quantum Field Theory, ed. N. Craigie, P. Goddard and W. Nahm (Springer-Verlag, 1982), p.157.

[9] M. Green and J. Schwarz, Phys. Lett. 149B (1984) 117; D. Gross, J.A. Harvey, E. Martinec and R. Rohm, Phys. Rev. Lett. 54 (1985) 502; Nucl. Phys. B256 (1985) 253.

[10] R.C. Brower, Phys. Rev. D6 (1972) 1655.

[11] P. Goddard and C.B. Thorn, Phys. Lett. 40B (1972) 235.

[12] C.B. Thorn, in Vertex Operators in Mathematics and Physics, ed. J. Lepowsky, S. Mandelstam and I. Singer (Springer-Verlag, 1984), p.411.

[13] A. Neveu and J.H. Schwarz, Nucl. Phys. B31 (1971) 86; P. Ramond, Phys. Rev. D3 (1971) 2415.

[14] P. Goddard and D. Olive, Kac-Moody and Virasoro Algebras in Relation to Quantum Physics, to be published in Journal of Modern Physics A.

[15] E. Corrigan, Durham preprints DTP 85/19 and 85/21.

[16] P. Goddard, D. Olive and A. Schwimmer, Phys. Lett. 157B (1985) 393; P. Goddard, W. Nahm, D. Olive and A. Schwimmer to appear.

NON-ABELIAN GAUGE FIELDS FROM SUPERSTRING COMPACTIFICATION

Leonardo Castellani
Istituto Nazionale di Fisica Nucleare, Sezione di Torino, Italy

ABSTRACT

The N = 1 closed superstring is compactified on a particular 6 - torus. Solitonic excitations annihilate the potential tachyon of the spinning string, and give rise to massless non-abelian gauge vectors. The massless states of the resulting 4-dimensional string coincide with those of N = 4, d = 4 supergravity coupled to a N = 4 non-abelian vector multiplet.

Superstrings[1] offer a promising framework for a finite, unified theory of gravity and matter. The discovery[2] that type I superstrings are finite and anomaly-free if the gauge group is SO(32) has sparked off a renewed interest in string models. With the addition of the recently constructed heterotic string[3], there are now five superstring theories. All of them live in ten space-time dimensions.

For phenomenological applications, the most attractive candidate seems to be the $E_8 \times E_8$ heterotic string[3,4]. Its compactification on Calabi-Yau manifolds (Ricci-flat with SU(3) holonomy) is argued to produce, in the field theory limit, N = 1 four dimensional supergravity coupled to E_6 vector multiplets[4].

Type II superstrings contain only closed, oriented strings: Yang-Mills interactions are absent, and it was deemed unlikely that they could appear upon compactification.

In this work I present a way to compactify the closed N = 1 superstring, such that the resulting four dimensional theory contains massless non-abelian gauge vectors. In other words, it is possible to generate a non-abelian gauge group upon $10 \rightarrow 4$ compactification, without need of preexisting Yang-Mills states in the ten dimensional string.

In view of this result, type IIB chiral superstrings may well become relevant for phenomenology, especially since they are known to be anomaly-free (at least in the field theory limit)[5].

As starting point, let us briefly review the superstring in the "old formalism" of the a, b, d oscillators of Neveu-Schwarz and Ramond[6,7]. This has the advantage of allowing a tachyonic state. The tachyon is now a desirable feature, since it can be annihilated by solitonic excitations of the compactified dimensions;

if the compact internal space is chosen appropriately, these solitons give rise, in the massless sector, to non-abelian gauge vectors[8,9,3].

The bosonic (Neveu-Schwarz) sector of the closed spinning string is described in terms of the oscillators a_n^μ, $\tilde{a}_n^\mu$ ($n \in \mathbb{Z}$) and b_r^μ, $\tilde{b}_r^\mu$ ($r \in \mathbb{Z} + 1/2$), satisfying the (anti)commutation relations

$$\left[a_m^\mu, a_{-m}^\nu\right] = \left[\tilde{a}_m^\mu, \tilde{a}_{-m}^\nu\right] = \eta^{\mu\nu}\delta_{mm} \qquad (1)$$

$$\left\{b_r^\mu, b_{-s}^\nu\right\} = \left\{\tilde{b}_r^\mu, \tilde{b}_{-s}^\nu\right\} = \eta^{\mu\nu}\delta_{rs} \qquad (2)$$

all other (anti)commutators vanishing; a_n^μ, b_r^μ refer to right-moving modes, $\tilde{a}_n^\mu$, $\tilde{b}_r^\mu$ to left-moving modes. The physical states $|\phi\rangle$ are defined by the infinite set of gauge conditions[6,7]

$$\begin{aligned} L_n|\phi\rangle &= \tilde{L}_m|\phi\rangle = 0 \qquad m>0\,,\ m\in\mathbb{Z} \\ G_r|\phi\rangle &= \tilde{G}_r|\phi\rangle = 0 \qquad r>0\,,\ r\in\mathbb{Z}+1/2 \\ (L_0-\tfrac{1}{2})|\phi\rangle &= (\tilde{L}_0-\tfrac{1}{2})|\phi\rangle = 0. \end{aligned} \qquad (3)$$

The last equation implies the mass shell condition :

$$\frac{1}{2}\alpha' m^2 = N + \tilde{N} - 1 \qquad (4)$$

and the constraint

$$N - \tilde{N} = 0 \qquad (5)$$

where N and $\tilde{N}$ are given by

$$N = \sum_{n=1}^{\infty} n\, a_{-n} \cdot a_n + \sum_{r=1/2}^{\infty} r\, b_{-r} \cdot b_r$$
$$\tilde{N} = \sum_{n=1}^{\infty} n\, \tilde{a}_{-n} \cdot \tilde{a}_n + \sum_{r=1/2}^{\infty} r\, \tilde{b}_{-r} \tilde{b}_r \qquad (6)$$

Thus the ground state $|0_{NS}\rangle|\tilde{0}_{NS}\rangle$ is a tachyon, with $\alpha' m^2 = -1$. In order to build a supersymmetric model, Gliozzi, Scherk and Olive[7] retained only the G = +1 sector of the spectrum, G being the conserved multiplicative quantum number :

$$G = (-1)^{\sum_r b_{-r} \cdot b_r - 1} = (-1)^{\sum_r \tilde{b}_{-r} \cdot \tilde{b}_r - 1} \qquad (7)$$

The last equality follows from the constraint (5). The G = +1 sector is tachyon-free.

The fermionic modes of the supersymmetric spinning string are described by the oscillators a_n^μ, $\tilde{a}_n^\mu$ and d_n^μ, $\tilde{b}_r^\mu$, the right-handed d_n^μ oscillators being integrally moded (Ramond type) and the left-handed $\tilde{b}_r^\mu$ half-integrally moded (Neveu-Schwarz type). The physical states satisfy[6,7]

$$\begin{aligned} &F_m |\phi\rangle = 0 \qquad m \geq 0, \qquad m \in \mathbb{Z} \\ &L_m |\phi\rangle = \tilde{L}_m |\phi\rangle \qquad m > 0 \\ &\tilde{G}_r |\phi\rangle = 0 \qquad r > 0, \qquad r \in \mathbb{Z} + 1/2 \\ &(\tilde{L}_0 - \tfrac{1}{2}) |\phi\rangle = 0, \quad L_0 |\phi\rangle = 0 \end{aligned} \qquad (8)$$

From the last two equations one deduces the mass-shell condition

$$\frac{1}{2} \alpha' m^2 = R + \tilde{R} - 1/2 \qquad (9)$$

and the constraint

$$R = \tilde{R} - 1/2 \tag{10}$$

with

$$\begin{aligned} R &= \sum_1^\infty n\, a_{-n} \cdot a_n + \sum_1^\infty n\, d_{-n} \cdot d_n \\ \tilde{R} &= \sum_1^\infty n\, \tilde{a}_{-n} \cdot \tilde{a}_n + \sum_{1/2}^\infty r\, \tilde{b}_{-r} \cdot \tilde{b}_r \end{aligned} \tag{11}$$

The spectrum given by (9) is tachyon-free, the state with $R = \tilde{R} = 0$ being removed by the condition (10).

Putting together the G = +1 bosonic sector and the "mixed" fermion sector (supplemented by the Majorana-Weyl condition) yields the supersymmetric spinning string[7]. Its massless modes combine into the N = 1, d = 10 supergravity multiplet. The graviton, the antisymmetric tensor and the scalar correspond to the symmetric traceless, antisymmetric, and trace part of the state:

$$b_{-1/2}^{\mu}\, \tilde{b}_{-1/2}^{\nu}\, |0_{NS}\rangle\, |\tilde{0}_{NS}\rangle \tag{12}$$

The gravitino and the spin 1/2 are contained in

$$\tilde{b}_{-1/2}^{\mu}\, |0_R\rangle\, |\tilde{0}_{NS}\rangle \tag{13}$$

where $|0_R\rangle$ is the spinorial Ramond vacuum.

Let us now examine the $10 \longrightarrow 4$ compactification of this model on a 6 - torus. The 6 compact string coordinates are given by

$$\begin{aligned} x^I(\tau,\sigma) &= x^I + p^I\tau + 2L^I\sigma + \\ &+ \frac{i}{2}\sum_{m\neq 0} a_m^I e^{-2im(\tau-\sigma)} + \frac{i}{2}\sum_{m\neq 0} \tilde{a}_m^I e^{-2im(\tau+\sigma)} \end{aligned} \tag{14}$$

where L^I counts how many times $x^I(\tau,\sigma)$ wraps around the I-th circle when σ varies from 0 to π. Separating right and left-moving modes :

$$X^I(\tau-\sigma) = \frac{x^I}{2} + \left(\frac{p^I}{2} - L^I\right)(\tau-\sigma) + \frac{i}{2}\sum_{m\neq 0} a_m^I e^{-2im(\tau-\sigma)}$$
$$\tilde{X}^I(\tau+\sigma) = \frac{x^I}{2} + \left(\frac{p^I}{2} + L^I\right)(\tau+\sigma) + \frac{i}{2}\sum_{m\neq 0} \tilde{a}_m^I e^{-2im(\tau+\sigma)} \tag{15}$$

For the bosonic sector (henceforth denoted by NS $\otimes$ $\widetilde{\mathrm{NS}}$), eqs. (4), (5) become ($\alpha' = 1/2$):

$$\frac{m^2}{4} = N + \tilde{N} - 1 + \frac{1}{2}(P^I)^2 + \frac{1}{2}(\tilde{P}^I)^2 \tag{16}$$

$$N + \frac{1}{2}(P^I)^2 = \tilde{N} + \frac{1}{2}(\tilde{P}^I)^2 \tag{17}$$

where $P^I \equiv p^I/2 - L^I$, $\tilde{P}^I \equiv p^I/2 + L^I$

For the "mixed" fermion sector (R $\otimes$ $\widetilde{\mathrm{NS}}$), eqs. (9), (10) become :

$$\frac{m^2}{4} = R + \tilde{R} - \frac{1}{2} + \frac{1}{2}(P^I)^2 + \frac{1}{2}(\tilde{P}^I)^2 \tag{18}$$

$$R + \frac{1}{2}(P^I)^2 = \tilde{R} + \frac{1}{2}(\tilde{P}^I)^2 - \frac{1}{2} \tag{19}$$

Now the crucial observation is that the operator

$$G = (-1)^{\sum_{1/2}^{\infty} b_{-r}\cdot b_r - 1} \tag{20}$$

is not necessarily equal to

$$\tilde{G} = (-1)^{\sum_{1/2}^{\infty} \tilde{b}_{-r}\cdot \tilde{b}_r - 1} \tag{21}$$

as in the non-compactified theory. Indeed because of solitonic excitations ($L^I \neq 0$), the constraint $N = \tilde{N}$ is modified as in (17), thus allowing $G \neq \tilde{G}$. Recalling that $G = +1(-1)$ selects states with an odd (even) number of b_r^μ oscillators, $G = +1$ (-1) corresponds to half-integer (integer) values for N. The same correspondance holds for $\tilde{G}$ and $\tilde{N}$. Hence we have the four possibilities:

$$\begin{array}{ll} G = \tilde{G} = +1 & N = 1/2, 3/2, 5/2, \ldots \\ & \tilde{N} = 1/2, 3/2, 5/2, \ldots \\[2ex] G = \tilde{G} = -1 & N = 0, 1, 2, \ldots \\ & \tilde{N} = 0, 1, 2, \ldots \\[2ex] G = +1, \ \tilde{G} = -1 & N = 1/2, 3/2, 5/2, \ldots \\ & \tilde{N} = 0, 1, 2, \ldots \\[2ex] G = -1, \ \tilde{G} = +1 & N = 0, 1, 2, \ldots \\ & \tilde{N} = 1/2, 3/2, 5/2 \ldots \end{array} \tag{22}$$

For solitonic states, the constraint (17) allows $N \in \mathbb{Z}$, $\tilde{N} \in \mathbb{Z} + \frac{1}{2}$ (or viceversa), provided that $(P^I)^2 - (\tilde{P}^I)^2$ is an odd integer.

The compactified model I want to propose has only two of the four sets of states in (22), namely the two sectors with G = +1.* The spectrum is therefore tachyon-free (the tachyon belongs to the $G = -1$, $\tilde{G} = -1$ sector), and contains massless solitonic states, in addition to the Kaluza-Klein reduced massless supergravity multiplet.

Using formulas (16) and (18), it is easy to verify that the following states are massless and satisfy the constraints (17) and (19) :

---

* This is a consistent choice, since the interaction of the two sectors does not produce states with G = - 1.

$NS \otimes \widetilde{NS}$ sector

$$b^{\mu}_{-1/2}\,\tilde{b}^{\nu}_{-1/2}\,|0_{NS}\rangle|\tilde{0}_{NS}\rangle \qquad \text{graviton , scalar , antisymmetric tensor (= pseudoscalar)}$$

$$\left.\begin{aligned} & b^{\mu}_{-1/2}\,\tilde{b}^{I}_{-1/2}\,|0_{NS}\rangle|\tilde{0}_{NS}\rangle \\ & b^{I}_{-1/2}\,\tilde{b}^{\mu}_{-1/2}\,|0_{NS}\rangle|\tilde{0}_{NS}\rangle \end{aligned}\right\} \qquad \text{6 + 6 Kaluza-Klein vectors}$$

$$b^{I}_{-1/2}\,\tilde{b}^{J}_{-1/2}\,|0_{NS}\rangle|\tilde{0}_{NS}\rangle \qquad 6 \times 6 \text{ scalars}$$

$$b^{\mu}_{-1/2}\,|0_{NS}\rangle|(\tilde{P}^{I})^2 = 1\rangle \qquad m(\Lambda_1)\text{vectors}$$

$$b^{I}_{-1/2}\,|0_{NS}\rangle|(\tilde{P}^{I})^2 = 1\rangle \qquad m(\Lambda_1)\text{scalars}$$

(23)

$R \otimes \widetilde{NS}$ sector

$$\tilde{b}^{\mu}_{-1/2}\,|0_{R}\rangle|\tilde{0}_{NS}\rangle \qquad \text{4 gravitini , 4 spin 1/2}$$

$$\tilde{b}^{I}_{-1/2}\,|0_{R}\rangle|\tilde{0}_{NS}\rangle \qquad 4 \times 6 \text{ spin 1/2}$$

$$|0_{R}\rangle|(\tilde{P}^{I})^2 = 1\rangle \qquad 4 \times m(\Lambda_1) \quad \text{spin 1/2}$$

(24)

The 6 - torus has been chosen so that the "momenta" P^I lie on a lattice that contains a finite sublattice Λ_1 of unit lenght; $m(\Lambda_1)$ is the number of points in Λ_1 .

The massless states in (23), (24) arrange themselves into the N = 4, d = 4 supergravity multiplet :

STATE	SPIN	# FIELDS
$b^{(\mu}\tilde{b}^{\nu)}\lvert 0_{NS}\rangle\lvert\tilde{0}_{NS}\rangle$	2	1
$\left(\tilde{b}^{\mu}\lvert 0_R\rangle\lvert\tilde{0}_{NS}\rangle\right)_{3/2}$	3/2	4
$b^{I}\tilde{b}^{\mu}\lvert 0_{NS}\rangle\lvert\tilde{0}_{NS}\rangle$	1	6
$\left(\tilde{b}^{\mu}\lvert 0_R\rangle\lvert\tilde{0}_{NS}\rangle\right)_{1/2}$	1/2	4
$b^{\mu}\tilde{b}^{\mu}\lvert 0_{NS}\rangle\lvert\tilde{0}_{NS}\rangle$	0^+	1
$b^{[\mu}\tilde{b}^{\nu]}\lvert 0_{NS}\rangle\lvert\tilde{0}_{NS}\rangle$	0^-	1

and into the N = 4 vector multiplet :

STATE	SPIN	# FIELDS
$b^{\mu}\tilde{b}^{I}\lvert 0_{NS}\rangle\lvert\tilde{0}_{NS}\rangle$, $b^{\mu}\lvert 0_{NS}\rangle\lvert(\tilde{P}^I)^2 = 1\rangle$	1	$1\times[m(\Lambda_1)+6]$
$\tilde{b}^{I}\lvert 0_R\rangle\lvert\tilde{0}_{NS}\rangle$, $\lvert 0_R\rangle\lvert(\tilde{P}^I)^2 = 1\rangle$	1/2	$4\times[m(\Lambda_1)+6]$
$b^{I}\tilde{b}^{J}\lvert 0_{NS}\rangle\lvert\tilde{0}_{NS}\rangle$, $b^{I}\lvert 0_{NS}\rangle\lvert(\tilde{P}^I)^2 = 1\rangle$	0	$6\times[m(\Lambda_1)+6]$

Notice that the Ramond 10-dimensional spinorial vacuum $\lvert 0_R\rangle$ splits into four 4-dimensional spinors, $\lvert 0_R\rangle$ being Majorana-Weyl in 10 dimensions.

A non-abelian symmetry (generated by Frenkel-Kac vertex operators) is obtained by choosing Λ_1 to be an integral, simply laced weight lattice[8,9]. This selects Λ_1 to be the weight lattice of the vector representation of SO(2r). Then $n(\Lambda_1) = 2r$ and there exist $r\,(2r-1)$ vertex operators transforming the $2r$ $|(P^I)^2 = 1\rangle$ states in the vector representation of SO(2r).

The $r\,(2r-1)$ vertex operators are bilinear in $2r$ anticommuting operators associated to the lattice Λ_1. They close the algebra of SO(2r), and commute with the super-Virasoro generators so to map physical states into physical states.

In the $10 \rightarrow 4$ compactified model of this work, $r = 6$ and Λ_1 is the six-dimensional weight lattice of the 12 of SO(12). The 12 massless vector states $|(P^I)^2 = 1\rangle$ transform in the vector representation of SO(12). The question is whether they can be interpreted as gauge fields, belonging to the adjoint representation of a local symmetry group $\mathbb{G}$.

This happens if $\mathbb{G}$ is a subgroup of SO(12), such that the 12 of SO(12) branches into the adjoint of $\mathbb{G}$. Then the SO(12) vertex operators corresponding to the $\mathbb{G}$- subalgebra act irreducibly on the 12 vectors, and transform them in the adjoint.

The remarkable fact is that there exist a (maximal) subgroup SU(3) X SU(2) of SO(12) for which

$$\underline{12} \xrightarrow{SU(3)\times SU(2)} (8,1) + (1,3) + (1,1) \qquad (25)$$

where the first (second) entry in the parentheses refers to the SU(3) (SU(2)) irreps. We see that the vector of SO(12) branches into the adjoint of SU(3) X SU(2) . The (1,1) singlet can be thought as gauging an extra U(1) local symmetry. The 12 massless vectors therefore do belong to the adjoint of SU(3) X SU(2) X U(1).

The SU(3) X SU(2) subgroup in (23) is embedded in SO(12) as

$$SO(12) \supset SO(11) \supset SU(2) \times SO(8) \supset SU(3) \times SU(2)$$

where SU(3) is maximally embedded in SO(8) so that

$$\underline{8} \text{ (of } SO(8)) \xrightarrow{SU(3)} \underline{8} \text{ (of } SU(3))$$

Note that for SU(3) X SU(2) X U(1) to appear as a symmetry, the compactified dimensions must be at least six. Indeed SO(2r) with $r < 6$ never branches into the adjoint of a subgroup SU(3) X SU(2).

The compactification scheme we have discussed can in principle be applied also to type II superstrings, provided one uses an equivalent "old formalism" in terms of a, b, d oscillators, considering both the mixed fermi sectors $R \otimes \widetilde{NS}$ and $\widetilde{R} \otimes NS$. This yields the two massless gravitini necessary for N=2 supersymmetry. Symmetrization between left and right-moving modes brings back the model to the N = 1 superstring (with an "old formalism" describing unoriented closed strings). Notice that in ref. [7], and in the present Letter, the a, b, d oscillators describe oriented closed strings. The asymmetry is due to the choice of the $R \otimes \widetilde{NS}$ sector. A symmetrized $[(R \otimes \widetilde{NS}) + (\widetilde{R} \otimes NS)]_{symm}$ is probably more correct, since we know that type I superstrings are unoriented.

There is a number of other questions to be addressed, one of which is the conjectured N = 4 supersymmetry of the model. The massless states are consistent with N = 4. Furthermore, each massive level contains an equal number of bose and fermi states. This is obvious for the $(P^I)^2 = (\tilde{P}^I)^2 = 0$ states, since they can be reassembled in the d = 10 supermultiplets of the N = 1 closed superstring. Bose - fermi matching between solitonic states is not difficult to prove, recalling that

i) the closed string states $NS \otimes \widetilde{NS}$, $R \otimes \widetilde{NS}$ are tensorial products of the open string states NS, $\widetilde{NS}$, R.

ii) for open strings, the Ramond (Majorana-Weyl) states $\{R\}$ precisely match the Neveu-Schwarz $G = +1$ states $\{NS\}$

iii) the action of soliton creation operators $e^{i\varphi^I p^I}$ does not alter this matching, although it does change the $(\text{mass})^2$ of a state. The correspondance of bose and fermi $(P^I)^2 = 0$ states at a given $(\text{mass})^2 = M$ is just reproduced for $(P^I)^2 \neq 0$ states with $(\text{mass})^2 = M + 1/2\,(P^I)^2$

This strongly hints at the existence of an $N = 4$, $d = 4$ superstring, yet to be formulated in an explicitly supersymmetric formalism .

ACKNOWLEDGEMENTS

I wish to acknowledge useful discussions with P. Fré, R. D'Auria and F. Gliozzi.

NOTE

The SO(12) or SU(3) X SU(2) X U(1) symmetries of the compactified model are really only symmetries of the massless states. For a discussion on symmetries preserved by interactions, and on nonabelian gaugings, see ref. [10] : There we show that if D dimensions of the closed N = 1 supersymmetric string are compactified on a torus, solitonic excitations can enlarge the $[U(1)]^D$ Kaluza-Klein gauge symmetry to $[SU(2) \times U(1)]^D$. For example, if D = 6, the 12 + 6 + 6 vectors in (23) gauge the group $[SU(2) \times U(1)]^6$. In the same paper [10] , the "old formalism is extended to obtain the N = 2 supersymmetric string. In its compactification to four dimensions, we find that massless solitons break N=8 to N=4 supersymmetry.

REFERENCES

[1] M. B. Green and J. H. Schwarz, Nucl. Phys. B 181 (1981) 502, B 198 (1982) 252, B 198 (1982) 441; Phys. Lett 109B (1982) 444; M. B. Green, J. H. Schwarz and L. Brink, Nucl. Phys. B 198 (1982) 474.

[2] M. B. Green and J. H. Schwarz, Phys. Lett. 149 B (1984) 117, 151 B (1985) 21.

[3] D. Gross, J. Harvey, E. Martinec and R. Rohm , Nucl. Phys. B 256 (1985) 253.

[4] P. Candelas, G. T. Horowitz, A. Strominger and E. Witten, Nucl. Phys. B 258 (1985) 46.

[5] L. Alvarez-Gaumé and E. Witten, Nucl. Phys. B 234 (1983) 269.

[6] A. Neveu and J. H. Schwarz, Nucl. Phys. B 31 (1971) 86; Phys. Rev. D4 (1971)1109.
P. Ramond, Phys. Rev. D3 (1971) 2415.

[7] F. Gliozzi, J. Scherk and D. Olive, Nucl. Phys. B 122 (1977) 253.

[8] I. B. Frenkel and V. G. Kac, Inv. Math. 62 (1980) 23.

[9] P. Goddard and D. Olive, DAMTP preprint (1983).

[10] L. Castellani, R. D' Auria, F. Gliozzi and S. Sciuto, Phys. Lett. 168B (1986) 47.

INTRODUCTION TO GAUGE COVARIANT STRING FIELD THEORY*

P. West**
CERN, Geneva

ABSTRACT

A local gauge covariant formulation of free strings is given and their interactions are discussed.

The work discussed in this contribution was found in collaboration with A. Neveu and H. Nicolai and is contained in references [1] and [2]. Before beginning to covariantly quantize an extended object, namely the string, it will be instructive to consider the corresponding path from the classical to the second quantized point particle.

The Point Particle

The trajectory of a classical point particle in space time is parameterized by its proper time τ and is given by $x^{\mu}(\tau)$. The path it takes is so as to minimise the action

$$-m \int d\tau \sqrt{(-\dot{x}^{\mu}\dot{x}^{\nu}\eta_{\mu\nu})} = \tfrac{1}{2}\int d\tau \left\{ V^{-1}\dot{x}^{\mu}\dot{x}^{\nu}\eta_{\mu} - m V \right\} . \qquad (1)$$

* Lectures given also at the Scottish Universities Nato Summer School, Edinburgh, August 1985.

** Permanent address: Mathematics Department, King's College, The Strand, London WC2, U.K.

where $\dot{x}^{\mu} = \frac{dx^{\mu}}{d\tau}$. These actions are invariant under reparameterizations of the proper time $\tau \to f(\tau)$ with the transformation of x^{μ} being $\delta x^{\mu} = f(\tau)\dot{x}^{\mu}$.

Let us now give a Hamiltonian treatment of the first action of equation (1), although the same results can be found from the second action. The canonical momentum is given by

$$P_{\mu} = \frac{\delta A}{\delta \dot{x}^{\mu}(\tau)} = \frac{m\,\dot{x}^{\mu}}{\surd(-\dot{x}^{\mu}\dot{x}^{\nu}\eta_{\mu\nu})} \tag{2}$$

and the equation of motion is of the form

$$\partial_{\tau} P_{\mu} = 0 \quad . \tag{3}$$

Due to the reparameterization invariance, the system is constrained by

$$\phi \equiv P_{\mu}^{2} + m^{2} = 0 \quad . \tag{4}$$

and the Hamiltonian vanishes;

$$H = P^{\mu}\dot{x}^{\mu} - L = 0 \quad . \tag{5}$$

The method of dealing with such a system was given in reference [3] and we now apply this in outline to the point particle. The reader who wishes to read further details is encouraged to consult reference [4]. We take the Hamiltonian to be proportional to the constraints, i.e.

$$H = v(\tau)\,(P^{\mu^{2}} + m^{2}) \tag{6}$$

where $v(\tau)$ is an arbitrary function of τ. One may verify that in this case there are no further constraints and that H generates time translations or reparameterizations in the sense that

$$\frac{\partial x^{\mu}}{\partial \tau} = \left\{ x^{\mu}(\tau), H \right\} = 2\, v(\tau)\, P^{\mu} \quad . \tag{7}$$

The fundamental Poisson brackets vanish except for

$$\left\{ x^{\mu} \, , \, P^{\nu} \right\} = i\hbar\, \eta^{\mu\nu} \quad . \tag{8}$$

To quantize the theory we make the usual transition from Poisson brackets to commutators, which are represented by the repacements

$$x^{\mu} \rightarrow x^{\nu} \quad ; \quad P^{\mu} \rightarrow - i\hbar \frac{\partial}{\partial x^{\mu}} \quad . \tag{9}$$

The constraints then become

$$\phi = (-\partial^{2} + m^{2}) \quad . \tag{10}$$

To find the second quantized field theory, we consider the state to be described by a field $\psi(x^{\mu}, \tau)$ and we impose the constraint

$$\phi\, \psi = 0 = (-\partial^{2} + m^{2})\, \psi \quad . \tag{11}$$

We also impose the Schrodinger equation

$$i\hbar \frac{\partial \psi}{\partial \tau} = H\, \psi \quad . \tag{12}$$

The right-hand side of this equation vanishes and we find that ψ is independent of τ. In the second quantized theory, there is in any case more than one particle and so the concept of more than one proper time is problematical.

The action that leads to the above Klein-Gordon equation is

$$A = \int d^4 x\, \psi(-\partial^2 + m^2)\psi \tag{13}$$

and we may use it to weight the Feynman path integral that can then be used to find the Greens function of the second quantized theory.

We note that the original reparameterization invariance of the proper time which was so important for determining the form of the classical action is absent in the second quantized theory; its only remnant being the field equation itself. Since we performed a Hamiltonian quantization with respect to the proper time this is only to be expected, but one might consider if one could second quantize in such a way as to maintain this invariance. It is the above steps that we now wish to repeat for the string.

<u>The Bosonic String</u>

The bosonic string, whose length is parameterized by σ sweeps out in time τ to a two-dimensional surface parameterized by $\xi^\alpha = (\tau,\sigma)$ in a space-time x^μ according to the function $x^\mu(\xi)$. This trajectory is so as to give the minimum area and so its action [5] is given by

$$A = -\frac{1}{2\pi\alpha'}\int d^2\xi \ \sqrt{(-\det \partial_\alpha x^\mu \partial_\beta x^\nu \eta_{\mu\nu})} \tag{14}$$

where α' has the dimensions of $(\text{mass})^2$. It is invariant under arbitrary reparameterizations of the two-dimensional surface $(0 \leqslant \sigma \leqslant \pi)$.

$$\xi^\alpha \to \xi^\alpha + f^\alpha(\xi) \quad , \quad x^\mu \to x^\mu + \xi^\alpha \partial_\alpha x^\mu \tag{15}$$

as well as invariants under the Poincaré group transformations acting on the space time x^μ.

The canonical momentum is given by

$$p_\mu = \frac{\delta A}{\delta \frac{\partial x^\mu}{\partial \tau}} = \frac{\partial_\tau x^\mu (x'^\nu)^2 - x'^\mu(\partial_\tau x^\nu x'^\nu)}{2\pi\alpha' \sqrt{(-\det(\partial_\alpha x^\mu \partial_\beta x^\nu \eta_{\mu\nu}))}} \tag{16}$$

where $x'^{\mu} = \frac{dx^{\mu}}{d\sigma}$.

Due to the invariance mentioned above, we have the constraints

$$P_{\mu}^{2} + \frac{1}{2\pi\alpha'}(x'^{\mu})^{2} = 0 \tag{17}$$

$$x'^{\mu} P_{\mu} = 0 \quad . \tag{18}$$

It is convenient to mathematical extend the range of σ from $-\pi$ to π by requiring

$$x^{\mu}(\sigma) = \begin{cases} x^{\mu}(\sigma) + & 0 \leqslant \sigma \leqslant \pi \\ x^{\mu}(-\sigma) & -\pi \leqslant \sigma \leqslant 0 \end{cases} \tag{19}$$

that is, $x^{\mu}(\sigma) = x^{\mu}(-\sigma)$. Using this extension, the above constraints can then be written as

$$(P^{\mu})^{2} \equiv \left[p^{\mu} + \frac{x'^{\mu}}{\sqrt{(2\pi\alpha')}}\right]^{2} = 0 \quad . \tag{20}$$

It will be advantageous to take the Fourier transform of these constraints and so we can define

$$L_{n} \equiv \frac{\pi\alpha'}{2} \int_{-\pi}^{\pi} d\sigma \, (P^{\mu}) \, e^{-in\sigma} \quad . \tag{21}$$

One finds that the L_n's obey the algebra [6].

$$\{L_{n} , L_{m}\} = (n-m) L_{n+m} \quad . \tag{22}$$

Since the Hamiltonian vanishes we take it to be proportional to the constraints, i.e.

$$H = \sum_{n=-\infty}^{\infty} C_n L_n \quad . \tag{23}$$

One may verify that the L_n's are the generators of two-dimensional conformal transformations in the world sheet of the string. Generally, the conformal group has only a finite number of generators, but for two dimensions only it is an infinite dimensional algebra which corresponds to making an arbitrary analytic transformation in $z = \tau + i\sigma$. Clearly, the two-dimensional flat metric which can be written as $dz\,d\bar{z}$ scales under $z \to f(z)$. The emergence of the two-dimensional conformal group rather than the original two-dimensional general coordinate group is presumably related to the choice of τ as the time to be used in the Hamiltonian approach. As we shall see, the Virasoro algebra and hence the two-dimensional conformal group play an important role in the second quantized gauge covariant theory.

The fundamental Poisson brackets of the theory are

$$\begin{aligned} \left\{x^{\mu}(\sigma)\ ,\ p^{\nu}(\sigma')\right\} &= \delta(\sigma-\sigma') \\ \{x^{\mu}(\sigma)\ ,\ x^{\nu}(\sigma')\} = 0 &= \{P^{\mu}(\sigma)\ ,\ P^{\nu}(\sigma')\} \end{aligned} \quad . \tag{24}$$

To quantize the theory we replace these relations by commutator relations which are represented by the changes

$$x^{\mu}(\sigma) \to x^{\mu}(\sigma) \quad ; \quad P^{\mu}(\sigma) \to -i\hbar \frac{\delta}{\delta x^{\mu}(\sigma)} \quad . \tag{25}$$

Making these replacements in the generators of the constraints, we find that

$$L_n = \frac{\pi\alpha'}{2} \int_{-\pi}^{\pi} d\sigma\, e^{-in\sigma} \quad \hat{P}^{\mu}(\sigma)\, \hat{P}^{\nu}(\sigma)\eta_{\mu\nu} \tag{26}$$

where

$$\hat{\mathbf{P}}^{\mu}(\sigma) = -i\frac{\delta}{\delta x^{\mu}(\sigma)} + \frac{1}{2\pi\alpha'}\frac{dx^{\mu}}{d\sigma} . \qquad (27)$$

We now impose the following constraints on the functional $\psi[x^{\mu}(\sigma)]$, which, like the particle, is τ independent.

$$\begin{aligned} (L_0 - 1/\alpha)\psi &= 0 \\ L_n \psi &= 0 \qquad n \geqslant 1 . \end{aligned} \qquad (28)$$

These are not entirely what we might naively expect. The $-1/\alpha$ in the first equation corresponds to the possibility of their being, due to L_0 not being uniquely defined by the classical theory, a normal ordering constant. We shall see that this is fixed to be $-1/\alpha$ by requiring 26 dimensional Lorentz invariance. In the second equation we do not require all the L_n's vanish on ψ , since this would imply that ψ itself would vanish due to the central term in the Virasoro algebra which we will discuss shortly. However, the latter equation implies that

$$(\psi, L_n \psi) = (L_{-n}\psi, \psi) = 0 \qquad (29)$$

as $L_n^+ = L_{-n}$. In this way one recovers in the classical limit that all the L_n's vanish in accord with equation (20). This procedure is the same in the Gupta Bleuther formulations of quantum electrodynamics.

The necessity of the constraints of equation (28) is guaranteed by the following theorem.

Theorem [7]

Equations (28) describe a ghost-free set of on-shell states provided the dimension D of space-time is less than or equal to 26.

In fact for $D < 26$ there are other problems and for the remainder of this contribution we will take $D = 26$. It may be helpful to recall the distinction between a ghost and a tachyon: for a scalar with action

$$\int dx\,(-c\,(\partial_\mu A)^2 - d\,A^2) \tag{30}$$

we say it is a ghost if $c < 0$ and a tachyon if $d < 0$. We will see that the open bosonic string does indeed possess a tachyon.

We are now in a position to specify what requirements a second quantized gauge covariant formulation of strings must satisfy. We must demand that there be an action whose equations of motion imply equations (18). Of course, we may have to make some gauge choices and we expect $L_n \psi = 0 \; n \geqslant 1$ to be the result of the gauge choices, while $(L_0 - 1)\psi = 0$ is the remaining equation of motion. (We will no longer explicitly write the α''s.) We further expect the action to be local in that the free theory should contain no more than two space-time derivatives. The actions we will obtain will contain the fields of Yang-Mills and gravity for the open and closed bosonic strings respectively; they will therefore possess these corresponding gauge invariances and so we must find that the gauge symmetries of the string will contain these particular symmetries.

To quantize this system we will use this action after appropriate gauge fixing and ghosts to weight a Feynman path integral. The vacuum-to-vacuum amplitude is given generally by

$$\int D\psi \exp i/\hbar\, S \tag{31}$$

where the action S is of the generic form

$$S = \int (D x^\mu) f(\psi(x^\mu(\sigma))) \tag{32}$$

+ gauge fixing + ghost + source terms.

Up until recently, the second quantization of strings has either been performed with constraints being present or carried out in a given gauge, such as the light-cone gauge [8], where the constraints have been solved. Quantization using BRS techniques of the linearized theory in a given gauge has been discussed in Ref. [9]. It is possible that one can, in principle, discover all the properties of a theory by quantizing in a given gauge. However, without the ability to use all the wisdom acquired with second quantization, this may by difficult. An example is the computation of anomalies which are absent in the light-cone gauge; however, a careful search of Lorentz invariance would show that it is violated when gauge anomalies are present. Also, the whole subject of the non-perturbative semi-classical phenomena and spontaneous symmetry breaking has up to now only been developed in the gauge invariant framework. It is also possible that the finiteness properties of strings may become particularly apparent in a covariant formulation, as they did in the case of supersymmetric theories.

Another advantage of obtaining a second quantized field theory of strings is that it will help us to understand what strings are. One of the remarkable developments of modern physics is that the theories relevant to nature are almost entirely determined by symmetries. For example, a theory possessing local gauge invariance realized on a vector potentials A_μ and which must be no more than second order in derivatives in the action, can only be on the Yang-Mills action. Similarly, Einstein's action is determined uniquely by general coordinate transformations realized on the metric and supergravity is controlled by local supersymmetry. The string theories are unique up to distinctions about being open or closed, supersymmetric or bosonic and they also contain the local symmetries mentioned above. It is natural to suppose that the string is also completely determined by a symmetry. A knowledge of this symmetry would explain the many wonderous cancellations found in string theory, as well as, hopefully, lead to many more. Clearly, possessing an invariant action under a set of transforming fields whose algebra is known would make it much easier to guess the principle which underlies this symmetry, and hence string theory itself.

Oscillator Formalism

In order to analyse the Virasoro conditions of equation (28), it is useful to reexpress the quantities discussed above in terms of creation and annihilation oscillators. We may write

$$x^{\mu}(\sigma) = \sum_{n=-\infty}^{\infty} x_n^{\mu}\, e^{in\sigma} \tag{33}$$

where $x_0^{\mu} = x^{\mu}$ and $x_{-n}^{\mu} = (x_n^{\mu})$ as a result of equation (19). The form of the expansion for $x^{\mu}(\sigma)$ is such as to obey the boundary conditions for the open string

$$\frac{dx^{\mu}}{d\sigma}\Big|_{\sigma=0} = \frac{dx^{\mu}}{d\sigma}\Big|_{\sigma=\pi} = 0 \;. \tag{34}$$

We may use the chain rule to rewrite the functional derivations

$$\frac{\delta}{\delta x^{\mu}(\sigma)} = \sum_{n=-\infty}^{\infty} \frac{\partial x_n^{\nu}}{\delta x^{\mu}(\sigma)} \frac{\partial}{\partial x_n^{\nu}} = \frac{i}{2\pi} \sum_{n=-\infty}^{\infty} e^{in\sigma} \frac{\partial}{\partial x_n^{\mu}} \tag{35}$$

Let us define

$$\hat{P}^{\mu}(\sigma) = \frac{1}{\pi(2\alpha')^{1/2}} \sum_{n=-\infty}^{+\infty} \alpha_n^{\mu}\, e^{in\sigma} \;. \tag{36}$$

Using equation (7) we find that

$$\alpha_n^{\mu} = i\,\Big(\Big(\frac{\alpha'}{2}\Big)^{1/2} \frac{\partial}{\partial x_n^{\mu}} + n(2\alpha')^{-1/2}\, x_n^{\mu}\Big) \;. \tag{37}$$

From the reality of $\hat{P}^{\mu}(\sigma)$ we find that $\alpha_{-n}^{\mu} = \alpha_n^{\mu +}$. The α's commutation relations are

$$\begin{aligned} [\alpha_n^{\mu}, \alpha_m^{\nu}] &= 0 \\ [\alpha_n^{\mu}, \alpha_m^{\nu +}] &= n\, \delta_{n,m}\, \eta^{\mu\nu} \end{aligned} \tag{38}$$

for $n,m > 0$. The Virasoro operators [6] can be expressed in terms of the α_n's by

$$L_n = \tfrac{1}{2} \sum_{m=-\infty}^{+\infty} \alpha_m^{\mu} \alpha_{n-m}^{\mu} \tag{39}$$

where L_o is understood to be normal ordered. They obey the modified algebra

$$[L_n, L_m] = (n-m)L_{n+m} + \frac{26}{12} n(n^2-1)\, \delta_{n-m} \;. \tag{40}$$

We find in particular that

$$L_o = \tfrac{1}{2} \left[\alpha_o^{\mu}\right]^2 + \sum_{m=1}^{\infty} \alpha_m^{\mu+} \alpha_m^{\mu} \,, \tag{41}$$

$$L_1 = L_{-1}^{+} = \alpha_o^{\mu} \alpha_1^{\mu} + \sum_{m=1}^{\infty} \alpha_m^{\mu+} \alpha_{m+1}^{\mu} \,, \tag{42}$$

where

$$\alpha_o^{\mu} = -i \left(\frac{\alpha'}{2}\right)^{\frac{1}{2}} \frac{\partial}{\partial x^{\mu}} \;.$$

We may write the state of the string in occupation number basis in terms of the creation operators $\alpha_n^{\mu+}$ by

$$\Psi[X^{\mu}(\sigma)] = \Big\{ \phi(x) + A_{\mu}^{1} \alpha_1^{\mu+} + h_{\mu\nu} \alpha_1^{\mu+} \alpha_1^{\nu+} + \alpha_2^{\mu+} A_{\mu}^{2} + \ldots \Big\} \langle x^{\mu}(\sigma)|0\rangle \tag{43}$$

The vacuum satisfies the equation

$$\alpha_n^{\mu} \langle x^{\mu}(\sigma)|0\rangle = 0, \; n \geq 1. \tag{44}$$

The vacuum of equation (44) is of the form

$$<x^{\mu}(\sigma)|0> = \prod_{n=1} c_n \exp(-\frac{n}{2\alpha'} x_n^{\mu} x_n^{\nu} \eta_{\mu\nu}) \quad . \tag{45}$$

The action of the $\alpha_n^{\mu+}$ on $<x^{\mu}(\sigma)|0>$ produces the well-known complete set of Hermite polynomials. In terms of component fields, we find (28) has as consequences

$$(\partial^2 + \frac{1}{\alpha'}) \phi(x) = 0 = [\partial^2 - (\ell-1)\frac{1}{\alpha'}] A_{\mu}^{\ell} = (\partial^2 - \frac{1}{\alpha'}) h_{\mu\nu} \; ... \; etc. \tag{46}$$

as well as

$$\partial^{\mu} A_{\mu}^{1} = 0 = \partial^{\mu} h_{\mu\nu} + A_{\nu}^{(2)} = 2\partial^{\mu} A_{\mu}^{(2)} + h_{\nu}^{\nu} \tag{47}$$

The theorem of reference [7] is remarkable in the sense that the presence of a_n^{o+} could lead to many ghost states which are forbidden by the Virasoro constraints of equation (28).

We now wish to find an action which does not have the constraints of equation (28) but instead possesses gauge invariances that allow the constraints of equation (29) to arise as a gauge choice upon the equation of motion. Consequently, we expect an infinite number of gauge invariances, one for every one of the constraints which generate the conformal group. This can be achieved mass level by mass level by successively releasing the constraints on ψ.

At the first level, we release the constraint $L_1\psi = 0$ but ψ is still subject to

$$L_1^2 \psi = L_3 \psi = L_2 L_1 \psi = .. = 0 \quad . \tag{48}$$

Consider now the transformation of ψ

$$\delta\psi = L_{-1}\Lambda_1 \quad , \tag{49}$$

where Λ_1 is subject to

$$L_1\Lambda_1 = L_2\Lambda_1 = 0 \quad . \tag{50}$$

Using the form of L_{-1} given in equation (42), we find that this invariance contains the transformation $\delta A^1_\mu = \partial_\mu \Lambda_1(x)$ which is the Abelian transformation expected for a linearized Yang-Mills theory.

An action invariant under $\delta\psi = L_{-1}\Lambda_1$ is given by

$$\tfrac{1}{2}\,(\psi\ ,\ (L_o - 1 - \tfrac{1}{2}L_{-1}L_1)\psi) \quad . \tag{51}$$

The equation of motion is given by

$$(L_o - 1 - \tfrac{1}{2}\,L_{-1}L_1)\psi = 0 \quad . \tag{52}$$

Explicitly testing this, we find that

$$(L_o - 1 - \tfrac{1}{2}\,L_{-1}L_1)L_{-1}\Lambda_1$$

$$= (L_oL_{-1} - L_{-1} - L_{-1}L_o)\Lambda_1 = 0 \quad , \tag{53}$$

since $L_1\Lambda_1 = 0$.

Equation (52) was probably known to a few people in the old hey-day of string theory, but has been rediscovered more recently [10],[11].

The projector P of a string field onto the physical states of equation (28) was given in [12]. The projector is an object that has the property $PL_{-n} = 0$ and at lowest order it is given by

$$P = 1 - \tfrac{1}{2}\,L_{-1}\,\frac{1}{L_o}\,L_1 \quad . \tag{54}$$

We note that at this level the equation of motion is given by

$$(L_o - 1)\, P\psi = 0 \ . \qquad (55)$$

It has been proposed recently [11],[13] that equation (55) is the correct equation of motion of all levels of the string, and this speculation has been encouraged by the above coincidence for the spin one at the first level. However, P has been formally computed for all levels, and (55) is explicitly non-local at the second level and more and more so for higher levels. One's suspicions are further aroused by the fact that P can be constructed in an arbitrary space-time dimension, and that $D = 26$ is not particularly favoured. The clearest way to show that (55) is not the right generalization is to consider the first level of the closed bosonic-orientated string. At this level, the covariant degrees of freedom are described by a single symmetric field $h_{\mu\nu}$. The Virasoro condition implies that it satisfies

$$\partial^2 h_{\mu\nu} = 0 \quad , \quad \partial^{\mu} h_{\mu\nu} = 0 \ . \qquad (56)$$

These equations tell us that at this level the closed string contains only a spin two and a spin zero. The generalization of (55) for this level is

$$\partial^2 R^{\rho}_{\mu} R^{\lambda}_{\nu} h_{\rho\lambda} = 0 \quad , \qquad (57)$$

where

$$R^{\rho}_{\mu} = (\delta^{\rho}_{\mu} - \frac{\partial_{\mu}\partial^{\rho}}{\partial^2}) \ . \qquad (58)$$

The reader immediately recognizes that this is a non-local equation, which does not admit a Hamiltonian formulation. Making it local by multiplication by ∂^2 leads to additional states.

In fact, it is known [14] that there is no Lorentz invariant, gauge invariant, local action constructed from $h_{\mu\nu}$ alone, which describes both spin two and spin zero. The only way to achieve this is to introduce another, scalar field to describe the spin zero; this is the well-known Einstein + massless scalar action. As we shall see, rather than the spin one, the first level of the closed string illustrates the general pattern, namely, for a local formulation, the theory naturally requires supplementary fields, as we shall now demonstrate. In fact, only for spins 0, 1 and ½ does multiplication by ∂^2 lead to a local field equation. Higher spins must be treated differently and the relation between their projectors and field equations is more subtle.

<u>Gauge Covariant Formulation of the Free String</u>

Let me examine whether any supplementary fields are required at the second level of the bosonic open string. To count the number of on-shell states at this level we could examine the Virasoro constraints at this level. The Virasoro constraints, however, possess on-shell gauge invariance that must be chosen before the on-shell states become apparent. This is clear even at the first level for which the conditions are $\partial^2 A_\mu = 0$ and $\partial^\mu A_\mu = 0$. As is well known, it is the additional on-shell invariance $\delta A_\mu = \partial_\mu \Lambda$ where $\partial^2 \Lambda = 0$ which allows the reduction to two rather than three on-shell states. A much faster method is to use the fact that the constraints are solved in the light core gauge and so one only has the on-shell states corresponding to oscillator α_n^i $i = 1$ to 24. In this case ψ has the expansion

$$(\phi(x) + (\alpha_1^i)^+ A_i^1 + h_{ij} (\alpha_1^i)^+ (\alpha_1^j)^+ + (\alpha_2^i)^\dagger B_2^i \ldots) \langle x^\mu(\sigma)|0\rangle \quad (59)$$

At the first level we have (D-2) states corresponding to the $(D-2)$ states contained in the vector representation of SO(D-2), which is the little group for massless particles in D dimension. Since the massive states must belong to representations of the little group SO(D-1). This demonstrates that the normal-order constant in equation (58) was chosen correctly. Any other choice would violate Lorentz invariance in D dimensions since (D-2) states can not carry a representation of SO(D-1).

At the second level we have

$$\frac{(D-2)(D-1)}{2} + (D-2) = \frac{D(D-1)}{2} - 1 \tag{60}$$

on-shell states. These we can only identify with the second rank traceless symmetric representation of SO(D-1). We shall refer to this as "pure spin two".

"Pure-spin two" is described on-shell by the field $h_{\mu\nu} = h_{\nu\mu}$ subject to the equations

$$\partial^{\mu} h_{\mu\nu} = h^{\mu}_{\mu} = (\partial^2 - m^2) h_{\mu} = 0 \quad . \tag{61}$$

The projector is well known [15]. It involves terms of the form $\partial_{\mu}\partial_{\lambda}\partial_{\rho}\partial_{\sigma}/(\partial^2)^2$ and hence multiplication (∂^2) does not lead to a local field equation. In fact, there is no way to describe in a Lorentz covariant way only a massive spin-two particle in terms of only $h_{\mu\nu} = h_{\nu\mu}$ subject to $h^{\mu}_{\mu} = 0$. The correct equations of motion involve the introduction of a supplementary field ϕ are are given by

$$(-\partial^2+m^2)h_{\mu\nu} + (\partial_{\mu}\partial^{\rho}h_{\rho\nu} + \partial_{\nu}\partial^{\rho}h_{\rho\mu}) - 2\,\frac{\delta_{\mu\nu}}{D}\,\partial^{\rho}\partial^{\lambda}h_{\rho\lambda}$$

$$= \frac{D-2}{D-1}\left(\partial_{\mu}\partial_{\nu}\phi - \frac{\delta_{\mu\nu}}{D}\,\partial^2\phi\right) \quad , \tag{62}$$

$$\partial^{\mu}\partial^{\nu}h_{\mu\nu} = \left(\partial^2 - \frac{D}{D-2}\,m^2\right)\phi \quad ,$$

where D is the dimension of space time.

Indeed, these coupled equations lead to the desired result, namely

$$0 = \partial^{\mu}h_{\mu\nu} = \phi = (\partial^2-m^2)h_{\mu\nu} \quad . \tag{63}$$

The above equations illustrate a more general method [16] of introducing additional field in order to propagate higher spin fields. At the second level the string contains the fields $h_{\mu\nu}$ and A^2_μ, however A^2_μ is gauge away leaving $h_{\mu\nu}$ traceless and so we require one extra supplementary field ϕ to implement the massive spin-two field equation. This field ϕ will be the lowest component of a supplementary string field

$$X^{(2)}[(x^\mu(\sigma)] = (\phi(x) + \ldots) \langle x^\mu(\sigma)|0\rangle \tag{64}$$

We will now find the gauge covariant action at the next level. Starting from the action of equation (51) and releasing the constraints of equation (48) of the first level, we subject ψ and X to

$$L_3\psi = L_2L_1^2\psi = L_1^3\psi = \ldots = 0 \ , \quad L_1X^{(2)} = L_2X^{(2)} = \ldots = 0 \ . \tag{65}$$

The most general expression of the correct order is

$$(L_0 - 1 - \tfrac{1}{2}L_{-1}L_1 - \tfrac{1}{4}\gamma L_{-2}L_2)\psi + (L_{-1}^2 + \tfrac{3}{2}\beta L_{-2})X^{(2)} = 0 \ . \tag{66}$$

$$(L_1^2 + \tfrac{3}{2}\beta L_2)\psi = (a L_0 + 8)X^{(2)} \tag{67}$$

The use of the same β in (65) and (66) is required by demanding that the equations of motion follow from an action. An alternative way of searching for a gauge invariance is to demand that L_1 on equation (66) should vanish when we use equation (67). Carrying this out and using constraints of equation (65) we find

$$-\tfrac{1}{2}L_{-1}(L_1^2 + \tfrac{3}{2}\gamma L_2)\psi + (4 L_{-1}L_0 + 2L_{-1})X^{(2)} + \tfrac{9}{2}\beta L_{-1}X^{(2)} = 0 \ . \tag{68}$$

To eliminate ψ by equation (67) requires $\gamma = \beta$ and we find

$$-\tfrac{1}{2} L_{-1}(aL_0 + b)X^{(2)} + L_{-1}(4L_0 + 2 + \tfrac{9}{2}\beta)\, X^{(2)} = 0 \;. \tag{69}$$

Hence we conclude that $a = 8$ and $b = 4 + 9\beta$. Applying L_2 in a similar way fixes $\beta = 1$, and we find the equations of motion

$$\begin{aligned} &(L_0 - 1 - \tfrac{1}{2}\sum_{n=1}^{2} \frac{L_{-n}L_n}{n})\psi + (L_{-1}^2 + \tfrac{3}{2}L_{-2})X^{(2)} = 0 \,, \\ &(L_1^2 + \tfrac{3}{2}L_2)\psi = (8L_0 + 13)X^{(2)} \;. \end{aligned} \tag{70}$$

This system of equations is in fact invariant under the gauge transformations

$$\delta_1\psi = L_{-1}\Lambda_1 \;,\; \delta_1 X^{(2)} = \tfrac{1}{2} L_1\Lambda_1 \;,\; \delta_2\psi = L_{-2}\Lambda_2,\; \delta_2 X^{(2)} = \tfrac{3}{2}\Lambda_2 \;, \tag{71}$$

with

$$L_2\Lambda_1 = L_1^2\Lambda_1 = \ldots = 0 \;, L_1\Lambda_2 = L_2\Lambda_2 = \ldots = 0 \;. \tag{72}$$

We stress that equations (70) are Λ_2 invariant only for $D = 26$. It may be possible with the introduction of further supplementary fields, to relax this condition. The corresponding action is given by

$$\begin{aligned} &\tfrac{1}{2}(\psi,(L_0 - 1 - \tfrac{1}{2}L_{-1}L_1 - \tfrac{1}{4}L_{-2}L_2)\psi) \\ &+(\psi,(L_{-1}^2 + \tfrac{3}{2}L_{-2})X^{(2)} - \tfrac{1}{2}(X^{(2)},(8L_0 + 13)X^{(2)}) \;. \end{aligned} \tag{73}$$

This completes the second level.

Before constructing the action to all orders, it will be instructive to rewrite the second-level result in a kind of first-order form. Completing the squares in the $L_1\psi$ and $L_2\psi$ terms in the action, we may rewrite equation (73) as

$$\begin{aligned} &\tfrac{1}{2}(\psi,(L_0 - 1)\psi) - \tfrac{1}{4}(L_1\psi - 2L_{-1}X^{(2)},\, L_1\psi - 2L_{-1}X^{(2)}) \\ &- \tfrac{1}{8}(L_2\psi - 6X^{(2)},\, L_2\psi - 6X^{(2)}) - 2(X^{(2)},(L_0 + 1)X^{(2)}) \end{aligned} \tag{74}$$

where we have at this level $L_1 \chi^{(2)} = 0$. We now introduce the auxiliary fields $\phi^{(1)}$ and $\phi^{(2)}$ to rewrite the action as

$$
\begin{aligned}
&(\psi,(L_0-1)\psi) + (\phi_1, L_1\psi + L_{-1}\,\xi^{(1,1)})) \\
&+ (\phi_2, L_2\psi + 3\,\xi^{(1,1)}) + (\phi_1,\phi_1) \\
&+ 2\,(\phi^2,\phi^2) - \tfrac{1}{2}(\xi^{(1,1)}, (L_0+1)\,\xi^{(1,1)})
\end{aligned}
\tag{75}
$$

where $\xi^{(1,1)} = -\tfrac{1}{2}\chi^{(2)}$.

The construction up to the sixth level was given in [1] and the all-orders construction which is now presented was given in [2]. We first write down the final-field equations which we will then justify.

The field equations are:

<u>field</u>		<u>field equation</u>
ψ	:	$(L_0-1)\psi+L_{-1}\phi^{(1)}+L_{-2}\phi^{(2)} = 0$
$\phi^{(1)}$	:	$L_1\psi = -2\phi^{(1)}-L_{-1}\xi^{(1,1)}-L_{-2}\zeta^{(2,1)}$
$\phi^{(2)}$	:	$L_2\psi = -4\phi^{(2)}-L_{-1}\zeta^{(1,2)}-L_{-2}\xi^{(2,2)}-3\zeta^{(1,1)}$
$\zeta^{(1,1)}$	:	$L_1\phi^{(1)} = (L_0+1)\zeta^{(1,1)}-3\phi^{(2)}$
$\zeta^{(2,1)}$	:	$L_2\phi^{(1)} = (L_0+2)\zeta^{(1,2)}$
$\zeta^{(1,2)}$	:	$L_1\phi^{(2)} = (L_0+2)\zeta^{(2,1)}$
$\zeta^{(2,2)}$	:	$L_2\phi^{(2)} = (L_0+3)\zeta^{(2,2)}$

(76)

This pattern is easily established: we require a ψ equation that begins with $(L_0 - 1)\psi$. We also expect, due to the gauge invariance $\delta\psi = L_{-1}\Lambda_1$ and $\delta\psi = L_{-2}\Lambda_2$, that L_1 and L_2 acting on the ψ equation give zero provided we use the other field equations. Hence we require $L_1\psi$ and $L_2\psi$, which are determined by the field equations of the new fields $\phi^{(1)}$ and $\phi^{(2)}$. However, this implies terms in the action of the form $(\phi^{(1)}, L_1\psi) + (\phi^{(2)}, L_2\psi)$ and so $\phi^{(1)}$ and $\phi^{(2)}$ occur in the ψ equation as above. Now, however, we require $L_1\phi^{(1)}$, $L_2\phi^{(1)}$, $L_1\phi^{(2)}$ and $L_2\phi^{(2)}$, which are determined by the field equations of $\zeta^{(1,1)}$, $\zeta^{(2,1)}$, $\zeta^{(1,2)}$ and $\zeta^{(2,2)}$ respectively. Remarkably, no new fields are required. We then write down the most general equations of this form and of the correct order. The arbitrary constants are now determined by the application of L_1 and L_2 to the ψ equation.

We leave the application of L_1 to the reader and consider the application of L_2 in more details: we find that

$$(L_0+1)L_2\psi + 3L_1\phi^{(1)} + L_{-1}L_2\phi^{(1)} + (4L_0+D/2)\phi^{(2)} + L_{-2}L_2\phi^{(2)} = 0 \quad . \tag{77}$$

Substituting for $L_2\psi$, $L_1\phi^{(1)}$, $L_2\phi^{(1)}$ and $L_2\phi^{(2)}$, we find that

$$(D - 26)\,\phi^{(2)} = 0 \quad . \tag{78}$$

Consequently, we discover that this system only exists in the critical dimension $D = 26$.

Given the field equations, we can search for the full gauge invariance, which one may easily check, given by

$$\delta\psi = \sum_{n=1}^{z} L_{-n}\Lambda_n \;, \quad \delta\phi^{(n)} = -(L_0+n-1)\Lambda_n \;, \tag{79}$$

$$\delta\zeta^{(n,m)} = -L_m\Lambda_n - (2m+n)\Lambda_{m+n} \quad \textit{for } m,n = 1,2 \quad . \tag{80}$$

it is straightforward to write down an action from which the above equation follows:

$$
\begin{aligned}
&\tfrac{1}{2}\,(\psi,(L_o-1)\psi) + \sum_{n=1}^{2} (\phi^{(n)},\, L_n\psi) \\
&+ \sum_{n,m=1}^{2} (L_n\phi^{(m)},\, \zeta^{(n,m)}) + \sum_{n=1}^{2} n(\phi^{(n)},\phi^{(n)}) \qquad (81)\\
-\tfrac{1}{2} &\sum_{n,m=1}^{2} (\zeta^{(n,m)},(L_o+n+m-1)\,\zeta^{(m,n)}) + \sum_{n,m=1}^{2} (2n+m)(\phi^{(m\mp n)},\, \zeta^{(n,m)}) \ .
\end{aligned}
$$

In fact, one can find free-gauge covariant formulation of all known strings in this way: the open and closed bosonic string, and the open and closed superstring theories and the hetoric string. We refer the reader to reference [2] for these other formulations.

One can extend the result for the open bosonic string given above so that it contains an infinite number of supplementary fields [17]. This will have the advantage that the generators of the Virasoro algebra will appear on a more equal footing. This is achieved by introducing the fields

$$\psi,\ \phi^{n},\ \zeta^{n,m} \qquad n,m = 1,2,\ \ldots\ \infty \tag{82}$$

and the action is the same as that of equation (80) except now all the sums run from 1 to ∞. This action is invariant under the transformations of equation (79) except the sums also run from 1 to ∞.

Interactions

In this section, we will demonstrate how to construct the interactions for the open bosonic string and carry out the first step of this procedure. To find the interacting theory from its linearized form, we will employ a generalization of the Noether method. In this method, one starts with the linearized theory which possesses Abelian local invariances and a similar number of rigid invariances. One then makes the rigid invariances local and finds an invariant non-linear action order by order in the coupling constant. This is achieved by knitting together the rigid and local Abelian invariances.

For the linearized theory given in the previous section, the linearized Abelian invariances are the transformations with parameters $\Lambda^{(1)}$, $\Lambda^{(2)}$, $\Lambda_{(3)}$, etc. The rigid invariances are the usual rigid rotations of the fields under the group G. In fact, there are an infinite set of such invariances, since we can make an independent rotation at every mass level. The first such invariance is the one which rotates all mass levels in the same way; it is given by

$$\delta\psi = [\ \Omega^{(1)}, \psi] \tag{83}$$

where $\Omega^{(1)}$ is independent of $x^{\mu}(\sigma)$. Let us denote the nth rigid rotation by $\Omega^{(n)}$. $\Omega^{(n)}$ is an infinitesimal anti-Hermitian matrix in the fundamental representation of the group acting on the Chan-Paton indices carried by the anti-Hermitian matrix ψ . The first step is to make this rigid invariance into a local one, which is most easily achieved by working in the operator formalism. We regard ψ as a ket $|\psi\rangle$ and promote $\Omega^{(1)}$ $[\Omega^{(n)}]$ to be a ket $|\Omega^{(1)}\rangle$ $[|\Omega^{(n)}\rangle]$. The variation of $|\psi\rangle$ is given by

$$\delta|\psi_b\rangle = \langle\Omega_a|\langle\psi_c|T_1 + \langle\psi_a|\langle\Omega_c|T'_2 \tag{84}$$

where T_1 and T'_2 are operators in the form

$$T_i = f_i(\alpha_a,\alpha_b,\alpha_c)|o_a\rangle|o_b\rangle|o_c\rangle \tag{85}$$

which are to be determined. The labels a, b and c correspond to the three "legs" of T_i which are illustrated in Fig. 1:

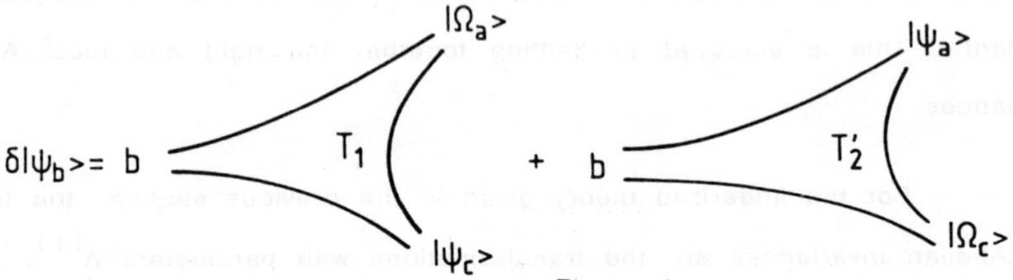

Fig. 1

There also exist similar equations for the supplementary fields $|\phi^n\rangle$. In this sense, the method we are using represents a generalization of the usual Noether method.

Upon substituting the now-local variations of $|\psi\rangle$ of equation (84), and those for $|\phi^{(n)}\rangle$ into the linearized action, we will obtain an expression which is linear in $\langle\Omega^{(n)}|$ and bilinear in $|\psi\rangle$ and $|\phi^{(n)}\rangle$. The action of the L_n which occur in the linearized action does not, in general, obey the Liebnitz rule. Hence the T's will have to be very special operators, so that the action of the L_n produces further L_n's on the other legs.

In the discussion which follows, we will only refer to $\Omega^{(1)}$, but the extension to $\Omega^{(n)}$ is in principle clear. The variation of the linearized action can then be written in the form

$$\langle\Omega^{(1)}|\,L_n S^{(n)} \tag{86}$$

where $S^{(n)}$ is an expression bilinear in $|\psi\rangle$ and $|\phi^{(n)}\rangle$. This variation of the linearized action can be cancelled by the addition of a term of order g to the action. The variation of this term at order g^0 results from only the variation of the fields under the Abelian invariance whose parameter $\Lambda^{(1)}$ is identified by

$$\Lambda^{(1)}\rangle = \frac{1}{g}|\Omega^{(1)}\rangle \quad . \tag{87}$$

For example, a term of the form

$$\langle\, \Omega^{(1)}|\, L_1 S^{(1)} \tag{88}$$

is cancelled by the addition of the term of the form

$$g\langle\, \psi|\, S^{(1)} \quad . \tag{89}$$

The variations of this new action at order g are then cancelled by adding new terms to the action and possibly new terms to the transformation laws. This is repeated until one obtains an action invariant to all orders in g .

From the light-cone formulation of the interacting string field theory and the geometrical interpretation of interactions such as splitting, joining and rearrangement of strings, one can expect at most quartic interactions for open strings and cubic ones for closed strings also in a gauge covariant formulation. However, the vertices will necessarily involve the supplementary fields $\phi^{(n)}$. The occurrence of interactions at least quadratic in $\phi^{(n)}$ results in the practical inability to integrate out the supplementary fields in the path integral. Such terms are inevitable; for example, there exists in the closed string sector the usual graviton-massless scalar coupling, even before taking the field theory limit, hence a non-zero $\phi^{(1)}\psi\phi^{(1)}$ interaction term. This shows that the non-local formulation of the free string theory of

[13] can definitely not be generalized to the interacting case.

The above procedure can be carried out level by level, and we will now find the order g interactions at the first level. We begin with the action

$$\tfrac{1}{2} \langle \psi \,|(L_0 - 1 - \tfrac{1}{2} L_{-1} L_1)\,|\, \psi \rangle \quad . \tag{90}$$

Using the variation of $|\psi\rangle$ of equation (84), we find

$$\langle \psi_b |(L_0 - 1 - \tfrac{1}{2} L_{-1} L_1)^{\dagger} \left\{ \langle \psi_a | \langle \Omega_c | T_1 + \langle \Omega_a | \langle \psi_c | T_2' \right\} \quad . \tag{91}$$

This must be expressible in the form

$$\langle \Omega_a | L_1^{a} \langle \psi_b | \langle \psi_c | W \tag{92}$$

which can cancel with the variation of the term

$$-\frac{1}{3} g \langle \psi_a | \langle \psi_b | \langle \psi_c | W \quad . \tag{93}$$

Comparing equations (91) and (92), we require the operators T_i and W to satisfy

$$L_1^{a} W = (L_0 - 1 - \tfrac{1}{2} L_{-1} L_1)_b \, T_1$$

$$+ (L_0 - 1 - \tfrac{1}{2} L_{-1} L_1)_c \, T_2' \quad . \tag{94}$$

where

$$T_2'(a,b,c) = T_2(c,a,b) \tag{95}$$

Long ago, a cyclically symmetric three-point vertex was written down [18] in the form

$$V = \exp - \sum_{\substack{n=1 \\ m=o}}^{\infty} (\alpha^a_{-n}\alpha^b_{-m} + \alpha^b_{-n}\alpha^c_{-m} + \alpha^c_{-n}\alpha^a_{-m}) \frac{(-1)^m \Gamma(n)}{\Gamma(m+1)\Gamma(n-m+1)} |0\rangle . \tag{96}$$

It is instructive to evaluate V, taking the external states to contain only the massless Yang-Mills vectors; one finds the result

$$k_1 \cdot \epsilon_3\, \epsilon_1 \cdot \epsilon_2 + k_2 \cdot \epsilon_1\, \epsilon_2 \cdot \epsilon_3 + k_3 \cdot \epsilon_2\, \epsilon_1 \cdot \epsilon_3$$

$$+ k_1 \cdot \epsilon_3\, k_2 \cdot \epsilon_1\, k_3 \cdot \epsilon_2 . \tag{97}$$

The last term, of course, disappears in the zero slope limit. This vertex differs from one which is antisymmetric under the interchange of any two external lines by longitudinal terms which are

$$\begin{aligned} \epsilon_1^\mu \epsilon_2^\nu \epsilon_3^\rho [&\tfrac{1}{2} k_1^\mu \delta_{\nu\rho} + \tfrac{1}{2} k_2^\nu \delta_{\mu\rho} + \tfrac{1}{2} k_3^\rho \delta_{\mu\nu} \\ &- \tfrac{1}{2}(k_1^\mu k_3^\nu k_2^\rho + k_3^\rho k_2^\mu k_1^\nu + k_2^\nu k_1^\rho k_3^\nu) \\ &+ \tfrac{1}{2} k_1^\mu k_2^\nu k_3^\rho] . \end{aligned} \tag{98}$$

We therefore take our operators to be of the form

$$\begin{aligned} W &= [1 + \alpha(L^1_{-1} + L^b_{-1} + L^c_{-1}) + \beta(L^b_{-1}L^c_{-1} + L^a_{-1}L^b_{-1} + L^c_{-1}L^a_{-1}) \\ &\qquad + \delta\, L^a_{-1}L^b_{-1}L^c_{-1}]\, V \\ T_1 &= \gamma(1 + \alpha' L^c_{-1})\, V \\ T_2 &= \epsilon(1 + \alpha'(L^b_{-1} + L^c_{-1} + L^a_{-1}))V \end{aligned} \tag{99}$$

These are the most general operators, taking into account the constraints for the first level written in the previous section:

$$L_1|\Omega\rangle = L_2|\psi\rangle = L_1^2|\psi\rangle = \ldots = 0 . \tag{100}$$

Using the result (ref. [19])]

$$(L_1^a - L_o^a + L_{-1}^b - L_o^b + L_o^c)\, V = 0 \quad , \tag{101}$$

and the analogous results obtained by cyclically permuting a, b and c, we can evaluate L_1^a on W and fix the above coefficients. One finds

$$W = [1 - \tfrac{1}{2}(L_{-1}^c + L_{-1}^b + L_{-1}^c) + \tfrac{1}{2}(L_{-1}^a L_{-1}^b + L_{-1}^b L_{-1}^c + L_{-1}^c L_{-1}^a) - \tfrac{1}{2} L_{-1}^a L_{-1}^b L_{-1}^c]\, V \quad ,$$

$$T_1 = V \quad ,$$
$$T_2 = -(1 - L_{-1}^b + L_{-1}^c + L_{-1}^a) V . \tag{102}$$

At this level, T'_2 happens to be cyclically symmetric, and the permutation (102) is irrelevant. It may become relevant in next order.

Hence the order g result at the first level is

$$\tfrac{1}{2} \langle \psi | L_o - 1 - \tfrac{1}{2} L_{-1} L_1 | \psi \rangle$$
$$- \frac{1}{3} g \langle \psi_a | \langle \psi_b | \langle \psi_c | W \quad . \tag{103}$$

This action is invariant to order g under

$$\delta|\psi_b\rangle = \frac{1}{g} L_{-1}^b |\Omega_b\rangle + \langle \psi_a | \langle \Omega_c | T_1 + \langle \Omega_a | \langle \psi_c | T_2 \tag{104}$$

Having found the transformation law of ψ at order g we can test the closure at this level. One finds that it does indeed close, yielding a composite Abelian gauge transformation. One general feature that emerges is that the structure constants are determined by the three Reggeon vertex operators and it is clear that the group structure underlying string will involve these vertex operators.

After these lectures were given, we received preprints from T. Banks and M. Peskin [SLAC preprint 3740 (1985)], W. Siegel and B. Zwiebach [U.C. Berkeley, preprint PTH 85/30 (1985)] and K. Itoh, T. Kugo, H. Kumitomo and H. Ooguri [Kyoto preprint HE-TH 85/04 (1985)], where material related to that presented here is discussed. The extension of the gauge covariant supersymmetric string formulation to include an infinite set of supplementary fields and so place the generators of the Ramond-Neveu-Schwarz algebra on a more equal footing is given in a preprint by Neveu and West (CERN preprint TH.4267/85).

More recently, there has been subtantial progress . For an elegant and very useful gauge covariant formulation of the free theory, see H. Nicolai, A. Neveu and P.West, CERN preprint TH.4297/85 and for an account of the interacting open bosonic string, see A. Neveu and P. West, CERN preprint TH.4315/85 and H. Hata, K. Itoh, T. Kugo, H. Kumitomo and K. Ogawa, Kyoto preprint HE(TH)85/10 (1985) and E. Witten, Princeton preprint (1985).

References

[1] A. Neveu and P. West "Gauge covariant local formulation of bosonic strings. CERN preprint TH4200 (1985).

[2] A. Neveu, H. Nicolai and P. West. Gauge covariant local formulation of free strings and superstrings.

[3] P.A.M. Dirac Lectures in quantum mechanics. (Belfer Graduate School of Science, Yeshiva University; New York 1964).

[4] P. Goddard, J. Goldstone, C. Rebbi and C. Thorn. Nucl. Phys. B56, 109 (1973).

[5] Y. Nambu, Proc. Int. Conf. on Symmetries and Quark Modes. Detroit 1969. Gordon and Breach, New York (1970).

[6] M. Virasoro Phys. Rev. D1, 2933 (1970).

[7] R.C. Brower, Phys. Rev. D6, 1655 (1972);
P. Goddard and C.B. Thorn, Phys. Lett. 40B, 235 (1972).

[8] E. Cremmer and J.-L. Gervais, Nucl. Phys. B90, 410 (1975);
M. Kaku and K. Kikkawa, Phys. Rev. D10, 1110, 1823 (1974).

[9] W. Siegel, Phys. Lett. 148B, 556 (1984); 149B, 157, 162 (1984).

[10] S. Raby and P.C. West, unpublished, January 1985.

[11] T. Banks and M. Peskin, Proceedings of the Symposium on Anomalies, Geometry and Topology, Argonne Nat. Lab. March 1985; M. Kaku and J. Lykken, ibid.

[12] C. Brower and C.B. Thorn, Nucl. Phys. B31, 163 (1971).

[13] D. Friedan, Univ. of Chicago preprint EFI-85-27 (April 1985).

[14] P. van Nieuwenhuizen, Nucl. Phys. B60, 478 (1973).

[15] H. van Dam and M. Veltman, Nucl. Phys. B22, 397 (1970).

[16] L.P.S. Singh and C.R. Hagen, Phys. Rev. D9, 898 (1974).

[17] A. Neveu, J. Schwarz and P. West Gauge symmetries of the free bosonic string. CERN preprint TH 4200 (1985).

[18] L. Caneschi, A. Schwimmer and G. Veneziano, Phys. Lett. 30B, 351 (1969).

COVARIANT PERTURBATION THEORY FOR SUPERSYMMETRIC σ-MODELS

K.S. Stelle
The Blackett Laboratory
Imperial College
London SW7 2BZ.

ABSTRACT

A manifestly covariant background field formalism for the N=2 supersymmetric non-linear σ-model is presented. The formalism allows the symmetries of the model to be exploited to the full in the discussion of the ultra-violet divergences in the quantum theory. This proves the cohomological triviality of the metric counterterms at the $\ell \geq 2$ loop orders. The formalism confirms the finiteness of models with Ricci-flat metrics through the three-loop order. However, it seems unlikely that these cancellations will persist to higher orders. This general analysis is borne out by a study of the supercurrent structure. It is shown that while there is a component axial U(1) current which obeys an Adler-Bardeen theorem, this current is not in the supercurrent multiplet and its existence cannot therefore be used to prove conformal invariance at the quantum level.

This paper is a report of work done in collaboration with P.S. Howe and G. Papadopoulos.

The conformal properties of two dimensional supersymmetric non-linear σ-models are important for determining the ground states of superstring theories. In the case of a supersymmetry-preserving compactification of a superstring theory where the torsion-free spin connection is identified with a non-zero ground state gauge connection, the relevant σ-models have N=2 supersymmetry.[1] The subject is also of intrinsic interest within the general context of ultraviolet cancellations in supersymmetric field theories.

Previous discussions[2,3] of the ultraviolet problem for d=2 σ-models have concentrated on an analysis of the counterterm structures expected on the basis of the models geometrical structure. On the other hand, it is known from studies of higher dimensional supersymmetric theories that there can be cancellations of divergences otherwise allowable by the symmetry properties of counterterms. An example of this is the finiteness of the N=4, d=4 super Yang-Mills theory despite the existence of the classical action as a potential counterterm.[3-10] These otherwise "miraculous" cancellations may be most easily understood in a formalism which maintains manifestly the maximal possible symmetry.[6-9] Indeed, there are no "miraculous" cancellations that cannot be understood in this way.[11]

In this paper we apply these methods to the N=2, d=2 non-linear super σ-models. The most convenient way of controlling the geometrical structure of quantized σ-models is the background field method.[12] For a bosonic σ-model with action

$$I_{N=0} = \int d^2x \; g_{ij}(\phi_T)\partial_\mu\phi_T^i \; \partial^\mu\phi_T^j \quad , \tag{1}$$

the total field ϕ_T^i is split into background and quantum parts as follows: let $\phi^i(s)$ be the geodesic starting from the background field $\phi_B^i = \phi^i(s=0)$ and passing through the total field at proper length s=1 : $\phi^i_T = \phi^i(s=1)$. The quantum field ξ^i is the tangent vector $\frac{d\phi^i}{ds}(s=0)$ to this geodesic at ϕ^i_B. Explicitly, we must solve the system

$$\frac{d^2\phi^i}{ds^2} + \Gamma^i_{jk}\frac{d\phi^j}{ds}\frac{d\phi^k}{ds} = 0 .$$

$$\frac{d\phi^i}{ds}(s=0) = \xi^i , \quad \phi^i(0) = \phi^i_B , \tag{2}$$

The solution is

$$\phi^i_T = \phi^i_B + \xi^i - \frac{1}{2}\Gamma^i_{jk}\xi^j\xi^k + \ldots \quad . \tag{3}$$

Substituting this expansion into (1), one finds that the coefficients of all orders in the quantum field ξ^i are constructed from geometrical tensors that are functions of the background field. The covariance of this expansion may easily be seen by using normal coordinates, for example. This construction is easily generalized to N=1, d=2 supersymmetry. In this case ϕ^i is a superfield, and the action is

$$I_{N=1} = \int d^2x d^2\theta \; g_{ij}(\phi^i_T) D^\alpha \phi^i_T D_\alpha \phi^j_T \tag{4}$$

where D_α is the normal N=1 Majorana supercovariant spinorial derivative.

In N=2 supersymmetry, the basic superfield ϕ^a is an N=2 chiral superfield,

$$\bar{D}_\alpha \phi^a = 0 \quad , \tag{5}$$

where D_α is now a Dirac spinor, and

$$\{D_\alpha, \bar{D}_\beta\} = i\gamma^\mu_{\alpha\beta}\partial_\mu \tag{6}$$

$$\{D_\alpha, D_\beta\} = \{\bar{D}_\alpha, \bar{D}_\beta\} = 0 \quad . \tag{7}$$

The holomorphic index a runs from 1 to n and the σ-model manifold is an n complex-dimensional Kähler manifold. The N=2 action is just the full superspace integral of the Kähler potential,[13]

$$I_{N=2} = \int d^2x d^4\theta \; K(\phi,\bar{\phi}) \qquad . \qquad (8)$$

We would like to make a covariant background-quantum split, but there is a problem. The geodesic equation is incompatible with the chirality condition on ϕ^a, because $\Gamma_{bc}{}^a$ depends on both ϕ^a and $\bar{\phi}^{\bar{a}}$. Another aspect of this problem is the absence of holomorphic normal coordinates, again because $\Gamma_{bc}{}^a$ depends on both holomorphic and anti-holomorphic coordinates.

This problem can be circumvented by the introduction of an unconstrained complex prepotential $X^a(s)$ defined along the (non-geodesic) curves $\phi^a(s)$ which are again taken to connect the background field $\phi_B^a = \phi^a(s=0)$ with the total field $\phi_T^a = \phi^a(s=1)$,

$$\phi^a(s) = \bar{D}^2 X^a(s) \qquad ; \qquad \bar{D}^2 = \bar{D}^\alpha \bar{D}_\alpha \qquad . \qquad (9)$$

The field $X^a(s)$ no longer has a nice geometrical interpretation, but we can define its transformation under holomorphic coordinate reparameterizations of the type

$$\phi^a \to \phi'^a = f^a(\phi) = h^a{}_b(\phi)\phi^b \qquad (10)$$

to be

$$X^a(s) \to X'^a(s) = \int_0^s f^a{}_{,b} \frac{dX^b}{dt} dt + X'^a(0) \qquad (11)$$

where

$$X'^a(0) = h^a{}_b(\phi_B) X^b(0) \qquad .$$

Equation (11) ensures that (10) is reproduced and also that $\frac{dX^a}{ds}$ transforms as a vector under the transformation (10). Then we can replace the geodesic equation (2) by the equation of parallel transport for $\frac{dX^a}{ds}$:

$$\frac{d^2X^a}{ds^2} + \Gamma^a_{bc}(\phi,\bar{\phi})\frac{dX^b}{ds}\frac{d\phi^c}{ds} = 0 \quad . \tag{12}$$

Equation (12) is to be solved subject to the initial conditions

$$X^a(s=0) = X^a_B \quad ; \quad \frac{dX^a(s=0)}{ds} = \xi^a \quad . \tag{13}$$

where X^a_B is the background prepotential and ξ^a is the quantum field which is again an unconstrained superfield. Explicitly,

$$X^a_T = X^a_B + \xi^a - \frac{1}{2}\Gamma^a_{bc}(\phi_B,\bar{\phi}_B)\xi^b\bar{D}^2\xi^c + \ldots \tag{14}$$

We observe that all higher order terms in the expansion (14) involve the background field via ϕ_B (not X_B) so that the substitution of (14) into the action (8) will yield a Lagrangian which depends on ϕ^a_B and the quantum field ξ^a. Despite the non-existence of normal coordinates there is no difficulty in showing that this expansion is reparameterization covariant with the coefficients at all orders in the quantum field ξ^a being constructed from geometrical tensors that are functions of the background field ϕ^a_B. The leading terms in the expansion are given by

$$I_{N=2}[\phi_B, \xi] = \int d^2x d^4\theta \; \{ K(\phi_B, \bar{\phi}_B) + (g_{a\bar{b}} \; \xi^a \; \bar{D}^2 \; \bar{\phi}^b + c.c.) + g_{a\bar{b}} \; \bar{D}^2 \; \xi^a \; D^2 \; \bar{\xi}^b + \ldots \} \tag{15}$$

where $g_{a\bar{b}} = \partial_a \partial_{\bar{b}} K(\phi_B, \bar{\phi}_B)$ is the Kähler metric.

The above procedure is therefore adequate to establish a covariant background field expansion but the introduction of the unconstrained prepotential means that the action (15) now has a quantum (pre-) gauge invariance which has to be gauge-fixed and accompanied by Fadeev-Popov ghosts in the usual way. The pregauge transformation is given implicitly by

$$\delta \left(\xi^a - \frac{1}{2} \Gamma^a_{bc}(\phi_B, \bar{\phi}_B)\, \xi^b \bar{D}^2 \xi^c + \ldots \right) = \bar{D}^\alpha \Lambda^a_\alpha$$

$$\delta \phi^a_B = 0 \qquad (16)$$

Such a transformation leaves the total field ϕ^a_T invariant. Since the action is a functional of ϕ^a_T, (16) is obviously an invariance of it.

The ghost structure of the model is actually somewhat complicated due to the fact that covariant gauge-fixing functions must be used, which in turn requires the introduction of Nielsen-Kallosh ghosts. In addition, the Fadeev-Popov ghost action is itself invariant under a new gauge transformation involving an even larger superfield parameter than the original transformation (16). Indeed, the strict maintenance of covariance would involve an infinite sequence of ghosts which would all couple to the background field. This sequence must be terminated in order to define the quantum theory, but since only the first generation ghosts couple to the quantum field and hence contribute to all orders in perturbation theory, it is possible to do this at the cost of losing manifest covariance at the one-loop level only. We defer a fuller discussion of this point to a later publication [14] but remark that the procedure can indeed be carried through satisfactorily, along similar lines to the supersymmetric Yang-Mills case [9].

It is worth remarking that, whereas the geodesic equation (2) cannot be modified, the parallel transport equation can be. For example, we could replace (12) by

$$\frac{d^2X^a}{ds^2} + \Gamma^a_{bc}\frac{dX^b}{ds}\frac{d\phi^c}{ds} = k\, R^a{}_{bc\bar{d}}\frac{d\phi^c}{ds}\frac{d\bar{\phi}^{\bar{d}}}{ds}\frac{dX^b}{ds} \tag{17}$$

Equation (17) would be an equally good equation on which to base the background-quantum split and should lead to identical results. Since the term on the right-hand side of (17) modifies (14) it will alter the pregauge transformation (16) and this will in turn affect the structure of the ghost sector of the theory. Thus, the effect of adding terms to the right-hand side of equation (12) is to modify the Lagrangian expansion (15), this modification being compensated by a concomitant modification of the ghost action. However, it could be that an appropriate choice of such additional terms would simplify the expansion and indeed this turns out to be the case up to order ξ^4 if one uses equation (17) with $k = -1/3$.

To summarize: it is possible to set up a manifestly N=2 supersymmetric reparameterization covariant background field formalism for N=2 non-linear σ-models based on the parallel transport equation (12) or some non-minimal extension such as (17), provided that one is prepared to introduce the unconstrained quantum prepotential ξ^a. As a consequence, the Feynman rules constructed from the expanded action (15) will yield manifestly covariant divergences $\Gamma^{Div}(\phi_B)$. These can be removed by counterterms which are integrals over the full superspace of globally defined scalar functions of the background field that are constructed from the Riemann tensor and its covariant derivatives. These scalars correspond to metric counterterms which are symmetric rank two Kähler tensors whose corresponding two-forms are exact, i.e. they belong to the trivial cohomology class in $H^{1,1}(M)$. The only exception to this rule is at one loop where the method fails due to the ghost structure of the theory. This fits in with the well-known result that the one-loop metric counterterm is given by the Ricci tensor, the corresponding Ricci form being cohomologically non-trivial.

We emphasize that the possible counterterms do not all vanish by inspection. One may have hoped, for example, that a judicious choice of non-minimal terms on the right-hand side of the parallel transport equation would yield a Lagrangian which would define a manifestly finite theory. This would have been the case if the background fields entered into the vertices of the theory only via the connection $\Gamma^a_{\alpha b} \equiv D_\alpha \phi^c \Gamma^a_{bc}$, since by power counting all divergences correspond to dimension zero counterterms and can therefore not involve D's. Indeed, the dangerous vertices in the expansion can readily be identified by setting the background field to be constant. Up to order ξ^4, and using equation (17) with $k = -1/3$, one finds

$$I = \int d^2 x \, d^4\Theta \; \{ \; K(\phi_B, \bar{\phi}_B) + g_{a\bar{b}} \, \bar{D}^2 \, \xi^a \, D^2 \bar{\xi}^{\bar{b}} + \frac{1}{6} R_{a\bar{b}c\bar{d}} \, (\phi_B, \bar{\phi}_B)$$

$$\bar{D}^2 \, \xi^a \, \bar{D}^2 \, \xi^c \, D^2 \, \bar{\xi}^{\bar{b}} \, D^2 \, \bar{\xi}^{\bar{d}} \quad + \text{terms in } D_\alpha \phi^a_B \; \} \quad . \qquad (18)$$

The four-point vertex in (18) allows one to construct potentially divergent diagrams such as those of figs. 1, 2 and 3 :

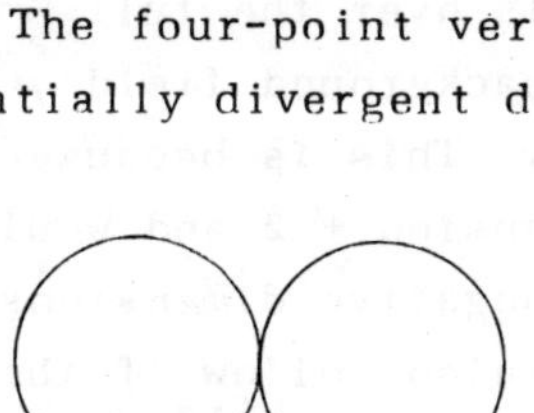

Fig. 1

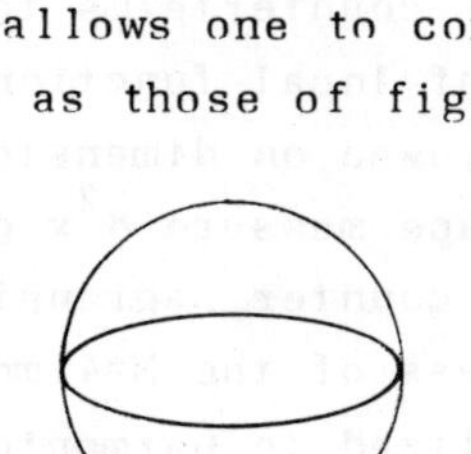

Fig. 2

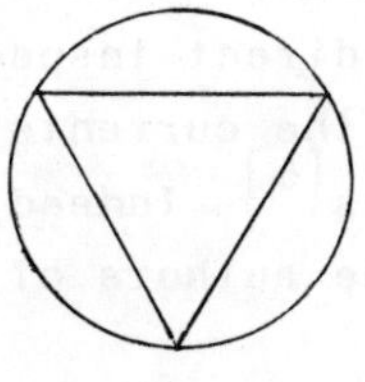

Fig. 3

The two-loop graph of fig. 1 requires a counterterm proportional to the curvature scalar, R, and hence vanishes in the Ricci-flat case. The three-loop graph of fig. 2 is the only 3-loop graph which can have a $1/\varepsilon$ divergence.[15] The corresponding counterterm would be $(R_{a\bar{b}c\bar{d}})^2$ but the coefficient must vanish as can be seen by appealing to universality of the N=2 expansion and finiteness of the N=4 model. Since $(\text{Riemann})^2$ does not vanish in the hyperKähler case, which is known to be finite by other arguments, this coefficient must be zero for all N=2 theories. Thus we can conclude that Ricci-flat N=2 σ-models are finite through three loops but there are no grounds for arguing that this state of affairs should persist to all orders. Indeed the diagram of fig. 3 has been analysed in conventional perturbation theory and it appears to be divergent.[16]

We note that our N=2 formalism does not supply an immediately obvious explanation of the finiteness[3] of the N=4 models. However, it is clear that a fully N=4 supersymmetric formulation of these theories would lead to such a conclusion. The background fields in such a formalism would presumably have to be dimension zero scalar superfields (for example, if one could construct these models in terms of the relaxed hypermultiplet[7]), so if the Feynman rules required all counterterms to be integrals over the full N=4 superspace of local functions of these background fields, none would be allowed on dimensional grounds. This is because the N=4 superspace measure $d^2x\, d^8\theta$ has dimension + 2 and would require the counter Lagrangian to have negative dimensions. The finiteness of the N=4 models should also follow if they can be quantized in harmonic superspace language [17].

Although it is not possible to establish the finiteness of Ricci flat N=2 models by direct inspection, one should also consider arguments based on the currents of the model as were made for N=4 super Yang-Mills.[5] Indeed, such an argument has recently been proposed by the authors of ref. [18]. However these authors invoked an Adler-Bardeen type theorem for the axial U(1) symmetry of the N=2 models without placing this current into its supersymmetric context. We conclude this paper by giving a brief analysis of the supercurrent which bears out the general conclusions that can be drawn from the N=2 expansion above.

The classical supercurrent is a real vector superfield J_μ obeying the constraint

$$(\gamma^\mu)_\alpha{}^\beta D_\beta J_\mu = 0 \qquad (19)$$

Its components are

$$j_\mu = J_\mu|, \quad S_{\mu\alpha} = D_\alpha J_\mu|, \quad t_{\mu\nu} = (\gamma_\mu)^{\alpha\beta} [D_\alpha, \bar{D}_\beta] J_\nu| \qquad (20)$$

where the bar notation indicates the superfield evaluated at $\theta=\bar{\theta}=0$. j_μ is the U(1) vector current, $S_{\mu\alpha}$ the supersymmetry current and $t_{\mu\nu}$ the energy-momentum tensor. All three are conserved and obey the conformal conditions

$$t^\mu{}_\mu = \gamma^\mu S_\mu = \varepsilon^{\mu\nu} \partial_\mu j_\nu = 0 \,. \qquad (21)$$

The last of equations (21) implies that there is also a conserved axial vector current

$$j^5_\mu = \varepsilon_\mu{}^\nu j_\nu \qquad (22)$$

At the quantum level, one expects the conservation law (20) to develop an anomalous right-hand side. From the structure of N=2 d=2 supergravity in superspace [14], there are two possible minimal anomaly multiplets :

$$(\gamma^\mu D)_\alpha J_\mu = \bar{D}_\alpha \bar{C} \qquad (23)$$

or

$$(\gamma^{\mu}D)_{\alpha} J_{\mu} = D_{\alpha} L \tag{24}$$

where C is a chiral superfield, $\bar{D}_{\alpha}C=0$, and L is a real linear superfield, $D^2L = \bar{D}^2L = 0$. Both forms yield conserved supersymmetry and translational currents, but the conformal anomalies differ. For the chiral case we have

$$\gamma^{\mu} S_{\mu},\ t^{\mu}{}_{\mu},\ \partial^{\mu} j_{\mu} \neq 0 \tag{25}$$

whereas in the linear case one finds

$$\gamma^{\mu} S_{\mu},\ t^{\mu}{}_{\mu},\ \varepsilon^{\mu\nu} \partial_{\mu} j_{\nu} \neq 0 \tag{26}$$

Since in this case, the vector current is conserved, this is the natural form. In practice the anomaly superfields are related to the counter Lagrangian ΔK by

$$C \sim \bar{D}^2 \Delta K \quad \text{or} \quad L \sim \bar{D}^{\alpha} D_{\alpha} \Delta K \tag{27}$$

and one can shift from one form to the other by a suitable redefinition of J_{μ} :

$$J_{\mu} \rightarrow J_{\mu} + k(\gamma_{\mu})^{\alpha\beta} [D_{\alpha}, \bar{D}_{\beta}] \Delta K \tag{28}$$

where k is a real number. If the original J_{μ} was reparameterization invariant, then the redefined one will be also provided that ΔK is a globally defined scalar. Furthermore, if we demand that the anomaly equation itself be reparameterization invariant, then the chiral form of the anomaly is not allowed in the case where ΔK is not globally defined. Hence, in the general N=2 case we can define a reparameterization invariant component axial current by

$$j'^{5}_{\mu} = \varepsilon_{\mu}{}^{\nu} j_{\nu} + k(\gamma_5\gamma_{\mu})^{\alpha\beta} [D_{\alpha}, \bar{D}_{\beta}] \tilde{\Delta}K| \tag{29}$$

where the tilde indicates that one should reject non globally defined contributions. The current j'^{5}_{μ} does obey an Adler-Bardeen theorem but is not the dual of the vector current, nor does it sit in the natural supercurrent multiplet(i.e. the one containing the conserved vector current). In the Ricci flat case, ΔK is globally defined so that we can define a conserved axial current, but again it does not sit in the natural supercurrent and hence one cannot conclude from its existence that the trace of the energy-momentum tensor vanishes.

The main technical result of this paper is the proof of the cohomological triviality of the counterterms for N=2, d=2 super σ-models at $\ell\geqslant 2$ loops. Since the detailed Feynman rules do not make manifest any further power-counting results, and Adler-Bardeen type considerations do not appear to be applicable, these results cast doubt upon the suggestion [3] that Ricci-flat N=2 σ-models are ultraviolet finite to all orders. The issue now appears to hang on the outcome of detailed four-loop calculations. [15,16]

In considering possible divergent structures at high loop order, account must be taken of the universality [3] property of the σ-model counterterms. All super σ-models with extended supersymmetry are just N=1 models with additional restrictions on the target manifold. Thus, all counterterms must be able to be written without the use of any special tensors such as ε-symbols or complex structures, which do not always exist for general Riemannian manifolds. Moreover, all counterterms with non-vanishing coefficients must vanish themselves when the manifold is restricted to one known to be finite at the loop order in question. Thus, all counterterms written in N=1 form must vanish for hyperKähler (N=4) manifolds and at $\ell\geqslant 2$ loops for locally symmetric ($\nabla_i R_{jklm}=0$) Kähler manifolds (for an instanton-based discussion of the locally symmetric N=2 case, cf. [19]). The four-loop calculation of reference [15] produces a correction to the Kähler potential of the form

$$\Delta_4 K \sim R_{a\bar{a}b\bar{b}}\, R^{c\bar{a}d\bar{b}}\, R_{c\ d}^{\ a\ b} - R_{a\bar{a}b\bar{b}} R^{b\bar{b}c\bar{c}} R_{c\bar{c}}^{\ \ a\bar{a}} \quad . \tag{30}$$

This satisfies the requirement that it vanish when the manifold is hyperKähler, because then the holonomy group is SU(2)xUsp(n/2) and there is just one independent R^3 contraction, which cancels between the two terms. The metric counterterm $\partial_a \partial_{\bar{b}} \Delta_4 K$ can also be written without using the complex structure in this case.[20] Thus all the consistency conditions for this counterterm to be allowable are satisfied.

Even if general Ricci-flat N=2 models diverge at higher loop orders, it may turn out that restrictions short of imposing the full N=4 supersymmetry may be sufficient for finiteness. For example, there may be special cancellations in models of specific dimensionality. An example of such an accident is the circumstance that all four-dimensional Ricci-flat compact Kähler manifolds are in fact hyperKähler. There are no hyperKähler 6-dimensional manifolds, but an important question remains whether there can be any similar accidents for the 6-dimensional manifolds with SU(3) holonomy that are relevant for string compactification.

ACKNOWLEDGEMENTS

We would like to thank P.K.Townsend and C.M. Hull for stimulating conversations.

REFERENCES

1. P. Candelas, G. Horowitz, A. Strominger and E. Witten, Nucl. Phys. B256 (1985) 46.

2. D. Friedan, Phys. Rev. Lett. 45 (1980) 1057; Ann. Phys. (N.Y.) 163 (1985), 318.

3. L. Alvarez-Gaumé and D.Z. Freedman, Phys. Rev. D22 (1980), 846; Comm. Math. Phys. 80 (1981), 443.

C.M. Hull, Nucl. Phys. B260 (1985), 182.

L. Alvarez-Gaumé and P. Ginsparg, Comm. Math. Phys. 102 (1985), 311.

4. O.V. Tarasov and A.A. Vladimirov, Phys. Lett. 96B (1980), 94;

M.T. Grisaru, M. Rocek and W. Siegel, Phys. Rev. Lett. 45 (1980) 1063;

W. Caswell and D. Zanon, Phys. Lett. 100B (1980), 152.

5. M.F. Sohnius and P.C. West, Phys. Lett. 100B (1981) 245.

6. M.T. Grisaru and W. Siegel, Nucl. Phys. B201 (1982) 292.

7. P.S. Howe, K.S. Stelle and P.K.Townsend, Nucl. Phys. B214 (1983) 519.

8. P.S. Howe, K.S. Stelle and P.C. West, Phys. Lett. 124B (1983) 55.

9. P.S. Howe, K.S. Stelle and P.K. Townsend, Nucl.Phys. B236 (1984) 125.

10. S. Mandelstam, Nucl. Phys. B213 (1983), 149;

L. Brink, O. Lindgren and B.E.W. Nielsen, Nucl.Phys. B212 (1983), 401; Phys. Lett. 123B (1983), 328.

11. P.S. Howe and K.S. Stelle, Phys. Lett. 137B (1984), 175;

N. Marcus and A. Sagnotti, Phys. Lett. 135B (1984) 85; Nucl. Phys. B256 (1985), 77.

12. J. Honerkamp, Nucl. Phys. B36 (1972), 130;

L. Alvarez-Gaumé, D.Z. Freedman and S. Mukhi, Ann. Phys. (N.Y.) 134 (1981), 85.

13. B. Zumino, Phys. Lett. B87 (1979).

14. P.S. Howe, G. Papadopoulos and K.S. Stelle, in preparation.

15. D. Zanon, Harvard preprint HUTP-85/A064 (1985).

16. D. Zanon, private communication.

17. A. Galperin, E. Ivanov, V. Ogievetsky and P.K. Townsend, Dubna preprint E2-85-732 (1985).

18. L. Alvarez-Gaumé, S. Coleman and P. Ginsparg, Harvard Preprint HUTP-85/A037 (1985).

19. V. Novikov, M. Shifman, A. Vainshtein and V. Zakharov, Phys. Lett. 139B (1984) 389;

A.Y. Morozov, A.M. Perelomov and M.A. Shifman, Nucl. Phys. B248 (1984), 279.

20. C.N. Pope, M.F. Sohnius and K.S. Stelle, in preparation.

SUPERSTRING COMPACTIFICATIONS WITH TORSION AND SPACETIME SUPERSYMMETRY

C. M. Hull

DAMTP, Silver Street, Cambridge CB3 9EW, England.

Abstract

The light-cone gauge action for the heterotic superstring in an arbitrary background with non-vanishing metric, torsion, dilaton and Yang-Mills fields is considered and the conditions for the existence of Killing spinors and hence for space-time supersymmetry are derived. It is shown that the ususal Green-Schwarz and Ramond-Neveu-Schwarz formulations of the superstring are equivalent for backgrounds admitting Killing spinors, but not for more general backgrounds. For compactifying backgrounds that lead to four-dimensional Minkowski space, N=1 supersymmetry requires the internal manifold to be an almost complex space with a connection with torsion whose holonomy group is SU(3) and with a non-constant dilaton field appearing in the ten-dimensional metric as a "warp-factor". There are many such supersymmetric spaces, but most do not correspond to sigma-models that are perturbatively conformally invariant. The extra conditions thus needed for these spaces to be solutions of the string field equations are described. Anomalies, sigma-model finiteness, phenomenology and type II superstrings are briefly discussed.

The effective physics of superstring theories [1,2] is greatly affected by the choice of vacuum, which should be a solution of some (as yet unknown) string field equations. One approach to finding such vacua is to seek backgrounds M in which it is possible to define a consistent string theory. This requires in particular that the appropriate two-dimensional non-linear sigma-model defined on M be conformally invariant [3-10]. Conversely, conformal invariance should imply that all tadpole expectation values vanish and hence that the string field equations are satisfied [4,5]. For the bosonic string this leads to group manifold solutions [3] while for the superstring this leads to the Calabi-Yau compactifications [4,5] and certain generalisations of these with torsion [8,9].

Superstring vacua of the form $M = M_4 \times K$ (where M_n is n-dimensional Minkowski space and K is a compact six-manifold), that lead to an effective four-dimensional theory with an unbroken N = 1 supersymmetry may be phenomenologically attractive. An alternative approach to superstring compactification has been to seek such supersymmetric backgrounds [4,5], again leading, in the case of vanishing torsion, to Calabi-Yau spaces. Although Calabi-Yau compactifications have a number of attractive features, no fully realistic scenario has yet emerged, so that it is desirable to seek a more general class of solutions. Here the analysis of [5] will be extended to find the conditions for a superstring background with torsion to admit Killing spinors, so that on compactification there is some unbroken supersymmetry. The treatment here is slightly more general than that presented in [9].

To illustrate the relevant ideas, consider the closed bosonic string $X^\mu(\sigma,\tau)$ ($\mu=0,\ldots,d-1$) moving in a background d-dimensional space with metric $g_{\mu\nu}(X)$. The action in conformal gauge is

$$S = \frac{1}{4\pi\alpha'}\int d\sigma\, d\tau\ \frac{1}{2} g_{\mu\nu}(X)\, \partial_+ X^\mu \partial_- X^\nu \quad (1)$$

where $\partial_{\pm}=\partial/\partial\tau\pm\partial/\partial\sigma$ and α' is the inverse string tension. This action is invariant under the general co-ordinate transformation

$$X^{\mu} \to X^{\mu} + \xi^{\mu}(X), \quad g_{\mu\nu} \to g_{\mu\nu} - 2\nabla_{(\mu}\xi_{\nu)} \tag{2a}$$

corresponding to the local reparameterisation invariance of string theory. This is to be contrasted with the transformation

$$X^{\mu} \to X^{\mu} + \xi^{\mu}(X), \quad g_{\mu\nu} \to g_{\mu\nu} + \xi^{\rho} g_{\mu\nu,\rho} \tag{2b}$$

which is a symmetry of the action only if ξ^{μ} is a Killing vector.

$$\nabla_{(\mu}\xi_{\nu)} = 0. \tag{2c}$$

When this holds (2b), unlike (2a), is a "proper" symmetry of the 2-dimensional field theory, with a corresponding Noether charge, and could be gauged. The Killing vectors correspond to the symmetries of the background, e.g. if $M = M_d$, then there are $\frac{1}{2}d(d+1)$ Killing vectors corresponding to the generators of the Poincare group ISO(d-1,1), while for a general background, this will be "spontaneously broken" to a smaller group. Similarly, the superstring coupled to a general supergravity background [11] will be formally invariant under local supersymmetry transformations analogous to (2a). However, there may also be a number of "proper" supersymmetries of the background analogous to (2b) corresponding to solutions of appropriate "Killing spinor" equations. For example, as will be seen, on $M = M_4 \times T^6$ (where T^6 is the 6-torus) the heterotic superstring has 16 Killing spinors leading to N = 4 supersymmetry in four dimensions, whereas $M = M_4 \times K$ with K a Calabi-Yau manifold has 4 Killing spinors so that the vacuum spontaneously breaks the N = 4 local supersymmetry of the full theory down to N = 1.

One is then led to consider the supersymmetry analogues of (2b),(2c). In flat space, the supersymmetry transformations corresponding to (2b) have been given in light-cone gauge by Green and Schwarz [1] and shown to satisfy the appropriate super-Poincare algebra. The generalisations of these for a non-trivial background will be presented here and the corresponding "Killing spinor" equations derived. The conditions for unbroken N = 1 supersymmetry obtained in this way are different when there is torsion from those obtained previously by considering part of the low-energy field theory [4,19]. As will be seen, in backgrounds that admit Killing spinors, it is possible to relate space-time and world-sheet supersymmetries and hence the Green-Schwarz and Ramond-Neveu-Schwarz formulations of superstrings in that background. The supersymmetric spaces found are generalisations of the Calabi-Yau solutions [4,5] that have non-vanishing torsion and dilaton field. However, the requirement of conformal invariance places extra restrictions on these backgrounds, leading to essentially the solutions found earlier [8,9] by requiring conformal invariance to lowest order in perturbation theory.

The heterotic superstring theory [2] is formulated in 10-dimensional space-time and can have as its gauge group either $Spin(32)/\mathbb{Z}_2$ or $E_8 \times E_8$. The bosonic degrees of freedom $X^\mu(\sigma,\tau)$ give the position of the world-sheet (parameterised by σ,τ) in space-time (with coordinates $X^\mu, \mu=0,1,\ldots,9$). In light-cone coordinates $X^\mu = (X^+,X^-,X^i)$, ($i=1,\ldots,8$) the 8 transverse coordinates X^i are the physical degrees of freedom, transforming (in M_{10}) as an 8-vector under the transverse Lorentz group $SO(8) \subset SO(9,1)$. The superpartner of X is $S(\sigma,\tau)$, a 16-component Majorana-Weyl spinor of $SO(9,1)$. Under the transverse $SO(8)$, $S \to (S_p, S_{\dot q})$ where $p,\dot q = 1,\ldots,8$ label the two inequivalent spinor representations of $SO(8)$. The 16x16 gamma matrices of $SO(8)$ have 8x8 blocks

$$\Gamma^i = \begin{pmatrix} 0 & \gamma^i_{p\dot q} \\ \gamma^i_{\dot r s} & 0 \end{pmatrix}$$

$$\{\Gamma^i, \Gamma^j\} = 2\delta^{ij}$$

where $\gamma^i_{p\dot{q}}$ are the "triality coefficients" relating the three inequivalent 8-dimensional representations of $SO(8)$ [15]. In the light-cone gauge [1,2], the condition $S_{\dot{q}} = 0$ is imposed, so that 8 components S_p remain. In the fermionic representation [2], the remaining degrees of freedom are the anti-commuting $\psi^A(\sigma,\tau)$, $(A=1,\ldots,32)$ transforming as a scalar under $SO(8)$, a left-handed spinor under world-sheet Lorentz transformations ($SO(1,1)$) and as a 32 of Spin $(32)/\mathbb{Z}_2$ or as a $16\oplus16$ under the $SO(16)\times SO(16)$ subgroup of $E_8\times E_8$. (It transforms non-linearly under the remainder of $E_8\times E_8$). In flat space (M_{10} or $M_4\times T^6$), the light-cone gauge action for the heterotic string is

$$S = \frac{1}{4\pi\alpha'}\int d\sigma d\tau\left[\frac{1}{2}\delta_{ij}\partial_+X^i\partial_-X^j + \frac{i}{2}S^p\partial_+S^p + \frac{i}{2}\psi^A\partial_-\psi^A\right] \quad (3)$$

where again $\partial_\pm = \partial/\partial\tau \pm \partial/\partial\sigma$ and α' is the inverse string tension. Thus on-shell the S^p are right-moving ($\partial_+S = 0$) and the ψ are left-moving ($\partial_-\psi = 0$).

The action (3) is invariant under a space-time supersymmetry parameterised by a 16-component $SO(9,1)$ Majorana-Weyl spinor η, which decomposes under the transverse $SO(8)$ as $\eta = (\delta_p, \epsilon_{\dot{q}})$. Under world-sheet Lorentz transformations, ϵ is a left-handed Majorana-Weyl spinor and δ is a right-handed one. The supersymmetry transformations are

$$\delta_\delta X^i = 0, \qquad \delta_\delta S_p = -2i\alpha'\sqrt{p^+}\,\delta_p \quad (4)$$

$$\delta_\epsilon X^i = (p^+)^{-\frac{1}{2}}\epsilon^{\dot{q}}\gamma^i_{\dot{q}p}S^p, \qquad \delta_\epsilon S_p = -i(p^+)^{-\frac{1}{2}}(\partial_-X^i)\gamma^i_{\dot{q}p}\epsilon^{\dot{q}} \quad (5)$$

with $\delta_\eta\psi = 0$ and where p^+ is a component of the 10-momentum p^μ. The commutator of two such transformations gives both a space-time translation ($\delta X^i = a^i$, $\delta S = 0$) and a world-sheet translation ($\delta(\tau-\sigma) = \xi^-$, $\delta(\tau+\sigma) = 0$)

$$[\delta_{\eta_1}, \delta_{\eta_2}] X^i = a^i + \xi^- \partial_- X^i \tag{6}$$

$$[\delta_{\eta_1}, \delta_{\eta_2}] S_\rho = \xi^- \partial_- S_\rho$$

where

$$a^i \equiv -2i\, \gamma^i_{\rho\dot{q}} (\delta_1^\rho \epsilon_2^{\dot{q}} + \delta_2^\rho \epsilon_1^{\dot{q}}), \quad \xi^- \equiv -\frac{2i}{p^\tau} \epsilon_1{}^{\dot{q}} \epsilon_{2\dot{q}}.$$

Thus each of the 8 components of $\epsilon_{\dot{p}}$ parameterises a world-sheet super-symmetry, i.e. there is (0,8) world sheet supersymmetry. (In 2 dimensions, the super-algebra of type (p,q) is that generated by p right-handed Majorana-Weyl super-charges and q left-handed ones [12].)

Consider now a background space-time of the form $M = M_2 \times L$ where L is some 8-dimensional space with positive-definite metric $g_{ij}(X)$. (Later this will be restricted to be of the form $L = \mathbb{R}^2 \times K$, so that $M = M_4 \times K$. L need not be compact.) Suppose that the anti-symmetric tensor gauge field and Yang-Mills field have background expectation values $b_{ij}(X)$, $A_i^M(X)$, respectively (M labels the adjoint of the gauge-group). Then the light-cone-gauge action for the heterotic string in such a background is, to lowest order in α', [2,5,6,10,12]

$$S = \frac{1}{4\pi\alpha'} \int d\sigma d\tau\, \frac{1}{2} \Big[\partial_+ X^i \partial_- X^j (g_{ij} + b_{ij}) + i S^\rho (D_+ S)^\rho \tag{7}$$

$$+ i \psi^A (D_- \psi)^A + \frac{1}{2} F_{ijAB}\, \sigma^{ij}_{pq}\, S^p S^q \psi^A \psi^B + O(\alpha') \Big]$$

where $F_{ijAB} = F^M_{ij} (T^M)_{AB}$ is the Yang-Mills field strength in the representation defined by the matrices $(T^M)_{AB}$.

$$(D_-\psi)^A = \partial_-\psi^A + A_i^{AB}(\partial_- X^i)\psi^B \tag{8}$$

$$(D_+ S)^p = \partial_+ S^p + \tfrac{1}{2}\left(\omega_i^{ab} - T_i^{ab}\right)\sigma_{ab}^{pq} S^q$$

$$T_{ijk} = \tfrac{3}{2}\, b_{[ij,k]}\,, \qquad T_i^{ab} = T_{ijk}\, e^{aj} e^{bk} \tag{9}$$

$e_i^a(X)$ are some orthonormal frames on L ($e_i^a e_j^b \delta_{ab} = g_{ij}$ and ω_i^{ab} is the torsion-free metric spin-connection, $de^a + \omega^{ab} \wedge e^b = 0$). In these frames, $\gamma^a_{p\dot{q}}$ are constant numerical matrices, $\gamma^i_{p\dot{q}} = e_a^i \gamma^a_{p\dot{q}}$ and

$$\sigma^{ab}_{pq} = \gamma^{[a}_{\dot{r}[p}\gamma^{b]}_{q]\dot{r}}\,, \qquad \sigma^{ab}_{\dot{p}\dot{q}} = \gamma^{[a}_{r[\dot{p}}\gamma^{b]}_{\dot{q}]r} \tag{10}$$

As will be seen later, the metric g_{ij} here is related to the conventionally scaled one by a conformal transformation. The action (7) is intimately related to the (0,1) supersymmetric non-linear sigma model [12], as will be seen.

Classically (i.e. to lowest order in α') (7) is formally invariant under local G rotations

$$\delta\psi^A = \Lambda^A{}_B(X)\,\psi^B\,, \qquad \delta A_i^{AB} = \partial_i\Lambda^{AB} + 2A_i^{C[A}\Lambda^{B]C} \tag{11}$$

and local transverse Lorentz rotations (SO(8))

$$\delta S^p = \tfrac{1}{2}\Theta^{ab}(X)\,\sigma_{ab}^{pq} S^q\,, \qquad \delta\omega_i^{ab} = \partial_i\Theta^{ab} + 2\omega_i^{c[a}\Theta^{b]c} \tag{12}$$

analagous to the general coordinate transformations (2a). However, these symmetries are anomalous [12]. These anomalies can be cancelled if b_{ij} transforms under these symmetries as [12, 13, 14]

$$\delta b_{ij} = \alpha'\left[A_{[i}^{AB}\,\partial_{j]}\Lambda^{BA} - \tilde{\omega}_{[i}^{ab}\,\partial_{j]}\Theta^{ba}\right] \tag{13}$$

where $\omega_i^{ab} = \omega_i^{ab} + \kappa_i^{ab}$ and κ_i^{ab} is any convariant "contorsion" tensor. Then

$$H = db + \alpha'\left[\frac{1}{30}\mathrm{Tr}\left(A_\wedge F - \frac{1}{3}A_\wedge A_\wedge A\right) - \mathrm{tr}\left(\tilde{\omega}_\wedge \tilde{R} - \frac{1}{3}\tilde{\omega}_\wedge\tilde{\omega}_\wedge\tilde{\omega}\right)\right] \quad (14)$$

defines a gauge invariant field strength H_{ijk}, with the differential form notation

$$H = \frac{1}{6} H_{ijk}\, dX^i{}_\wedge dX^j{}_\wedge dX^k, \qquad b = \frac{1}{2} b_{ij}\, dX^i{}_\wedge dX^j \quad (15)$$

$$A = A_i\, dX^i, \quad \tilde{\omega} = \tilde{\omega}_i\, dX^i, \quad \tilde{R} = d\tilde{\omega} + \tilde{\omega}_\wedge\tilde{\omega}$$

with Lie algebra indices suppressed. Here tr denotes a trace in the fundamental representation while Tr denotes an adjoint trace. As T_{ijk} is no longer gauge invariant, it should be replaced by $\frac{1}{2}H_{ijk}$ in (8). H satisfies the Bianchi identity

$$dH = \alpha'\left[\frac{1}{30}\mathrm{Tr}(F_\wedge F) - \mathrm{tr}\, R_\wedge R\right] \quad (16)$$

Although (14),(16) depend on the choice of $\tilde{\omega}$, they have an integrability condition that is independent of the choice of connection and depends only on the topology [23]. Indeed, for there to be a 3-form H satisfying (16), it is necessary and sufficient that the Pontriagin class (the cohomology class defined by $tr(R_\wedge R)$) and the second Chern class (the cohomology class represented by $\frac{1}{30}Tr(F_\wedge F)$) are equal.

The tensor H_{ijk} can be interpreted as a torsion on L. Thus it is convenient to define the connections and curvatures with torsion*

$$R_{(\pm)\,jkl}^{\;i} = \partial_k \Gamma_{(\pm)\,l\,j}^{\;i} + \Gamma_{(\pm)\,k\,m}^{\;i}\,\Gamma_{(\pm)\,l\,j}^{\;m} - (k \leftrightarrow l)$$

---

* *Note that the conventions here are the opposite of those of [8,9]. To compare with [8,9], interchange right-movers with left-movers and* $\omega_{(+)}$ *with* $\omega_{(-)}$*, etc..*

$$\Gamma_{(\pm)j}{}^{i}{}_{k} = \left\{ {i \atop j\,k} \right\} \pm \frac{1}{2} H^{i}_{jk} \qquad (17)$$

$$H^{i}_{jk} = g^{il} H_{jkl}\,, \qquad \omega_{(\pm)i}{}^{ab} = \omega_i^{ab} \mp \frac{1}{2} H_i^{ab} = -e^{jb}\left(e^{a}_{j,i} - \Gamma_{(\pm)i}{}^{k}{}_{j}\, e^{a}_{k}\right)$$

and corresponding covariant derivatives on L, $\nabla_{(\pm)i}$. As will be seen, a natural choice for $\tilde{\omega}$ is $\tilde{\omega}_i^{ab} = \omega_{(+)i}^{ab}$, $\kappa_i^{ab} = -\frac{1}{2}H_i^{ab}$. Then H appears on both sides of (14) and can be calculated perturbatively in α'.

As a motivation for the choice $\tilde{\omega} = \omega_{(+)}$, consider the light-cone gauge action for the Green-Schwarz superstring [1] in a background field configuration [6, 7, 8, 10]

$$S = \frac{1}{4\pi\alpha'} \int d\sigma d\tau \, \frac{1}{2} \Big[(g_{ij} + b_{ij})\, \partial_+ X^i \partial_- X^j + i S^p (D_+ S^p) \qquad (18)$$

$$+ i S^{\dot{q}} (D_- S)^{\dot{q}} + \frac{1}{4} R_{(+)ijkl}\, \sigma^{ij}_{\dot{p}\dot{q}}\, \sigma^{kl}_{rs}\, S^{\dot{p}} S^{\dot{q}} S^r S^s \Big]$$

with (8), (9) and (17) and

$$(D_- S)^{\dot{q}} = \partial_- S^{\dot{q}} + \frac{1}{2}\left(\omega_i^{ab} - T_i^{ab}\right) \sigma_{ab}^{\dot{p}\dot{q}} S^{\dot{q}} \qquad (19)$$

This is anomaly-free and so requires no Chern-Simons corrections. Let ι: $SO(8) \to G$ denote an embedding of the tangent space group into the Yang-Mills group. Then comparing (7) and (18), it can be seen that if the Yang-Mills connection A_i^{AB} and the image $\iota(\omega_i^{ab} - T_i^{ab})$ are equal, then (7) is the same classically as (18), with ψ^A $(A = 1, \ldots, 8)$ identified with the $S^{\dot{p}}$, together with the action for 24 free fields ψ^A $(A = 9, \ldots, 32)$ that do not couple to the background. However, if the anomaly and the corrections (13, 14) are taken into account, T is not covariant so that $\omega - T$ no longer transforms as a connection. This suggests replacing T with $\frac{1}{2}$H and identifying A_i^{AB} with $\iota(\omega_i^{ab} - \frac{1}{2} H_i^{ab} \equiv \omega_{(+)i}^{ab})$. This would, in general, lead not to (18) but to (18) plus extra terms coming from Cherns-Simons terms in (14). However, with the special choice $\tilde{\omega}_i^{ab} = \omega_{(+)i}^{ab}$ in (14), then on identifying A_i^{AB} and $\iota(\omega_{(+)i}^{ab})$ the Chern-Simons terms in H (14) cancel and (7) again becomes precisely (18) plus the action for 24 free fields. The importance of the identification of A and $\iota(\omega_{(+)})$ is that it is sufficient (and possibly necessary) to cancel the 2-loop sigma-model β-function for the metric [8, 9].

Thus taking $\tilde{\omega}$ as $\omega_{(+)}$ (possibly plus some fermion bilinear and dilaton dependent terms) seems a prefered choice for string theory. However, for generality, $\tilde{\omega}$ will be kept arbitrary in the following analysis. The conditions to be obtained for space-time supersymmetry for the heterotic string will immediately give the corresponding conditions for the Green-Schwarz superstring on taking $\tilde{\omega} = \omega_{(+)}$, $\iota(\omega_{(+)}) = A$.

Consider now the question of space-time supersymmetry. In flat space, the action (3) is invariant under (4), (5) corresponding to $N = 1$ supersymmetry in 10 dimensions or, on toroidal compactification, to $N = 4$ supersymmetry in four dimensions. For most background field configurations, the action (7) will not be invariant under any such supersymmetries. For any given compactifying background, the number of unbroken supersymmetries present in the lower dimension will depend on the number of (possibly X^I-dependent) supersymmetries of the type (4), (5) under which the action (7) for this background is invariant. Thus one is led to consider the conditions for (7) to be invariant (at least to lowest order in α') under supersymmetries of the form (3), (4) for some spinor fields (Killing spinors) $\delta^p(X), \epsilon^{\dot{q}}(X)$ on L. Consider first the δ-supersymmetries (4). The action (7) will be invariant under (4) for some $\delta^p(X)$ provided δ satisfies

$$\left(D_+^{(-)}\delta\right)^p \equiv \partial_+\delta^p + \tfrac{1}{2}\,\omega_{(-)i}^{\;ab}\left(\sigma_{ab}^{pq}\right)\left(\partial_- X^i\right)\delta^q = 0 \qquad (20)$$

and

$$F^{M}_{ij}\left(\sigma^{ij}_{pq}\right)\delta^{q} = 0 \tag{21}$$

For (20) to be satisfied for all $X^{i}(\sigma,\tau)$ it is necessary that δ be a covariantly constant spinor on L with respect to the connection with torsion $\omega_{(-)}$

$$\nabla_{(-)i}\,\delta^{p} \equiv \partial_{i}\delta^{p} + \frac{1}{2}\,\omega_{(-)i}^{\ \ ab}\left(\sigma^{pq}_{ab}\right)\delta^{q} = 0. \tag{22}$$

Consider next the ϵ-supersymmetries (5). Comparison with [12] and covariance requirements suggests modifying (5) to become

$$\delta_{\epsilon} X^{i} = (p^{+})^{-\frac{1}{2}}\,\epsilon^{\dot q}\,\gamma^{i}_{\dot q p} S^{p}$$

$$\delta_{\epsilon} S^{p} + \frac{1}{2}\left(\delta_{\epsilon}X^{i}\right)\hat{\omega}_{i}^{ab}\,\sigma^{pq}_{ab} S^{q} = -i\epsilon^{\dot q}\,\gamma^{a}_{\dot q p}\,e_{ai}\left(\partial_{-}X^{i}\right) \tag{23}$$

$$\delta_{\epsilon}\Psi^{A} + \left(\delta_{\epsilon}X^{i}\right)A_{i}^{AB}\,\Psi^{B} = 0$$

for some spin-connection $\hat{\omega}_{i}^{ab}$. Using identities such as

$$\sigma^{pq}_{bc}\,\gamma^{a}_{\dot p q} + \sigma^{\dot p \dot q}_{bc}\,\gamma^{a}_{p \dot q} = \delta^{a}_{[b}\,\gamma_{c]\dot p p} \tag{24}$$

it can be shown that the action (7) is invariant under (23) only if $\hat{\omega}_{i}^{ab} = \omega_{i}^{ab}$, the minimal (torsion-free) spin-connection, and

$$\nabla_{(-)i}\,\epsilon^{\dot p} = 0 \tag{25}$$

$$F^{M}_{ij}\,\sigma^{ij}_{\dot p \dot q}\,\epsilon^{\dot q} = 0 \tag{26}$$

which can be combined with (21), (22) to give a condition on the 16-spinor $\eta = (\delta, \epsilon)$

$$\nabla_{(-)i}\,\eta \equiv \partial_i \eta + \tfrac{1}{2}\,\omega_{(-)i}{}^{ab}\left(\Gamma_{[a}\Gamma_{b]}\right)\eta = 0 \tag{27}$$

$$F^{M}_{ij}\,\Gamma^{[i}\Gamma^{j]}\,\eta = 0 \tag{28}$$

These, then, are the analogues of (2c). However, just as it is possible to make the action invariant under general coordinate transformations and local tangent space transformations by defining the transformations of the metric as (2a) and of the spin-connection as (12), it is possible to make the action invariant under N=1 local supersymmetry for arbitrary backgrounds by introducing a gravitino field χ_i and gaugino field β^M transforming under local supersymmetry with arbitrary parameter $\eta(X)$ as $\delta\chi_i=\nabla_{(-)i}\eta+\ldots$ $\delta\beta^M=(F^M_{ij}\Gamma^i\Gamma^j)\eta+\ldots$ Then using the Noether method a locally supersymmetric action can be obtained, presumably corresponding to a light-cone gauge component form of the actions of [11].

It is convenient to write

$$\epsilon^{\dot p}(X) = \epsilon\,\alpha^{\dot p}(X) \tag{29}$$

where ϵ is a constant anti-commuting Majorana-Weyl world-sheet spinor (and space-time scalar) and $\alpha^{\dot p}(X)$ is a commuting spinor on L (and world-sheet scalar). From (25), $\sum_{\dot p}\alpha^{\dot p}\alpha^{\dot p}$ is a constant which may be normalised to unity. Then

$$M^{a}{}_{q} = \alpha^{\dot p}\,\gamma^{a}_{\dot p q} \tag{30}$$

is a covariantly constant orthogonal 8x8 matrix ($MM^T = 1$) that can be used to convert SO(8) spinor indices into vector indices, so that an SO(8) vector (and world-sheet spinor) is defined in terms of the SO(8) spinor (and SO(1,1) spinor) S^q by

$$S^a \equiv \alpha^{\dot p}\gamma^a_{\dot p q} S^q \,, \qquad S^i = e^i{}_a S^a \tag{31}$$

Using the identity (24) and (25),(26) the action (7) can then be rewritten as

$$S = \frac{1}{4\pi\alpha'}\int d\sigma\, d\tau\, \tfrac{1}{2}\Big[\partial_+ X^i \partial_- X^j (g_{ij} + b_{ij}) \tag{32}$$

$$+ i\,\psi^A D_-\psi^A + i\, S^a\big(\partial_+ S^a + \omega_i^{\ ab}(\partial_+ X^i) S^b\big) + \tfrac{1}{2} F_{ab\,AB} S^a S^b \psi^A \psi^B + O(\alpha')\Big]$$

This is precisely the light-cone action for the Ramond-Neveu-Schwarz formulation of the heterotic string [2] in a non-trivial background, i.e. the (0,1) supersymmetric non-linear sigma-model [12], with world-sheet supersymmetry [12]

$$\delta X^i = \epsilon S^i \,, \qquad \delta S^i = -i\epsilon\, \partial_- X^i \tag{33}$$

$$\delta\psi^A + A_i^{\ AB}(\delta X^i)\psi^B = 0$$

(Note that $\delta S^i = \delta S^i + (^i_{jk})\, \delta X^j S^k$ is covariant as the connection term cancels.) However, using (24-26),(29) and (31) this becomes precisely the transformation (23) giving an immediate check of the invariance of (7) under (23). Note that (32) and (7) are only equivalent if (25),(26) hold. Strictly speaking (7) is only exactly invariant under (23) if $dH = 0$, as will be the case if $\omega_{(+)} = \tilde\omega$, $A = \iota(\omega_{(+)})$. Otherwise the change in (7) under (23) is proportional to $\alpha' \times \frac{1}{4\pi\alpha'}$, coming from the variation of the Chern-Simons terms in (14). There is some evidence that this change is cancelled by one-loop effects [14].

Suppose then that (7) is invariant (to lowest order in α') under (23) for m independent Killing spinors $\epsilon^{\dot p}(X)$, $\epsilon^{(J)\dot p}(X)$ ($J = 1,\ldots,m-1$, $m \leq 8$) satisfying (25),(26) with corresponding commuting normalised spinors $\alpha^{\dot p}$.

$\beta^{(J)\dot{p}}$ and matrices (30) and $M^{(J)a}{}_q \equiv \beta^{(J)\dot{p}} \gamma^a_{\dot{p}q}$. A representation of the $\gamma^a_{\dot{p}q}$ can be chosen such that one of the m matrices $M, M^{(J)}$ is the identity, $M^a{}_q$, say, while the other m−1 are anti-symmetric $M^{(J)} = -(M^{(J)})^T$ and satisfy

$$\{M^{(I)}, M^{(J)}\} = -2\delta^{IJ} \tag{34}$$

(see e.g. [15]). Then defining

$$J^{(J)}_{ij} \equiv \beta^{(J)}_{\dot{p}} \sigma^{\dot{p}\dot{q}}_{ij} \alpha_{\dot{q}} = M^{(J)a}{}_q \delta^{qb} e^a_{[i} e^b_{j]} \tag{35}$$

and choosing the normalisation $\sum_{\dot{p}} \beta^{(J)}_{\dot{p}} \beta^{(K)}_{\dot{p}} = \delta^{JK}$ it follows that the tensors $J^{(I)}_{ij}$ satisfy

$$J^{(I)}{}_i{}^j J^{(J)}{}_j{}^k + J^{(J)}{}_i{}^j J^{(I)}{}_j{}^k = -2\delta^{IJ}\delta_i{}^k \tag{36}$$

so that each $J^{(I)}$ is an almost complex structure on L ($J^2 = -1$) while from (35), $J_{ij} = -J_{ji}$. This is equivalent to

$$J_i{}^j J_k{}^l g_{jl} = g_{ik} \tag{37}$$

so that the metric is hermitian. As a result of (25) the almost complex structure is covariantly constant with respect to the connection with torsion (17)

$$\nabla_{(-)i} J^{(I)}{}_{jk} = 0 \tag{38}$$

Equation (26) implies $\beta(F.\sigma)\alpha=0$ and $\beta[(F.\sigma),\sigma_{ij}]\alpha=0$ which can be rewritten as

$$J^{ij} F^M_{ij} = 0 \tag{39}$$

$$J_{[i}{}^{k} F^{M}_{j]k} = 0 \tag{40}$$

respectively. The condition (38) implies that for each J_{ij} (suppressing the I index)

$$H_{ijk} = -\tfrac{1}{2} N_{ijk} - \tfrac{1}{12} J_{i}{}^{m} J_{j}{}^{n} J_{k}{}^{p} J_{[mn,p]} \tag{41}$$

where

$$N_{ij}{}^{k} = g^{km} N_{ijm} = J^{m}{}_{i} J^{k}_{[j,m]} - i(\leftrightarrow j) \tag{42}$$

is the Nijenhuis tensor. From (41), N_{ijk} must be totally antisymmetric

$$N_{ijk} = N_{[ijk]} \tag{43}$$

and the Bianchi identity (16) gives an additional constraint on the right-hand-side of (41).

Suppose now that $m \geqslant 2$ and consider the commutator of the supersymmetry transformation (23) with parameter $\epsilon^{\dot{p}}(X)$ and that with parameter $\epsilon^{(J)}_{\dot{p}}(X)$, say, satisfying (25,26) with corresponding almost complex structure $J_{ij} = J^{(J)}_{ij}$. Then (6) is replaced by

$$[\delta_\epsilon, \delta_{\epsilon^{(J)}}] X^i = b^i \tag{44}$$

$$b^i \equiv J^{l}{}_{j} N^{i}_{lk} \epsilon^{\dot{p}} \epsilon^{\dot{q}}_{(J)} \gamma^{j}_{\dot{p}p} \gamma^{k}_{\dot{q}q} S^p S^q \tag{45}$$

so that the commutator of two "world-sheet supersymmetry" transformations has become a space-time translation. Note that as $\epsilon, \epsilon_{(J)}$ are Killing spinors satisfying (25),(26), the space-time translation $\delta X^i = b^i$ must be an invariance of the action (7), so that b^i must be a Killing vector satisfying

$$\nabla_{(i} b_{j)} = 0 \tag{46}$$

$$\partial_{[l}\left(T_{ij]}{}^{k} b_k\right) = 0 \tag{47}$$

for all values of the fermion field $S^{\rho}(\sigma,\tau)$. Clearly, (43), (46) and (47) are extremely restrictive if N^k_{ij} is non-zero. If (42) vanishes, the algebra (6) is regained, and $J^i{}_j$ is a complex structure. L is a complex manifold and it is possible to choose complex co-ordinates $X^i \to (z^{\alpha}, \bar{z}^{\bar{\beta}})$, $(\alpha,\bar{\beta} = 1,\ldots,4)$, $\bar{z}^{\bar{\alpha}} = (z^{\alpha})^*$ in which the complex structure is constant and diagonalised $J^{\alpha}{}_{\beta} = i\delta^{\alpha}{}_{\beta}$, $J^{\alpha}{}_{\bar{\beta}} = 0$ (and complex conjugate equations). Then equations (37), (39) and (40) become

$$g_{\alpha\beta} = 0, \quad g_{\bar{\alpha}\bar{\beta}} = 0, \quad g^{\alpha\bar{\beta}} F_{\alpha\bar{\beta}}{}^{M} = 0 \tag{48}$$

$$F_{\alpha\beta}{}^{M} = 0, \quad F_{\bar{\alpha}\bar{\beta}}{}^{M} = 0 \tag{49}$$

so that (49) implies that A is a connection for a holomorphic vector bundle. (41) becomes

$$H_{\alpha\beta\bar{\gamma}} = g_{\alpha\bar{\gamma},\beta} - g_{\beta\bar{\gamma},\alpha}, \quad H_{\alpha\beta\gamma} = 0 \tag{50}$$

Defining the fundamental 1-form

$$J = \tfrac{1}{2} J_{ij}\, dX^i \wedge dX^j = i\, g_{\alpha\bar{\beta}}\, dz^{\alpha} \wedge d\bar{z}^{\bar{\beta}} \tag{51}$$

(50) can be re-written

$$H = i(\partial J - \bar{\partial} J) \tag{50a}$$

where ∂ ($\bar{\partial}$) denotes (anti-) holomorphic exterior differentiation (see e.g. [16]). Then taking the curl of this gives, using (16),

$$i\partial\bar{\partial}H = \alpha'\left[\frac{1}{30}\mathrm{Tr}(F\wedge F) - \mathrm{tr}(\tilde{R}\wedge\tilde{R})\right] \tag{52}$$

If the connections are suitably identified so that the right hand side vanishes, the equation can be rewritten in terms of the metric as

$$g_{\alpha[\bar{\beta},\bar{\gamma}]\delta} - g_{\delta[\bar{\beta},\bar{\gamma}]\alpha} = 0 \tag{53}$$

and locally this can be solved in terms of a local "potential" $k = k_\alpha dz^\alpha$, $\bar{k} = (k)^*$ [12]

$$g_{\alpha\bar{\beta}} = \partial_\alpha \bar{k}_{\bar{\beta}} + \partial_{\bar{\beta}} k_\alpha \,, \quad J = i(\partial\bar{k} - \bar{\partial}k) \tag{54}$$

If $H_{ijk} = 0$, then L is Kahler while if the vector

$$V^i \equiv J^{ij} J^{kl} H_{jkl} = J^{ij} \nabla_k J^k{}_j \tag{55}$$

vanishes, L is semi-Kahler. This vector (55) has great importance for the following discussion. If $dH \neq 0$, there will be corrections to (53), (54) proportional to α'.

Then, to summarise, the light-cone action for the heterotic string in a fixed background, given to lowest order in α' by (7), will be invariant under the supersymmetry transformation with parameter η given, to lowest order in α', by (4),(23), provided that the spinor η satisfies, to lowest order in α', the equations (27),(28). If the string theory is conformally invariant, these equations may be exact. More generally, the corrections to these equations will be small if $r^2/4\pi\alpha'$ is sufficiently large, where r is some typical length scale (e.g. radius) of L. (On rescaling the metric

$g_{ij} \to e^{\Phi/2} g_{ij}$ where Φ is the dilaton expectation value, this condition becomes the requirement that $e^{\Phi/2} r^2/4\pi\alpha'$ be large.)

The number of covariantly constant spinors satisfying (27) is determined by the holonomy group H_- of the connection $\Gamma^i_{(-)jk}$ (17). Necessarily $H_- \subseteq O(8)$ and if H_- is $O(8)$ (or $SO(8)$), as it will be for most manifolds, there will be no solutions of (27). If $H_- \subseteq Spin^+(7)$ there will be at least one right-handed covariantly constant spinor satisfying (22) while if $H_- \subseteq Spin^-(7)$ there will be one solution of (25). (The superscripts ± are used to distinguish inequivalent embeddings in SO(8) in the usual way.) There will be two solutions of (22) if $H_- \subseteq SU(4)^+$ and two of (25) if $H_- \subseteq SU(4)^-$. If there is one solution of (22) and one of (25), then $H_- \subseteq Spin^+(7) \cap Spin^-(7) = G_2$ and there is also a covariantly constant vector (as $G_2 \subset SO(7)$) so that there is a flat direction in L (i.e. $L = L' \times \mathbb{R}$ or $L = L' \times S^1$ for some L'). More generally, given any solution δ of (22) and ϵ of (25) there is a covariantly constant vector proportional to $\epsilon^{\dot p} \gamma^i_{\dot p q} \delta^q$ and a corresponding flat direction. The first integrability condition for (25) is

$$R_{(-)ab\,ij} \left(\sigma^{ab}_{\dot p \dot q}\right) \epsilon^{\dot q} = 0 \tag{56}$$

If there are at least two solutions $\epsilon, \epsilon^{(J)}$ of (25) then multiplying (56) by $\epsilon^{(J)}$ and factoring out the Grassman variables gives, using (35)

$$C_{ij} \equiv J^l{}_k \, R_{(-)}{}^k{}_{l\,ij} = 0 \tag{57}$$

i.e. the U(l) part of the curvature vanishes so that $H_- \subseteq SU(4)^-$. The cohomology class of all 2-forms of the form $C + d\alpha$, where $C \equiv \frac{1}{2} C_{ij} dX^i \wedge dX^j$ and α is a 1-form, is the first Chern class so that, from (57), a necessary integrability condition for there to be 2 solutions of (25) is the topological restriction that L have vanishing first Chern class. In the same way, if there are two solutions of (22) there is, again, an almost complex structure and L must again have vanishing first Chern class.

Of particular interest are spaces of the form $L = \mathbb{R}^2 \times K$, $M = M_4 \times K$ for some 6-space K, giving a compactification to four-dimensional Minkowski space, M_4. There are then two flat directions in L (X^7 and X^8, say) and two corresponding covariantly constant vectors ($\partial/\partial X^7$, $\partial/\partial X^8$) so that $H_- \subseteq O(6)$. For there to be a solution of (25) it is necessary that $H_- \subseteq SU(3)$. In fact there are then two solutions ϵ_1, ϵ_2 of (25). There will then also be two solutions of (22), given by $\delta_1^q = \epsilon_1^{\dot p}\gamma^8_{\dot p q}$, $\delta_2^q = \epsilon_2^{\dot p}\gamma^8_{\dot p q}$. (Note that $\epsilon_1^p\gamma^7_{pq}$, $\epsilon_2^{\dot p}\gamma^7_{pq}$ are linearly dependent on these).

On compactifying to four dimensions a 16-component Majorana-Weyl spinor of SO(9,1) decomposes into four 4-component Majorana spinors of SO(3,1). If H_ is trivial, that is, if K if flat (e.g. $K = T^6$) and $F^M_{ij} = 0$ there will be 16 solutions of (27),(28) and the vacuum will be invariant under N=1 10-dimensional supersymmetry or N=4 4-dimensional supersymmetry. If K has SU(3) holonomy, there will be 4 solutions of (27), $\delta_1, \delta_2, \epsilon_1, \epsilon_2$ which combine into a single Majorana spinor in 4 dimensions. Then if these solutions also satisfy (28), the vacuum will be invariant under N=1 4-dimensional supersymmetry so that the N=4 local supersymmetry of the theory is spontaneously broken by the vacuum down to N=1 supersymmetry. Consider now equation (28). Any solution of (27) satisfies the integrability condition

$$R_{(-)kl\,ij}\left(\Gamma^k\Gamma^l\right)\eta = 0 \tag{58}$$

The curvature tensors (17) with totally anti-symmetric torsion $\pm\frac{1}{2}H$ satisfy the identity

$$R_{(+)ijkl} - R_{(-)klij} = -2\,H_{[ijk,l]} \qquad (59)$$

where the right-hand side, given by the Bianchi identity (16), is of order α'. Then if, for some embedding $\iota : SO(8) \to G$, the Yang-Mills connection satisfies $A_i^{AB} = \iota(\omega_{(+)i}^{ab})$, then $F_{ij}^{AB} = \iota(R_{(+)abij})$ so that, from (59), (58) would imply (28) up to terms of order α'. If further $\bar{\omega}_i^{ab} = \omega_{(+)i}^{ab}$ in (16) then on identifying A and $\omega_{(+)}$, the right-hand side of (59) vanishes and (58) implies (28) exactly. Then for any geometry admitting a covariantly constant spinor (27), there is at least one choice of the Yang-Mills field such that (28) is satisfied. Suppose now that there are two solutions of (25). If the torsion vanishes, H = 0, then (43) vanishes so that L is a complex manifold. Further, (38) then implies that L is Kahler, while (57) implies that the Ricci-tensor R_{ij} vanishes, as it is simply related to the U(1) part of the curvature (57). Indeed, in complex co-ordinates,

$$C_{\alpha\beta} = 0\,, \qquad C_{\alpha\bar{\beta}} = i\,R_{\alpha\bar{\beta}} \qquad (60)$$

If L is compact or $L = \mathbb{R}^2 \times K$ and K is compact, then, by Yau's theorem, [17], for any given choice of complex structure and cohomology class of the Kahler form, there is a unique Kahler Ricci-flat metric, provided the first Chern class vanishes. Further, by the Uhlenbeck-Yau theorem, [18], there is an essentially unique choice of the Yang-Mills field for any given holomorphic stable Yang-Mills bundle such that (39),(40) hold. If the Yang-Mills bundle is topologically equivalent to the tangent bundle times a trivial bundle, then $A = \iota(\omega)$. These are the Calabi-Yau compactifications of Candelas et al [4,5].

Suppose now that $H \neq 0$ but that (43) still vanishes. Then if (55) vanishes, L is semi-Kahler and using (50),(53) it is found that (60) still holds but with $R_{ij} \equiv R^k_{(-)lkj}$ now the Ricci-tensor with torsion. More generally, if (55) is non-zero, (60) becomes [8]

$$C_{\alpha\beta} = \partial_{[\alpha} V_{\beta]} \tag{61}$$

$$C_{\alpha\bar{\beta}} = R_{(-)\alpha\bar{\beta}} + \nabla_{(\alpha} V_{\bar{\beta})} + \partial_{[\alpha} V_{\bar{\beta}]} + \frac{1}{2} H_{\alpha\bar{\beta}\gamma} V^{\gamma} \tag{62}$$

If (61) vanishes then, at least locally, there are real scalars A(X), B(X) such that

$$V_{\alpha} = \partial_{\alpha}(A + iB) \tag{63}$$

Then (62) becomes

$$\partial_{\alpha} \partial_{\bar{\beta}} \log\left(e^{-A} \det[g_{\gamma\bar{\delta}}]\right) = 0 \tag{64}$$

so that the condition for SU(n) holonomy is that (62) and (64) vanish, implying that coordinates can be chosen such that

$$A = \log\det[g_{\gamma\bar{\delta}}] + A_0 \tag{65}$$

for some constant A_0. However, as detg is not globally defined in general, the constant A_0 can be different in different coordinate patches. On the other hand, exp(−A)detg <u>is</u> well-defined globally and can be expressed in a coordinate independent way in terms of the globally defined covariantly constant spinors, or, equivalently, the covariantly constant holomorphic n-form that necessarily exists on spaces of SU(n) holonomy (cf. [24]). For such a space, there is at least one configuration of the Yang-Mills field for which (28) will be satisfied, that in which A and $\iota(\omega_{(+)})$ are identified. It seems plausible that the solution of (39) will again be unique on these generalisations of Calabi-Yau manifolds with torsion (as it is when H=0), for any stable holomorphic Yang-Mills bundle. If $\tilde{\omega} = \omega_{(+)}$ then dH

= 0 for these solutions and these spaces can have arbitrary size, i.e. if the fields are rigidly rescaled or α' is changed, (16),(27) and (28) will still have solutions. For large enough spaces, the corrections to these equations of order α' are negligible. For almost all other values of the Yang-Mills field (or if $\tilde{\omega} \neq \omega_{(+)}$), $dH \neq 0$ and as the left and right hand sides of (14,16) scale differently, the size of L or K would be fixed.

It seems likely that, as in the torsion free case [4,5], given a stable holomorphic vector bundle over a complex manifold, there will be a choice of metric, torsion and Yang-Mills connection such that (16) holds and there are two Killing spinors satisfying (25),(26) provided only that certain topological integrability conditions are satisfied, namely that the first Chern class vanishes and that the second Chern class equals the Pontriagin class. However, not all such backgrounds are exact solutions of the string field equations. For example, there is a contribution to the 2-loop metric β-function and trace anomaly of the heterotic non-linear sigma model that is quadratic in the curvature tensor and Yang-Mills field strength [6,8,9,10]. This in fact vanishes if $A = \iota(\omega_{(+)})$ [8,9]. However, for this to also vanish for some other choice of Yang-Mills field A', say, satisfying (39,40) it is necessary that the "stress-energy" tensor $F^M_{ik}F^M_{jk}$ for A' equal that for $\iota(\omega_{(+)})$. This is very restrictive and will not hold for most solutions of (39,40). Considering the effects of the ambiguities inherent in the definitions of β-functions and trace anomalies [21] does not appear to change the conclusion that most supersymmetric spaces do <u>not</u> satisfy the string field equations obtained by requiring conformal invariance [6]. However, if the square of the radius, r, of the internal manifold is large compared with α' the 2-loop corrections to trace anomaly will be small, so that these spaces are in some sense approximate solutions- the field equations are satisfied up to terms of order α'/r^2.

Remarkably, however, if $A = \iota(\omega_{(+)})$, the (0,2) supersymmetric non-linear sigma model [12] defined on a complex manifold satisfying (48),(49),(50),(53) with SU(n) holonomy ($C_{ij}=0$) is completely finite to at least two-loop order [8,9] and there is some evidence that the finiteness persists to all orders, so these may be exact solutions. Indeed, if it is assumed that the two ϵ-supersymmetries (23) remain symmetries of the renormalised theory, then (25),(26) must have solutions in the remormalized background. However, if H = 0, then by the Yau theorem [17] and the Uhlenbeck-Yau theorem [18], for a given compact manifold with vanishing first Chern class and complex structure on that manifold, there is a unique metric and Yang-Mills field in a given topological class such that (25),(26) have two solutions ϵ, ϵ^{1}. Then if the equations (25),(26) were to hold in the renormalised theory, it would be necessary that the renormalisation of the metric and Yang-Mills field be trivial, i.e. the sigma-model must be finite [20]. Thus proving the finiteness of these models is equivalent to proving that the classical symmetry (23) is a symmetry of the quantum theory. Similarly, if $H \neq 0$, this argument again greatly restricts the possible counterterms if the symmetry (23) is maintained in the quantum theory. However, care is needed with such simple arguments. They seem to imply that all sigma-models defined on spaces that admit Killing spinors are finite, which is incorrect. It seems that there must be anomalies in the space-time supersymmetry, analagous to the gravitational and Yang-Mills anomalies discussed earlier, that can presumably be cancelled by a Green-Schwarz mechanism. Such an anomaly cancellation would lead to modifications of the supersymmetry transformation rules and action. These should, however, cancel when the Yang-Mills and spin connections are identified, $A = \iota(\omega_{(+)})$, as they do for the gravitational and gauge anomalies. This is because with this identification, the heterotic superstring (7) is equivalent to the Type II superstring (18) (plus free fields) which should be anomaly-free. Then only when $A = \iota(\omega_{(+)})$ would it seem possible to conclude that the heterotic sigma model with supersymmetric target space is finite.

Finiteness is not sufficient for conformal invariance, however [21]. If, for these spaces, the vector (55) is the gradient of some scalar $\Phi(X)$,

$$V_i = -2\partial_i \Phi \tag{66}$$

then the $\partial X \partial X$ part of the one-loop trace anomaly can be cancelled by adding a finite local counterterm (the "Fradkin-Tseytlin" term [7]) to the effective action [21]. The scalar $\Phi(X)$ is then interpreted as the vacuum expectation value of the dilaton field (the massless scalar mode of the heterotic string). Then for conformal invariance it is necessary that the vector v defined in terms of the torsion and complex structure by (55) be curl-free, $v_{[i,j]} = 0$. This will again not be true for all supersymmetric spaces. Further, if the dilaton field is to be globally defined, it is necessary that the 1-form $v = v_i dX^i$ be an element of the trivial first cohomology class on L. (This will, of course, automatically be the case if $H^1(L)$=0.) Then from above A=-2Φ, B=0 so that the dilaton field is given in terms of the metric by (65). It should be emphasised that many supersymmetric spaces will be ruled out as string vacua as they will have non-trivial B.

Perturbative conformal invariance also requires that the order $(\alpha')^2$ contribution to the central charge of the super-Virasoro algebra, c, vanishes [6]

$$c = R - \frac{1}{12}H^2 + 4\nabla^2\Phi - 4(\nabla\Phi)^2 \tag{67}$$

Then the string field equations should be satisfied, at least to lowest orders in α' [6]. If v^i=0, Φ is constant and L is semi-Kahler and from (61, 66, 67) H must also vanish and L is then, in fact, Kahler. Thus non-zero torsion requires a non-zero dilaton field if the theory is to be conformally invariant order by order in α' and further that it be given by

(65). There is, of course, also the possibility of a non-perturbative solution [8].

A constant expectation value of Φ can be absorbed into the string coupling constants [23]. For non-constant Φ, the field equations obtained by requiring conformal invariance can be derived from a 10-dimensional action [6] $\int d^{10}x \sqrt{(g^{10})} e^{-2\Phi} R_{10} + \ldots$ where the 10-dimensional metric is $g^{10}_{\mu\nu}(x,y) = \mathrm{diag}\ [\eta_{mn}, g_{ij}(y)]$ where x^m, η_{mn} are the co-ordinates and metric on M_4 $(m,n = 0,\ldots,3)$ and $y^i, g_{ij}(y)$ are the co-ordinates and metric on K $(i,j = 4,\ldots,9)$. To obtain the conventional action in 10-dimensions, it is necessary to conformally rescale the metric $g_{10} \to e^{½\Phi} g_{10}(x,y)$ so that it becomes a "warped product" of the form diag $(e^{½\Phi}\eta_{mn}, \hat{g}_{ij})$. For four-dimensional Lorentz invariance of the vacuum, it is necessary that the dilaton be independent of x^m, $\Phi = \Phi(y)$. On dimensionally reducing to four dimensions, it is necessary to make a further rescaling of the four-dimensional metric $e^{½\Phi}\eta_{mn} \to \eta_{mn}$ must be made in order to obtain the conventional four dimensional Hilbert action $\int d^4x \sqrt{g^4}\ R_4 + \ldots$ Thus in four dimensions the physics is indeed that of Minkowski space, in spite of the "warp factor".

Complex manifolds K with SU(3) holonomy with non-zero torsion H, dilaton Φ and Yang-Mills field A satisfying (16,48-50,52,57) lead to unbroken N=1 supersymmetry in 4-dimensions and are solutions of the string field equations if, further, (66) holds, c=0 and the Yang-Mills potential and the spin-connection are identified, at least to lowest order in α'. It is possible that these are exact solutions and that the supersymmetry is also exact. Examples of such spaces were considered in [9].

Witten has argued that it should be possible to deform a Calabi-Yau space so as to obtain a new space with $H \neq 0$ satisfying the string field equations and maintaining unbroken N=1 supersymmetry in four dimensions [20], corresponding to moving in flat directions in the scalar potential of the effective four dimensional theory. If the supersymmetry (4), (23) is to be maintained, the geometry of the deformed spaces must be as discussed above. If the field equations are to be satisfied order by order in α', (so that the solutions are of arbitrary size) then dH = 0 (as the left and right hand sides of (14, 16) are of different orders in α') and A and $\tilde{\omega}$ must be identified. For the 2-loop trace anomaly to cancel, A must be identified with $\omega_{(+)}$. Although the holonomy group H_- of $\omega_{(-)}$ is SU(3) the holonomy group $H_+ \subseteq O(6)$ is not so restricted. The gauge group G will then be broken down to the subgroup $\hat{G}$ that commutes with $\iota(H_+)$. If $H_+ = O(6)$ then SO(32) would be broken to SO(26) while $E_8 \times E_8$ would be broken to $E_8 \times SO(10)$ (under the standard embeddings $O(6) \to$ G that lead to dH = 0). The possibility of an SO(10) gauge group is quite attractive - it would lead to families in the <u>16</u> (spinor) representation of SO(10). By the arguments of [4, 5, 22], the number of families is given half the Euler number of K. If, as suggested above, conformal invariance requires $A = \iota(\omega_{(+)})$ this will be the only possible Yang-Mills configuration and the gauge group $E_8 \times \hat{G}$ must contain $E_8 \times SO(10)$ as a subgroup so that SO(10) is the minimal grand unified group. More general supersymmetric spaces with $A \neq \iota(\omega_{(+)})$ could have more general gauge groups such as SU(5), as in [20], but most would not be exact solutions of the string field equations and it is as yet unknown whether there can be <u>any</u> such more general solutions. It would be of interest to see whether making a different choice for $\tilde{\omega}$ could lead to any different solutions. It would also be of interest to extend the analysis of [24] to investigate how the masses and Yukawa couplings change in the presence of torsion.

Finally, it is of interest to seek solutions that are almost complex, with (42) non-vanishing. The classic example of a six-dimensional almost complex manifold is the six-sphere $S^6 \approx G_2/SU(3)$, which has an almost complex structure whose Nijenhuis tensor (42) is non-vanishing but totally anti-symmetric (43). The connection $\Gamma_{(-)}$ with torsion (41) has holonomy SU(3) and (37),(38) are satisfied with (55) vanishing, so that the dilaton field is trivial. In particular, this implies the existence of two solutions to (25) so that if the Yang-Mills potential is set equal to the spin-connection, as before, the covariantly constant spinors would also satisfy (26) and so it would seem that the six-sphere should be a supersymmetric background. Unfortunately, the torsion (41) is not closed, $dH \neq 0$, so that the Yang-Mills potential cannot be set equal to the spin-connection if the Bianchi identity (16) is to hold and it is unknown whether any other Yang-Mills ansatz (or choice of $\tilde{\omega}$) might work - any such solution would be non-pertubative and higher order corrections to the equations consid ered here would be important. However, the fact that the six-sphere comes so close to being a supersymmetric compactifying space suggests that there may indeed be almost complex spaces meeting the requirements discussed here, in addition to the complex spaces considered above.

Further details will be given elsewhere. Some of the results obtained here have been found independently by Andy Strominger.

References

[1] J. H. Schwarz, Phys. Rep. 89 (1982) 223

M. B. Green, Surveys in High Energy Physics 3 (1983) 127

[2] D. J. Gross, J. A. Harvey, E. Martinec and R. Rohm Phys. Rev. Lett. 54 (1983) 502 ; Nucl. Phys. B256 (1985) 253

[3] D. Friedan and S. Shenker, unpublished lectures at the Apsen Summer Institute (1984)

D. Nemeschansky and S. Yankielowicz, Phys. Rev. Lett. 54 (1985) 620

E. Bergshoeff, S. Randjbar-Daemi, Abdus Salam, H. Sarmadi and E. Sezgin, ICTP Trieste preprint (1985)

S. Jain, R. Shankar and S. R. Wadia, Phys. Rev. D32 (1985) 2713

[4] P. Candelas, G. Horowitz, A. Strominger and E. Witten, Nucl. Phys. B528 (1985) 46

[5] P. Candelas, G. Horowitz, A. Strominger and E. Witten, to appear in proceedings of the Argonne Conference

[6] C. G. Callan, D. Friedan, E. Martinec and M. Perry, Nucl. Phys. B262 (1985) 593

[7] E. S. Fradkin and A. A. Tseytlin, Phys. Lett. 158B (1985) 316 and Nucl. Phys. B261 (1985) 1

[8] C. M. Hull, Nucl. Phys. B267 (1986) 266

[9] C. M. Hull, DAMPT preprint (1985) to appear in "Superstrings Anomalies and Supergravity", Ed. by G. W. Gibbons, S. W. Hawking and P. K. Townsend

[10] A. Sen, Fermilab preprints (1985)

[11] E. Witten, Princeton preprint (1985)

M. Grisaru, P. Howe, L. Mezincescu, B. Nilsson and

P. K. Townsend, Phys. Lett. 162B (1985) 116

[12] C. M. Hull and E. Witten, Phys. Lett. 160B (1985) 398

[13] C. M. Hull, Phys. Lett. 167B (1986) 51

[14] A. Sen, Phys. Lett. 166B (1986) 300

[15] M. B. Green and J. H. Schwarz, Nucl. Phys. B218 (1983) 43, Appendix A and Nucl. Phys. B219 (1983) 43 Appendix A

[16] S. S. Chern, "Complex Manifolds Without Potential Theory", 2nd ed., Springer Verlag, (1979)

[17] S. T. Yau, Proc. Natl. Acad. Sci. 74 (1977) 1798

[18] K. Uhlenbeck and S. T. Yau, unpublished

[19] I. Bars, Phys. Rev. D33 (1986) 383

[20] E. Witten, Princeton pre-print (1985)

[21] C. M. Hull and P. K. Townsend, DAMTP pre-print (1985), to appear in Nucl. Phys. B.

[22] I. Bars and M. Visser, USC pre-print (1985)

[23] E. Witten, Phys. Lett. 149B (1984) 351

[24] A. Strominger and E. Witten, Commun. Math. Phys. 101 (1985) 341, A. Strominger, UCSB pre-prints (1985)

Spacetime supersymmetric particles and strings in background fields

P.K. Townsend.

Department of Applied Mathematics & Theoretical Physics.

University of Cambridge.

Silver Street.

Cambridge CB3 9EW.

England.

Abstract

A pedagogical review of "superparticles" and "superstrings" (as against "spinning" particles or strings) is presented. An account is given of recent progress in the construction of actions for superparticles and superstrings in electromagnetic and gravitational backgrounds.

1. Introduction

The action for a free bosonic string can be rewritten as a two-dimensional (d=2) conformally invariant field theory of D scalar fields in interaction with d=2 gravity. Spin degrees of freedom are incorporated by supersymmetrization, but this can be done in essentially two ways. In one way, historically the first, the action is a two-dimensional superconformally invariant field theory of D scalar supermultiplets interacting with d=2 supergravity [1]. This is commonly referred to as the "spinning" string formulation. It has manifest d=2, or worldsheet, supersymmetry. Remarkably, the spectrum is that of a D-dimensional, or spacetime, supersymmetric theory if (i) one takes the spacetime dimension D to be the critical dimension D=10, and (ii) one truncates the spectrum by requiring that all "physical states" have positive "G-parity" [2]. One suspects that there is another formulation in which the spacetime supersymmetry is manifest, and indeed there is. In this, later version, of the supersymmetric string, the coordinates are taken to be those of a D-dimensional superspace [3]. The action can be considered as a kind of d=2 σ-model with superspace as the target space [4]. The D-dimensional supersymmetry is then a consequence of the super Poincaré isometry group of flat superspace, but d=2 supersymmetry is absent. This is commonly referred to as the "superstring" formulation.

The spinning string and the superstring appear to be complementary formulations of the same theory, each having advantages and disadvantages. One advantage of the superstring formulation is that, possessing a manifest D-dimensional supersymmetry, it is relatively straightforward to generalize it to include interactions with external fields of super-Yang-Mills or supergravity theories. There are two reasons why we might be interested in such generalizations. If we think of the superstring as a unified theory of elementary particles and their interactions then we are presumably limited to the critical dimension D=10. But the D=10 background spacetime cannot be flat for a realistic theory, and it is presumably given by some solution of an effective low-energy supergravity theory. So we need the generalization of the superstring to background geometries that are solutions of supergravity equations. The other reason is that if there was an epoch in which the universe was governed by an effective N=1 D=4 supergravity arising as the low-energy limit of a D=10 superstring theory there would be cosmic superstrings around [5] which should couple consistently to supergravity. In this case we think of the D=4 superstring as a classical extended object interacting with D=4 supergravity. Moreover, in this case one requires a consistent coupling that allows for the back-reaction of the string on the background geometry, i.e. the full dynamical system of superstrings and supergravity fields.

The principal aim of this work is to review some recent progress in the coupling of D=10 and D=4 superstrings to supergravity. For D=10 the action for the type I superstring in a background supergravity configuration was found by Witten [6] who showed that consistency requires that at least some subset of the supergravity field equations be satisfied. Furthermore these field equations do not include the string as a source and do not include the back-reaction of the string. It would therefore appear that a formulation in D=10 of the coupled dynamical system of superstrings and supergravity is not possible. Fortunately we require only a fixed background because all perturbations about this background should be contained in the modes of the D=10 superstring. Unfortunately, the N=1 D=10 supergravity is anomalous; what we really want is the heterotic superstring in an N=1 Einstein–Yang–Mills background, but this is, as yet, an unsolved problem. The coupling of the type IIB superstring to N=2 chiral supergravity has been obtained, however [7]. Consistency of the (classical) N=1 D=4 superstring in a supergravity background does not imply the supergravity field equations. It does imply constraints on the supergeometry that reduce the field content to that of an off-shell supergravity [8]. the most natural choice appears to be a version of the U(1) invariant "new–minimal" supergravity in which the local U(1) invariance is "compensated" by a chiral superfield [9, 10]. This is called 16+16 supergravity and is also the version of supergravity that occurs most naturally as the low energy limit of the D=10 superstring after compactification on a Calabi Yau space, although it is in fact equivalent to minimal supergravity coupled to a linear superfield [11].

But before delving further into the details of these superstring matters we shall take a step backward to examine some of the issues in the simpler setting of superparticle actions. Superparticles were apparently first discussed by Casalbuoni [12], but the action of relevance here is that of Brink and Schwarz [13]. It can be considered as a one-dimensional σ-model with flat superspace as the target manifold. It has a peculiar fermionic gauge-invariance [14] that is also a feature of the superstring action. This gauge invariance allows the action to be reduced to a form of a free (one- or two-dimensional) field theory. (It is also essential if the spectrum of the superstring is to be supersymmetric as it allows the elimination of superfluous fermionic degrees of freedom). The associated commutator algebra closes only "on-shell" i.e. when the field equations are satisfied, but for the superparticle at least, auxiliary fields can be found to close the algebra. (The action with auxiliary fields given in this work was found in collaboration with M. Grisaru and P. Howe.) This algebra is then no longer so closely tied to the particular action of ref (13), and one can expect other actions to exist that realize the same algebra. In fact, the new superparticle action of Siegel [15] does just this. The algebra is really more fundamental than the particular model that realizes it. One can expect the analogous algebra for the superstring, which is much more complicated [3,16], to play a crucial role in the future. But although the Brink-Schwarz (B-S) superparticle action with auxiliary fields is simpler in some respects, it complicates the construction of the action for a superparticle interacting with super-Maxwell or supergravity fields. Of course, without auxiliary fields the fermionic gauge transformations are modified by the interactions and this is also what happens in the superstring case.

The coupling of a particle to electromagnetism is achieved via the interaction term $\int_\Gamma A = \int_\Gamma A_m \dot{x}^m dt$, i.e. the integral along the worldline Γ of the pull-back of the Maxwell one-form A. This is essentially a one-dimensional Wess-Zuminó (W-Z) term. The coupling of a string to an antisymmetric tensor field is achieved via the similar interaction term $\int_S B = \int_S B_{mn} \dot{x}^m x'^m \, dt \, d\sigma$, i.e. the integral over the world sheet S of the pull-back of the two-form potential B. The extension to the superparticle and superstring is achieved by the analogous line or surface integrals of superspace one or two-form potentials, respectively. Their respective field strengths F=dA and H=dB are closed two- and three forms in superspace. The interaction terms can be written in a manifestly gauge invariant way in terms of these field strength forms as $\int_D F$ and $\int_B H$ where $\partial D = \Gamma$ and $\partial B = S$.

The essential difference between the bosonic and supersymmetric cases is that for the latter the preservation of the fermionic gauge invariance mentioned above constrains the field strengths F and H. The principal differences between the superstring and superparticle is that the 3-form H of the former has a flat superspace limit whereas the 2 form F of the latter vanishes in this limit.

2. Particles and Superparticles

The standard action for a free relativistic point particle of mass μ is

$$I_1 = -\mu \int dt \left\{ \left(\dot{x}^m \dot{x}^n \eta_{mn} \right)^{\frac{1}{2}} \right\} \tag{2.1}$$

where η_{mn} = diag(+, −, −, ..., −) is the D-dimensional Minkowski metric. This can be re-expressed as the one-dimensional generally covariant field theory

$$I_2 = -\frac{1}{2} \int dt \left\{ V^{-1} \dot{x}^m \dot{x}^n \eta_{mn} + \mu^2 V \right\} \tag{2.2}$$

where V is the "einbein" on the worldline. Eliminating V in I_2 through its field equation reproduces I_1. One can rewrite I_2 in first order form as

$$I_3 = -\int dt \left\{ p_m \dot{x}^m - \frac{1}{2} V \left(p^2 - \mu^2 \right) \right\} \tag{2.3}$$

The one-dimensional general coordinate transformations are

$$\delta x^m = \lambda \dot{x}^m \quad , \quad \delta p_m = \lambda \dot{p}_m \quad , \quad \delta V = \frac{d}{dt}(V\lambda) \tag{2.4}$$

This can be combined with the trivial invariance of I_3 under

$$\delta x^m = \lambda \frac{\delta I_3}{\delta p_m} = -\lambda(\dot{x}^m - Vp^m) \quad , \quad \delta p_m = -\lambda \frac{\delta I_3}{\delta x^m} = -\lambda \dot{p}_m \tag{2.5}$$

to yield the new set of transformations

$$\delta_\xi x^m = \xi p^m \quad , \quad \delta_\xi p_m = 0 \quad , \quad \delta_\xi V = \dot{\xi} \tag{2.6}$$

with the new parameter $\xi = V\lambda$. The advantage of this is that the transformations (2.6) are those of an Abelian group, with gauge field V. Remarkably, the commutator algebra for both (2.4) and (2.6) are closed off-shell. Notice also that there is clearly no problem with the massless limit in either I_2 or I_3. In the following we shall be concerned with massless particles.

The B-S superparticle action is obtained by the simple replacement

$$\dot{x}^m \to \omega^m \equiv \dot{x}^m + \tfrac{1}{2}\bar{\theta}\Gamma^m\dot{\theta} \tag{2.7}$$

which is invariant under the rigid ($\dot{\epsilon} = 0$) supersymmetry transformation

$$\delta x^m = -\tfrac{1}{2}\bar{\epsilon}\Gamma^m\theta \tag{2.8}$$

$$\delta\theta = \epsilon$$

Here we are supposing that θ is Majorana, so that what follows will be valid as written for those dimensions D for which Majorana spinors exist, in particular for D=4 and D=10. Slight modifications are required for other dimensions. Making the substitution (2.7) in (2.3) we arrive at the superparticle action

$$I_2 = -\frac{1}{2}\int dt \left\{ V^{-1} \omega^m \omega^n \eta_{mn} \right\} \tag{2.9}$$

which can be written in first order form as

$$I_3 = -\int dt \left\{ \pi_m \omega^m - \frac{1}{2} V \pi^2 \right\} . \tag{2.10}$$

This has the fermionic gauge invariance [14] $(\eta = \eta(t))$,

$$\delta_\eta x^m = \frac{i}{2} \bar{\eta} \not{\pi} \Gamma^m \Theta \quad , \quad \delta_\eta \Theta = -i \not{\pi} \eta \tag{2.11}$$

$$\delta_\eta \pi_m = 0 \quad , \quad \delta_\eta V = 2i \bar{\eta} \dot{\Theta} \quad ,$$

which is easily seen from the variation $\delta \omega^m = i \bar{\eta} \not{\pi} \Gamma^m \dot{\Theta}$ that follows from (2.11). The commutator of two η-transformations yields a transformation of the type (2.6), which is now

$$\delta_\xi x^m = \xi \pi^m \quad , \quad \delta_\xi \pi_m = 0, \quad \delta_\xi V = \dot{\xi} \quad , \quad \delta_\xi \Theta = 0 . \tag{2.12}$$

But this algebra closes only if the V and x^m-field equations are used. To close the algebra "off-shell" we introduce two auxiliary fermion fields χ and d. The action becomes

$$I_4 = -\int dt \left\{ \pi_m \omega^m - \frac{1}{2} V \pi^2 + i \bar{\chi} d \right\} \tag{2.13}$$

and the ξ and η-transformations become

$$\delta x^m = \xi \pi^m + \frac{i}{2} \bar{\eta} \not{\pi} \Gamma^m \Theta + \bar{\eta} \Gamma^m d$$

$$\delta \pi_m = 0$$

$$\delta V = \dot{\xi} + 2i \bar{\eta} \dot{\Theta} + i \bar{\eta} \chi$$

$$\delta \Theta = -i \not{\pi} \eta$$

$$\delta\chi = i\not\pi\eta$$
$$\delta\alpha = \tfrac{1}{2}\pi^2\eta \qquad (2.14)$$

and the commutator algebra is

$$[\delta_{\eta_1}, \delta_{\eta_2}] = \delta_\xi = 2\bar{\eta}_1\not\pi\eta_2$$
$$[\delta_\eta, \delta_\xi] = [\delta_{\xi_1}, \delta_{\xi_2}] = 0 . \qquad (2.15)$$

The transformations (2.14) also commute with supersymmetry. The importance of this fermionic gauge invariance is that it ensures that the supersymmetric extension of a free particle is again a free particle. The equations of motion imply

$$\not\pi\dot{\Theta} = 0 \quad , \quad \pi^2 = 0 \qquad (2.16)$$

These equations do not imply $\dot{\Theta} = 0$, but we can find an η-gauge for which this is true. We can also choose the ξ-gauge e=1, and the π and x-equations then reduce to $\pi^m = \dot{x}^m$, $\dot{\pi}^m = 0$, i.e. a free massless particle.

The quantization of the B-S superparticle is not at all straightforward. For this reason Siegel has recently proposed a new action [15]. He introduces a new spinor field ψ with ξ and η-transformations

$$\delta\psi = \dot{\eta} \qquad (2.17)$$

i.e. ψ is a gauge field for fermionic gauge transformations. Observe now that $(\chi + \dot{\Theta} + i\not\pi\psi)$ is ξ and η-inert, as well as being supersymmetric, so that we may simply substitute $-(\dot{\Theta} + i\not\pi\psi)$ for χ in the action (2.13) without affecting the algebra (2.15) or supersymmetry. The resulting action is

$$I_{Siegel} = -\int dt \left\{ \Pi_m \omega^m - i\bar{d}\dot{\theta} - \tfrac{1}{2}V\pi^2 - \bar{\psi}\not{\pi}d \right\} . \qquad (2.18)$$

We now have three sets of fields, the superspace coordinates (x^m, θ^μ), their conjugate momenta (Π_m, d_μ) and the gauge fields (V, ψ_μ) associated with the algebra (2.15). Actually d is not conjugate to θ because $\dot{\theta}$ also appears in ω^m. Consequently it is

$$d - \tfrac{i}{2}\not{\pi}\theta \qquad (2.19)$$

that is conjugate to θ and it is this quantity that becomes $\partial_{\bar{\theta}}$ upon canonical quantization. Hence, on quantizing,

$$d \rightarrow D = \frac{\partial}{\partial\bar{\theta}} + \tfrac{1}{2}\not{p}\theta \qquad (2.20)$$

which is the spinor covariant derivative that commutes with the supersymmetry charge $Q = \partial_{\bar{\theta}} - \frac{1}{2}\not{p}\theta$. Evidently the algebra of the constraints $\pi^2 \approx 0$, $\not{\pi}d \approx 0$, that follow from (2.18) is a simple consequence of the algebra of spinor derivatives. Imposing these constraints on a wavefunction $\phi(x, \theta)$ yields a set of linearized superfield equations. For D=4 these equations are equivalent to those of a linear multiplet, while for D=10 they are equivalent to the linearized D=10 supergravity field equations.

It will prove convenient to rewrite the action of (2.10) in terms of Z^M, the superspace coordinates (x^m, θ^μ) and the flat superspace vielbein $e_M{}^A$ (and its inverse $e_A{}^M$). Thus

$$I_3 = -\int dt \left\{ e_t^a \pi_a - \tfrac{1}{2}V\pi^2 \right\} \qquad (2.21)$$

where

$$e_t{}^A \equiv \dot{Z}^M e_M{}^A = \left(e_t{}^a = \omega^m \delta_m^a \,,\; e_t{}^\alpha = \dot{\theta}^\mu \delta_\mu^\alpha \right) . \qquad (2.22)$$

With the definition

$$\delta e^A \equiv \delta z^M e_M{}^A , \tag{2.23}$$

the fermionic gauge transformations of (2.11) can be written as

$$\delta_\eta e^a = 0 \quad , \quad \delta_\eta e^\alpha = -i(\not{p}\eta)^\alpha$$

$$\delta_\eta \Pi_a = 0 \quad , \quad \delta_\eta V = 2i\,\bar{\eta}_\alpha e_t{}^\alpha . \tag{2.24}$$

The invariance of the action follows from the identity

$$\delta(e_t{}^A) = \frac{d}{dt}(\delta e^A) - \delta e^B e_t^C t_{CB}{}^A \tag{2.25}$$

where $t_{CB}{}^A$ is the flat superspace torsion $\left(t_{\gamma\beta}{}^a = -(C\Gamma^a)_{\gamma\beta}\right.$ being the only non-zero component$\left.\right)$.

The coupling to a Maxwell supermultiplet is achieved by the addition of

$$I_{int.} = \int dt\{\dot{z}^M A_M(z)\} \equiv \int dt\{e_t{}^A A_A(z)\}, \tag{2.26}$$

to the action. The variation of this new term takes the form

$$\delta I_{int.} = \int dt\left\{\delta e^A e_t^B F_{BA}\right\} \tag{2.27}$$

where

$$F_{BA} = D_B A_A - (-1)^{ab} D_A A_B - t_{BA}{}^C A_C \tag{2.28}$$

are the usual components of the field strength two-form F associated to the superspace potential one-form $A = dz^M A_M$. Because $\delta_\eta e^a = 0$, this variation is particularly simple, viz.

$$\delta_\eta I_{int} = \int dt \left\{ \delta_\eta e^\alpha e_t^\beta F_{\beta\alpha} + \delta e^\alpha e_t^b F_{b\alpha} \right\} \tag{2.29}$$

The first term cannot be cancelled, so that the constraint

$$F_{\alpha\beta} = 0 \tag{2.30}$$

is necessary if the interaction is to preserve the fermionic gauge invariance. Given (2.30), the Bianchi identity $dF = 0$ implies that

$$F_{b\alpha} = (C\Gamma_b W)_\alpha \tag{2.31}$$

where W_α is the field strength superfield of the super-Maxwell field theory. The second term in (2.29) can now be reduced to

$$\int dt \left\{ i(\bar{\eta} \not{\pi} \Gamma_b W) e_t^b \right\} \tag{2.32}$$

But this can be cancelled by the further variations

$$\delta'_\eta V = -2iV\bar{\eta}W \quad , \quad \delta'_\eta \pi^a = i\bar{\eta}\not{\pi}\Gamma^a W , \tag{2.33}$$

so that $I_3 + I_{int}$. is fermionic gauge invariant provided (2.30) holds. The reason that these additional variations are required is that we chose to work with the action I_3, i.e. <u>without</u> the auxiliary fields of I_4. As the algebra of η-transformations closes only on-shell, the transformation laws must depend on the interactions. Despite this complication, the calculation described above is simpler because $\delta_\eta e^a = 0$. If we were to include the auxiliary fields then $\delta_\eta e^a \neq 0$, but presumably one would arrive in the end at the same result.

Remarkably, the constraint $F_{\alpha\beta} = 0$ is just what is needed to reduce the large number of components of Ω_M to those of the super-Maxwell theory. For both D=4 and D=10 it implies that all components of F_{AB} can be expressed in terms of the single gauge invariant superfield W_α, whose lowest component is the spinor of the Maxwell supermultiplet. But for D=10 $F_{\alpha\beta} = 0$ also implies the field equations, whereas for D=4 it does not. This requires a different interpretation of the term $I_{int.}$ in D=4 and D=10, as mentioned in the introduction.

To couple the superparticle to gravity we have simply to replace the flat superspace vielbein $e_M{}^A$ by the curved superspace vielbein $E_M{}^A$ [6]. Thus, we write

$$I_3 = -\int dt \left\{ E_t{}^a \pi_a - \tfrac{1}{2} V \pi^2 \right\} \tag{2.34}$$

The fermionic gauge transformation (2.24) is similarly generalized to

$$\delta_\eta E^a = 0 \quad , \quad \delta_\eta E^\alpha = -i(\not{\pi}\eta)^\alpha \tag{2.35}$$

but we leave open for the moment the form of the transformations for π_a and V, anticipating that, as previously, they will receive additional contributions. We shall need the identity

$$\delta(E_t{}^A) = \frac{d}{dt}(\delta E^A) - \delta E^B E_t{}^C T_{CB}{}^A - E_t{}^B \delta z^M \Omega_{MB}{}^A \quad , \tag{2.36}$$

which generalizes (2.25) to curved superspace. $T_{CB}{}^A$ are the components of the torsion two form

$$T^A = \tfrac{1}{2} E^C E^B T_{BC}{}^A \quad , \tag{2.37}$$

and $\Omega_{MB}{}^A$ of the connection one-form

$$\Omega_B{}^A = dz^M \Omega_{MB}{}^A \tag{2.38}$$

of the curved superspace. We shall take the superspace structure group to be the Lorentz group (possibly times a U(1)-factor) so that $\Omega_B{}^A$ is non-zero only if both indices are either bosonic or fermionic. Using (2.36) we obtain

$$\delta I_3 = \int dt \Big\{ \delta E^\beta E_+{}^\gamma T_{\gamma\beta}{}^a \Pi_a + \delta E^\beta E_t{}^c T_{c\beta}{}^a \Pi_a \tag{2.39}$$

$$+ E_t{}^b \Pi^a \delta z^M \Omega_{Mba} + \tfrac{1}{2}\delta V \Pi^2 - \left(E_t{}^a - V\Pi^a\right)\delta\Pi_a \Big\},$$

which can be rewritten as

$$\delta I_3 = \int dt \Big\{ -i\left[(\Gamma_c\eta)^\beta E_+{}^\gamma T_{\gamma\beta}{}^a + V(\not\eta)^\beta T_{c\beta}{}^a\right]\Pi_a \Pi^c \tag{2.40}$$

$$+ \left[\Pi^a \delta z^M \Omega_{Mca} - i(\not\eta)^\beta T_{c\beta}{}^a \Pi_a - \delta\Pi_c\right]\left(E_t{}^c - V\Pi^c\right) + \tfrac{1}{2}\delta V \Pi^2 \Big\}.$$

Clearly we should choose

$$\delta\Pi_c = \Pi^a\left[\delta z^M \Omega_{Mca} - i(\not\eta)^\beta T_{c\beta}{}^a\right] \tag{2.41}$$

which vanishes for flat superspace in agreement with (2.24). In order that the remaining terms in (2.40) may be cancelled by a choice of δV we require that the term in the first square bracket be proportional to δ_c^a. In fact, as T cannot depend on $\dot\theta$, but only on θ, each of the two terms in this bracket must separately be proportional to δ_c^a. For D=10 this requires that

$$T_{\gamma\beta}{}^a = -\phi\left(C\Gamma^a\right)_{\gamma\beta}, \quad \eta_{c(a} T_{b)\beta}{}^c = \eta_{ab}(CT)_\beta \tag{2.42}$$

where ϕ is a scalar superfield and T_α a spinor superfield. Then δI_3 vanishes if

$$\delta V = 2i\left[\phi\,\bar{\eta}_\gamma E_t^{\ \gamma} - V(\bar{\eta}\not{\!T})\right] . \tag{2.43}$$

But if ϕ and T^β were non-zero their lowest components would have to be sub-canonical dimension auxiliary fields. As a consequence, they must vanish in an on-shell D=10 supergravity background.

For D=4 one requires almost the same constraints as (2.42), the only difference being that $T_{\gamma\beta}{}^a$ may include a $(C\Gamma_5\Gamma^a)_{\gamma\beta}$ term times some pseudoscalar sub-canonical dimension auxiliary superfield. But in almost all versions of off-shell D=4 supergravity (and certainly for the three irreducible off-shell versions) there are no sub-canonical scalar or pseudoscalar fields. This means that we may set $\phi = 1$ in (2.42) without any essential loss of generality.

3. Strings and superstrings

The Nambu-Goto action for a free massless relativistic string is

$$S_1 = -\int d^2\xi\left\{\left[-\det(\partial_i x^m\,\partial_j x^n\,\eta_{mn})\right]^{\frac{1}{2}}\right\} , \tag{3.1}$$

which can be re-expressed as the two-dimensional generally covariant field theory

$$S_2 = -\frac{1}{2}\int d^2\xi\left\{\sqrt{\gamma}\,\gamma^{ij}\,\partial_i x^m\,\partial_j x^n\,\eta_{mn}\right\} \tag{3.2}$$

Here γ^{ij} is the d=2 worldsheet inverse metric and $\gamma = -\det\gamma_{ij}$. As $\det(\sqrt{\gamma}\,\gamma^{ij}) = -1$, only two of the components of γ_{ij} actually appear in the action. As a result the action is d=2 Weyl invariant. A d=2 cosmological constant would not preserve this invariance and, in fact, would lead to an inconsistent field equation, so that it is omitted. As it stands, $\sqrt{\gamma}\,\gamma^{ij}$ can be eliminated from (3.2) by its field equation to recover (3.1).

The first order form of the action is

$$S_3 = -\int dt \int d\sigma \left\{ p_m \dot{x}^m - L(p^2 + x'^2) - S(p_m x'^m) \right\} , \tag{3.3}$$

but we shall deal exclusively with the first order action S_2 in the following.

The superstring is obtained in two steps. Firstly one makes the replacement

$$\partial_i x^m \rightarrow \partial_i x^m + \tfrac{1}{2}\bar{\theta}\Gamma^m \partial_i \theta , \tag{3.4}$$

as for the superparticle. At this stage the action can be viewed as a two-dimensional σ-model with flat superspace as its target manifold, and can be written as $(e_i{}^A \equiv \partial_i z^M e_M{}^A)$

$$S_2^{(+)} = -\frac{1}{2}\int d^2\xi \left\{ \sqrt{\gamma}\, \gamma^{ij}\, e_i{}^a e_j{}^b \eta_{ab} \right\} \tag{3.5}$$

But in d=2 there is also the possibility of parity violating kinetic terms, so we should also consider[3]

$$S_2^{(-)} = -\frac{1}{2}\int d^2\xi \left\{ \varepsilon^{ij} e_i{}^A e_j{}^B b_{BA}(z) \right\} , \tag{3.6}$$

where $b = \frac{1}{2} e^A e^B b_{BA}$ is some two-form in flat superspace. Under a variation of z^M,

$$\delta S_2^{(-)} = -\frac{1}{2}\int d^2\xi \left\{ \varepsilon^{ij} \delta e^A e_i{}^B e_j{}^C h_{CBA} \right\} \tag{3.7}$$

where $h = \frac{1}{6} e^A e^B e^C h_{CBA} = db$ is a closed three form in superspace.

The volume term in (3.7) can be rewritten as

$$-\frac{1}{6}\int_{B_3} \delta h^* \tag{3.8}$$

where B_3 is a three-dimensional ball with the d=2 worldsheet as its boundary, and h^* is the pull-back of the 3 form h to B_3. Now we want $S_2^{(-)}$ to be super invariant, and we see from (3.8) that, up to surface terms, it will be if h is. The obvious choice for h is,

$$h = -\frac{1}{2} e^\beta e^\alpha e^c (\Gamma_c)_{\alpha\beta} \equiv \frac{1}{2} e^\beta e^\alpha e^c t_{\alpha\beta}{}^d \eta_{dc} \tag{3.9}$$

where $t_{\alpha\beta}{}^c$ is the non-vanishing component of the flat superspace torsion, but the closure of h requires the identity

$$(C\Gamma^a)_{\alpha\beta}(C\Gamma_a)_{\gamma\delta} + (C\Gamma^a)_{\gamma\alpha}(C\Gamma_a)_{\beta\delta} + (C\Gamma^a)_{\beta\gamma}(C\Gamma_a)_{\alpha\delta} = 0 \tag{3.10}$$

which is valid only for D=3, 4, 6, 10.

A term of the form (3.6) also arises in the context of d=2 (bosonic) σ-models with a group manifold as the target space. In that context this term is known as a "Wess–Zumino term" and the three form h has a direct interpretation as the parallelizing torsion of the manifold. But although the analogy of the superstring action to a bosonic σ-model is quite close, it is misleading to think of the superspace three form as related to the torsion of superspace for several reasons, viz. (i) the torsion form is a two form not a three-form, (ii) in curved superspace the torsion two-form is not closed, (iii) even in flat D=4 or D=10 superspace the three-form (3.9) is not closed for N > 1. As we shall see later the superspace formulation of certain supergravity theories introduces in a natural way a closed three form in superspace and the three-form h should be considered as its flat superspace limit.

The importance of $S_2^{(-)}$ is that without it the action would not possess a fermionic gauge invariance of the type discussed previously for the superparticle, and without this invariance the supersymmetrization of a free string would not yield a free superstring. We shall now proceed with the demonstration that $S_2 = S_2^{(+)} + S_2^{(-)}$ is fermionic gauge invariant. As for the superparticle it is convenient to express the gauge transformations in terms of $\delta e^A \equiv \delta z^M e_M{}^A$, thus

$$\delta e^a = 0 \,, \qquad \delta e^\alpha = -i(\not{e}_i \eta^i)^\alpha$$
$$\delta \gamma_{ij} = -4i\, \bar{\eta}_{(i\alpha}\, e_{j)}{}^\alpha \tag{3.11}$$

where η^i is a D-dimensional spinor and a d=2 dimensional <u>self-dual</u> vector, i.e.

$$\sqrt{\gamma}\, \eta^i = \varepsilon^{ij} \gamma_{jk} \eta^k \tag{3.12}$$

Using the result, analogous to that of (2.25), that

$$\delta(e_i{}^A) = \partial_i(\delta e^A) - \delta e^B e_i{}^C t_{CB}{}^A \tag{3.13}$$

we have that

$$\delta S_2^{(+)} = i \int d^2\xi \left\{ \sqrt{\gamma}\, \gamma^{ij} (\bar{\eta}^k \not{e}_k \not{e}_i)_\alpha e_j{}^\alpha + \frac{i}{2}\, \delta(\sqrt{\gamma}\gamma^{ij})\, e_i{}^a e_j{}^b \eta_{ab} \right\} \tag{3.14}$$

while

$$\delta S_2^{(-)} = i \int d^2\xi \left\{ \varepsilon^{ij} (\bar{\eta}^k \not{e}_k \not{e}_i)_\alpha e_j{}^\alpha \right\} . \tag{3.15}$$

Adding (3.14) to (3.15) and making use of (3.12) we have

$$\delta S_2 = \frac{1}{2} \int d^2\xi \left\{ i(\sqrt{\gamma}\gamma^{ij} + \varepsilon^{ij})\left(\gamma^{kl} + \frac{\varepsilon^{kl}}{\sqrt{\gamma}}\right)(\bar{\eta}_l \not{e}_k \not{e}_i)_\alpha e_j{}^\alpha - \delta(\sqrt{\gamma}\gamma^{ij})\, e_i{}^a e_j{}^b \eta_{ab} \right\}$$

$$= i \int d^2\xi \left\{ \sqrt{\gamma} \left(2 \gamma^{j(i} \gamma^{k)l} - \gamma^{ik} \gamma^{jl}\right)\left(\bar{\eta}_{l\alpha} e_j{}^\alpha - \frac{i}{4} \delta\gamma_{lj}\right) e_i{}^a e_k{}^b \eta_{ab} \right\} \tag{3.16}$$

the last step following by repeated use of (3.12). Substituting now the γ_{ij} transformation law in (3.11) we find that

$$\delta S_2 = 0 \tag{3.17}$$

We shall pass by the problem of coupling the superstring to super–Maxwell or super Yang–Mills fields and proceed directly to the supergravity coupling. The first step will, of course, be to replace $e_m{}^A$ by the full superspace vielbein $E_m{}^A$, as for the superparticle. But in order to generalize $S_2^{(-)}$ we shall have to postulate the existence of a two form superspace potential B, the analogue of the flat superspace form b, with superspace field strength closed 3–form H = dB. In addition, a possible complication arises if the background supergravity fields include scalars because these will be the $\Theta = 0$ components of scalar superfields and some scalar function of these may multiply the kinetic (i.e. parity preserving) term in the superstring action. Such factors cannot appear in the Wess-Zumino (i.e. parity violating) term in the action however, essentially because if H is a closed 3 form a function of scalars times H will not be. These considerations lead us to consider a curved superspace generalization of the superstring action of the form [6]

$$S = -\frac{1}{2}\int d^2\xi \left\{ e^{k} \sqrt{\gamma}\, \gamma^{ij} E_i{}^a E_j{}^b \eta_{ab} + \varepsilon^{ij} E_i{}^B E_j{}^C B_{AB} \right\} \tag{3.18}$$

where k is some scalar superfield. We similarly generalize the fermionic gauge transformations of (3.11) to

$$\delta E^a = 0 \quad , \quad \delta E^\alpha = -i(\not{E}_i \eta^i)^\alpha \tag{3.19}$$

$$\delta\gamma_{ij} = -4i\,\bar{\eta}_{\alpha(i} E_{j)}{}^\alpha$$

To evaluate δS we need the identity analogous to (2.36), which is

$$\delta(E_i{}^A) - \partial_i(\delta E^A) - \delta E^B E_i{}^C T_{CB}{}^A - E_i{}^B \delta E^C \Omega_{CB}{}^A \quad , \tag{3.20}$$

but the Ω–dependent terms in δS vanish separately because $\Omega_{CB}{}^a = 0$ unless $B = b$ and because $\Omega_{cba} = -\Omega_{cab}$. Thus

$$\delta S = -\tfrac{1}{2}\int d^2\xi \Big\{ \delta\big(e^k \sqrt{\gamma}\gamma^{ij}\big) E_i^{\ a} E_j^{\ b} \eta_{ab} - 2e^k \sqrt{\gamma}\gamma^{ij} E_i^{\ a} \delta E^{\beta} E_j^{\ C} T_{C\beta}^{\ \ a}$$
$$+ \varepsilon^{ij}\delta E^{\gamma} E_i^{\ B} E_j^{\ A} H_{AB\gamma} \Big\} . \qquad (3.21)$$

Using the specific form of δE^{β} of (3.19) we find

$$\delta S = -\tfrac{1}{2}\int d^2\xi \Big\{ \delta\big(e^k \sqrt{\gamma}\gamma^{ij}\big) E_i^{\ a} E_j^{\ b} \eta_{ab}$$
$$+ 2i\Big[e^k\sqrt{\gamma}\gamma^{ij} E_i^{\ a} E_k^{\ b} (\Gamma_b \eta^k)^{\beta} E_j^{\ \gamma} T_{\gamma\beta}^{\ \ a} - \varepsilon^{ij}(\Gamma_b\eta^k)^{\gamma} E_k^{\ c} E_j^{\ a} E_i^{\ \beta} H_{a\beta\gamma}\Big]$$
$$+ i\Big[2e^k\sqrt{\gamma}\gamma^{ij} E_i^{\ a} E_j^{\ c} E_k^{\ b} (\Gamma_b\eta^k)^{\beta} T_{c\beta}^{\ \ a} - \varepsilon^{ij}(\Gamma_c\eta^k)^{\gamma} E_k^{\ C} E_i^{\ b} E_j^{\ a} H_{ab\gamma}\Big]$$
$$- i\varepsilon^{ij}(\Gamma_c\eta^k)^{\gamma} E_k^{\ c} E_i^{\ \beta} E_j^{\ \alpha} H_{\alpha\beta\gamma} \Big\} \qquad (3.22)$$

Clearly we require $H_{\alpha\beta\gamma} = 0$. The $\alpha\beta\gamma\delta$ component of the $dH = 0$ Bianchi identity then implies that $H_{a\beta\gamma}$ is proportional to $h_{a\beta\gamma}$ as given in (3.9) and we must take the proportionality factor to be e^k. We also need $T_{\gamma\beta}^{\ \ a} = -(C\Gamma^a)_{\gamma\beta}$ (we again ignore the possibility of subcanonical scalar superfields), and this requires $H_{a\beta\gamma} = -e^k(C\Gamma_a)_{\beta\gamma}$. The terms in the first square bracket in (3.22) then cancel with the $\delta(\sqrt{\gamma}\gamma^{ij})$ variation, as in the flat superspace case. The terms in the second square bracket must then cancel against either further variations of $\sqrt{\gamma}\gamma^{ij}$ or the variation of e^k in the first term. For this to happen there must be a contraction on two of the indices abc. To arrange this we are forced to set

$$\eta_{c(b} T_{a)\alpha}^{\ \ \ c} = \eta_{ab}(CT)_{\alpha}$$
$$H_{ab\gamma} = e^k (C\Gamma_{ab}H)_{\gamma} \qquad (3.23)$$

where T_{α} and H_{α} are two spinor superfields. The second square bracket in (3.22) can now be reduced to

$$\Big[-2ie^k\sqrt{\gamma}\big(\bar{\eta}^j \not{E}^k H - \tfrac{1}{2}\gamma^{jk}\bar{\eta}^i \not{E}_i H\big) E_k^{\ a} E_j^{\ b}\eta_{ab} \qquad (3.24)$$
$$+ ie^k\sqrt{\gamma}\big(\bar{\eta}^i\not{E}_i H - 2\bar{\eta}^i \not{E}_i T\big) E_k^{\ a} E_j^{\ b} \eta_{ab}\gamma^{jk}\Big]$$

The first term can be cancelled against the further variation

$$\delta' \gamma_{ij} = 4i\, \bar{\eta}_{(i} \not{E}_{j)} H \tag{3.25}$$

whereas the second term must cancel against the δe^k variation which is

$$-\frac{1}{2}\int d^2\xi \left\{ e^k \sqrt{\gamma}\gamma^{ij} E_i{}^a E_j{}^b \eta_{ab}\left(\delta E^b \partial_b k + \delta E^\alpha D_\alpha k\right)\right\} \tag{3.26}$$

$$= -\frac{i}{2}\int d^2\xi \left\{ e^k \sqrt{\gamma}\gamma^{ij} E_i{}^a E_j{}^b \eta_{ab}\left(\bar{\eta}^k \not{E}_k D k\right)\right\}$$

Hence we require [8]

$$2H_\alpha - 4T_\alpha - D_\alpha k = 0 \ . \tag{3.27}$$

To summarize, the action (3.18) is invariant under the fermionic gauge transformations

$$\delta E^a = 0 \quad , \quad \delta E^\alpha = -i\left(\not{E}_i \eta^i\right)^\alpha \tag{3.28}$$

$$\delta \gamma_{ij} = -4i\, \bar{\eta}_{\alpha(i} E_{j)}{}^\alpha + 4i\, \bar{\eta}_{(i} \not{E}_{j)} H \ ,$$

provided that we impose the constraints

$$T_{\alpha\beta}{}^c = -\left(C\Gamma^c\right)_{\alpha\beta} \quad , \quad \eta_{c(b} T_{a)\alpha}{}^c = \eta_{ab}\left(CT\right)_\alpha \tag{3.29}$$

$$H_{\alpha\beta\gamma} = 0 \quad , \quad H_{\alpha\beta c} = -e^k\left(C\Gamma_c\right)_{\alpha\beta} \quad , \quad H_{ab\gamma} = e^k\left(C\Gamma_{ab}\right)_{\gamma\delta} H^\delta \ ,$$

with T^α and H^α related by (3.27). To establish the consistency of the superstring action (3.18) in a background satisfying (3.29) now requires only a check that the constraints are consistent with the Bianchi identities of the torsion and of H and indeed they pass the test.

Finally we shall apply this result to two cases of interest. D=10 supergravity and 16+16 D=4 supergravity. Both have as their component physical fields

$$\left\{ e_\mu{}^m \, ; \; \psi_\mu \, ; \; A_{\mu\nu} \, ; \; \lambda \, ; \; \sigma \right\} \tag{3.30}$$

For D=10 we can impose a chirality condition on all spinors and choose $k = \sigma$. $T_\alpha = 0$ and $H_\alpha = \frac{1}{2} D_\alpha k$ to obtain the action of Witten [6]. Alternatively we can choose $k = 0$ but $T_\alpha = \frac{1}{2} H_\alpha = D_\alpha \sigma$. In this case there is no e^k factor in the action. This is what happens if one takes the original constraints of Nilsson [17]. For any choice of the constraints they imply the D=10 supergravity field equations.

For D=4, we have $T = D\sigma \neq 0$. Again, one can choose $k = 0$, in which case $H_\alpha = 2D_\alpha \sigma$. In fact

$$H = -\frac{1}{2} E^\beta E^\alpha E^a (\Gamma_a)_{\alpha\beta} + E^b E^a E^\alpha (\Gamma_{ab} D\sigma)_\alpha + E^c E^b E^a H_{abc} \tag{3.31}$$

where H_{abc} is the field strength of the antisymmetric tensor field. For D=4, however, these constraints do not imply the supergravity field equations.

References

1. L. Brink, P. Di Vecchia, and P. Howe, Phys. Lett. 65B, (1976), 471; S. Deser and B. Zumino, Phys. Lett. 65B, (1976), 369; P. Howe, J. Phys. A. Math. Gen. 12, (1979), 393.

2. F. Gliozzi, J. Scherk, and D. Olive, Phys. Lett. 65B, (1976), 282; Nucl. Phys. B122, (1977), 253.

3. M.B. Green and J.H. Schwarz, Phys. Lett. 136B, (1984), 367; Nucl. Phys. B243, (1984), 285.

4. M. Henneaux and L. Mezincescu, Phys. Lett. 152B, (1985), 340.

5. E. Witten, Phys. Lett. 153B, (1985), 243.

6. E. Witten, Princeton preprint, (1985)

7. M. Grisaru, P. Howe, L. Mezincescu, B. Nilsson, and P.K. Townsend, Phys. Lett. 162B, (1985), 116.

8. E. Bergshoeff, E. Sezgin, and P.K. Townsend, Trieste preprint (1985).

9. A.S. Galperin, V.I. Ogievetsky and E.S. Sokatchev, J. Phys. A. Math. Gen. 15, (1982), 3785.

10. G. Giradi, R. Grimm, M. Muller, and J. Wess, Z. Phys. C26 (1984) 123; R. Grimm, M. Muller and J. Wess, Z. Phys. C26 (1984) 427; W. Lang, J. Louis, and B. Ovrut, Phys. Lett. 158B, (1985), 40.

11. W. Siegel, Maryland preprint 86-48 (1985).

12. R. Casalbuoni, Phys. Lett. 62B, (1976), 49.

13. L. Brink and J.H. Schwarz, Phys. Lett. 100B, (1981), 310.

14. W. Siegel, Phys. Lett. 128B, (1983), 397.

15. W. Siegel, Class. Quantum Grav. 2, (1985), L95.

16. W. Siegel, Berkeley preprint (1985).

17. B. Nilsson, Nucl. Phys. B188, (1981), 176.

STRUCTURE OF HETEROTIC σ-MODELS COUPLED TO CONFORMAL SUPERGRAVITY

E. Sezgin
International Centre for Theoretical Physics, Trieste, Italy.

ABSTRACT

The structure of (1,0) and (2,0) supersymmetric σ-models with Wess-Zumino term coupled to d = 2 conformal supergravity is described. The locally supersymmetric (4,0) model with heterotic fermions is discussed. Implications of the heterotic σ-models for heterotic strings propagating in a curved background are elucidated.

1. INTRODUCTION

We define the heterotic σ-models to be those with (p,q) type supersymmetry [1], generated by p left-handed real <u>or</u> pseudoreal Majorana-Weyl charges and q left-handed ones, with $p \neq q$.

The (p,q) superalgebra with real Majorana-Weyl supercharges exists in 2 mod 8 dimensions. Its anticommutators are

$$\begin{aligned} \{Q_{\alpha i}, Q_{\beta j}\} &= \gamma^{\mu}_{\alpha\beta}\, \delta_{ij}\, P_\mu & i&=1,\dots,p \\ \{Q^{\alpha r}, Q^{\beta s}\} &= \gamma^{\mu\,\alpha\beta}\, \delta^{rs}\, P_\mu & r&=1,\dots,q \\ \{Q_{\alpha i}, Q^{\beta s}\} &= \delta_\alpha^\beta\, Z_i{}^s & \alpha&=1,\dots,2^{d/2-1} \end{aligned} \tag{1.1}$$

where Z_i^s are the central charges. In $d = 2$ the (p,0) algebra admits a superconformal extension with one of the following bosonic subalgebras [2]

$$\begin{aligned} &O(2,1) \oplus U(p) & p&\neq 2 \\ &O(2,1) \oplus SU(2) & & \\ &O(2,1) \oplus O(p) & p&=1,2,\dots \\ &O(2,1) \oplus O(3) \oplus Sp(p/2) & p&=2,4,\dots \\ &O(2,1) \oplus G_2 & & \end{aligned} \tag{1.2}$$

So far only actions based on O(2,1) [3], O(2,1) ⊕ SO(2) [4], and O(2,1) ⊕ SU(2) [5] have been constructed. In fact, it has been argued that these are the only ones which can admit an extension to an infinite dimensional (Virasoro) superconformal algebra [6].

The (p,q) superalgebra with pseudoreal Majorana-Weyl spinors exists in 6 mod 8 dimensions. Its anticommutators are

$$\begin{aligned} \{Q_{\alpha i}, Q_{\beta j}\} &= \gamma^{\mu}_{\alpha\beta}\, \Omega_{ij}\, P_\mu & i&=1,\dots,p \\ \{Q^{\alpha r}, Q^{\beta s}\} &= \gamma^{\mu\,\alpha\beta}\, \Omega^{rs}\, P_\mu & r&=1,\dots,q \\ \{Q_{\alpha i}, Q^{\beta s}\} &= \delta_\alpha^\beta\, Z_i{}^s & \alpha&=1,\dots,2^{d/2-1} \end{aligned} \tag{1.3}$$

where $\Omega_{ij} = -\Omega_{ji}$ and $\Omega_{rs} = -\Omega_{sr}$ are the invariant tensors of $Sp(p/2)$ and $Sp(q/2)$, respectively. In $d = 6$ the $(p,0)$ case admits a superconformal extension with bosonic subalgebra $O(6,2) \oplus Sp(p/2)$ [2].

In this paper we shall focus our attention to the description of the (1,0), (2,0) and (4,0) heterotic σ-models in $d = 2$, and their coupling to conformal supergravity. Our main motivation for considering these models is that, when free of all possible anomalies, they describe consistent first quantized spinning strings propagating in the σ-model target manifold.

Generically, the $(p,0)$ models contain the following off-shell supermultiplets

$$(e_\mu^m\,,\ \psi_{\mu L}\,,\ A_\mu)\,,\quad (\phi\,,\ \chi_R)\,,\quad (\lambda_L\,,\ F) \tag{1.4}$$

where e_μ^m $(\mu,m = 1,2)$ is the zweibein, $\psi_{\mu L}$ is the (left-handed) gravitino, A_μ is absent in the $(1,0)$ model, the gauge field of $U(1)$ in the $(2,0)$ model, and of $SU(2)$ in the $(4,0)$ model. The scalars ϕ are the string coordinates and χ_R their fermionic partners. The heterotic fermions, λ_L, have no physical bosonic partners, as F is the auxiliary field, and among other things, they are needed for the cancellation of the $d = 2$ Lorentz anomaly [7].

The fields of the conformal supergravity multiplet have no dynamics. They act as Lagrange multipliers whose equations of motion are nothing but the Virasoro constraints, essential for a consistent ghost-free description of the string.

Actions for the $(p,0)$ models, generically, consist of three parts: (a) An action which is the supersymmetric extension of the kinetic term $\partial_\mu\phi^i\partial^\mu\phi^j g_{ij}$, where g_{ij} is the metric on the scalar manifold, (b) an action which is the supersymmetric extension of the Wess-Zumino term $\varepsilon^{\mu\nu}\partial_\mu\phi^i\partial_\nu\phi^j B_{ij}$. (Only for $p = 1$ (a) and (b) are separately supersymmetric, see below (3.13)). (c) and an action which is the supersymmetric extension of the kinetic term for the heterotic fermions. The curl of B_{ij} has the interpretation of torsion. The fact that B_{ij} can be set equal to zero without losing supersymmetry is in contrast to the case of Green-Schwarz superstring actions [8] where a Wess-Zumino term defined in superspace [9] is necessary for Green-Schwarz-Siegel invariance [8],[10].

One can also add to the action the Poincaré supersymmetric extension of the Fradkin-Tseytlin term $\Omega(\phi)R^{(2)}$, where Ω is the dilaton and $R^{(2)}$ is the $d = 2$ scalar curvature [11]. Since this term is conformally noninvariant it should be treated as a one loop counterterm which contributes to the trace anomaly [12].

In the absence of the Wess-Zumino term, the geometry of the scalar manifold is restricted by supersymmetry to be [13] (a) Riemannian in the (1,0) model (b) Kahler in the (2,0) model and (c) hyper-Kahler in the (4,0) model. These geometries are not altered by going from rigid supersymmetry to local supersymmetry *), essentially due to the fact that there is no gravitino kinetic term in $d = 2$ [5]. (When there is one, the variation $\bar{\psi}_\mu \gamma^{\mu\nu\rho}[D_\nu, D_\rho]\varepsilon$ gives information on the curvature of the holonomy group which is related to the automorphism group of the superalgebra).

In the presence of the Wess-Zumino term the geometry of the globally supersymmetric (p,0) models is modified [14] by the introduction of a totally antisymmetric torsion G_{ijk} as follows

$$\Gamma_{+ij}{}^{k} = \left\{ {k \atop ij} \right\} + G_{ij}{}^{k} \tag{1.5}$$

where $G_{ijk} = G_{ij}{}^{\ell} g_{\ell k}$ is given by (at least locally)

$$G_{ijk} = \frac{1}{2}\left(B_{ij,k} + B_{ki,j} + B_{jk,i} \right) \tag{1.6}$$

In the (1,0) model there is no further condition on G_{ijk}. In the (2,0) model, an extra condition on G_{ijk} arises due to the fact that the complex structure $J_i{}^j$ is covariantly constant with respect to the torsionful connection [14]

$$\partial_i J_j{}^{k} - \Gamma_{+ij}{}^{\ell} J_\ell{}^{k} + \Gamma_{+i\ell}{}^{k} J_j{}^{\ell} = 0 \tag{1.7}$$

Together with the integrability condition for $J_i{}^j$ (i.e. the vanishing of the Nijenhuis tensor [15]),(1.7) implies [14],[15]

$$G_{ijk} = 3\, J_{[i}{}^{m} J_j{}^{n} G_{k]mn} \tag{1.8a}$$

$$= \frac{3}{2} J_{[i}{}^{\ell} J_{jk]\ell} \tag{1.8b}$$

---

*) In Ref.[5] it has been argued that the quaternionic geometry is allowed in the (4,4) model coupled to conformal supergravity.

where

$$J_{ijk} \equiv J_{ij,k} + J_{ki,j} + J_{jk,i} \tag{1.8c}$$

and (1.8b) follows from (1.7) and (1.8a). If the heterotic fermions, λ^r ($r = 1,\ldots,\ell$), are coupled to a gauge connection $A_i^{\ rs}$, then the supersymmetry of the action requires that [1]

$$J_{[i}^{\ \ \ell} F_{j]\ell}^{\ \ \ rs} = 0 \tag{1.9}$$

where F is the Yang-Mills curvature of A.

In the (4,0) model, the restriction (1.7) applies to each one of the three complex structures $J^I{}_i{}^j$ ($I = 1,2,3$) satisfying an $SU(2)$ algebra $J^I J^J = -\delta^{IJ} + \varepsilon^{IJK} J^K$. One consequence of this is [14]

$$G_{ijk}\,\delta^{IJ} = 3\,J^I{}_{[i}{}^m\, J^J_j{}^n\, G_{k]mn} \tag{1.10a}$$

$$= \frac{3}{2}\, J^I_{[i}{}^{\ell}\, J^J_{jk]\ell} + 3\,\varepsilon^{IJK} J^K_{[i}{}^{\ell}\, G_{jk]\ell} \tag{1.10b}$$

where

$$J_{ijk}{}^I \equiv J^I_{ij,k} + J^I_{ki,j} + J^I_{jk,i} \tag{1.10c}$$

In the case of coupling the heterotic fermions to a Yang-Mills connection, supersymmetry imposes the condition

$$J^I{}_{[i}{}^{\ell}\, F_{j]\ell}{}^{rs} = 0 \tag{1.11}$$

Coupling of the (1,0) and (2,0) models with Wess-Zumino term to conformal supergravity does not modify the geometry described above [16]. Whether this will be the case also for the locally supersymmetric (4,0) model with Wess-Zumino term remains to be seen.

We now turn to a more detailed description of the $(p,0)$ models. We shall elaborate most on the $(1,0)$ model, since it is the only one in which the scalar manifold is allowed to be a group manifold. Further, the group manifold seems to be the only known manifold which leads to a non-perturbative and nontrivial fixed point of the β-function [17] thereby allowing the possibility of reducing the critical dimension for strings [18], [19], [20]. We shall elaborate on this at some length in Sec. 2. The $(2,0)$ and the $(4,0)$ models will be discussed in Sec. 3 and Sec. 4 respectively.

2. THE (1,0) MODEL

The supermultiplets of the $(1,0)$ model are

$$\begin{array}{lll} \text{Conformal} & e_\mu^m\ ,\ \psi_\mu & (\mu, m = 1,2\ ,\ \gamma_5\psi_\mu = \psi_\mu) \\ \text{Matter} & \phi^i\ ,\ \chi^i & (i=1,\dots,n\ ,\ \gamma_5\chi = -\chi) \\ \text{Heterotic} & \lambda^r & (r=1,\dots,\ell\ ,\ \gamma_5\lambda = \lambda) \end{array} \tag{2.1}$$

All the fermions are Majorana-Weyl. The scalars ϕ^i are real and they parametrize an arbitrary Riemannian manifold, M. The heterotic fermions take value in a vector bundle over M, with fibre metric $G_{rs}(\phi)$, connection $A_i{}^{rs}(\phi)$, and structure group $G \subseteq SO(\ell)$ [1]. Locally one can take $G_{rs} = \delta_{rs}$. Then the Lagrangian for the $(1,0)$ model reads *) [16] (the string tension parameter, α', can be introduced through the scaling $\mathcal{L} \to 2\pi\alpha'\mathcal{L}$).

---

*) Conventions: $\eta_{mn} = (+\,-)$, $\gamma_m\gamma_n = \eta_{mn} + \gamma_5\epsilon_{mn}$, $\gamma_5 = \sigma_3$, $\epsilon^{01} = 1$, $C^T = -C$, $\gamma_\mu^T = -C\gamma_\mu C^{-1}$. Symmetry: $\bar{\chi}\gamma^{\nu_1,\dots,\nu_n}\lambda = (-1)^n\,\bar{\lambda}\gamma^{\nu_n,\dots,\nu_1}\chi$.

Fierz identity: $\chi\bar{\psi} = -\frac{1}{2}(\bar{\psi}\chi) - \frac{1}{2}(\bar{\psi}\gamma_5\chi)\gamma_5 - \frac{1}{2}(\bar{\psi}\gamma^\lambda\chi)\gamma_\lambda$.

$$e^{-1}\mathcal{L}^{(1,0)} = \frac{1}{2} g^{\mu\nu}\partial_\mu\phi^i\partial_\nu\phi^j g_{ij} + \frac{1}{2}\varepsilon^{\mu\nu}\partial_\mu\phi^i\partial_\nu\phi^j B_{ij}$$

$$+\frac{i}{2}\bar{\chi}^i\gamma^\mu D_\mu\chi^j g_{ij} + \bar{\psi}_\mu\gamma^\nu\gamma^\mu\chi^i\partial_\nu\phi^j g_{ij}$$

$$-\frac{i}{3}G_{ijk}\bar{\psi}_\mu\chi^i\bar{\chi}^j\gamma^\mu\chi^k + \frac{i}{2}\bar{\lambda}^r\gamma^\mu D_\mu\lambda_r$$

$$+\frac{1}{8}F_{ijrs}\bar{\chi}^i\gamma^\mu\chi^j\bar{\lambda}^r\gamma_\mu\lambda^s \tag{2.2}$$

where

$$D_\mu\chi^i = \partial_\mu\chi^i - \frac{1}{2}\omega_\mu(e,\psi)\chi^i + \Gamma_{+jk}{}^i\partial_\mu\phi^j\chi^k \tag{2.3a}$$

$$D_\mu\lambda^r = \partial_\mu\lambda^r + \frac{1}{2}\omega_\mu(e,\psi)\lambda^r + A_i{}^r{}_s\partial_\mu\phi^i\lambda^s \tag{2.3b}$$

and

$$\Gamma_{\pm ij}{}^k = \left\{ {k \atop ij} \right\} \pm G_{ij}{}^k \tag{2.4}$$

$$G_{ij}{}^l g_{lk} \equiv G_{ijk} = \frac{1}{2}(B_{ij,k} + B_{ki,j} + B_{jk,i}) \tag{2.5}$$

In terms of the covariantly constant vielbein fields, $V_i{}^a$, on the scalar manifold, the spin connection reads

$$\omega_{\pm i}{}^{ab} = -V^{jb}\left(\partial_i V_j{}^a - \Gamma_{\pm ij}{}^k V_k{}^a\right)$$

$$= \omega_i{}^{ab}(V) \pm G_i{}^{ab} \tag{2.6}$$

where $G_{iab} = G_{ijk}V^k_a V^j_b$. The supercovariant spin connection $\omega_\mu{}^{mn} \equiv -\omega_\mu\varepsilon^{mn}$ is given by

$$\omega_\mu(e,\psi) = -e^{-1}\varepsilon^{\rho\sigma}\left(e_\mu^{\ m}\partial_\rho e_{\sigma m} + i\,\bar{\psi}_\rho\gamma_\mu\psi_\sigma\right) \tag{2.7}$$

Note that ω_μ drops out in the action, but <u>not</u> in the spinor field equations.

The action of $\mathcal{L}^{(1,0)}$ is invariant under *)

$$\delta e_\mu^{\ m} = -2i\,\bar{\varepsilon}\gamma^m\psi_\mu - \Lambda_D e_\mu^{\ m} + \Lambda_M\,\varepsilon^{mn} e_\mu^{\ n}$$

$$\delta\psi_\mu = \left(\partial_\mu + \tfrac{1}{2}\omega_\mu(e,\psi)\right)\varepsilon + \gamma_\mu\eta - \tfrac{1}{2}\Lambda_D\psi_\mu - \tfrac{1}{2}\Lambda_M\psi_\mu$$

$$\delta\phi^i = -\bar{\varepsilon}\chi^i$$

$$\delta\chi^i = i\gamma^\mu\varepsilon\left(\partial_\mu\phi^i + \bar{\psi}_\mu\chi^i\right) + \tfrac{1}{2}\Lambda_D\chi^i + \tfrac{1}{2}\Lambda_M\chi^i$$

$$\delta\lambda^r = \bar{\varepsilon}\chi^i A_i^{\ rs}\lambda_s + \tfrac{1}{2}\Lambda_D\lambda^r - \tfrac{1}{2}\Lambda_M\lambda^r \tag{2.8}$$

where $(\varepsilon,\eta,\Lambda_D,\Lambda_m)$ are the parameters of supersymmetry, special (conformal) supersymmetry, dilatation and Lorentz transformations, respectively.

The algebra closes off-shell on all the fields but λ^r, on which it closes on-shell. For off-shell closure on λ^r, one introduces the auxiliary fields F^r as follows:

$$\delta F^r = \frac{\delta\mathcal{L}}{\delta\lambda_r}\,\varepsilon$$

$$\delta_{extra}\lambda^r = F^r\varepsilon \tag{2.9}$$

$$\mathcal{L}_{extra} = F^2$$

---

*) The dilatation gauge field b_μ has been set equal to zero by use of conformal boost transformations. It can easily be reintroduced by an η-transformation with parameter $\gamma^\mu b_\mu\varepsilon$. Although in this way it appears in the transformation rules, it drops out in the action.

In any event, the nontrivial part of the conformal superalgebra is given by

$$[\delta_{\epsilon_1}, \delta_{\epsilon_2}] = \delta_{g.c.}(\xi^\lambda) + \delta_M(-\xi^\lambda \omega_\lambda) + \delta_\epsilon(-\xi^\lambda \psi_\lambda)$$

$$[\delta_\eta, \delta_\epsilon] = \delta_D(2i\bar{\epsilon}\eta) + \delta_M(2i\bar{\epsilon}\eta) + \delta_\eta(-i\bar{\epsilon}\eta\gamma^\lambda\psi_\lambda) \tag{2.10}$$

where $\xi^\lambda = 2i\bar{\epsilon}_1\gamma^\lambda\epsilon_2$.

The Lagrangian given in (2.2) can be interpreted as describing a spinning string propagating in gravity (g_{ij}), tordion (B_{ij}) and Yang-Mills $(A_{ir}{}^s)$ background. The dilaton, $\Omega(\phi)$, can also be coupled to the string [11] as $\Omega(\phi)R^{(2)}$ where $R^{(2)}$ is the $d = 2$ scalar curvature. This coupling is not conformally invariant, however interpreted as a counterterm it does serve a purpose; it contributes to the trace anomaly [12]. The supersymmetric extension of this coupling is

$$e^{-1}\mathcal{L} = \Omega(\phi)R^{(2)} - 4\bar{\chi}^i\gamma^{\mu\nu}D_\mu\psi_\nu\,\Omega(\phi)_{,i} \tag{2.11}$$

where D_μ <u>and</u> $R^{(2)}$ contain the supercovariant spin connection $\omega_\mu(e,\psi)$. In proving the supersymmetry of (2.11), it is useful to recall that $eR^{(2)} = -2\epsilon^{\mu\nu}\partial_\mu\omega_\nu$, and that $\delta\omega_\mu(e,\psi) = -4i\bar{\epsilon}\gamma^\lambda D_{[\lambda}\psi_{\mu]}$.

In order to couple the Yang-Mills field to string we introduced new fermionic coordinates, namely λ^r. Another way to achieve this coupling is to let ϕ^i be the coordinates of the direct product of spacetime with a compact internal space, and to perform dimensional reduction with suitable ansatze for g_{ij}, B_{ij} and Ω. This programme has been carried out for group manifolds in Refs. 21 and 22. The ansatze these authors consider is

$$g_{ij}(x,y) = \begin{pmatrix} g_{\alpha\beta}(x) + A_\alpha^I A_\beta^J g_{IJ}(y) & A_\alpha^I g_{IJ}(y) \\ g_{IJ}(y) A_\beta^J & g_{IJ}(y) \end{pmatrix} \tag{2.12}$$

$$B_{ij}(x,y) = \begin{pmatrix} B_{\alpha\beta}(x) + A^I_{[\alpha} V^J_{\beta]}\, g_{IJ}(y) & V^I_\alpha\, g_{IJ} \\ -\, g_{IJ}(y)\, V^J_\beta & B_{IJ}(y) \end{pmatrix} \tag{2.13}$$

$$\Omega(x,y) = \text{Constant} + \phi(x) \tag{2.14}$$

where (x^α, y^I) are the coordinates of (spacetime, group manifold), and

$$A^I_\alpha \equiv A_\alpha{}^a(x)\, L^I{}_a(y) + B_\alpha{}^a(x)\, R^I{}_a(y)$$

$$V^I_\alpha \equiv A_\alpha{}^a(x)\, L^I{}_a(y) - B_\alpha{}^a(x)\, R^I{}_a(y) \tag{2.15}$$

Here, a = 1,..., dim G, and (L^I_a, R^I_a) are the (left, right) killing vectors on G. The gauge fields $(A^a_\alpha,\ B^a_\alpha)$ are those of (G_L, G_R). Substitution of (2.12) - (2.14) into the bosonic part of the action (2.2) yields a $G_L \times G_R$ invariant gauged σ-model with Wess-Zumino term in d = 2 [21],[22] *). Insertion of (2.12)-(2.14) into the full action (2.2) yields the d = 2 supersymmetric extension of this gauged σ-model. Note that the supersymmetry transformation laws (2.8) remain unchanged while the gauge transformation laws of A^a_α, B^a_α and $B_{\alpha\beta}$ are deduced from a special general coordinate transformation in $d + d_G$ dimensions characterized by $\xi^i(x,y) = (0,\ \alpha^a(x)L^i_a(y) + \beta^a(x)R^i_a(y))$ with α^a and β^a arbitrary [21],[22]. Note also that the dimensional reduction of the λ^r-dependent sector of our action (2.2) with a suitable ansatz for $A_I{}^{rs}(x,y)$ would lead to the coupling of the scalar fields to the spinning string in d dimensions.

---

*) One can also gauge the isometries of the internal space which depend on the coordinates of the two dimensional spacetime by introducing a d = 2 vector multiplet [16]. However, in this case $B_{IJ}(y)$ must have vanishing Lie derivative with respect to the gauged isometries.

Considering the case $d = 10$ and $G = E_8 \times E_8$, one may establish a relation between the model described above, and the coupling of the heterotic string to its massless sector [21],[22]. Although there are interesting results obtained in this direction, the question of how the fermionic sector of the heterotic string arises is somewhat mysterious in this approach. A remarkable scheme to explain the fermionic sector starting from $d = 26$ and compactifying nontrivially to $d = 10$, is due to Casher et al. [23] according to which one identifies the transverse $d = 10$ spacetime group, $SO(8)_{spacetime}$, with the $SO(8)$ diagonal subgroup of $SO(8)_{internal} \times SO(8)_{spacetime}$, and choose the embedding $SO(8)_{internal} \subset G_R$ to be such that the adjoint of G_R contains the 8_s representation of $SO(8)_{internal}$. In this way, for example the G_R gauge field $B^a_\alpha(x)$ will contain $B_\alpha{}^m(x)$, where α is M_d spacetime index, and m is an M_d spinor index, which then can be interpreted as the $d = 10$ gravitino.

An alternative approach to the problem of how to generate the fermionic sector of the heterotic string might be to elevate the bosonic coordinates $\phi^i(\sigma,\tau)$ to superspace coordinates $z^M(\sigma,\tau)$, and consequently the bosonic fields $g_{ij}(\phi)$, $B_{ij}(\phi)$ and $\Omega(\phi)$ to the superfields $g_{ij}(z)$ and $\Omega(z)$, respectively, in $d + d_G$ bosonic dimensions [24]. One may then look for a Green-Schwarz-Siegel type world-sheet local supersymmetries which may put constraints on the geometry of the superspace, thereby reducing the number of physical fields contained in these superfields which, in general, have enormous number of components. This procedure would be analogous to that of Witten's [25] describing the Green-Schwarz superstring action [8] in $d = 10$, $N = 1$ curved superspace.

Focusing our attention to the (1,0) model described in (2.2) we now discuss the conditions under which it describes a consistent first quantized spinning string. First of all, let us consider the special case when the scalar manifold is $M_d \times G$, where M_d is a d-dimensional Minkowski spacetime and G is a group manifold. At least at the string tree level, the consistency of a string model propagating in $M_d \times G$ requires that in the first quantized theory the super Virasoro algebra holds (to insure conformal invariance) with a fixed constant value of the central extension (to insure Lorentz invariance in M_d).

The salient features of checking the super Virasoro symmetry in the (1,0) model is as follows [26]. Firstly, on $M_d \times G$ the action, (2.2), decomposes into the sum of an action on M_d and one on G. For notational convenience let the action (2.2) be the one on G. Hence it will be understood that the total action is (2.2) plus the standard Neveu-Schwarz-Ramond action [3] (with chiral fermions) on M_d.

Since the third Betti number of a group manifold is nonvanishing, the field B_{ij} is not globally well-defined. Demanding that the ambiguity in its definition does not lead to a multivalued action requires the rescaling of B_{ij} as [27]

$$B_{ij} \longrightarrow \frac{k\alpha'}{4a^2} B_{ij} \quad , \qquad k \in \mathbb{Z} \tag{2.16}$$

where a is the radius of G, and the curl of the new B_{ij} is proportional to the structure constants f_{ijk} of G (See below). In the superconformal gauge ($g_{\mu\nu} = e^{\rho}\, \eta_{\mu\nu}$, $\psi_\mu = \gamma_\mu \zeta$), and background choice $A_i^{\ rs} = 0$, it follows that the field equations for the physical fields ϕ^i, χ^i and λ^r are

$$\partial_\mu \partial^\mu \phi^i + \left\{ {}^{\ i}_{jk} \right\} \partial_\mu \phi^j \partial^\mu \phi^k - \frac{k\alpha'}{4a^2} G_{jk}{}^{i}\, \partial_\mu \phi^j \partial_\nu \phi^k \varepsilon^{\mu\nu} = 0 \tag{2.17}$$

$$\gamma^\mu \left(\partial_\mu \delta^i_k + \left(\left\{ {}^{\ i}_{jk} \right\} + \frac{k\alpha'}{4a^2} G_{jk}{}^{i} \right) \partial_\mu \phi^j \right) e^{-\rho/4} \chi^k = 0 \tag{2.18}$$

$$\gamma^\mu \partial_\mu \left(e^{\rho/4} \lambda^r \right) = 0 \tag{2.19}$$

In terms of the left killing vectors L_i^a (satisfying $[L_i, L_j] = f_{ij}^{\ \ k} L_K$) the metric on G is $g_{ij} = L_i^a L_j^a$. Hence,

$$\left\{ {}^{\ i}_{jk} \right\} = - L_k^a L^i_{a,j} + \frac{1}{2a} f_{jk}{}^{i} \tag{2.20}$$

Note that $f_{ijk} = f_{abc} L_i^a L_j^b L_k^c$ where f_{abc} are the structure constants of G. Substituting (2.20) into (2.18) and setting $G_{ijk} = -a^{-1} f_{ijk}$, we find the term $(-1/2a + k\alpha'/4a^2) f_{jk}{}^{i}$ which vanishes by choosing

$$k\alpha' / 2a^2 = 1 \quad , \qquad k \in \mathbb{Z} \tag{2.21}$$

With this condition, defining $\chi^a = \chi^i L_i^a e^{-\rho/4}$, the χ-field equation (2.18) reduces to $\gamma^\mu \partial_\mu \chi^a = 0$. As for the ϕ field equation, we first set $G_{ijk} = -a^{-1} f_{ijk}$ in (2.17) and then substitute for $\{{}^{\,i}_{jk}\}$ and $f_{jk}{}^i$ as obtained from (2.20). Recalling (2.21) we thus find the result $\partial_\mu J_L^{\mu a} = 0$, where $J_L^{\mu a} = (\eta^{\mu\nu} - \varepsilon^{\mu\nu}) L_i^a \partial_\nu \phi^i$. In fact this is just the conserved current corresponding to left translations, $\delta\phi^i = \alpha^a L_a^i$, on G. There is also a conserved current corresponding to right translations, $\delta\phi^i = \beta^a R_a^i$, on G. One can easily obtain the conservation law, for example, from (2.18) by making use of (2.20) with L_i^a replaced by R_i^a, and f_{ijk} by $-f_{ijk}$, or by starting directly from $\partial_\mu J_L^\mu = 0$ and substituting the standard relation $L_i{}^a = m^a{}_b R_i$, where $m^a{}_b(\phi)$ is the adjoint representation. In the latter case, one uses the relation $(m^{-1}\partial_i m)_{ab} = 2R^a_{[i,j]} R^{jb}$. Either method leads to the result: $\partial_\mu J_R^{\mu a} = 0$ where $J_R^{\mu a} = (\eta^{\mu\nu} + \varepsilon^{\mu\nu}) R_i^a \partial_\nu \phi^i$.

To summarize, the field equation for physical fields are $\partial_\mu J_L^\mu = 0$, $\partial_\mu J_R^\mu = 0$, $\partial_\mu \chi^a = 0$ and $\partial_\mu \tilde{\lambda}^r = 0$, where $\tilde{\lambda}^r = e^{\rho/4} \lambda^r$. The field equations for (Minkowski)$_d$ variables are, of course, the usual free field equations. The next step is to find the constraint equations, which are nothing but the graviton and gravitino field equations, which we denote by $T_{\mu\nu} = 0$ and $S_\mu = 0$, respectively. Quantizing our system in the light cone gauge, and recalling that $J^{\mu a}_{L,R}$ obey the Kac-Moody algebra [17], one finds that the Virasoro generators L_n, L_n and F_n, which are essentially the Fourier components of T_{++}, T_{--} and S_+, respectively, obey the super Virasoro algebra given by

$$[L_n, L_m] = (n-m) L_{n+m} + \frac{c}{12} n(n^2-1) \delta_{m+n,0}$$

$$\{F_n, F_m\} = 2 L_{n+m} + \frac{1}{3} c \left(n^2 - \frac{1}{4}\right) \delta_{m+n,0}$$

$$[L_n, F_m] = \left(\frac{1}{2} n - m\right) F_{n+m} \qquad (2.22)$$

$$[\tilde{L}_n, \tilde{L}_m] = (n-m) \tilde{L}_{n+m} + \frac{\tilde{c}}{12} n(n^2-1) \delta_{m+n,0}$$

where the central extensions are given by

$$c = (d-2) + \frac{d_G}{1+\frac{C_A}{2k}} + \frac{1}{2}(d-2) + \frac{1}{2}d_G \tag{2.23}$$

$$\tilde{c} = (d-2) + \frac{d_G}{1+\frac{C_A}{2k}} + \frac{1}{2}\ell \tag{2.24}$$

Here d_G is the dimension of the group manifold G, C_A is the second Casimir eigenvalue in the adjoint representation, k is any positive integer, and ℓ is the number of heterotic fermions λ^r. In (2.23) the contributions to c by M_d-bosons, G-bosons, M_d-fermions and G-fermions, and in (2.24) the contributions to $\tilde{c}$ by M_d-bosons, G-bosons and heterotic fermions are shown separately. The fact that (2.22) is established shows that the system has quantum conformal invariance. However, since L_n and F_n are obtained in the light cone gauge one must check the Lorentz invariance of the system. Since the Lorentz generators involve L_n and F_n, understandably the central extensions c and $\tilde{c}$ play an important role, and in fact it turns out that the Lorentz invariance holds provided that (a) $c = 12$ and $\tilde{c} = 24$ (b) the constraint $(L_0 - \frac{1}{2}) = (\tilde{L}_0 - 1)$ holds and, (c) in the zero transverse momentum frame the mass^2 is given by $(\frac{1}{4})\, \alpha' M^2 = (L_0 - \frac{1}{2})$, where the operators L_0 and $\tilde{L}_0$ are defined by

$$\begin{aligned} L_0 &= N_B^d + N_B^G + N_F^d + N_F^G + \varepsilon(d-2) + \varepsilon' d_G \\ \tilde{L}_0 &= \tilde{N}_B^d + \tilde{N}_B^G + N_F^{Het.} + \varepsilon'' \ell \end{aligned} \tag{2.25}$$

where ε, ε', and ε'' are (0)16 for (anti)periodic boundary conditions for the M^d-fermions, G-fermions and heterotic fermions, respectively. Furthermore, N's are the number operators with subscripts denoting bosons/fermions, and superscripts standing for M_d, G or heterotic.

The conditions $c = 12$ and $\tilde{c} = 24$ stated above, in view of (2.23) and (2.24) imply the following two conditions both of which must be satisfied simultaneously:

$$d = 10 - \frac{2}{3}\,\frac{d_G}{1+\frac{C_A}{2k}} - \frac{1}{3}\,d_G \qquad (2.26)$$

and

$$d = 26 - \frac{d_G}{1+\frac{C_A}{2k}} - \frac{1}{2}\,\ell \qquad (2.27)$$

We recall that k is any positive integer and C_A is given by

G	SU(n)	SO(2n+1)	Sp(n)	SO(2n)	G_2	F_4	E_6	E_7	E_8
C_A	2n	4n-2	2n+2	4n-4	8	18	24	36	60

(2.28)

For simple Lie groups the solution of (2.26) has been already given in Ref.[20]. Clearly, (2.27) will also be satisfied for those solutions, in such a way that ℓ will be fixed. Thus the solutions to (2.26) and (2.27) for simple groups can be summarized as follows [26]

d	8	8	6	5	4	4	3	2
G	SO(3)	SU(2)	SU(3)	SO(5)	SO(5)	SU(3)	SU(4)	SO(5)
k	1	2	1	1	2	5	1	7
ℓ	33	33	36	37	36	34	40	34

(2.29)

We shall neither discuss the mass spectrum, nor the question of modular invariance here. However, we wish to emphasize the $d = 6$ solution, since anomaly free supergravity theories do exist in six dimensions. It is concievable that the 36 heterotic fermions can give rise to a simply laced, rank 18 group, H, via the Frenkel-Kac mechanism. In that case, the symmetry group would be a rank 26-d = 20 group, SU(3) x H. The only other solution in (2.29) having this property is the one in $d = 3$. In that case one expects the rank 26-d = 23 group SU(4) x H where rank H = 20. In connection with the question of whether heterotic strings can exist in lower than 10 dimensions, and in particular, in 3, 4 and 6 dimensions, see [28],[29],[30].

We now consider the case when the target manifold of our (1,0) model is of the form $M_d \times K$ where K is a compact manifold of arbitrary dimensions which, however, is not a group manifold. First of all it has been recently shown that if K is a homogeneous manifold other than a group manifold, then the σ-model does not admit the Kac-Moody algebra as a current algebra [31]. This means that for homogeneous K we cannot go through the analysis of the type described above to determine the exact, nonperturbative central extension of the Virasoro algebra underlying the (1,0) model. Nevertheless, one may try to establish the Virasoro algebra directly without the benefit of the Kac-Moody algebra of currents. Nonperturbative results are hard to obtain, however, due to the presence of self-interactions. In the perturbative approach, one has to show that the (1,0) model is free from super-Weyl anomalies (dilatation and conformal supersymmetry), super-Lorentz anomalies (Lorentz, supersymmetry), and σ-model anomalies [32] (due to $A_i^{rs}(\phi)$ and $\omega_{+i}{}^{ab}$) [1]. Assuming that the super-Weyl and super-Lorentz anomalies form separate multiplets, one then only has to worry about the dilatation, Lorentz and σ-model anomalies [33]. The Lorentz anomaly is easily deduced from the work of Alvarez-Gaume and Witten [34], and vanishes for *)

$$\ell = d + d_K + 22 \tag{2.30}$$

where ℓ is the number of heterotic fermions, d and d_K are the dimensions of M_d and K, respectively.

The σ-model anomaly, on the other hand, is given by $-\frac{1}{2}\alpha'(\mathrm{tr}\,\Lambda F_{ij} - \mathrm{tr}\,\theta R_{ij})\,\partial_\mu\phi^i\partial_\nu\phi^j\,\varepsilon^{\mu\nu}$ where Λ and θ are the gauge parameters, and F_{ij} is the curvature of $A_i^{rs}(\phi)$, while R_{ij} is the

*) Note that (2.30) is satisfied by all the solutions listed in (2.29).

$H \subseteq SO(d_K)$ valued curvature of the torsionful connection $\omega_{+i}{}^{ab}$ defined in (2.6). These anomalies can be cancelled by attributing to B_{ij} the transformation law $\delta B_{ij} = \alpha'(\mathrm{tr}\,\Lambda F_{ij} - \mathrm{tr}\,\theta R_{ij})$ [1]. Hull has recently shown that the anomalies only change trivially if a covariant term is added to the gauge connection [35]. Thus, letting

$$\bar{\omega}_i{}^{ab} = \omega_{+i}{}^{ab} + \tau_i{}^{ab} \tag{2.31}$$

where $\tau_i{}^{ab}$ is an arbitrary covariant tensor, the 3-form field strength invariant under the modified transformation rules will be

$$H \equiv G + \alpha' \omega_{3Y}(A) - \alpha' \omega_{3L}(\bar{\omega}) \tag{2.32}$$

where $\omega_{3Y}(\omega_{3L})$ are the usual Yang-Mills (Lorentz) Chern-Simons forms. Replacing G by H in the (1,0) model given in (2.2) will clearly break supersymmetry since

$$\begin{aligned} \delta_\epsilon \mathcal{L}^{(1,0)}(G \to H) = {} & (G_{ijk} - H_{ijk})\, \varepsilon^{\mu\nu} \partial_\mu \phi^i \partial_\nu \phi^j \delta\phi^k \\ & - \tfrac{2}{3} i\, H_{[ijk,\ell]} \left(\partial_\mu \phi^i + \tfrac{1}{2} \bar{\psi}_\mu \chi^i\right) \bar{\chi}^j \gamma^\mu \chi^k \delta\phi^\ell \\ & + \text{total derivative} \end{aligned} \tag{2.33}$$

where we have used the fact that $\delta\phi^i = -\bar{\varepsilon}\chi^i$. The G-term in (2.33) comes from the variation of the Wess-Zumino term. Using (2.32) in (2.33) one finds

$$\begin{aligned} \delta_\epsilon \mathcal{L}^{(1,0)}(G \to H) = {} & \alpha' \left(\omega_{3Y}(A) - \omega_{3L}(\bar{\omega})\right)_{ijk} \varepsilon^{\mu\nu} \partial_\mu \phi^i \partial_\nu \phi^j \bar{\varepsilon}\chi^k \\ & + \tfrac{2}{3} i \alpha' \left(\mathrm{tr}\, F_{ij} F_{k\ell} - \mathrm{tr}\, R_{ij}(\bar{\omega}) R_{k\ell}(\bar{\omega})\right)\left(\partial_\mu \chi^i + \tfrac{1}{2} \bar{\psi}_\mu \chi^i\right) \bar{\chi}^j \gamma^\mu \chi^k \bar{\varepsilon}\chi^\ell \\ & + \text{total derivative} \end{aligned} \tag{2.34}$$

One particular way to restore supersymmetry is to choose $\tau_{iab} = -2\, G_{iab}$, and under an embedding of $H \subseteq SO(d_K)$ into the gauge group G, to set the resulting $\bar{\omega}_i{}^{ab} = \omega_{-i}{}^{ab}$ equal to the gauge connection $A_i{}^{rs}$ [36].

The dilatation anomaly, i.e. the conformal or trace anomaly, is especially important as it puts severe restrictions on the model. The absence of the dilatation anomaly is ensured by the vanishing of the trace of the renormalized stress energy tensor [12],[36],[37]

$$T = \beta_{(ij)}\, \partial_\mu \phi^i \partial_\nu \phi^j g^{\mu\nu} + \beta_{[ij]}\, \partial_\mu \phi^i \partial_\nu \phi^j \varepsilon^{\mu\nu} + \beta_i{}^{rs}\, \bar\lambda^r \gamma^\mu \lambda^s \partial_\mu \phi^i + c\, R^{(2)} = 0 \tag{2.35}$$

where the β's are the β-functions, and c is the central extension of the Virasoro algebra satisfied by the energy momentum tensor.

Thus, the model is conformally invariant provided that

$$\beta_{ij} = 0 \quad , \quad \beta_i{}^{rs} = 0 \quad , \quad c = 0 \tag{2.36}$$

The β's and the c have been computed by several authors. The result, part of which is still to be verified (see below), is [12],[36],[37]

$$\beta_{ij} = R_{ij}(\eta_+) - \nabla_i(\eta_-)\partial_j \Omega - 2\alpha' \, tr\left(F_{ik} F_j{}^k - R_{ik}(\eta_-) R_j{}^k(\eta_-)\right) + \cdots = 0$$

$$\beta^{j\,rs} = \nabla_i(\eta_+) F^{ij\,rs} + O(\alpha') = 0 \tag{2.38}$$

$$c = \frac{d + d_K - 10}{16\pi} + \frac{\alpha'}{2}\left(R - \frac{1}{3}H^2 - 4(\nabla\Omega)^2 + 4\nabla^2\Omega\right) + O(\alpha'^2) = 0 \tag{2.39}$$

In (2.39) R and ∇ are defined with respect to the Christoffel symbol, H is as defined in (2.32) with $\bar\omega = \omega_-$, and

$$\eta_\pm \equiv \omega_\pm \pm \alpha'\left(\omega_{3Y}(A) - \omega_{3L}(\omega_-)\right) \tag{2.40}$$

For the definition of $\omega_{\pm iab}$, see (2.6). The Ω-terms in (2.37) and (2.39) come from the Fradkin-Tseytlin term $\Omega R^{(2)}$ (see (2.11). The Chern-Simons terms occuring in $\beta_{(ij)} = 0$, in (2.38) and in (2.39) to our knowledge, have not been verified as yet. The ... in (2.37) refers to order α'^2 terms, and to order α' terms which vanish upon the use of the zeroth order in α' equation: $R_{ij}(\eta_+) - \nabla_i(\eta_-)\partial_j\Omega = 0$.

In summary, for the (1,0) model to describe a consistent (at the string tree level) heterotic string moving in $M_d \times K$, it is sufficient that the conditions (2.30), (2.37), (2.38), (2.39) and $A = \omega_-$ are satisfied. In particular, demanding that (2.37)-(2.39) are satisfied order by order in α' implies that $d + d_K = 10$ and $\ell = 32$. With those two conditions and $A = \omega_-$ satisfied, a special solution of (2.37)-(2.39) is obtained by setting $G = 0 = \Omega$ and by taking K to be a Ricci flat manifold. Two examples of such solutions are: (a) $d = 4$, K = Calabi Yau manifold, A = SU(3) gauge field [38] (b) $d = 6$, $K = K_3$, A = SU(2) gauge field [39]. It is important to realize that although case (a) involves a σ-model which has global (2,0) supersymmetry [38], as it is coupled to supergravity which has only (1,0) supersymmetry, the full system is still a (1,0) model. With this in mind, the σ-model sector of a (1,0) model can be chosen to have (4,0) global supersymmetry as well [36].

We conclude this section with the following remarks. Firstly, as is well known Eqs. (2.37)-(2.39) are closely related to the bosonic field equations of the Green-Schwarz modified N = 1, d = 10 supergravity. Furthermore, Eqs. (2.37) and (2.38), apart from the Ω-dependent term, do not depend on the fields of d = 2 conformal supergravity, and therefore they are just the β-functions of a globally supersymmetric σ-model. In such a model the conditions $\beta_i{}^{rs} = 0$ and $\beta_{ij} = \nabla_i(\eta_-)V_j + \partial_{[i}\lambda_{j]}$, where V_i and λ_i are arbitrary vectors, would suffice to insure on-shell finiteness [40]. However, from (2.37)-(2.39) it is clear that, although necessary, these conditions are not sufficient for the conformal invariance of the supersymmetric σ-model coupled to d = 2 conformal supergravity. Further conditions are $\lambda^i = 0$, $V_i = \partial_i\Omega$ and Eq. (2.39) [40].

3. THE (2,0) MODEL

Denoting the supersymmetry parameter of the (1,0) model by ε_1, the supersymmetry of the model can be extended to (2,0) by introducing a second supersymmetry parameter, ε_2, as follows [1],[13],[14]

$$\delta\phi^i = \bar{\varepsilon}_1 \chi^i + \bar{\varepsilon}_2 J^i{}_j \chi^j \quad , \quad i = 1, .., 2n \tag{3.1}$$

where $J^i{}_j$ is the complex structure on a Hermitian manifold, i.e. a manifold with $J_i{}^k J_j{}^\ell g_{k\ell} = g_{ij}$. Eq. (3.1) has the following local U(1) invariance

$$\delta\chi^i = -\Lambda J^i{}_j \chi^j \ , \quad \delta\epsilon_1 = -\Lambda\epsilon_2 \ , \quad \delta\epsilon_2 = \Lambda\epsilon_1 \tag{3.2}$$

In coupling the system to the (2,0) conformal supergravity, it is convenient to make the U(1) invariance manifest. To this end we work in a complex coordinate system ϕ^α ($\alpha = 1,\ldots,n$) in which the complex structure reads [15]

$$J_i{}^j = \begin{pmatrix} i\,\delta_\alpha{}^\beta & 0 \\ 0 & -i\,\delta_{\bar\alpha}{}^{\bar\beta} \end{pmatrix} \tag{3.3}$$

Hence, (3.1) becomes

$$\delta\phi^\alpha = \bar\varepsilon\,\chi^\alpha \tag{3.4}$$

with $\varepsilon_1 + i\varepsilon_2 \equiv \varepsilon$, and (3.2) now reads

$$\delta\chi^\alpha = i\Lambda\chi^\alpha \ , \qquad \delta\varepsilon = i\Lambda\varepsilon \tag{3.5}$$

Clearly, ε and χ^α are no longer Majorana-Weyl, but instead just Weyl spinors.

The scalar fields, ϕ^α, parametrize a Hermitian manifold with torsion, and Eqs. (1.5)-(1.8) hold. In particular, subsitituting (3.3) into (1.7) one finds (recall that $g_{\alpha\beta} = 0$)

$$\Gamma_{+\alpha\beta\gamma} = 0 \ , \qquad \Gamma_{+\bar\alpha\beta\gamma} = 0 \tag{3.6}$$

Using this information in (1.5) one now finds

$$G_{\alpha\beta\gamma} = 0 \ , \qquad G_{\bar\alpha\beta\gamma} = \frac{1}{2}\left(g_{\bar\alpha\beta,\gamma} - g_{\bar\alpha\gamma,\beta}\right) \tag{3.7}$$

Therefore,

$$\Gamma_{+\alpha\beta\bar\gamma} = g_{\alpha\bar\gamma,\beta} \ , \qquad \Gamma_{+\alpha\bar\beta\gamma} = \left(g_{\bar\beta\gamma,\alpha} - g_{\bar\beta\alpha,\gamma}\right) \tag{3.8}$$

Furthermore, from (1.6), and (1.9) we have

$$G_{\bar\alpha\beta\gamma} = \frac{1}{2}\left(B_{\bar\alpha\beta,\gamma} - B_{\bar\alpha\gamma,\beta}\right) \ , \qquad F_{\alpha\beta}{}^{rs} = 0 \tag{3.9}$$

We are now ready to present the (2,0) model which contains the supermultiplets

$$\begin{array}{llr} \text{Conformal} & e_\mu^m\,,\ \psi_\mu\,,\ A_\mu & \mu, m = 1,2 \\ \text{Matter} & \phi^\alpha\,,\ \chi^\alpha & \alpha = 1,\dots,n \\ \text{Heterotic} & \lambda^r & r = 1,\dots,\ell \end{array} \tag{3.10}$$

The spinors ψ_μ and χ^α are Weyl ($\gamma_5\psi_\mu = \psi_\mu$, $\gamma_5\chi = -\chi$), however λ^r is Weyl <u>and</u> Majorana ($\gamma_5\lambda^r = \lambda^r$). (See below (2.1)).

With (3.6)-(3.9) in mind, by Noether procedure we find the following Lagrangian for the (2,0) model [16]

$$\begin{aligned} e^{-1}\mathcal{L}^{(2,0)} = \; & g^{\mu\nu}\partial_\mu\phi^\alpha\partial_\nu\phi^{\bar\beta}g_{\alpha\bar\beta} + \varepsilon^{\mu\nu}\partial_\mu\phi^\alpha\partial_\nu\phi^{\bar\beta}B_{\alpha\bar\beta} \\ & + \Big(\tfrac{i}{2}\bar\chi_\beta\gamma^\mu D_\mu\chi^\beta + \bar\chi_\beta\gamma^\nu\gamma^\mu\psi_\nu\partial_\mu\phi^\beta \\ & + \tfrac{1}{2}\bar\chi_\beta\gamma^\nu\gamma^\mu\psi_\nu\bar\psi_\mu\chi^\beta + iG_{\alpha\beta\bar\gamma}\bar\psi_\mu\chi^\alpha\bar\chi^{\bar\gamma}\gamma^\mu\chi^\beta + \text{h.c.}\Big) \\ & + \tfrac{i}{2}\bar\lambda^r\gamma^\mu D_\mu\lambda_r + \tfrac{1}{4}F_{\bar\alpha\beta rs}\bar\chi^{\bar\alpha}\gamma^\mu\chi^\beta\bar\lambda^r\gamma_\mu\lambda^s \end{aligned} \tag{3.11}$$

where

$$\begin{aligned} D_\mu\chi^\beta = \; & \partial_\mu\chi^\beta - \tfrac{1}{2}\omega_\mu(e,\psi)\chi^\beta - iA_\mu\chi^\beta \\ & + \Gamma_{+\gamma\delta}{}^\beta\,\partial_\mu\phi^\gamma\chi^\delta + \Gamma_{+\bar\gamma\delta}{}^\beta\,\partial_\mu\phi^{\bar\gamma}\chi^\delta \end{aligned}$$

$$D_\mu \lambda^r = \partial_\mu \delta^r_s + \frac{1}{2}\omega_\mu(e,\psi)\lambda^r + A_\alpha{}^r{}_s\, \partial_\mu \phi^\alpha \lambda^s + A_{\bar\alpha}{}^r{}_s\, \partial_\mu \phi^{\bar\alpha} \lambda^s$$

$$\omega_\mu(e,\psi) = -e^{-1}\varepsilon^{\rho\sigma}\left(e_\mu^m \partial_\rho e_{\sigma m} + i\,\bar\psi_\rho \gamma_\mu \psi_\sigma\right) \tag{3.12}$$

The action of (3.11) is invariant under

$$\delta e_\mu^m = -i\,\bar\varepsilon \gamma^m \psi_\mu + \text{h.c.} - \Lambda_D e_\mu^m + \Lambda_M \varepsilon^{mn} e_\mu^n$$

$$\delta\psi_\mu = \left(\partial_\mu + \frac{1}{2}\omega_\mu(e,\psi) - i A_\mu\right)\varepsilon + \gamma_\mu \eta - \frac{1}{2}\Lambda_D \psi_\mu - \frac{1}{2}\Lambda_M \psi_\mu + i\Lambda\psi_\mu$$

$$\delta A_\mu = -\frac{1}{2} e^{-1}\varepsilon^{\rho\sigma}\bar\varepsilon\gamma_\mu D_\rho \psi_\sigma + \frac{1}{2}\bar\psi_\nu \gamma_\mu \gamma^\nu \eta + \text{h.c.} + \frac{1}{2}(g_{\mu\nu} - \varepsilon_{\mu\nu})\partial^\nu \Lambda$$

$$+ \frac{1}{2}(g_{\mu\nu} + \varepsilon_{\mu\nu})\partial^\nu \Lambda'$$

$$\delta\phi^\alpha = -\bar\varepsilon \chi^\alpha$$

$$\delta\chi^\alpha = i(\partial_\mu \phi^\alpha + \bar\psi_\mu \chi^\alpha)\gamma^\mu \varepsilon + \frac{1}{2}\Lambda_D \chi^\alpha + \frac{1}{2}\Lambda_M \chi^\alpha + i\Lambda\chi^\alpha \tag{3.13}$$

$$\delta\lambda^r = \bar\varepsilon \chi^\alpha A_\alpha{}^r{}_s\, \lambda^s + \bar\chi^{\bar\alpha}\varepsilon\, A_{\bar\alpha}{}^r{}_s\, \lambda^s + \frac{1}{2}\Lambda_D \lambda^r - \frac{1}{2}\Lambda_M \lambda^r$$

where $(\varepsilon, \eta, \Lambda_D, \Lambda_m, \Lambda, \Lambda')$ are the parameters of local supersymmetry, conformal supersymmetry, dilatation, Lorentz, U(1) and U(1)' transformations, respectively. The fact that λ^r is inert under U(1) can be seen from the commutator $[\delta_\varepsilon, \delta_\eta]$ on λ^r, (See (3.14)), while the presence of the $\chi^2\psi_\mu{}^2$ term in the action can be understood by noting that its η-variation cancels the η-variation of A_μ.

It is important to realize that, unlike in the (1,0) case, the G-dependent part of (3.11) is not separately supersymmetric, since in the variation of the G-independent action one encounters $g_{\bar\alpha[\beta,\gamma]}$, which vanishes for a Kahler manifold but is proportional to G (see 3.7)) for a hermitian manifold with torsion G.

The existence of the gauge field A_μ which gauges <u>two</u> U(1) groups is a phenomenon peculiar to two dimensions. It can be best understood from the analysis of the superconformal Lie algebra underlying the model [16],[41]. A consequence of two U(1) invariances is that A_μ has no degrees of freedom. In analogy with e_μ^m and ψ_μ, A_μ plays the role of a Lagrange multiplier, yielding the constraint $\bar{\chi}_\beta \gamma^\mu \chi^\beta = 0$ needed in the quantization of the model.

The nontrivial part of the superconformal algebra resulting from (3.12) is

$$[\delta_{\epsilon_1}, \delta_{\epsilon_2}] = \delta_{g.c.}(\xi^\lambda) + \delta_M(-\xi^\lambda \omega_\lambda) + \delta_\epsilon(-\xi^\lambda \psi_\lambda) + \delta_\Lambda(-\xi^\lambda A_\lambda)$$

$$[\delta_\eta, \delta_\epsilon] = \delta_D(i\bar{\epsilon}\eta + h.c.) + \delta_M(i\bar{\epsilon}\eta + h.c.) + \delta_\Lambda(-\bar{\epsilon}\eta + h.c.) + \delta_\eta(-i\eta\bar{\psi}_\lambda\gamma^\lambda\epsilon) \qquad (3.14)$$

where $\xi^\lambda = (i\bar{\epsilon}_1\gamma^\lambda\epsilon_2 + h.c.)$. Notice the absence of the Λ'- transformations in (3.14) [16].

We conclude this section with remarks on the string applications of the (2,0) model. As mentioned earlier, a <u>globally</u> supersymmetric (2,0) σ-model can occur in a (1,0) model, playing an important role in finding solutions to the conditions (2.37) and (2.38) [36],[38]. In fact, with $A = \omega_-$ (see Sec. 2), a d_K-dimensional (2,0) manifold (Eqs.(1.5)-(1.7)) with $SU(d_K/2)$ holonomy, and satisfying $\partial_{\bar{\beta}} \Gamma_{+\alpha\gamma}{}^\gamma = 0$, already fulfills the conditions (2.37) and (2.38), to lowest order in α' [36]. In that case, there remains only (2.30) and (2.39) to satisfy for an (1,0) model based (order α') consistent string theory. Note that G_{ijk} need not be vanishing in this situation.

Insisting that the full locally supersymmetric (2,0) model describes a spinning string (with U(1) symmetry), one runs into trouble. This is mainly due to the fact that, in (2.39), the expression $(d + d_K - 10)$ gets replaced by $(d + d_K - 2)$ [42]. Therefore in a perturbative (in α') scheme there does not exist a consistent U(1) spinning string. In a nonperturbative scheme, one must find a nontrivial exact solution of (2.39). This is not easy, however, since an (2,0) manifold can never be a group manifold, and consequently the utility of Kac-Moody symmetry is not available here [31].

4. THE (4,0) MODEL

Extension of the (1,0) supersymmetry to (4,0) proceeds naturally by introduction of three additional supersymmetry parameters, ε^I (I = 1,2,3), as follows [13],[14]

$$\delta_\varepsilon \phi^i = \bar{\varepsilon}\chi^i + \bar{\varepsilon}^I J^{I\,i}{}_j \chi^j \qquad i=1,\ldots,4n \tag{4.1}$$

We recall that $J^{I\,i}{}_j$ are the three complex structures obeying an SU(2) algebra $J^I J^J = -\delta^{IJ} + \varepsilon^{IJK} J^K$, and are covariantly constant with respect to the connection $\Gamma_{+ij}{}^k$ which contains the torsion G_{ijk} (See Eqs. (1.5), (1.6), (1.7) and (1.9)). Recall also that as the torsion is switched off the scalar manifold becomes hyper-Kahler.

Next, we observe that Eq. (4.1) is invariant under the following transformations,

$$\delta_\Lambda \chi^i = \Lambda^I J^{I\,i}{}_j \chi^j$$

$$\delta_\Lambda \varepsilon = \Lambda^I \varepsilon^I$$

$$\delta_\Lambda \varepsilon^I = -\Lambda^I \varepsilon + \varepsilon^{IJK} \Lambda^J \varepsilon^K \tag{4.2}$$

These transformations form the SU(2) subgroup of an $SO(4) \approx SU(2) \times \widetilde{SU(2)}$. To see this, let us first introduce a notation which makes the SO(4) transformations manifest:

$$\varepsilon^P = (\varepsilon, \varepsilon^I) \qquad P=1,\ldots,4 \qquad I=1,2,3$$

$$J^{P\,i}{}_j = (\delta^i_j, J^{I\,i}{}_j)$$

$$\Lambda^{PQ} = (\Lambda^I, -\varepsilon^{IJK}\Lambda^K) \tag{4.3}$$

Since Λ^I describes $\widetilde{SU(2)}$ only, we have

$$\Lambda^{PQ} = -\frac{1}{2}\,\varepsilon^{PQRS}\,\Lambda^{RS} \qquad (4.4)$$

In this notation (4.1) and (4.2) becomes

$$\delta_\epsilon \phi^i = \bar{\varepsilon}^P J^{Pi}{}_j \chi^j$$

$$\delta_\Lambda \chi^i = \Lambda^I J^{Ii}{}_j \chi^j$$

$$\delta_\Lambda \varepsilon^P = \Lambda^P{}_Q \varepsilon^Q \qquad (4.5)$$

Next, in order to make the $SU(2) \times \widetilde{SU(2)}$ invariance manifest, we make use of the $SO(4)$ γ-matrices:

$$\{\gamma^P, \gamma^Q\} = 2\,\delta^{PQ}$$

$$\gamma^P = \begin{pmatrix} 0 & \sigma^P \\ \sigma^{P\dagger} & 0 \end{pmatrix} \quad , \quad \sigma^P = (1, i\sigma^I) \quad , \qquad (4.6)$$

and define the supersymmetry parameter $\varepsilon^{A\dot{A}}$ and the spinor $\chi^{a\dot{A}}$ as follows

$$\varepsilon^P = \sigma^P_{A\dot{A}}\,\varepsilon^{A\dot{A}}$$

$$J^{Pi}{}_j \chi^j = \frac{1}{2}\,\sigma^{P\dot{A}A}\,\chi^a_{\dot{A}}\,V^i_{aA} \qquad (4.7)$$

Here, $a = 1,\ldots,2n$ labels $Sp(n)$, $A = 1,2$ labels $SU(2)$, $\dot{A} = 1,2$ labels $\widetilde{SU(2)}$ and V^i_{aA} are the $4n$-beins on the scalar manifold satisfying

$$g_{ij} V^i_{aA} V^j_{bB} = \Omega_{ab}\,\Omega_{AB} \;, \qquad V^i_{a\dot{A}} V^{j\,aB} + i \leftrightarrow j = g^{ij}\,\delta^B_A$$

$$V^i_{aA} V^{j\,bA} + i \leftrightarrow j = \frac{1}{n}\, g^{ij}\,\delta^b_a \tag{4.8}$$

where $\Omega_{ab}(\Omega_{AB})$ are the antisymmetric invariant tensors of $Sp(n)$ $(Sp(1))$. The Weyl spinor $\chi^{a\dot{A}}$ is real in the sense that

$$\chi^{a\dot{A}} = \Omega^{ab}\,\Omega^{\dot{A}\dot{B}}\, C\,\bar{\chi}^T_{b\dot{B}} \;, \qquad \gamma_5\,\chi^{a\dot{A}} = -\chi^{a\dot{A}} \tag{4.9}$$

Substituting (4.7) into (4.5) one finds

$$\begin{aligned} \delta_\epsilon \phi^i &= -V^i_{aA}\,\bar{\epsilon}^{A\dot{A}}\,\chi^a_{\dot{A}} \\ \delta_\Lambda \chi_{a\dot{A}} &= -i\,\Lambda^I\,\sigma^I{}_{\dot{A}}{}^{\dot{B}}\,\chi_{a\dot{B}} \\ \delta_\Lambda \epsilon_{A\dot{A}} &= -i\,\Lambda^I\,\sigma^I{}_{\dot{A}}{}^{\dot{B}}\,\epsilon_{A\dot{B}} \end{aligned} \tag{4.10}$$

These are precisely the transformation rules resulting from the chiral truncation of the (4,4) model of Ref. [5].

Using the notation (4.10) we now describe the (4,0) model without a Wess-Zumino term, but with heterotic fermions. The model contains the following multiplets

Conformal	$e^m_\mu,\ \psi^{A\dot{A}}_\mu,\ B^I_\mu$	$A,\dot{A}=1,2\ ,\ I=1,2,3$
Matter	$\phi^i\quad \chi^{a\dot{A}}$	$i=1,..,4n\ ,\ a=1,..,2n$
Heterotic	λ^r	$r=1,..,\ell$

(4.11)

where, we recall, A labels the global $SU(2)$, $\dot{A}$ labels the local $\widetilde{SU(2)}$, and B^I_μ is the $SU(2)$ gauge field. The spinors ψ_μ and χ are doubly-symplectic-Majorana and Weyl (see (4.9)), and λ^r is an ordinary Majorana-Weyl spinor.

Chiral truncation of the (4,4) model [5] and addition of the heterotic fermions yields the following result *)

$$e^{-1}\mathcal{L}^{(4,0)} = \frac{1}{2} g^{\mu\nu}\partial_\mu\phi^i\partial_\nu\phi^j g_{ij} + \frac{i}{2}\bar{\chi}^a\gamma^\mu D_\mu\chi_a$$

$$-\bar{\chi}^a\gamma^\nu\gamma^\mu\psi_\nu^A\partial_\mu\phi^i V_{iaA} + \frac{1}{2}\bar{\chi}^a\gamma^\nu\gamma^\mu\psi_{\nu A}\bar{\psi}_\mu^A\chi_a$$

$$+\frac{i}{2}\bar{\lambda}^r\gamma^\mu D_\mu\lambda_r + \frac{1}{8}F_{ijrs}V^i_{aA}V^{jA}_b\bar{\chi}^a\gamma_\mu\chi^b\bar{\lambda}^r\gamma^\mu\lambda^s \tag{4.12}$$

where

$$D_\mu\chi^a = \partial_\mu\chi^a - \frac{1}{2}\omega_\mu(e,\psi)\chi^a - \frac{i}{2}B_\mu^I\sigma^I\chi^a + \partial_\mu\phi^i\omega_{+i}{}^{ab}\chi_b$$

$$D_\mu\lambda^r = \left(\partial_\mu\delta^r_s + \frac{1}{2}\omega_\mu(e,\psi)\delta^r_s + \partial_\mu\phi^i A_i{}^r{}_s\right)\lambda^s$$

$$\omega_\mu(e,\psi) = -e^{-1}\varepsilon^{\rho\sigma}\left(e_\mu^m\partial_\rho e_{\sigma m} + \frac{i}{2}\bar{\psi}_\rho^A\gamma_\mu\psi_{\sigma A}\right) \tag{4.13}$$

Throughout we have suppressed the $\widetilde{SU(2)}$ indices $(\dot{A} = 1,2)$ on $\psi_\mu^{A\dot{A}}$ and $\chi^{a\dot{A}}$, e.g. $\bar{\chi}^a \not{D}\chi_a = \bar{\chi}^{a\dot{A}}\not{D}\chi_{a\dot{A}}$. Note also that the heterotic fermions are inert under $SU(2)\times\widetilde{SU(2)}$.

The action of (4.12) is invariant under

---

*) Symmetry properties: $\bar{\chi}^{aA}\lambda^{bB} = \bar{\lambda}^{bB}\chi^{aA}$, $\bar{\chi}^{aA}\gamma_\mu\lambda^{bB} = -\bar{\lambda}^{bB}\gamma_\mu\chi^{aA}$. Fierz identity for spinors of the same chirality: $\lambda^{A\dot{A}}\bar{\chi}^{B\dot{B}} = -\frac{1}{4}(\bar{\chi}^B\gamma_\mu\lambda^A)\gamma_\mu\Omega^{\dot{A}\dot{B}} - \frac{1}{4}(\bar{\chi}^B\gamma_\mu\sigma^I\lambda^A)\times\gamma_\mu(\sigma^I)^{\dot{A}\dot{B}}$. For spinors of the opposite chirality $-\frac{1}{4}(\gamma^\mu)(\gamma_\mu)$ is replaced by $-\frac{1}{2}$.

$$\delta e_\mu^m = -i\bar{\varepsilon}^A \gamma^a \psi_{\mu A} - \Lambda_D e_\mu^m + \Lambda_m \varepsilon^{mn} e_\mu^n$$

$$\delta \psi_\mu^A = D_\mu \varepsilon^A + \gamma_\mu \eta^A - \frac{1}{2}\Lambda_D \psi_\mu^A - \frac{1}{2}\Lambda_m \psi_\mu^A + \frac{i}{2}\sigma^I \Lambda^I \psi_\mu^A$$

$$\delta B_\mu^I = -e^{-1}\varepsilon^{\rho\sigma}\bar{\varepsilon}^A \sigma^I \gamma_\mu D_\rho \psi_{\sigma A} + \bar{\psi}_\nu^A \gamma_\mu \gamma^\nu \sigma^I \eta_A$$

$$+\frac{1}{2}(g_{\mu\nu} - \varepsilon_{\mu\nu}) D^\nu \Lambda^I + \frac{1}{2}(g_{\mu\nu} + \varepsilon_{\mu\nu}) D^\nu \Lambda'^I$$

$$\delta \phi^i = -V_{aA}^i \bar{\varepsilon}^A \chi^a$$

$$\delta \chi^a = i(\partial_\mu \phi^i V_i^{aA} + \bar{\psi}_\mu^A \chi^a)\gamma^\mu \varepsilon_A - \delta\phi^i \omega_i{}^{ab}(V)\chi_b$$

$$+\frac{1}{2}\Lambda_D \chi^a + \frac{1}{2}\Lambda_m \chi^a + \frac{i}{2}\sigma^I \Lambda^I \chi^a$$

$$\delta \lambda^r = \bar{\varepsilon}^A \chi^a V_{aA}^i A_i{}^r{}_s \lambda^s + \frac{1}{2}\Lambda_D \lambda^r - \frac{1}{2}\Lambda_m \lambda^r \tag{4.14}$$

where

$$D_\mu \varepsilon^A = \partial_\mu \varepsilon^A + \frac{1}{2}\omega_\mu(e,\psi)\varepsilon^A - \frac{i}{2}B_\mu^I \sigma^I \varepsilon^A \tag{4.15}$$

and $(\varepsilon^{A\dot{A}}, \eta^{A\dot{A}}, \Lambda_D, \Lambda_m, \Lambda^I, \Lambda'^I)$ are the parameters of the local supersymmetry, conformal supersymmetry, dilatation, Lorentz local $\widetilde{SU(2)}$ and local $\widetilde{SU(2)}'$ transformations, respectively. Furthermore, the supersymmetry of the λ^r-dependent part of the actions requires that

$$\left(V_{aA}^i V^{kaB} - i\leftrightarrow k\right) F_{jk\,rs} - i\leftrightarrow j = 0 \tag{4.16}$$

In fact, this condition arises in the cancellation of the $F\chi^3 \partial\phi$ terms. Provided that the complex suctuctures are represented as follows

$$J_{ij\,A}{}^B \equiv -\frac{i}{2}\sigma^I{}_A{}^B J_{ij}^I = \frac{1}{2}\left(V_{iaA} V_j^{aB} - i\leftrightarrow j\right) \tag{4.17}$$

Eq.(4.16) reduces to Eq.(1.11).

Notice that although the (4,4) scalar multiplet of Ref. [5] is on-shell, the (4,0) scalar multiplet obtained by its chiral truncation is off-shell. This is similar to the situation in the (1,0) and (2,0) models discussed earlier. The phenomenon of two SU(2) summetries gauged by one SU(2) gauge field is a nonabelian generatization of the similar phenomenon in the (2,0) model, and again, because of it the SU(2) gauge field has no degrees of freedom.

The nontrivial part of the conformal superalgebra which follows from (4.14) is

$$[\delta_{\epsilon_1}, \delta_{\epsilon_2}] = \delta_{g.c.}(\xi^\lambda) + \delta_M(-\xi^\lambda \omega_\lambda) + \delta_\epsilon(-\xi^\lambda \psi_\lambda^A) + \delta_\Lambda(-\xi^\lambda B_\lambda^I)$$

$$[\delta_\eta, \delta_\epsilon] = \delta_D(i\bar{\epsilon}^A \eta_A) + \delta_M(i\bar{\epsilon}^A \eta_A) + \delta_\Lambda(-2\bar{\epsilon}^A \sigma^I \eta_A) + \delta_\eta(i\eta^B \bar{\epsilon}^A \gamma^\lambda \psi_{\lambda B}) \tag{4.18}$$

where $\xi^\lambda = i\bar{\epsilon}_1^A \gamma^\lambda \epsilon_{2A}$.

We now turn to the question of constructing the (4.0) locally supersymmetric Wess-Zumino action. Working with the notation (4.1) and (4.2), the Wess-Zumino term can be introduced via Eq.(1.5) and (1.6), upon which the torsion G_{ijk} must satisfy the condition (1.8), as a consequence of (1.7). Since the globally supersymmetric (4,0) model with torsion satisfying these conditions already exist [14], what remains to be shown is that the coupling of the model to the (4,0) conformal supergravity containing ψ_μ and ψ_μ^I (I = 1,2,3) is consistent with these conditions. This question is currently under study [43]. In any event, we expect that the (4,0) model will have a structure similar to that of the (1,0) model. For example, the analogue of the $G\psi\chi^3$ term occurring in (2.2) will now be

$$G_{ijk}\bar{\psi}_\mu \chi^i \bar{\chi}^j \gamma^\mu \chi^k + 3\, G_{ijk} \bar{\psi}_\mu^I \chi^i \bar{\chi}^j \gamma_\mu J^{Ik}{}_\ell \chi^\ell \tag{4.19}$$

The reader can verify that (4.19) is invariant under the $\widetilde{SU(2)}$ transformations, (4.2), provided that the constraint (1.10a) is used.

Finally, we remark on the string application of the (4,0) model. Since hyper-Kahler (or quaternionic) manifolds are not group manifolds, exact computation of the central extension of the super-Virasoro symmetry underlying the (4,0) model is difficult. Its perturbative evaluation, on the other hand, gives the result $c = (d + d_K + 2)/16\pi$ (see (2.39)), at the lowest order [44], which is, of course, physically unacceptable. Nevertheless, the (4,0) model is useful in that its globally supersymmetric sector can be coupled to the (1,0) conformal supergravity. In fact, an example of such a model which is conformal invariant (i.e. satisfies (2.36)) at least to three loops has been provided by Hull [36] and involves a manifold K which is a product of a two torus with a four dimensional compact hyper-Kahler manifold.

ACKNOWLEDGMENT

I wish to thank Eric Bergshoeff for numerous discussions, and in particular for his collaboration on the contents of Sec. 4.

REFERENCES

[1] C.M. Hull and E. Witten, Phys. Lett. 160B, 398 (1985).

[2] W. Nahm, Nucl. Phys. B135, 149 (1978).

[3] S. Deser and B. Zumino, Phys. Lett. 65B, 369 (1976); L. Brink, P. di Vecchia and P. Howe, Phys. Lett. 65B, 471 (1976).

[4] L. Brink and J.H. Schwarz, Nucl. Phys. B121, 285 (1977).

[5] M. Pernici and P. van Nieuwenhuizen, "A covariant action for the SU(2) spinning string as a hyper-Kahler or quaternionic nonlinear σ-model", Stony Brook preprint (1985).

[6] P. Ramond and J.H. Schwarz, Phys. Lett. 64B, 75 (1976).

[7] D.J. Gross, J.A. Harvey, E. Martinec and R. Rohm, Nucl. Phys. B256, 253 (1985).

[8] M. Green and J.H. Schwarz, Phys. Lett. 136B, 367 (1984); Nucl. Phys. B243, 285 (1984).

[9] M. Henneaux and L. Mezincescu, Phys. Lett. 152B, 340 (1985).

[10] W. Siegel, Phys. Lett. 128B, 397 (1983).

[11] F.S. Frakdin and A.A Tseytlin, "Quantum string theory effective action", Lebedev Institute preprint (1984).

[12] C.G. Callan, D. Friedan, E.J. Martinec and M.J. Perry, "Strings in background fields", Princeton preprint (1985).

[13] L. Alvarez-Gaume and D.Z. Freedman, Commun. Math. Phys. 80, 443 (1981).

[14] S.J. Gates Jr., C.M. Hull and M. Rocek, Nucl. Phys. B248, 157 (1984).

[15] K. Yano, Differential Geometry on Complex and Almost Complex Spaces (Pergamon, Oxford, 1965).

[16] E. Bergshoeff, E. Sezgin and H. Nishino, Phys. Lett. 166B, 141 (1986).

[17] E. Witten, Comm. Math. Phys. 92, 455 (1984).

[18] D. Nemeschansky and S. Yankielowicz, Phys. Rev. Lett. 54, 620 (1984).

[19] S. Jain, R. Shankar and S.R. Wadia, Phys. Rev. D32, 2713 (1985).

[20] E. Bergshoeff, S. Randjbar-Daemi, Abdus Salam, H. Sarmadi and E. Sezgin, Nucl. Phys. B269, 77 (1986).

[21] M.J. Duff, B.E.W. Nilsson and C.N. Pope, Phys. Lett. 163B, 343 (1985); M.J. Duff, B.E.W. Nilsson, N.P. Warner and C.N. Pope, "Kaluza-Klein approach to the heterotic string (II)" CERN preprint (January 1986).

[22] R.I. Nepomechie, "Non-abelian symmetries from higher dimensions in string theories", Seattle preprint (December 1985).

[23] A. Casher, F. Englert, H. Nicolai and A. Taormina, Phys. Lett. 162B, 121 (1985).

[24] S.J. Gates Jr. and H. Nishino, "d = 2 superfield supergravity, local $(\text{supersymmetry})^2$, and nonlinear σ-models", Maryland preprint (1985).

[25] E. Witten, "Twister-like transform in ten dimensions", Princeton preprint (1985).

[26] E. Sezgin, "Lowering the critical dimension for heterotic strings", to appear in the Proceedings on Superstrings, Supergravity and Unified Theories, Trieste, June-July 1985.

[27] E. Bergshoeff, H. Sarmadi and E. Sezgin, "Critical dimensions for strings in a curved background", to appear in the Proceedings of the Cambridge Workshop on Supersymmetry and its Applications, June-July, 1985.

[28] C. Vafa and E. Witten, "Bosonic string algebras", Princeton preprint, (1985).

[29] K.S. Narain, "New heterotic string theories in uncompactified dimensions < 10" Rutherford preprint (1985).

[30] W. Nahm, "A classification of open string models", Bonn preprint (December 1985).

[31] W. Ogura and A. Hosoya, "Kac-Moody algebra and nonlinear σ-models", Osaka preprint (September 1985).

[32] G. Moore and P. Nelson, Phys. Rev. Lett. 53, 1519 (1984).

[33] E.S. Fradkin and A.H. Tseytlin, "Anomaly-free two-dimensional chiral supergravity-matter models and consistent string theories", Lebedev Institute preprint (1985).

[34] L. Alvarez-Gaume and E. Witten, Nucl. Phys. B234, 269 (1984).

[35] C.M. Hull, "Anomalies, ambiguities and superstrings", Cambridge preprint (1986).

[36] C.M. Hull, "Heterotic σ-models and nonlinear strings", Cambridge preprint (1985).

[37] A. Sen, "σ-model approach to the heterotic string theory" and references therein, SLAC preprint (September 1985).

[38] P. Candelas, G. Horowitz, A. Strominger and E. Witten, Nucl. Phys. B258, 46 (1985).

[39] M.B. Green, J.H. Schwarz and P.C. West, Nucl. Phys. B254, 327 (1985).

[40] C.M. Hull and P.K. Townsend, "Finiteness and conformal invariance in non-linear σ-models", Cambridge preprint (December 1985).

[41] P. van Nieuwenhuizen, "Connections between supergravity and strings", Stony Brook preprint (1985).

[42] M. Ademollo et al., Nucl. Phys. B111, 77 (1976); E.S. Fradkin and A.A. Tseytlin, Phys. Lett. 106B, 63 (1981); P.Bouwkengt and P. van Nieuwenhuizen, "Critical dimensions of N = 1 and N = 2 spinning string derived from Fujikawa's approach", Stony Brook preprint, (1985).

[43] E. Bergshoeff and E. Sezgin, work in progress.

[44] P. van Nieuwenhuizen, private communication.

Part Five: Anomalies

ANOMALIES IN SUPERSYMMETRIC GAUGE THEORIES

M. Tonin

Dipartimento di Fisica dell'Università di Padova
Istituto Nazionale di Fisica Nucleare - Sezione di Padova

* * *

Abstract

A cohomological method is presented to get a formulation of the consistent anomalies in N=1, D=4 supersymmetric gauge theories, both for theories coupled to external supergravity (local susy) and for theories in flat superspace (rigid susy). The result is the A.B.B.J., non supersymmetric, chiral anomaly coupled to a supersymmetric partner required for consistency. This result is extended to six dimensions. In the rigid case and for D=4 an equivalent, supersymmetric formulation is also obtained.

At the end the relation between N=1,D=4 superconformal anomalies and the renormalization theorem for supersymmetric theories, is briefly discussed.

1. Introduction

Recently the non abelian gauge anomalies in Supersymmetric Yang-Mills Theories (S.Y.M.T's) have received considerable attention. A theorem of Piguet, Schweda and Sibold [1] assures us that the consistent, chiral anomaly in N=1, D=4 S.Y.M.T'.s can be exhibited in a supersymmetric invariant form. Indeed this theorem states that in N=1, D=4 susy theories described by unconstrained superfields, the rigid supersymmetry is anomaly free. Moreover Piguet and Sibold [2] have shown that this anomaly exists and is unique. However Ferrara et al. [3] proved that the susy invariant anomaly is non polynomial in the components of the superconnection.

In addition to these general results, explicit calculations [4] of the regularized chiral superfield determinant have been performed. They lead to closed, supersymmetric but rather complicated expressions for this anomaly in agreement with [3].

Moreover the structure of the anomaly has been determined through algebraic, cohomological methods by using the extended transgression formula in superspace [5]-[9] or by exploiting the relevant Wess-Zumino consistency condition [1], [10], [11]. In particular in [10], [11] the anomaly is given in the Wess-Zumino gauge (in 4 as well as in 6 dimensions).

The abelian gauge anomaly in susy theories has been known for long time. On the contrary the structure of the non abelian anomaly, Δ is an algebraic problem only recently solved. The reason for this delay is now clear: from the theorems of ref. [1], [2] (see also [12]) it is correct to search for the non abelian anomaly in a susy invariant form; however the theorem of ref. [3] tells us that it is impossible to obtain Δ as a simple expression in terms of the vector superfield V(z) (more precisely, as a polynomial in $e^{\pm V(z)}$ and their superspace derivatives). Indeed the susy invariant formulas derived through explicit calculation [4] or cohomological methods [5], [7], [9] look very involved.

In [6]-[8] a different strategy was adopted: one gives up, <u>at first</u>, looking for a supersymmetric invariant gauge anomaly; instead one formulates and solves a coupled cohomological problem for gauge invariance and supersymmetry. In this way the answer is non supersymmetric, but polynomial and involves the usual A.B.B.J. anomaly and a supersymmetric partner required for consistency [6]. This result can be generalized [8] to curved superspace i.e. for S.Y.M.T'.s coupled to (external) supergravity. Afterward, in the case of flat superspace, one can derive [7] an equivalent supersymmetric invariant, non polynomial expression for the anomaly (similar to the one of ref. [5]). This is done by adding to the action a suitable, local counterterm which cancels the supersymmetric partner.

In this talk I shall follow the approach developed in [6]-[8]. Since the derivation of the chiral anomaly is not more difficult in curved superspace (with local supersymmetry) than in flat superspace (with rigid supersymmetry) the local case is treated at first (in sect. 4). In section 5 the result is specialized to the rigid case and a supersymmetric invariant version of the anomaly is presented. In sect. 6 the extension to higher dimensions is discussed and the anomaly in six dimension is derived [13]. Sections 2 and 3 are premised in order to make the talk selfcontained. In sect. 2 I recall the relation between anomalies and cohomologies and I summarize the method, based on the extended transgression formula, to solve the gauge cohomology in ordinary space-time. This method [14]-[17] is by now well known but is described here again since, when generalized to superspace, it will be the main tool to obtained the results of sections 4,5,6. In sect. 3 the formulation of S.Y.M.T'.s in superspace is reviewed. The last section deals with a different subject. It contains a comment on the N=1, D=4 superconformal anomalies and their relation with the non renormalization theorem for supersymmetric theories.

All the results presented here have been obtained in collaboration with L. Bonora and P. Pasti.

2. Anomalies and Cohomologies.

A nilpotent operator Ω with domain on a space $\mathcal{S}$ (coboundary operator) defines a cohomological problem:

$$\Omega X = 0$$
$$\Omega^2 = 0 \quad ; \quad X \in \mathcal{S} \tag{1}$$

The elements of $\mathcal{S}$ are called cochains. The solutions of (1) are called cocycles. The coboundaries (or trivial cocycles) are the cocycles

$$X = \Omega Y \quad ; \quad Y \in \mathcal{S}$$

and a cocycle is non trivial if it is not a coboundary. Two cocycles are equivalent if they differ by a coboundary and the cohomological classes are the equivalence classes w.r.t. this equivalence relation.

To any Lie algebra (or superalgebra) $\mathcal{G}$ one can associate a cohomology through the BRST construction. The corresponding coboundary operator is the generator of the BRST transformations (or BRST operator) on the space of functionals of the fields φ and the F.P. ghosts.

The BRST construction is completely general: it holds for local transformations (gauge transformation or diffeomorphisms) where the ghosts are fields as well as for rigid transformations where the ghosts are constant parameters; for ordinary transformations (with anticommuting ghosts) as well as far supersymmetric transformations (with commuting ghosts).

One is interested in the BRST cohomology on the space of <u>local</u> functionals of fields and ghosts.

These functionals can be filtered by means of the ghost number. Let us call $\mathcal{S}_n$ the space of local functionals of fields and ghosts with ghost-number n.

The BRST operator is a map from $\mathcal{S}_n$ to $\mathcal{S}_{n+1}$. The n--cocycles are the cocycles belonging to $\mathcal{S}_n$. The anomalies are the non trivial cohomological classes of 1-cocycles.

Indeed let us consider a theory invariant <u>at the classical</u> level under a Lie group (or supergroup) symmetry, say, R. We shall call c_R the related F.P. ghosts and Ω_R the corresponding nilpotent BRST operator. If $I[\varphi] \in \mathcal{S}_0$ is the classical action, the invariance under R gives: $\Omega_R I[\varphi] = 0$

<u>At the quantum level,</u> let us suppose at first that R

is a rigid symmetry or, if local, that the gauge fields (or vielbeins) and the ghosts are external unquantized fields. We denote by $\Gamma[\varphi]$ the renormalized effective action. If the regularization preserves R, one gets the Ward identity $\Omega_R \Gamma[\varphi] = 0$ and the symmetry is preserved at the quantum level. If the regularization breaks R one gets the modified Ward identity

$$\Omega_R \Gamma[\varphi] = \hbar q \Delta_R [c_R, \varphi] + O(\hbar^2) \qquad (2)$$

where q is a constant and Δ_R is a local functional of φ, linear in c_R so that $\Delta_R \in \mathcal{S}_1$. From the nilpotency of Ω_R one has the <u>Wess-Zumino consistency condition</u>

$$\begin{aligned} &\Omega_R \Delta_R = 0 \\ &\Omega_R^2 = 0 \ ; \quad \Delta_R \in \mathcal{S}_1 \end{aligned} \qquad (3)$$

which defines a cohomological problem.

If Δ_R is a coboundary i.e. $q \Delta_R = \Omega_R C[\varphi]$; $C[\varphi] \in \mathcal{S}_0$ then by adding to $\Gamma[\varphi]$ the counterterm $-\hbar C[\varphi]$ one can replace eq. (2) with

$$\Omega_R \Gamma' \equiv \Omega_R (\Gamma - \hbar C) = 0 + O(\hbar^2)$$

and the symmetry is preserved (at least at one loop).

If $\Delta_R \neq \Omega_R C[\varphi]$, $\forall C[\varphi] \in \mathcal{S}_0$ is a non trivial cocycle, R is broken at the quantum level: in this case Δ_R is an anomaly.

Since all the regularizations are equivalent a sufficient condition for the absence of the anomalies is that al least one, consistent, symmetry preserving regularization exists.

Two anomalies which are equivalent from the mathematical point of view (i.e. which belong to the same cohomological class) are also equivalent from a physical point of view. Indeed they can be transformed into each other by adding a finite counterterm to the action.

When the gauge and ghost fields are quantized, eq. (2) must be replaced by the <u>modified</u> Slavnov-Taylor identity. However the conclusion, eq. (3), and the subsequent remarks still hold.

The previous discussion extends easily to the case where more symmetries, say S and T, which exist at the classical levels, are broken by the regularization.

Let us call c_S and c_T, the F.P. ghosts and Ω_S and Ω_T, the BRST operators associated with S and T respectively. By choosing suitably the BRST transformations of c_S (c_T) under T (under S) one can make $\Omega_S+\Omega_T$ nilpotent: $(\Omega_S+\Omega_T)^2=0$ so that

$$\begin{aligned} \Omega_S^2 &= 0 \\ \Omega_T^2 &= 0 \\ \Omega_S\Omega_T+\Omega_T\Omega_S &= 0 \end{aligned} \tag{4}$$

Now the modified Ward identity is

$$(\Omega_S+\Omega_T)\,\Gamma[\varphi] = \hbar q\left(\Delta_S[c_S,\varphi]+\Delta_T[c_T,\varphi]\right)+O(\hbar^2) \tag{5}$$

where Δ_S and Δ_T are local functionals of φ linear in c_S and c_T respectively. The Wess-Zumino consistency conditions becomes

$$(\Omega_S+\Omega_T)(\Delta_S+\Delta_T) = 0 \tag{6}$$

which implies

$$\Omega_S\Delta_S = 0 \tag{6a}$$

$$\Omega_T\Delta_T = 0 \tag{6b}$$

$$\Omega_S\Delta_T+\Omega_T\Delta_S = 0 \tag{6c}$$

Eq. (4) and (6) define the coupled cohomological problem for S and T. Δ_S and Δ_T obey the usual W-Z consistency conditions, eqs. (6a) and (6b), but are not necessarily invariant under T and S respectively. However they must comply with the mixed consistency condition eq. (6c).

Let us consider in more detail the case of chiral gauge symmetry in the ordinary 4-dimensional space-time.

The BRST transformations are

$$\delta\psi(x) = -c(x)\psi(x) \tag{7a}$$

$$\delta A(x) = -dc(x)-[A(x),c(x)]_+ \equiv -Dc(x) \tag{7b}$$

$$\delta c(x) = -c(x)c(x) \tag{7c}$$

where $\psi(x)$ are the matter fields, $A(x) \equiv dx^m A_m(x) \in \mathcal{G}$ is the connection one-form, $c(x) \in \mathcal{G}$ are the anticommuting F.P. fields, d the exterior differential and D the exterior covariant differential; $\mathcal{G}$ is the Lie algebra of the gauge group. Moreover $F = dA + AA$ will denote the curvature two form and Ω_G the relevant BRST operator.

Our problem is to find the gauge anomalies i.e. the non trivial classes of the cohomology problem

$$\begin{aligned} &\Omega_G \Delta_G = 0 \\ &\Omega_G^2 = 0 \; ; \quad \Delta_G \in \mathcal{S}_1 \end{aligned} \tag{8}$$

Becchi, Rouet and Stora [18] have shown that the gauge cohomology (for simple gauge groups) has, at most, one non trivial class in $\mathcal{S}_1$ i.e. one anomaly modulo trivial cocycles.

A simple way to find the anomaly is provided by the method of the extended transgression formula proposed by Bonora and Cotta Ramusino [14], Stora [15] and Zumino, Zee, Wu [16].

The extended transgression formula results from the combination of two simple identities:

a) <u>The curvature identity</u>: Let us define

$$\hat{A} = A + kc \, , \qquad \hat{d} = d + k\delta \, , \qquad \hat{F} = \hat{d}\hat{A} + \hat{A}\hat{A}$$

where k is a real number. Then the curvature identity[15], [19] is

$$\hat{F} = F \tag{9}$$

Indeed by expanding eq. (9) in powers of k, one gets at the order $(k)^0$ the definition of the curvature and at the orders $(k)^1$ and $(k)^2$ del BRST transformations eqs. (7b) (7c).

b) <u>The transgression (or Chern-Weil) formula</u>: Let us define

$$A_t = tA \quad ; \quad F_t = dA_t + A_t A_t \quad ; \quad D^{A_t} = d + [A_t, \;]_\pm$$

where $0 \le t \le 1$ and the $\pm$ in the definition of the covariant differential D^{A_t} depends on the grading of the $\mathcal{G}$-valued quantity on which it operates. Then

$$sTr(F, F, F) = 3\, d \int_0^1 dt \, sTr(A, F_t, F_t) \tag{10}$$

(sTr means symmetrized trace)

Eq. (10) can be verified immediatly if one remarks that i) when d acts on $Tr(A,F_t,F_t)$, it can be replaced by D^{A_t}

ii) $D^{A_t} F_t = 0$ from the Bianchi identity

iii) $D^{A_t} A = \frac{dF_t}{dt}$.

Eq. (10) is purely algebraic. Since $\hat{d}$, $\hat{A}$, $\hat{F}$, obey the same algebraic relations as d, A, F, eq. (10) holds also if d, A, F are replaced by $\hat{d}, \hat{A}, \hat{F}$. Then taking into account the curvature identity (warning: $\hat{F}_t \neq F_t$) one gets the extended transgression formula

$$sTr(F,F,F) = 3\hat{d}\int_0^1 dt\, sTr(\hat{A},\hat{F}_t,\hat{F}_t) \tag{11}$$

Let us define

$$Q_5 \equiv 3\int_0^1 dt\, sTr(\hat{A},\hat{F}_t,\hat{F}_t) = \sum_{p=0}^{5} (k)^p Q_{5-p}^p (c,A) \tag{12}$$

where Q_{5-p}^p are (5-p)-forms with ghost number p. Then expanding eq. (11) in powers of k one gets the identities

$$
\begin{array}{llll}
\text{a)} & sTr(F,F,F) = dQ_5^0 & \text{e)} & 0 = \delta Q_2^3 + dQ_1^4 \\
\text{b)} & 0 = \delta Q_5^0 + dQ_4^1 & \text{f)} & 0 = \delta Q_1^4 + dQ_0^5 \\
\text{c)} & 0 = \delta Q_4^1 + dQ_3^2 & \text{g)} & 0 = \delta Q_0^5 \\
\text{d)} & 0 = \delta Q_3^2 + dQ_2^3 & &
\end{array}
\tag{13}
$$

Eqs. (13a) and (13b) vanish trivially since they concern 6-forms and 5-forms in 4 dimensions. Eq. (13c), integrated over space time, gives

$$\Omega_G \Delta_G = 0$$

where

$$\Delta_G = \int d^4x\, Q_4^1(c,A)$$

Then Δ_G is a 1-cocycle of the gauge cohomology eq. (8). Moreover it is a non trivial 1-cocycle. The proof is by reductio ad absurdum. Indeed the triviality of Δ_G would imply:

$$Q_4^1 = \delta \chi_4^0 + d \chi_3^1 \tag{14}$$

Eqs. (14) and (13d) yield

$$d\left(Q_3^2 - \delta \chi_3^1\right) = 0$$

so that

$$Q_3^2 = \delta \chi_3^1 + d \chi_2^2$$

and so on. At the end one gets

$$Q_0^5 = \delta \chi_0^4 \tag{15}$$

$\chi_{4-p}^{p}(c, A)$ are (4-p)-forms with ghost number p. However eq. (15) cannot hold since Q_0^5 is proportional to $Tr(c^5)$ and $Tr(c^4)$ vanishes. The expression of Q_4^1 can be obtained from eq. (12) and, in conclusion,

$$\Delta_G = \int d^4x \; sTr\left[dc, \left(A, dA + \frac{1}{2} A, A A\right)\right] \tag{16}$$

which is the usual A.B.B.J. anomaly.

3. S.Y.M.T. and S.U.G.R.A. in Superspace [20] .

The superspace Σ_D is spanned by the coordinates $z^M \equiv (x^m, \theta^\mu)$, ($m$=0,1,....D-1; μ=1,2,....N) where x^m are the coordinates of the D-dimensional space-time submanifold $\mathcal{M}_D$ and θ^μ are Grassmann variables. For D=4,6,10 N=4,8,16, respectively. Letters from the middle (the beginning) of the alphabeth denote curved superspace (tangent superspace) indices. Latin letters are reserved to the space-time sector, Greek letters to the spinor sector and Capital letters to the full superspace. The NW-SE summation convention for Greek indices is adopted.

Forms in superspace (superforms) can be defined as in ordinary space-time. If $e^A(z) = dz^M e_M{}^A(z)$ is a vielbeins basis, any n-superform

$$\psi(z) = dz^{M_1} \cdots dz^{M_n} \psi_{M_n \cdots M_1}(z)$$

can be expanded in this basis

$$\psi(z) = e^{A_1} \cdots e^{A_n} \psi_{A_n \cdots A_1}(z)$$

and $\psi_{A_m \cdots A_1}$ are the intrinsic components of ψ. Any (r,s)-superform is an (r+s)-superforms whose intrinsic components vanish except when r indices are Latin and s indices are Greek. Any n-superform ψ can be decomposed into (r,s)-superforms i.e.

$$\psi(z) = \sum_{r+s=n} \psi_{r,s}(z)$$

where

$$\psi_{r,s}(z) = \binom{n}{r} e^{a_1} \cdots e^{a_r} e^{\underline{\alpha}_1} \cdots e^{\underline{\alpha}_s} \psi_{\underline{\alpha}_s \cdots \underline{\alpha}_1 a_r \cdots a_1}$$

The symmetry properties of the curved (or intrinsic) components of ψ follow immediatdly from the rule that dx^m (and e^a) are Grassmann numbers and $d\theta^{\underline{\mu}}$ (and $e^{\underline{\alpha}}$) are c-numbers. To each superforms or superfield $\psi(z)$ one can assign a grading which is

$$q_\psi = n_1 + n_2 + n_3 \pmod 2$$

where n_1 is the number of free Greek indices, n_2 is the form degree and n_3 the ghost number of ψ. Two superforms or superfields anticommute if their gradings are both odd and commute otherwise.

The differential in Σ_D is denoted d_s :

$$d_s = dz^M \frac{\partial}{\partial z^M} = e^A D_A$$

The internal product of $\psi(z)$ w.r.t. the vector $h = h^A D_A$ is indicated, as usual, by

$$i_h \psi = n\, h^{A_1} e^{A_2} \cdots e^{A_n} \psi_{A_n \cdots A_1}$$

The integral of a D-superform ψ_D over $\mathcal{M}_D$ is defined by

$$\int_{\mathcal{M}_D} \psi_D = \frac{1}{D!} \int d^D x \; \varepsilon^{m_1 \cdots m_D} \psi_{m_D \cdots m_1} \tag{17}$$

In 4 dimensions: $\underline{\mu} = (\mu, \dot{\mu})$; $\mu, \dot{\mu} = 1,2$; $\underline{\alpha} = (\alpha, \dot{\alpha})$; $\alpha, \dot{\alpha} = 1,2$; the tangent superspace metric is the Minkowsky one η_{ab}, η^{ab}, with signature -2 in the space-time sector and the Ricci one i.e. $\varepsilon^{\alpha\beta} = -\varepsilon_{\alpha\beta} = \varepsilon_{\dot{\alpha}\dot{\beta}} = -\varepsilon^{\dot{\alpha}\dot{\beta}} = \begin{pmatrix} 0 & 1 \\ -1 & 0 \end{pmatrix}$ in the spinor sector; the rigid torsion is

$$i\sigma_{BC}{}^{A} = \begin{cases} i\sigma_{\beta\dot{\gamma}}{}^{a} \; ; \; i\sigma_{\dot{\beta}\gamma}{}^{a} \equiv i\sigma_{\gamma\dot{\beta}}{}^{a} \\ 0 \quad \text{otherwise} \end{cases}$$

where $\sigma^{a}_{\beta}{}^{\dot\gamma} = (1, \vec{\sigma})$ and $\vec{\sigma}$ are the Pauli matrices.

Supergravity is described by the vielbeins $e^A(z) = dz^M e_M{}^A(z)$ and the Lorentz-valued, spin connection $\omega_A{}^B(z) = dz^M \omega_{MA}{}^B = e^C \omega_{CA}{}^B$

The torsion T^A and the curvature $R_A{}^B$ are given by:

$$T^A = d_s e^A + e^B \omega_B{}^A = e^B e^C T_{CB}{}^A = \mathcal{D}_s e^A$$

$$R_A{}^B = d_s \omega_A{}^B + \omega_A{}^C \omega_C{}^B = e^D e^C R_{CDA}{}^B$$

with the constraints

$$T_{\underline{\beta}\underline{\gamma}}{}^a = i\,\sigma_{\underline{\beta}\underline{\gamma}}{}^a \quad ; \quad T_{bc}{}^a = T_{\underline{\beta}\underline{\gamma}}{}^{\underline{\alpha}} = T_{b\underline{\gamma}}{}^a = 0 \qquad (18)$$

$\mathcal{D}_s = e^A \mathcal{D}_A$ denotes the covariant differential w.r.t. the internal Lorentz transformations.

Supergauge transformations are a special composition of superdiffeomorphisms and field dependent Lorentz transformations in the tangent superspace. The associated BRST transformations are

$$\begin{aligned} \delta_s e^A &= e^B\left(\mathcal{D}_B \xi^A - 2\xi^C T_{CB}{}^A\right) = \mathcal{D}_s \xi^A - i_\xi T^A \\ \delta_s \omega_A{}^B &= -\, i_\xi R_A{}^B \\ \delta_s \xi^A &= -\tfrac{1}{2}\, i_\xi i_\xi T^A \;, \qquad \delta_s \psi = -\xi^A \mathcal{D}_A \psi \end{aligned} \qquad (19)$$

where $\phi(z)$ is a word scalar Lorentz covariant superfield and $\xi^A(z)$ are the F.P. ghosts so that $\xi^\alpha(\xi^a)$ have even (odd) grading.

The generator of these BRST transformations on the space of superfield functionals is called Ω_s. It is nilpotent only modulo fields dependent Lorentz transformations and, therefore, it is nilpotent on Lorentz invariant functionals. Only such functionals will be met in the following.

One can easily see that if $\psi(z)$ is a Lorentz invariant superform, independent of $\xi^A(z)$, the BRST transformations for $\psi(z)$ and $i_\xi \psi(z)$ are

$$\delta_s \psi = \mathcal{L}_\xi \psi \;; \quad \delta_s\left(i_\xi \psi\right) = \frac{1}{2}\left(\mathcal{L}_\xi i_\xi + i_\xi \mathcal{L}_\xi\right)\psi \qquad (20)$$

where $\mathcal{L}_\xi = d_s i_\xi - i_\xi d_s$ is the Lie derivative along ξ.

Supersymmetric Yang-Mills theories are described by a constrained $\mathfrak{g}$-valued superconnection and the constraints require that the intrinsic components of the curvature vanish when both indices are spinor like. There are different equivalent formulations of S.Y.M.T.s.

Framework (φ):
The superconnection $\varphi = e^A \varphi_A \in \mathfrak{g}$ is antihermitean: $\varphi^+ = -\varphi$
Then the gauge transformations

$$\varphi' = e^{-L} \varphi e^L + e^{-L} d_s e^L \; ; \; L(z) \in \mathfrak{g} \tag{21}$$

involve real gauge parameters: $L^+(z) = -L(z)$ and the BRST transformations associated with (21)

$$\begin{aligned} \delta_G \varphi(z) &= -d_s C(z) - [\varphi(z), C(z)]_+ \\ \delta_G C(z) &= -C(z)\, C(z) \quad ; \quad C(z) \in \mathfrak{g} \end{aligned} \tag{22}$$

involve real anticommuting F.P. superfields: $C^+ = -C$.
We shall call Ω^c_G the nilpotent generator of the BRST transf.(22) in the space of the functionals of superfields.

The curvature of φ

$$\mathcal{F} = d_s \varphi + \varphi\varphi = e^A e^B \mathcal{F}_{BA} \in \mathfrak{g}$$

fulfills the constraints

$$\mathcal{F}_{\alpha\beta} = 0 \tag{23}$$

These constraints can be solved in terms of the complex superfield $\mathcal{U}(z) \in \mathfrak{g}$:

$$\begin{aligned} \varphi_\alpha &= e^{-\mathcal{U}^+} D_\alpha e^{\mathcal{U}^+} \; , \quad \varphi_{\dot\alpha} = e^{\mathcal{U}} \bar{D}_{\dot\alpha} e^{-\mathcal{U}} \\ \varphi_a &= \frac{i}{4} (\sigma_a)^{\alpha\dot\beta} \left(D_\alpha \varphi_{\dot\beta} + \bar{D}_{\dot\beta} \varphi_\alpha + [\varphi_\alpha, \varphi_{\dot\beta}]_+ \right) \end{aligned} \tag{24}$$

Moreover $\mathcal{U}(z)$, $\mathcal{U}^+(z)$ transforms as

$$\begin{aligned} e^{\mathcal{U}'} &= e^{-L} e^{\mathcal{U}} e^{\Lambda} \\ e^{\mathcal{U}^{+\prime}} &= e^{-\bar\Lambda} e^{\mathcal{U}^+} e^{L} \end{aligned} \tag{25}$$

where $\Lambda(z) \in \mathcal{G}$ and $\bar{\Lambda}(z) = -\Lambda^+(z)$ are chiral and antichiral superfields respectively;

<u>Framework $(\omega, \bar{\omega})$:</u>

Let us define the superconnections

$$\begin{aligned} \omega &= e^{-\mathcal{U}} \varphi e^{\mathcal{U}} + e^{-\mathcal{U}} d_s e^{\mathcal{U}} \\ \bar{\omega} &= e^{\mathcal{U}^+} \varphi e^{-\mathcal{U}^+} + e^{\mathcal{U}^+} d_s e^{-\mathcal{U}^+} ; \quad \bar{\omega} = -\omega^+ \in \mathcal{G} \end{aligned} \tag{26}$$

and the real superfield $V(z) \in \mathcal{G}$ such that $e^V = e^{\mathcal{U}^+} e^{\mathcal{U}}$. One gets

$$\begin{aligned} \omega_\alpha &= e^{-V} D_\alpha e^V ; \quad \omega_{\dot{\alpha}} = 0 ; \quad \omega_a = \frac{i}{4} \sigma_a^{\alpha\dot{\beta}} \bar{D}_{\dot{\beta}} \omega_\alpha \\ \bar{\omega}_\alpha &= 0 ; \quad \bar{\omega}_{\dot{\alpha}} = e^V \bar{D}_{\dot{\alpha}} e^{-V} ; \quad \bar{\omega}_a = \frac{i}{4} \sigma_a^{\alpha\dot{\beta}} D_\alpha \bar{\omega}_{\dot{\beta}} \end{aligned} \tag{27}$$

The gauge transformations for ω and $\bar{\omega}$ are

$$\begin{aligned} \omega' &= e^{-\Lambda} \omega e^{\Lambda} + e^{-\Lambda} d_s e^{\Lambda} \\ \bar{\omega}' &= e^{-\bar{\Lambda}} \bar{\omega} e^{\bar{\Lambda}} + e^{-\bar{\Lambda}} d_s e^{\bar{\Lambda}} \end{aligned} \tag{28}$$

The associated BRST transformations involve the F.P. ghosts $\gamma(z) \in \mathcal{G}$ and $\bar{\gamma}(z) \in \mathcal{G}$ which are respectively chiral and antichiral, odd graded, superfields. We shall call Ω_G^γ and $\Omega_G^{\bar{\gamma}}$ the corresponding nilpotent BRST operators. In conclusion there are three independent groups of gauge transf. involving respectively the $\mathcal{G}$-valued superfields $L(z)$, $\Lambda(z)$, $\bar{\Lambda}(z)$.

It follows from eqs. (25) and (26) that the superfields $e^{\mathcal{U}}$ ($e^{-\mathcal{U}^+}$) play here the role of the vielbeins in gravitational theories.

Indeed they connect the gauge transformations with parameters $L(z)$ and $\Lambda(z)$ ($L(z)$ and $\bar{\Lambda}(z)$) just as the vielbeins connect general coordinates and Lorentz gauge transf. Let us define

$$\Omega_G = \Omega_G^c + \Omega_G^\gamma + \Omega_G^{\bar{\gamma}}$$

Ω_G is nilpotent and moreover $\Omega_G \Omega_S + \Omega_S \Omega_G = 0$ so that in the space of Lorentz invariant functionals $\Omega_G + \Omega_S$ is a nilpotent operator.

Let us consider a supersymmetric quantum gauge theory coupled to supergravity. According to the general discussion of sect. 2 the coupled cohomology for chiral gauge and supergauge gives

$$(\Omega_G + \Omega_S)(\Delta_G + \Delta_S) = 0$$

that is

$$\Omega_G \Delta_G = 0 \quad , \quad \Omega_S \Delta_S = 0$$
$$\Omega_G \Delta_S + \Omega_S \Delta_G = 0 \tag{29}$$

where Δ_G and Δ_S are local, Lorentz invariant functionals of the superfields, linear in the ghosts $C(z)$ and $\xi^A(z)$ respectively. In the next section we shall derive a non trivial cocycle of this cohomology (indeed the only one, modulo coboundaries, at least in the rigid case, according to Piguet, Sibold [2])

4. *Supersymmetric Gauge anomaly in D=4: the local case.*

The extended transgression formula (E.T.F.), being purely algebraic holds in superspace as well as in ordinary space-time.In superspace, and in framework (φ), reads

$$_s\mathrm{Tr}(\mathcal{F},\mathcal{F},\mathcal{F}) = \hat{d}_s \mathcal{Q}_5 \tag{30}$$

where

$$\mathcal{Q}_5 = 3\int_0^1 dt \; {}_s\mathrm{Tr}\left(\hat{\varphi}, \hat{\mathcal{F}}_t, \hat{\mathcal{F}}_t\right)$$
$$\hat{d}_s = d_s + \delta_G$$
$$\hat{\varphi} = \varphi + c \tag{31}$$
$$\hat{\mathcal{F}}_t = t\,\hat{d}_s\hat{\varphi} + t^2\hat{\varphi}\hat{\varphi}$$

Since $\mathcal{Q}_5$ is Lorentz invariant, $\hat{d}_s$ in eq. (30) can be replaced by

$$\hat{\mathcal{D}}_s = \mathcal{D}_s + \delta_G$$

($\mathcal{D}_s$ being the covariant differential w.r.t. the Lorentz connection)

There are two differences between eq. (30) and the correspondig formula in ordinary space-time.

The first difference is that the l.h.s. of the E.T.F. in the ordinary space-time vanishes being a 6-forms in 4D. Viceversa the l.h.s. of eq. (30) contains (6-s,s)-superforms which do not vanish for s=2,3 (for s>3 they vanish due to the curvature constraints).

Fortunately, in superspace, a new important relation holds, namely

$$ {}_{S}Tr\left(\mathcal{F},\mathcal{F},\mathcal{F}\right) = \hat{\partial}_S \,\chi_{4,1} \qquad (32)$$

where $\chi_{4,1}$ is the gauge invariant (4,1)-superform

$$\chi_{4,1}(z) = \frac{1}{6}\, S\ Tr\left(G,G,G\right) \qquad (33)$$

and

$$G = 2 e^{a}\, e^{\underline{\alpha}}\, \mathcal{F}_{\underline{\alpha} a} \qquad (34)$$

S is a linear map from (r,s) superforms to (r+1,s-2) superforms defined as follows:

$$S\, e^{\alpha} e^{\dot\beta} = S\, e^{\dot\beta} e^{\alpha} = -\frac{i}{4}\, e^{a}\, \sigma_a{}^{\alpha\dot\beta} \qquad (35)$$

on any couple of spinor-like vielbeins and gives zero otherwise.

With the definition

$$\tilde{Q}_5 = Q_5 - \chi_{4,1} \qquad (36)$$

eq. (30) becomes

$$\left(\partial_S + \delta_G\right)\tilde{Q}_5 = 0 \qquad (37)$$

Q_5 and $\tilde{Q}_5$ can be decomposed as

$$Q_5 = \sum_{p=0}^{5} Q^{p}_{5-p} \ , \quad \tilde{Q}_5 = \sum_{p=0}^{5} \tilde{Q}^{p}_{5-p} \qquad (38)$$

where Q^p_{5-p} and $\widetilde{Q}^p_{5-p}$ are (5-p)-superforms with ghost number p. Therefore eq. (37) contains many identities, one for each ghost sector. We remark that $\chi_{4,1}$ affects only $\widetilde{Q}^0_5$ and for $p \geq 1$ $\widetilde{Q}^p_{5-p} = Q^p_{5-p}$.

The second difference is that in ordinary space-time the first two identities with ghost number 0,1 vanish identically and only the one with ghost number 2 is used to get the anomaly. Viceversa in superspace also the first two identities are non trivial and we need the first three identities to solve the coupled cohomology, eq. (29). Indeed looking at the identities in the sectors with ghost number 0,1,2 one gets from eq. (37)

$$i_\xi i_\xi d_S \widetilde{Q}^0_5 = 0 \tag{39a}$$

$$i_\xi d_S Q^1_4 + \delta_G i_\xi \widetilde{Q}^0_5 = 0 \tag{39b}$$

$$d_S Q^2_3 + \delta_G Q^1_4 = 0 \tag{39c}$$

These eqs. can be rewritten as

$$\left(i_\xi \mathcal{L}_\xi + \mathcal{L}_\xi i_\xi\right) \widetilde{Q}^0_5 - d_S i_\xi i_\xi \widetilde{Q}^0_5 = 0 \tag{40a}$$

$$\mathcal{L}_\xi Q^1_4 - \delta_G i_\xi \widetilde{Q}^0_5 - d_S i_\xi Q^1_4 = 0 \tag{40b}$$

$$\delta_G Q^1_4 + d_S Q^2_3 = 0 \tag{40c}$$

Let us define

$$\Delta_G = \int_{\mathcal{M}_4} Q^1_4(z)\Big|_{\theta=\bar\theta=0} \quad ; \quad \Delta_S = -\int_{\mathcal{M}_4} i_\xi \widetilde{Q}^0_5(z)\Big|_{\theta=\bar\theta=0} \tag{41}$$

Then integrating eqs. (40) over $\mathcal{M}_4$ one has

$$\Omega_S \Delta_S = 0 \tag{42a}$$

$$\Omega_S \Delta_G + \Omega_G \Delta_S = 0 \tag{42b}$$

$$\Omega_G \Delta_G = 0 \tag{42c}$$

In conclusion

$$\Delta = \Delta_G + \Delta_S \tag{43}$$

is a non trivial 1-cocycle of the coupled cohomology for chiral gauge and supergauge. The non triviality is proved by the same argument as in ordinary gauge theories.

If Q_4^1 and $i_\xi \tilde{Q}_5^0$ are decomposed in (r,s)-superforms

$$Q_4^1 = \sum_{s=0}^{4} Q^1_{4-s,s} \quad , \qquad i_\xi \tilde{Q}_5^0 = \sum_{s=0}^{4} \left(i_\xi Q_5^0\right)_{4-s,s} \tag{44}$$

Δ_G and Δ_S , as defined in eq. (41), appear as sums of different contributions. The leading terms $\int_{M_4} Q^1_{4,0}$ and $\int_{M_4} \left(i_\xi \tilde{Q}_5^0\right)_{4,0}$ are proportional to the determinant of the space time vielbeins $e_m^a(x)$. The remaining ones are proportional to the gravitino field $e_m^\alpha(x)$. The same procedure can be carried out working in the framework (ω) or ($\bar{\omega}$). The corresponding results are obtained by simply replacing (φ, c) with (ω, γ) or ($\bar{\omega}, \bar{\gamma}$) in the above formulas. The anomalies in the framework (ω) and ($\bar{\omega}$) will be

$$\Delta^{\omega} = \Delta_G^{\omega} + \Delta_S^{\omega} \tag{45a}$$

$$\Delta^{\bar{\omega}} = \Delta_G^{\bar{\omega}} + \Delta_S^{\bar{\omega}} \tag{45b}$$

respectively. One has $\Delta^{\bar{\omega}} = \left(\Delta^{\omega}\right)^+$. To get a real quantity one must consider the semisum of eqs. (45a) and (45b)

$$\tilde{\Delta} = \frac{1}{2}\left(\Delta^{\omega} + \Delta^{\bar{\omega}}\right) \tag{46}$$

$\tilde{\Delta}, \Delta^{\omega}, \Delta^{\bar{\omega}}$ are functionals of the unconstrained superfield $V(z)$ and involve the F.P. ghosts $\gamma(z)$ and/or $\bar{\gamma}(z)$.

The anomalies $\Delta, \tilde{\Delta}, \Delta^{\omega}, \Delta^{\bar{\omega}}$ are equivalent and the equivalence can be easily verified by adopting the procedure of Bardeen and Zumino [21] for gravity. Indeed, as already remarked, the superfields e^{U} , e^{-U^+} play here the role of the vielbeins in gravitational theories. To be more precise let us define the Wess-Zumino action

$$I_{WZ}^{U} = \int_0^1 dy \int_{M_4} \bar{Q}_4^1(z, y)\Big|_{\theta=\bar{\theta}=0}$$

where $\bar{Q}_4^1$ is obtained from Q_4^1 by replacing $C(z)$ with $\mathcal{U}(z)$ and $\varphi(z)$ with

$$\varphi^y(z) = e^{-y\,\mathcal{U}(z)}\,\varphi(z)\,e^{y\,\mathcal{U}(z)} + e^{-y\,\mathcal{U}(z)}\,d_s\,e^{y\,\mathcal{U}(z)}$$

Then

$$\Omega_s I_{WZ}^{\mathcal{U}} = -\Delta_s + \Delta_s^{\omega}\ ;\quad \Omega_G^c I_{WZ}^{\mathcal{U}} = -\Delta_G\ ;\quad \Omega_G^{\gamma} I_{WZ}^{\mathcal{U}} = \Delta_G^{\omega}$$

so that

$$\Delta^{\omega} = \Delta + (\Omega_G + \Omega_s)\, I_{WZ}^{\mathcal{U}}$$

By defining $I_{WZ}^{-\mathcal{U}^+}$ in a similar way, one has

$$\Delta^{\bar{\omega}} = \Delta + (\Omega_G + \Omega_s)\, I_{WZ}^{-\mathcal{U}^+}$$

$$\tilde{\Delta} = \Delta + \frac{1}{2}(\Omega_G + \Omega_s)\left(I_{WZ}^{\mathcal{U}} + I_{WZ}^{-\mathcal{U}^+}\right)$$

(for further details see ref. [7]).

5. *Supersymmetric gauge anomaly in 4D: the rigid case.*

The anomalies, in the rigid case, are obtained from the local ones by replacing the vielbeins with the rigid vielbeins, by setting the spin superconnection to zero and by taking the F.P. ghosts ξ^A to be constant parameters. They will be specified by a superscript o on the corresponding symbols.

Rigid vielbeins are defined by the condition

$$d_s e^A = i\, e^B e^C\, \sigma_{BC}{}^A$$

which is solved by

$$e_M{}^A = \begin{pmatrix} \delta_m^a & 0 & 0 \\ \bar{\theta}^{\dot\nu}\sigma_{\dot\nu\mu}^a & \delta_\mu^\alpha & 0 \\ \theta^\nu \sigma_{\nu\dot\mu}^a & 0 & \delta_{\dot\mu}^{\dot\alpha} \end{pmatrix}$$

In particular the gravitino fields $e_m{}^\alpha(x)$ vanish. It follows that only the leading terms $Q_{4,0}^1$ and $(i_\xi Q_5^o)_{4,0}$ contribute to the anomalies Δ_G^o and Δ_S^o (or $\tilde{\Delta}_G^o$ and $\tilde{\Delta}_S^o$ etc.).

From the expressions of Q_4^1 and $\tilde{Q}_5^o$ which follow from eqs. (31)(33)(36) one has (+)

$$\Delta^0 = \Delta^0_G + \Delta^0_S \tag{47}$$

where

$$\Delta^0_G = \int_{M_4} sTr\left\{dc,\left[dA,A + \frac{1}{2}A,A\,A\right]\right\}\Big|_{\theta=\bar\theta=0} \tag{48}$$

$$\Delta^0_S = 6\int_{M_4}\int_0^1 dt\, sTr\left\{i_\xi\left[-(A,G_t,F_t) - \frac{1}{2}(\lambda F_t,F_t) + \frac{1}{36}S(G,G,G)\right]\right\}\Big|_{\theta=\bar\theta=0} \tag{49}$$

In eqs. (48) (49) the following notations have been used

$$A = \varphi_{1,0} = e^a\varphi_a\ ;\ \lambda = \varphi_{0,1} = e^{\underline{\alpha}}\varphi_{\underline{\alpha}}\ ;\ F = \mathcal{F}_{2,0}\ ;\ G = \mathcal{F}_{1,1}$$
$$F_t = (\mathcal{F}_t)_{2,0}\ ;\ G_t = (\mathcal{F}_t)_{1,1} \tag{50}$$

The anomaly Δ^0 is polynomial in the components A and λ of the superconnection and consists of two parts: the chiral gauge anomaly Δ^0_G (the usual A.B.B.J anomaly) which breaks supersymmetry and the supersymmetric partner Δ^0_S required for consistency. We remark that the ghost parameters ξ^a do not appear in $(i_\xi\, \mathcal{G}^0_5)_{4,1}$ and therefore in Δ^0_S since they multiply a 5-form in 4 dimensions. This was to be expected because Δ^0_G is invariant under space-time translations.

The theorem of ref. [1] implies that Δ^0_S is a coboundary of Ω_S that is

$$\Delta^0_S = \Omega_S C$$

for some local functional C. Then the new equivalent anomaly

$$\Delta'^0 = \Delta^0 - (\Omega_G + \Omega_S)C = \Delta^0_G - \Omega_G C \tag{51}$$

is a supersymmetric invariant gauge anomaly. The problem is the calculation of C. This problem has been solved in [7] to which we refer for the details of the derivation. In the framework $(\omega,\bar\omega)$ Δ'^0, as given by eq. (51), is

$$\Delta'^0[\Gamma,\bar\Gamma,V] = \frac{1}{8}\int d^4x\, d^4\theta\left\{-\frac{1}{2}sTr\left(\gamma, \bar D^{\dot\alpha}\omega^\beta, D_{\dot\alpha}\omega_\beta\right) - \right. \tag{52}$$
$$\left. -\int_0^1 dy\, \sigma^{a\alpha\dot\beta} P_{a\alpha\dot\beta}(\gamma,V;y)\right\} + h.c.$$

where

$$P(\gamma, V; y) = e^a e^\alpha e^\beta P_{a\alpha\beta}(\gamma, V; y) =$$
$$= 6\int_0^1 dt\, t(1-t)\, sTr\Big\{\big(\omega^y, \mathcal{F}_t^y, [\gamma^y, V]\big) - \quad (53)$$
$$-\big(\gamma^y, \mathcal{F}_t^y, [\omega^y, V]\big) + \big(V, \mathcal{F}_t^y, [\omega^y, \gamma^y]\big)\Big\}$$

and

$$\omega^y = e^{yV}\omega e^{-yV} + e^{yV} d_s e^{-yV}$$
$$\gamma^y = e^{yV}\gamma e^{-yV} + e^{yV}\delta_G^\gamma e^{-yV}$$
$$\mathcal{F}_t^y = t\, d_s\omega^y + t^2\omega^y\omega^y$$

δ_G^γ is the BRST variations related to the ghost γ.

$P(\gamma, V; y)$ is non polynomial in $e^{\pm V}$ in agreement with ref. [3]. Eq. (53) should be compared with the results of [4], [5], [9]. All these expressions for the anomaly must be equivalent due to the theorem of [2].

The difficulty to prove this equivalence lies in the fact that the local counterterms needed to transform an anomaly into another can be, a priori, highly non polynomial.

It is interesting to remark that the anomaly Δ^0, eqs.(47),(48), (49),restricted to the Wess-Zumino gauge, coincides with the anomalies obtained in [10],[11]. As noted in [10],[11] the susy partner in Wess-Zumino gauge cannot be avoided by adding a suitable local counterterm to the action. This is not in contraddiction with the theorem in [1]. Indeed, in order to maintain the Wess – Zumino gauge, the susy transformations must be implemented with suitable, compensating, gauge transformations and the latter are anomalous.

Coming back to SYMTs coupled to SUGRA,an open question is if, in the local case, the susy partner Δ_S is a coboundary of Ω_S or not. The fact that, up to now, no consistent, susy preserving, regularization has been found, could lead one to conjecture that the right answer is the negative one. However, Δ_S vanishes together with Δ_G . Therefore even if, in case, Δ_S turns out to be non trivial,it never affects a theory free from gauge anomalies.

6. *Supersymmetric gauge anomaly in higher dimensions.*

Extended transgression formulas in higher, even, dimensions hold in space-time as well as in superspace. In Σ_D, (D=2n-2), one has:

$$sTr(\mathcal{F}^n) = n\,\hat{d}_s \int_0^1 dt\; sTr(\hat{\varphi}, \hat{\mathcal{F}}_t^{n-1}) \equiv \hat{d}_s\, Q_{2n-1}$$
$$\left(sTr(\mathcal{F}^n) = s\,Tr(\underbrace{\mathcal{F},\ldots\mathcal{F}}_{n-\text{times}}), \text{etc}\right) \tag{54}$$

where the notations of sect. 3,4 are used. In particular φ is the $\mathfrak{g}$-valued gauge superconnection in D dimensions and $\mathcal{F}$ is its curvature with the constraints

$$\mathcal{F}_{\underline{\alpha}\underline{\beta}} = 0 \tag{55}$$

We are interested in supersymmetric theories in 6 and 10 dimensions. In 6 dimensions [22] the spinor indices are $\underline{\alpha} \equiv (\alpha, i)$; $\alpha = 1,..,4$; $i=1,2$; the space-time ones, a , are represented by an antisymmetric pair of Greek indices: $a = [\alpha, \alpha']$ - and the rigid torsion is $\sigma_{a\underline{\beta}\underline{\gamma}} = i\varepsilon_{\alpha\alpha'\beta\gamma}\;\varepsilon_{ij}$. In 10 dimensions, the supersymmetric charges form a left handed Weyl-Majorana spinor so that dotted spinor indices do not arise ($\underline{\alpha} = \alpha$; $\alpha = 1,\ldots,16$) and the rigid torsion is $\sigma^a{}_{\alpha\beta} = (\gamma^a C)_{\alpha\beta}$ (γ^a is a set of Dirac matrices in D=10 and C is the charge conjugation). A difference w.r.t. 4D is that here there is more than one invariant polynomial with n entries. In 6D, in addition to $Tr(\mathcal{F}^4)$ one has also $(Tr(\mathcal{F}^2))^2$ and in 10D in addition to $Tr(\mathcal{F}^6)$ one has $(Tr(\mathcal{F}^2))^3$; $Tr(\mathcal{F}^2)\,Tr(\mathcal{F}^4)$ and $(Tr(\mathcal{F}^3))^2$. For this new class of invariants, instead of the ETF corresponding to eq. (5.4) one can consider the simplified ETFs (which give equivalent answers)

$$sTr(\mathcal{F}^{m_1}) \ldots\ldots sTr(\mathcal{F}^{m_k}) =$$
$$= m_1\,\hat{d}_s \int_0^1 dt\; sTr(\hat{\varphi}, \hat{\mathcal{F}}_t^{m_1-1})\, sTr(\mathcal{F}^{m_2}) \ldots sTr(\mathcal{F}^{m_k}) \tag{56}$$

where $\sum m_i = n$; $m_i \geq 2$.

Moreover in these dimensions, also the Lorentz group is anomalous, so that supergravitational anomalies are expected. In addition, in D=10, the curvature constraints eq. (5.4), imply the field equations so that the anomaly to be found will be on shell. In principle, the same algebraic method developed in sect. 4 to get the anomaly for D=4, can be performed also in D=6,10. The only difficulty is to extend in D=6,10 the relation given by eq. (32) in 4D. This problem has been solved for D=6 [13]; for D=10 the work is in progress. Indeed in D=6 we have proved the relations

$$s\mathrm{Tr}\left(\mathcal{F}^4\right) = \frac{1}{5} d_s\left\{4\hat{S}\,\mathrm{Tr}\left(\mathcal{F}^4\right) + \hat{S}\,\mathrm{Tr}\left(G^4\right) - 4\hat{S}\hat{D}\hat{S}\,\mathrm{Tr}\left(G^4\right)\right\} \equiv \tag{57}$$

$$\equiv d_s \chi_7^{(A)}$$

$$\mathrm{Tr}\left(\mathcal{F}^2\right)\mathrm{Tr}\left(\mathcal{F}^2\right) = \frac{1}{5} d_s\left\{4\hat{S}\left(\mathrm{Tr}\left(\mathcal{F}^2\right)\right)^2 + \hat{S}\left(\mathrm{Tr}\left(G^2\right)\right)^2 - 4\hat{S}\hat{D}\hat{S}\left(\mathrm{Tr}\left(G^2\right)\right)^2\right\} =$$

$$\equiv d_s \chi_7^{(B)} \tag{58}$$

where $\hat{S} = \frac{1}{12} S$, $\hat{D} = \bar{\epsilon}^{\alpha} D_{\alpha}$ and G and S are defined in eqs (34)(35). Then the relevant ETFs can be written as:

$$\hat{d}_s \tilde{Q}_7^{(i)} = 0 \quad ; \quad (i = A, B) \tag{59}$$

where

$$\tilde{Q}_7^{(i)} = Q_7^{(i)} - \chi_7^{(i)} \tag{60}$$

$$Q_7^{(A)} = 4\int_0^1 dt\, s\mathrm{Tr}\left(\hat{\varphi}, \hat{\mathcal{F}}_t^3\right) \tag{61}$$

$$Q_7^{(B)} = 2\int_0^1 dt\; s\left[\mathrm{Tr}\left(\hat{\varphi}, \hat{\mathcal{F}}_t\right)\mathrm{Tr}\left(\mathcal{F}^2\right)\right] \tag{62}$$

The two anomalies $\Delta^{(A)}$ and $\Delta^{(B)}$ in D=6 follow from eqs.(59)-(64) by repeating the procedure of sect. 4. In the rigid case they are

$$\Delta^{0(i)} = \Delta_6^{0(i)} + \Delta_5^{0(i)} \tag{63}$$

where, using the notations introduced in sect. 5 eqs. (50) ,

$$\Delta_G^{0(A)} = 12 \int_{M_6} \int_0^1 dt\, (1-t)\, s\mathrm{Tr}\left[(dc, A, F_t^2) \right] \Big|_{\theta=\bar\theta=0} \tag{64a}$$

$$\Delta_S^{0(A)} = -4 \int_{M_6} \int_0^1 dt\, i_\xi\, s\mathrm{Tr}\Big[(\lambda, F_t^3) + 3(A, G_t, F_t^2) - \frac{4}{5} \hat{s}(F, G^3) + \frac{1}{5} \hat{s}\hat{D}\hat{s}(G^4) \Big] \Big|_{\theta=\bar\theta=0} \tag{64b}$$

$$\Delta_S^{0(B)} = \int_{M_6} \mathrm{Tr}(dcA)\,\mathrm{Tr}(F^2) \Big|_{\theta=\bar\theta=0} \tag{64c}$$

$$\Delta_S^{0(B)} = -4 \int_{M_6} \int_0^1 dt\, i_\xi \Big\{ \frac{1}{2}\left[\mathrm{Tr}(\lambda F_t) + \mathrm{Tr}(A G_t) \right] \mathrm{Tr}(F^2) + 2\,\mathrm{Tr}(A F_t)\,\mathrm{Tr}(GF) + - \frac{4}{5} \hat{s}\left[\mathrm{Tr}(FG)\,\mathrm{Tr}(G^2) \right] + \frac{1}{5} \hat{s}\hat{D}\hat{s}\left(\mathrm{Tr}(G^2) \right)^2 \Big\} \Big|_{\theta=\bar\theta=0} \tag{64d}$$

Eqs (63) (64), taken in Wess-Zumino gauge reproduce the result of [11].

7. N=1, D=4 superconformal anomalies.

This section deals with a different subject. Following [23], our aim is to point out the relation between superconformal anomalies and the supersymmetric non renormalization theorem. In Wess-Zumino models this relation has been also discussed by Tsao [24]. Renormalizable N=1, D=4 supersymmetric theories are invariant at the classical level and modulo soft terms under superconformal transformations. The superconformal algebra involves the following generators: P_a (translations), M_{ab} (Lorentz), D (dilatation), R (chirality), K_a (conformal), $Q_{\underline{\alpha}}$ (susy), $\hat{Q}_{\underline{\alpha}}$ (conformal susy).

They can be defined in terms of the improved stress energy tensor $\theta_{mn}(x)$, the spinor susy current $j_m^{\underline{\alpha}}(x)$ and the chiral current $j^5_{m}(x)$. $\theta_{mn}(x)$, $j_m^{\underline{\alpha}}(x)$ and $j^5_m(x)$ are the components of the real current supermultiplet [25] $V_{\alpha\dot\beta}(z) = \sigma^a_{\alpha\dot\beta} V_a(z)$ with the conservation equations

$$D_\alpha V^{\alpha\dot\beta} = 0 = \bar{D}_{\dot\beta} V^{\alpha\dot\beta} \tag{65}$$

which hold at the classical level modulo soft terms. Eq. (65) is broken by the anomalies and explicit model calculations give

$$\bar{D}_{\dot\beta} V^{\alpha\dot\beta}(z) = D^\alpha S(z) \quad ; \quad D_\alpha V^{\alpha\dot\beta}(z) = \bar{D}^{\dot\beta}\bar{S}(z)$$

where $S(z)$ is a chiral superfield and $\bar{S}(z) = S^+(z)$ is antichiral.

$$D_\alpha S\big|_{\theta=\bar\theta=0} \; ; \; \bar{D}_{\dot\alpha}\bar{S}\big|_{\theta=\bar\theta=0} \; ; \; (DDS - \bar{D}\bar{D}\bar{S})\big|_{\theta=\bar\theta=0} \; ; \; (DDS + \bar{D}\bar{D}\bar{S})\big|_{\theta=\bar\theta=0}$$

are the anomalies of $\hat{Q}_\alpha$, $\hat{Q}_{\dot\alpha}$, the chiral anomaly and the dilatation (or trace) anomaly respectively.

The result that the superconformal anomalies belong to chiral superfields can be obtained on general ground through cohomological methods [23]. In order to take care also of anomalies which are total derivatives (and this is always the case for the chiral anomaly [26]) one must work at the local level. Therefore we couple our renormalizable supersymmetric theory to external N=1 supergravity by the standard recipe. Then the theory at the classical level and modulo soft terms becomes invariant under a) superdiffeomorphisms, b) superLorentz transformations c) superWeyl transformations and we can study the coupled cohomology for these three symmetries.

The superWeyl gauge parameters are chiral, $\rho(z)$, and antichiral, $\bar\rho(z) = \rho^+(z)$, superfields. Let us define

$$\lambda(x) + i\alpha(x) = \rho(z)\big|_{\theta=\bar\theta=0} \; ; \; \xi_\alpha(x) = D_\alpha\rho(z)\big|_{\theta=\bar\theta=0} \; ; \; \bar\xi_{\dot\alpha}(x) = \bar{D}_{\dot\alpha}\bar\rho(z)\big|_{\theta=\bar\theta=0}$$

The parameters $\lambda(x), \alpha(x), \xi_\alpha(x), \bar\xi_{\dot\alpha}(x)$ are related, in the rigid limit, to dilatation D, chirality R and conformal susy $\hat{Q}_\alpha$, $\hat{Q}_{\dot\alpha}$. Dilatation and conformal susy receive contributions also from superdiffeomorphisms. We shall call $\sigma(z)$ $(\bar\sigma(z))$ the chiral (antichiral) F.P. ghosts related to $\rho(z)$ $(\bar\rho(z))$.

Under some technical assumptions it has been proved in [23] that, in 4D, and for theories free from gauge anomalies, superdiffeomorphisms and superLorentz transformations are not affected by anomalies. Therefore the anomalies of the coupled cohomology under study are merely the superWeyl ones and look as follows

$$\Delta_W = \int d^4x\, d^2\theta\, \sigma(z)\, S(z) + \int d^4x\, d^2\bar\theta\, \bar\sigma(z)\, \bar S(z)$$

$S(z)$ and $\bar S(z) = S^\dagger(z)$ are chiral and antichiral superfields which can be classified.

In conclusion the important result found by explicit calculations is confirmed : dilatation, chiral and susy conformal anomalies belong to the same chiral (antichiral) supermultiplet. This result allows us to get the relation between superconformal anomalies and the susy non renormalization theorem. This theorem says that chiral and antichiral action terms like

$$I = \int d^4x\, d^2\theta\, \mathcal{L}_K \quad ; \quad \bar I = \int d^4x\, d^2\bar\theta\, \bar{\mathcal{L}}_K$$

where $\mathcal{L}_K$ $(\bar{\mathcal{L}}_K)$ are genuine chiral (antichiral) superfields, do not require divergent counterterms. (By a genuine chiral (antichiral) superfield we means a chiral (antichiral) superfield which cannot be written as $DD\varphi$ $(\bar D\bar D\varphi)$) .

Therefore only action terms which are integrals over the full superspace, require renormalization.

As already noted, the chiral anomalies are total divergences so that

$$DDS - \bar D\bar D\bar S = \partial_m K^m \tag{66}$$

Eq. (66) can be satisfied only if

$$S = \bar D\bar D T$$

where T is a real superfield. Therefore the trace anomaly is

$$DDS + \bar D\bar D\bar S = \left(DD\bar D\bar D + \bar D\bar D DD\right)T = \int d^4\theta\, T$$

From the well known relation between trace anomalies (integrated over space-time) and divergent counterterms, the non renormalization theorem follows.

Acknowledgements

I am very grateful to S. Ferrara, L. Girardello, R. Grimm, E. Guadagnini, K. Konishi, M. Mintchev, O. Piguet and, especially, R. Stora, for fruitful discussions and communication of their results prior to publication. Special thanks are due to L. Bonora and P. Pasti for many useful discussions and for advices and help during the preparation of this report.

Footnote

(+) In the first paper of ref. [7] we wrote down inaccurately the coefficients of Δ^a_5, in eq. (4.14).

References

[1] O. Piguet, K. Sibold and M. Schweda, Nucl. Phys. B174, (1980), 183.
[2] O. Piguet and K. Sibold, Nucl. Phys. B247 (1984), 484.
[3] S. Ferrara, L. Girardello, O. Piguet and R. Stora, Phys. Lett. 157B (1985), 179.
[4] N.K. Nielsen, Nucl. Phys. B244 (1984), 499.
E. Guadagnini, K. Konishi and M. Mintchev, Phys. Lett. 157B (1985), 37.
M. Pernici and M. Riva, preprint ITP-5B-85-10.
K. Harada and K. Shituya, Tohoka, preprint TH-85-280.
R. Garreis, M. Schobl and J. Wess, Karlsruhe preprint, RATHEP-85-4.
[5] G. Girardi, R. Grimm and R. Stora, Phys. Lett. 156B (1985), 203.
[6] L. Bonora, P. Pasti and M. Tonin, Phys. Lett. 156B (1985), 341.
[7] L. Bonora, P. Pasti and M. Tonin, Nucl. Phys. B261, (1985) 241.
L. Bonora, P. Pasti and M. Tonin, Symposium on Anomalies Geometry and Topology, Chicago (1985).
[8] L. Bonora, P. Pasti and M. Tonin, preprint DFPD 21/85.
P. Pasti, Proceedings of the Workshop on Perturbative Methods, Montpellier 1985.
L. Bonora, Proceedings of 14th ICGTHP - Seul 1985.
[9] L.N. Mc Arthur and H. Osborn, preprint DAMPT 85-27.
[10] H. Itoyama, V.P. Nair and H.C. Ren, IAS preprint March 85.
H. Itoyama, V.P. Nair and H.C. Ren, FERMILAB-PUP 85-139.
[11] E. Guadagnini and M. Mintchev, Pisa preprint IFUP-TH 28/85.
[12] M. Porrati, Pisa preprint IFUP-TH 22/85.
[13] L. Bonora, P. Pasti and M. Tonin, in preparation.
[14] L. Bonora and P. Cotta Ramusino, Phys. Lett. 107B (1981), 87.
[15] R. Stora, Cargèse lectures (1983).
[16] B. Zumino, Les Houches lecture (1983).
B. Zumino, Y.S. Wu and A. Zee, Nucl. Phys. B239 (1984), 477.

[17] B. Zumino, Nucl. Phys. B253 (1985) 477.
L. Baulieu, Nucl. Phys. B241 (1984) 557.
L. Bonora, P. Pasti, Phys. Lett. 123B (1983) 75.
F. Langouche, T. Schucker and R. Stora, Phys. Lett. 145B (1984) 342.

[18] C. Becchi, A. Rouet and R. Stora, Ann. of Phys. 58 (1976) 287.
C. Becchi, A. Rouet and R. Stora, in "Field Theory Quantization and Statistical Physics", E. Trapeguei editor (1981).

[19] L. Bonora and M. Tonin, Phys. Lett. 98B, 48 (1981).

[20] J. Wess and Bagger "Supersymmetry and Supergravity" Princeton Univ. Press 1983.

[21] W.A. Bardeen and B. Zumino, Nucl. Phys. B244 (1984), 421.

[22] P.S. Howe, G. Sierra and T.K. Townsend, Nucl. Phys. B221 (1983),231.
J. Koller, Nucl. Phys. B222 (1983) 319.

[23] L. Bonora, P. Pasti and M. Tonin: Nucl. Phys. B252 (1985), 458.

[24] H.S. Tsao, Phys. Lett. B53 (1974) 351.

[25] S. Ferrara and B. Zumino, Nucl. Phys. B87 (1975), 207.

[26] V.S. Varadarajan: "Lie Group, Lie Algebra and their Representation" Springer Verlag, New York (1984).

INTERPLAY BETWEEN CHIRAL ANOMALY AND SUPERSYMMETRY

E. Guadagnini

and

M. Mintchev

Dipartimento di Fisica, Univesità di Pisa
Istituto Nazionale di Fisica Nucleare, Sezione di Pisa

Abstract: We discuss the structure of the chiral anomaly in supersymmetric gauge theories in both the manifestly SUSY covariant approach and the Wess-Zumino gauge. In four dimensions we derive the SUSY anomaly in the W-Z gauge from the SUSY invariant expression of the chiral anomaly. We give a prescription for the computation of the SUSY anomaly in higher dimensions and present the resulting expression in six dimensions.

Introduction

This talk is devoted to some aspects [1,2,3] of the chiral anomalies in supersymmetric (SUSY) gauge theories. More precisely, we consider the new features stemming from the relationship between chiral and SUSY transformations. We start with the four dimensional case and derive the anomaly content of the theory in the Wess-Zumino (W-Z) gauge from the chiral anomaly [4] obtained in the manifestly SUSY covariant approach. Afterwards we discuss a prescription for the computation of the anomalies in higher dimensions and present the resulting expression in six dimensions. Finally, we comment on some physical cosequences of the anomalous Ward identities in SUSY gauge theories.

It is well known that there are two ways of formulating SUSY gauge theories in four dimensions. The manifestly SUSY covariant approach, called in what follows ξ-SUSY picture, uses the vector superfield $V(x,\theta,\theta) = V^a(x,\theta,\theta)T^a$. The matter coupling has the form#1

$$S = \int d^8z\ \Phi_i^{+} (e^{2V})_{ij}\ \Phi_j \quad . \tag{1}$$

Here $\Phi = \{\Phi_i\} = \{(A + \sqrt{2}\,\theta\psi + \theta\theta F)_i\}$ is a chiral matter supermultiplet containing left-handed fermions ψ and transforming under an irreducible representation of a compact group G with hermitian generators T^a_{ij}. In this picture the rigid SUSY transformations, generated by $s(\xi) = \xi Q + \bar{\xi}\bar{Q}$ (see [5]), are not anomalous [6]. Let $\Gamma(V)$ be the regularized generating functional obtained from the action (1) after integration on Φ and Φ^+. The SUSY extension of the usual gauge transformations is

---

#1 In four dimensions we adopt the notation and conventions of Ref.5.

$$\delta_\Lambda e^{2V} = -i\Lambda^+ e^{2V} + ie^{2V}\Lambda \quad , \qquad (2)$$

where $\bar{D}\Lambda = 0$. By using the heat kernel technique one finds [4]

$$s(\xi)\Gamma(V) = 0 \quad , \qquad (3)$$

$$\delta_\Lambda \Gamma(V) = -i\int d^8z \; tr\{[\Lambda(z) - \Lambda^+(z)]G(z)\} \quad , \qquad (4)$$

where

$$G(z) = (1/32\pi^2)\int_0^{+\infty} dt \; \{ \partial^\mu[itLB_\mu L^2 - t^2L(\partial_\mu L)L] - 2t^2(\partial_\mu L)L(\partial^\mu L)$$
$$- 2LCL - iLB^\mu(\partial_\mu L) + it[(\partial_\mu L)LB^\mu L - LB^\mu L(\partial_\mu L)] - (1/2)LB_\mu LB^\mu L\}$$
$$(5a)$$

and

$$L = (e^{2V} + t)^{-1} \quad , \qquad B^\mu = -(1/2)(D^\alpha e^{2V})\sigma^\mu_{\alpha\dot{\alpha}}\bar{D}^{\dot{\alpha}} \quad ,$$

$$C = (1/16)(D^2e^{2V})\bar{D}^2 \quad . \qquad (5b)$$

The expression $\delta_\Lambda\Gamma$ can be given different forms [4,7-9] which are all equivalent due to the uniqueness theorem of Ref.10. It is worth stressing, however, that all of them are non-polinomial in e^{2V}. This is a characteristic feature of $\delta_\Lambda\Gamma$ which cannot be avoided modifying Γ by local counterterms [11]. The reason for this fact is in the presence of dimensionless components in V.

Although the recent progress in analizing $\delta_\Lambda\Gamma$ and the corresponding W-Z term [8,9,12,19], formulas like (5a) are always difficult to handle with and, which is more important, cannot be directly generalized to higher dimensions.

The second way of formulating SUSY gauge theories, called in what follows ε-SUSY picture, is to use the W-Z gauge [5]

$$S_{W-Z} = \int d^4x [\,|\partial_\mu A + iA_\mu A|^2 - i\overline{\Psi}\,\bar{\sigma}^\mu(\partial_\mu + iA_\mu)\Psi + F^+F$$

$$+ i\sqrt{2}(A^+T^a\overline{\Psi}\lambda^a - \overline{\lambda}^a T^a A\Psi) + D^a A^+ T^a A] \quad . \qquad (6)$$

In this picture, only the components (A_μ, λ, D) of V are present as external fields; in particular, the dimensionless components of V are absent. Consequently, the couplings with the external field in S_{W-Z} are of the standard type only.

We are interested now in the transformation properties of $\Gamma_{W-Z}(A_\mu, \lambda, D)$ under the local gauge transformations

$$\delta_\alpha A_\mu = \partial_\mu \alpha + i[A_\mu, \alpha] \equiv D_\mu \alpha \quad , \qquad (7a)$$

$$\delta_\alpha \lambda = i[\lambda, \alpha] \quad , \qquad (7b)$$

$$\delta_\alpha D = i[D, \alpha] \quad , \qquad (7c)$$

and under the ε-SUSY transformations

$$\delta_\varepsilon A_\mu = i\bar{\varepsilon}\bar{\sigma}_\mu\lambda - i\,\bar{\lambda}\bar{\sigma}_\mu\varepsilon \quad , \qquad \delta_\varepsilon\lambda = \sigma^{\mu\nu}\varepsilon F_{\mu\nu} + i\varepsilon D \quad , \qquad (8a)$$

$$\delta_\varepsilon D = -\varepsilon\sigma^\mu D_\mu\bar{\lambda} - D_\mu\lambda\,\sigma^\mu\bar{\varepsilon} \quad , \qquad (8b)$$

where

$$F_{\mu\nu} = \partial_\mu A_\nu - \partial_\nu A_\mu + i[A_\mu, A_\nu] \quad .$$

As easily seen, the transformations δ_α and δ_ε leave invariant the classical action S_{W-Z}. This is not the case for Γ_{W-Z}. It is well known [13] that

$$\delta_\alpha \Gamma_{W-Z} = \int d^4x\ \alpha^a(x)\,\mathcal{A}^a(x) \quad , \qquad (9)$$

$\mathcal{A}^a(x)$ being the chiral anomaly. In order to get some information on $\delta_\varepsilon\Gamma_{W-Z}$, we recall [14] that Eqs.(8) imply

$$\delta_{\varepsilon_1}\delta_{\varepsilon_2} - \delta_{\varepsilon_2}\delta_{\varepsilon_1} = -2i\varepsilon_{12}^{\nu}\,\hat{D}_{\nu} \quad . \tag{10}$$

Here the "covariant derivative" $\hat{D}_{\mu}$ is defined by

$$\hat{D}_{\mu}X(x) = \partial_{\mu}X(x) - \int d^4y\, A_{\mu}^{a}(y)[\delta/\delta\alpha^{a}(y)]X^{\alpha}(x)$$

and $X^{\alpha}(x) = X(x) + \delta_{\alpha}X(x)$. By definition, $\hat{D}_{\mu}$ coincides with the usual covariant derivative when applied to fields transforming covariantly under local gauge transformations, whereas $\hat{D}_{\mu}A_{\nu} = F_{\mu\nu}$. Combining Eq.(9) and Eq.(10) one finds the consistency condition

$$(\delta_{\varepsilon_1}\delta_{\varepsilon_2} - \delta_{\varepsilon_2}\delta_{\varepsilon_1})\Gamma_{W-Z} = 2i\varepsilon_{12}^{\nu}\int d^4x\, A_{\nu}^{a}(x)\,\mathcal{A}^{a}(x) \quad . \tag{11}$$

An immediate consequence of Eq.(11) is that the presence of a chiral gauge anomaly implies that the ε-SUSY transformations are anomalous. This fact has a simple explanation: roughly speaking, because of the chiral anomaly A_{μ}^{a} represents (for fixed a) actually three (two and one half in a more sophisticated counting) bosonic physical degrees of freedom which do not match with the two fermionic degrees of λ^{a}. Clearly, at this level the absence of chiral anomalies does not imply $\delta_{\varepsilon}\Gamma_{W-Z} = 0$.

Finally, the commutativity of δ_{α} and δ_{ε} gives rise to

$$\delta_{\varepsilon}\delta_{\alpha}\Gamma_{W-Z} = \delta_{\alpha}\delta_{\varepsilon}\Gamma_{W-Z} \quad . \tag{12}$$

The consistency conditions (11,12), supplemented by the standard W-Z consistency conditions [15] or, equivalently, by the knowledge of $\mathcal{A}^{a}(x)$, contain the whole information about the ε-SUSY anomaly. We will say that the couple $\{\delta_{\varepsilon}\Gamma_{W-Z}, \delta_{\alpha}\Gamma_{W-Z}\}$ is in a standard form when $\mathcal{A}^{a}(x)$ is given just by the standard Adler-Bell-Jackiw (A-B-J) expression (see e.g. [16]). Actually, this definition does not fix uniquely $\delta_{\varepsilon}\Gamma_{W-Z}$. One still has the freedom to modify $\delta_{\varepsilon}\Gamma_{W-Z}$ by adding gauge invariant local counterterms to Γ_{W-Z}.

It is worth stressing that the presence of ε-SUSY anomalies is not in contradiction with the absence of ξ-SUSY anomalies; the two types of SUSY transformations are in fact different. The ε-SUSY anomaly does not represent a new anomaly in the following sense: it is strictly related to the usual chiral anomaly in the ξ-picture and is actually contained already in $\delta_\Lambda \Gamma$. We elaborate on this point in the next section.

2. The ε-SUSY Anomaly in Four Dimensions

Instead of emploing the consistency conditions (11,12) to find $\delta_\varepsilon \Gamma_{W-Z}$, in four dimensions we use the already known form of $\delta_\Lambda \Gamma$, Eq.(4). We have some reasons for doing this. Firstly, we want to clarify the relationship between the two SUSY pictures. In particular, we will see that a naive reduction of the expression (4) to the W-Z gauge is incorrect. Secondly, we get a further insight into the structure of Eq.(4).

$\{\delta_\varepsilon \Gamma_{W-Z}, \delta_\alpha \Gamma_{W-Z}\}$ and $\delta_\Lambda \Gamma$ are related as follows. As well known [5], S_{W-Z} is obtained from S by the substitution

$$V \longmapsto V_{W-Z} = -\theta\sigma^\mu\bar{\theta}\, A_\mu(x) + i\,\theta\theta\bar{\theta}\bar{\lambda}(x) - i\,\bar{\theta}\bar{\theta}\theta\,\lambda(x)$$

$$+ (1/2)\,\theta\theta\bar{\theta}\bar{\theta}\, D(x) \quad . \tag{13}$$

The α-gauge transformations (7) on the components of V_{W-Z} are realized by a Λ-transformation (2) with

$$\Lambda_{(\alpha)}(y) = \alpha(y)\big|_{y_\mu = x_\mu + i\,\theta\sigma_\mu\bar{\theta}} \quad . \tag{14}$$

$\delta_{\Lambda_{(\alpha)}} V_{W-Z}$ is still in the W-Z gauge. Therefore

$$\delta_\alpha \Gamma_{W-Z}(A_\mu, \lambda, D) = \delta_{\Lambda(\alpha)} \Gamma(V_{W-Z}) \quad . \qquad (15)$$

Let us consider now the ε-SUSY transformations. The linear SUSY transformation $s(\varepsilon)$ applied to V_{W-Z} brings us out of the W-Z gauge. However, this can be compensated by a suitably choosen Λ-transformation with parameter

$$\Lambda_{(\varepsilon)}(y) = -2i\,\theta\sigma^\mu\bar{\varepsilon}A_\mu(y) - 2\theta\theta\,\bar{\varepsilon}\,\bar{\lambda}(y)\big|_{y_\mu = x_\mu + i\,\theta\sigma_\mu\bar{\theta}} \quad . \qquad (16)$$

Finally, as one can easily verify in components, one has

$$\delta_\varepsilon = s(\varepsilon) + \delta_{\Lambda(\varepsilon)} \quad . \qquad (17)$$

Consequently

$$\delta_\varepsilon \Gamma_{W-Z}(A_\mu, \lambda, D) = \delta_{\Lambda(\varepsilon)} \Gamma(V_{W-Z}) \quad , \qquad (18)$$

since $s(\varepsilon)\Gamma = 0$, [6].

At this point one can use Eqs.(4) and (5) to compute $\{\delta_\varepsilon\Gamma_{W-Z}, \delta_\alpha\Gamma_{W-Z}\}$. In order to give this couple of anomalies a standard form, we subtract from Γ_{W-Z} the local counterterm

$$C = (1/192\pi^2)\int d^4x\, \mathrm{tr}[12A_\mu\lambda\sigma^\mu\bar{\lambda} - 4(\partial_\mu A^\mu)^2 + 2(\partial_\mu A_\nu)(\partial^\mu A^\nu) + A_\mu A_\nu A^\mu A^\nu - 2(A_\mu A^\mu)^2 - 2F_{\mu\nu}F^{\mu\nu}] \quad . \qquad (19)$$

With this prescription one obtains

$$\delta_\varepsilon\Gamma_{W-Z} = i(1/48\pi^2)\int d^4x\, \mathrm{tr}\{(\bar{\varepsilon}\bar{\sigma}^\mu\lambda - \bar{\lambda}\bar{\sigma}^\mu\varepsilon)[3(\bar{\lambda}\bar{\sigma}_\mu\lambda) - \varepsilon_{\mu\nu\varrho\tau}(2A^\nu(\partial^\varrho A^\tau) + 2(\partial^\nu A^\varrho)A^\tau + 3iA^\nu A^\varrho A^\tau)]\} \quad , \qquad (20)$$

$$\delta_\alpha\Gamma_{W-Z} = (1/48\pi^2)\varepsilon^{\mu\nu\varrho\tau}\int d^4x\, \mathrm{tr}[\partial_\mu\alpha(2A_\nu\partial_\varrho A_\tau + iA_\nu A_\varrho A_\tau)] \quad . \qquad (21)$$

Eqs.(20,21) represent a standard form of the anomalies in the ε-picture. Several comments are in order:

(i) One can directly verify that the term

$$\varepsilon_{\mu\nu\rho\tau}\ \mathrm{tr}\{(\bar{\varepsilon}\bar{\sigma}^{\mu}\lambda - \bar{\lambda}\bar{\sigma}^{\mu}\varepsilon)[A^{\nu}(\partial^{\rho}A^{\tau}) + (\partial^{\nu}A^{\rho})A^{\tau}]\} \qquad (22)$$

appearing in $\delta_{\varepsilon}\Gamma_{W-Z}$ cannot be eliminated by any local counterterm. This confirms the previous conclusion based on the consistency condition (11), namely, that chiral anomalies imply ε-SUSY anomalies.

(ii) $\{\delta_{\varepsilon}\Gamma_{W-Z}, \delta_{\alpha}\Gamma_{W-Z}\}$ does not depend on the auxiliary field D. This fact has some consequences for the anomalies related to the ε-SUSY transformations defined without auxiliary fields.

(iii) We recall [4] that the A-B-J anomaly is contained in the $\theta\bar{\theta}$-component of $G^{a}(z)$. Eqs.(18) and (20) show that also the other components of $G^{a}(z)$ contain anomalous (i.e. not eliminable) terms.

(iv) $\delta_{\varepsilon}\Gamma_{W-Z}$ is proportional to $\mathrm{tr}T^{a}\{T^{b},T^{c}\} = d^{abc}$. Therefore, $d^{abc} = 0$ garantees the absence of both chiral and ε-SUSY anomalies. This is also the case in the ξ-picture [9].

(v) We have verified that (20) and (21) satisfies the consistency conditions (11),(12).

(vi) $\{\delta_{\varepsilon}\Gamma_{W-Z}, \delta_{\alpha}\Gamma_{W-Z}\}$, Eqs.(20,21), represent the minimal form of the anomalies in SUSY gauge theories since ε and α(x) are the minimal number of parameters associated with the symmetries of these theories.

(vii) Clearly, the ξ- and the ε-pictures are inequivalent at the quantum level. We have seen that the ε-SUSY transformations are anomalous because of the mismatch between bosonic and fermionic degrees of freedom. This is not the case for ξ-SUSY transformations since the whole superfield V contains the right degrees of freedom to describe a massive real (vector) supermultiplet. Just because of the difference in the number of external degrees of freedom, the ξ- and ε-SUSY transformations are different. So it would be incorrect to work out the expression (5) in the W-Z gauge and neglect the terms which are not proportional to the parameter $\alpha(x)$ of local gauge transformations.

We can summarize the content of this section as follows. In supersymmetric theories, whatever picture we adopt, the chiral anomalies involve new structures in addition to the standard A-B-J expression. One advantage of the ε-picture is that formulas (20,21) can be generalized to higher dimensions. This will be our main task in the next section.

3. ε-SUSY Anomalies in Higher Dimensions

In this section we describe a procedure for the computation of the ε-SUSY anomalies in d = 2n dimensions, which is based on the consistency conditions. Our conventions for the Γ-matrices are the following:[#2]

$$\{\Gamma_\mu,\Gamma_\nu\} = -2\eta_{\mu\nu} \quad , \quad \text{diag}\eta = (-,+,\ldots,+) \quad , \quad \Gamma_{d+1} = \Gamma_0\ldots\Gamma_{d-1}$$

The ε-SUSY transformations are defined by

$$\delta_\varepsilon A_\mu = i(\bar\lambda\Gamma_\mu\varepsilon - \bar\varepsilon\Gamma_\mu\lambda) \quad , \tag{23a}$$

$$\delta_\varepsilon\lambda = (1/2)\Gamma_{\mu\nu}\varepsilon F^{\mu\nu} \quad , \quad \delta_\varepsilon\bar\lambda = -(1/2)\bar\varepsilon\Gamma_{\mu\nu}F^{\mu\nu} \tag{23b}$$

and commute with δ_α, Eqs.(7).

#2 We work in the 2n-dimensional Minkowski space M_{2n}, since we need for n = 5 Majorana-Weyl spinors.

As well known [17], for n = 1, 2, 3 and 5 the algebra (23) closes on-shell (in practice, one needs $\not{D}\lambda = 0$ only), i.e.

$$\delta_{\varepsilon_1}\delta_{\varepsilon_2} - \delta_{\varepsilon_2}\delta_{\varepsilon_1} = -2i\varepsilon_{12}^{\nu}\,\hat{D}_{\nu} \quad , \tag{10}$$

where

$$\varepsilon_{12}^{\mu} = \bar{\varepsilon}_2\Gamma^{\mu}\varepsilon_1 - \bar{\varepsilon}_1\Gamma^{\mu}\varepsilon_2 \quad .$$

Therefore the consistency conditions (11,12) hold on-shell also in d = 2, 6 and 10 dimensions. Off-shell, Eq.(10) becomes:

$$\delta_{\varepsilon_1}\delta_{\varepsilon_2} - \delta_{\varepsilon_2}\delta_{\varepsilon_1} = -2i\varepsilon_{12}^{\nu}\,\hat{D}_{\nu}$$

$$+ \int d^{2n}x\ \{c_{2n}^{a}(x)[\delta/\delta\lambda^{a}(x)] + \bar{c}_{2n}^{a}(x)[\delta/\delta\bar{\lambda}^{a}(x)]\} \quad , \tag{24}$$

where $c_{2n}(x)$ and $\bar{c}_{2n}(x)$ vanish on-shell. For instance

$$c_4 = -(i/2)\varepsilon_{12}^{\mu}\Gamma_{\mu}\not{D}\lambda + i[\varepsilon_2(\bar{\lambda}\overleftarrow{\not{D}}\varepsilon_1) - \varepsilon_1(\bar{\lambda}\overleftarrow{\not{D}}\varepsilon_2)] \quad , \tag{25a}$$

$$c_6 = -i\varepsilon_{12}^{\mu}\Gamma_{\mu}\not{D}\lambda + i\varepsilon_2[(\bar{\lambda}\overleftarrow{\not{D}}\varepsilon_1) - (\bar{\varepsilon}_1\not{D}\lambda)]$$

$$- i\varepsilon_1[(\bar{\lambda}\overleftarrow{\not{D}}\varepsilon_2) - (\bar{\varepsilon}_2\not{D}\lambda)] \quad . \tag{25b}$$

Consequently, the consistency condition (11) gets modified and takes the form

$$(\delta_{\varepsilon_1}\delta_{\varepsilon_2} - \delta_{\varepsilon_2}\delta_{\varepsilon_1})\Gamma_{W-Z} = 2i\varepsilon_{12}^{\nu}\int d^{2n}x\ A_{\nu}^{a}(x)\mathcal{O}\!\!\!\mathcal{C}^{a}(x) + C_{2n}(\varepsilon_1,\varepsilon_2;\not{D}\lambda) \tag{26}$$

Here C_{2n} is obtained by acting on Γ_{W-Z} with the operator defined by the last two terms in the RHS of Eq.(24). By construction

$$C_{2n}(\varepsilon_1,\varepsilon_2;0) = 0 \quad . \tag{27}$$

In what follows we are not concerned with a particular model realization of the anomalies, but consider thr problem in its abstract form. More precisely, we are looking for a solution $\{{}^1\Omega_{2n}(\varepsilon),\ \omega^1_{2n}(\alpha)\}$ of the consistency conditions

$$\int_{M_{2n}} [\delta_{\varepsilon_1}{}^1\Omega_{2n}(\varepsilon_2) - \delta_{\varepsilon_2}{}^1\Omega_{2n}(\varepsilon_1) + 2i\omega^1_{2n}(\varepsilon^{\mu}_{12}A_{\mu})] = C_{2n}(\varepsilon_1,\varepsilon_2;\not{D}\lambda) \ , \tag{28}$$

$$\int_{M_{2n}} [\delta_{\alpha}{}^1\Omega_{2n}(\varepsilon) - \delta_{\varepsilon}\omega^1_{2n}(\alpha)] = 0 \quad . \tag{29}$$

where [16]

$$\omega^1_{2n}(\alpha) = n(n+1)\int_0^1 dt(1-t)\mathrm{Str}(d\alpha,A,F^{n-1}_t) \quad . \tag{30}$$

and C_{2n} satisfies Eq.(27). Here and in what follows we use the differential form notations adopted in Ref.16, however our A is hermitian, $F_t = tdA + it^2A^2$ and our ω^1_{2n} differs by a factor i^n.

In order to find ${}^1\Omega_{2n}$ we use the following procedure.#3 The first step consists in the integration of (29) in α, which leads to

$${}^1\Omega_{2n}(\varepsilon) = {}^1\omega_{2n}(\varepsilon) + {}^1\Lambda_{2n}(\varepsilon) \quad , \tag{31}$$

where ${}^1\omega_{2n}(\varepsilon)$ is a particular solution of (29) and ${}^1\Lambda_{2n}$ is a gauge invariant unknown function. In the second step one determines ${}^1\Lambda_{2n}$ and C_{2n} by using (28). Note that the general structure of ${}^1\Omega_{2n}$, described by Eq.(31), can be recognized already in Eq.(20), where

---

#3 As usual, we assume that the fields vanish at infinity.

$$^1\Lambda_4(\varepsilon) \sim \mathrm{tr}[(\bar{\varepsilon}\bar{\sigma}^\mu\lambda - \bar{\lambda}\bar{\sigma}^\mu\varepsilon)(\bar{\lambda}\bar{\sigma}_\mu\lambda)] \quad .$$

The form $^1\omega_{2n}(\varepsilon)$ has been found in a different context in Ref.18 and reads

$$^1\omega_{2n}(\varepsilon) = n(n+1)\int_0^1 dt\ t\ \mathrm{Str}(A,\delta_\varepsilon A,F_t^{n-1}) \quad . \tag{32}$$

By using the explicit form of $^1\omega_{2n}(\varepsilon)$ one obtains [1,3] from Eq.(28)

$$\int_{M_{2n}}[\delta_{\varepsilon_1}{}^1\Lambda_{2n}(\varepsilon_2) - \delta_{\varepsilon_2}{}^1\Lambda_{2n}(\varepsilon_1) + n(n+1)\mathrm{Str}(\delta_{\varepsilon_1}A,\delta_{\varepsilon_2}A,F^{n-1})] = C_{2n} \tag{33}$$

Let us make some remarks on the term C_{2n}:
(a) As already observed, $^1\Omega_4$ does not depend on the auxiliary field D. Therefore one can apply the ε-SUSY transformations (23) to the couple (20,21). In this case, Eq.(28) is satisfied with $C_4 = 0$.
(b) In higher dimensions the existence of a nontrivial term C_{2n} is an open possibility. In principle, this term may survive also when $\omega^1_{2n} = 0$.

We present now the result for $^1\Omega_6$. The term $^1\omega_6$ is easily derived from (32) and reads

$$^1\omega_6(\varepsilon) = \mathrm{tr}\{\delta_\varepsilon A[(dA)^2A + (dA)A(dA) + A(dA)^2$$
$$+ i(8/5)(A^3dA + dAA^3) + i(4/5)(A^2dAA + AdAA^2) - 2A^5]\} \quad . \tag{34}$$

For $^1\Lambda_6$ we get

$$^1\Lambda_6(\varepsilon) =$$

$$= 12\ \mathrm{Str}(T^a,T^b,T^c,T^d)F^a_{\mu\nu}(\bar{\lambda}^b\Gamma^\mu\lambda^c)(\bar{\lambda}^d\Gamma^\nu\varepsilon + \bar{\varepsilon}\Gamma^\nu\lambda^d) \equiv$$

$$\equiv 12\ \mathrm{Str}[F_{\mu\nu}(\bar{\lambda}\Gamma^{\mu}\lambda)(\bar{\lambda}\Gamma^{\nu}\varepsilon + \bar{\varepsilon}\Gamma^{\nu}\lambda)] \quad . \tag{35}$$

The proof that (35) satisfies Eq.(33) proceeds as follows: The variation of ${}^{1}\Lambda_6$ appearing in Eq.(33) contains two types of terms. The terms comming from the variation of λ and $\bar{\lambda}$ are

$$6\ \mathrm{Str}\{F_{\mu\nu}F_{\varrho\sigma}[(\bar{\lambda}\Gamma^{\mu\varrho\sigma}\varepsilon_1)(\bar{\lambda}\Gamma^{\nu}\varepsilon_2 + \bar{\varepsilon}_2\Gamma^{\nu}\lambda)$$
$$+ (\bar{\lambda}\Gamma^{\mu}\lambda)(\bar{\varepsilon}_2\Gamma^{\nu\varrho\sigma}\varepsilon_1) + \mathrm{h.c.}] - (1 \longleftrightarrow 2)\} \quad . \tag{36}$$

Using the Fierz identities for chiral fermions in six dimensions one can show that the expression (36) reproduces exactly the term $-12\ \mathrm{Str}(\delta_{\varepsilon_1}A, \delta_{\varepsilon_2}A, F^2)$. The terms obtained by varying $F_{\mu\nu}$ in (35) are

$$i12\ \mathrm{Str}\{(\bar{\lambda}\Gamma^{\mu}\lambda)[(D_{\mu}\bar{\lambda}\Gamma_{\nu}\varepsilon_1)(\bar{\lambda}\Gamma^{\nu}\varepsilon_2) + (D_{\mu}\bar{\lambda}\Gamma_{\nu}\varepsilon_1)(\bar{\varepsilon}_2\Gamma^{\nu}\lambda)$$
$$- (D_{\nu}\bar{\lambda}\Gamma_{\mu}\varepsilon_1)(\bar{\lambda}\Gamma^{\nu}\varepsilon_2) - (D_{\nu}\bar{\lambda}\Gamma_{\mu}\varepsilon_1)(\bar{\varepsilon}_2\Gamma^{\nu}\lambda) - \mathrm{h.c.}] - (1 \longleftrightarrow 2)\} \ .$$

Using again the Fierz identities, the above expression can be given the form

$$i\ \mathrm{Str}\{3(\bar{\lambda}\Gamma_{\mu}\lambda)[2(\bar{\varepsilon}_1\Gamma^{\mu}\lambda)(\bar{\varepsilon}_2\not{D}\lambda) + (\bar{\varepsilon}_2\Gamma_{\nu}\varepsilon_1)(\bar{\lambda}\Gamma^{\mu\nu}\not{D}\lambda)$$
$$- (\bar{\varepsilon}_2\Gamma^{\mu}\varepsilon_1)(\bar{\lambda}\not{D}\lambda)] + (5/4)(\bar{\lambda}\Gamma_{\mu\nu}\not{D}\lambda)(\bar{\lambda}\Gamma^{\mu}\varepsilon_1)(\bar{\lambda}\Gamma^{\nu}\varepsilon_2)$$
$$- \mathrm{h.c.} - (1 \longleftrightarrow 2)\} \quad . \tag{37}$$

The term (37) vanishes on-shell and, according to Eq.(33), its integral over M_6 is C_6.

Eqs.(34,35,37) represent a solution of the consistency conditions (28,29) and vanish when $\mathrm{Str}(T^a,T^b,T^c,T^d) = 0$. The uniqueness of this solution and the triviality (or not) of C_6

are still open problems.

4. Conclusions

Concerning the formal aspects of the problem, we have seen that the ε-picture (W-Z gauge) is useful in studying the structure of the chiral anomalies in supersymmetric theories. Presumably, it is the most convenient approach to this problem in the case of six and ten dimensions; at least, it provides a prescription for performing the calculations.

In four dimensions, the use of the ε-picture clarifies somehow the structure of the consistent chiral anomaly derived in the manifestly SUSY covariant approach.

By means of the consistency conditions (28,29), we were able to find the expression of the anomalies also in six dimensions. We stress once more that the ε-SUSY anomalies discussed in this talk are just a different manifestation of the chiral anomalies in supersymmetric theories.

We conclude with a remark on the physical implications of the anomalous Ward identities in SUSY gauge theories. Applying the methods of Ref.15, one can easily construct the supersymmetric W-Z term [12]. Even if the formal aspects are similar to the nonsupersymmetric case, the low energy physics of the Goldstone bosons presents new features. For instance, the presence of massless effective fermions, required by SUSY, modifies the anomalous interactions [12,19]. In general, the knowledge of the anomalies on fundamental level is not sufficient by itself to determine the anomalous vertices of the Goldstone bosons. Nevertheless, one can show [19] that it is still possible to describe these vertices in a compact way and that the description still has an intrisic geometrical meaning.

References

[1] H.Itoyama, V.P.Nair and H.C.Ren, IAS preprint, March 1985.

[2] B.Zumino, Univ. of California preprint, LBL-19794, 1985.

[3] E.Guadagnini and M.Mintchev, Pisa Univ. preprint, IFUP-TH 28/85, 1985.

[4] E.Guadagnini, K.Konishi and M.Mintchev, Phys. Lett. 157B (1985) 37.

[5] J.Wess and J.Bagger, Supersymmetry and Supergravity (Princeton Univ. Press, 1983).

[6] O.Piguet, K.Sibold and M.Schweda, Nucl. Phys. B174 (1980) 183.

[7] N.K.Nielsen, Nucl. Phys. B244 (1984) 499 and talk at this meeting.

[8] M.Pernici and F.Riva, Stony Brook preprint, ITP-SB-85-10.
R.Garreis, M.Scholl and J.Wess, Karlsruhe Univ. preprint, KA-THEP-85-4, 1985.
K.Harada and K.Shizuya, Phys. Lett. 162B (1985) 322.

[9] L.Bonora, P.Pasti and M.Tonin, Phys. Lett. 156B (1985) 341 and talk of M.Tonin at this meeting.
G.Girardi, R.Grimm and R.Stora, Phys. Lett. 156B (1985) 203.

[10] O.Piguet and K.Sibold, Nucl. Phys. B247 (1984) 484.
M.Porrati, Pisa Univ. preprint, IFUP-TH 22/85, 1985.

[11] S.Ferrara, L.Girardello, O.Piguet and R.Stora, Phys. Lett. 157B (1985) 179.

[12] E.Guadagnini, K.Konishi and M.Mintchev, Nucl. Phys. B262 (1985) 610.

[13] W.A.Bardeen, Phys. Rev. 184 (1969) 1848.

[14] P.van Nieuwenhuizen, Phys. Rep. 68 (1981) 189.

[15] J.Wess and B.Zumino, Phys. Lett. 37B (1971) 95.

[16] B.Zumino, W.Yang-Shi and A.Zee, Nucl. Phys. B239 (1984) 477.

[17] L.Brink, J.H.Schwartz and J.Scherk, Nucl. Phys. B121 (1977) 77.

[18] W.A.Bardeen and B.Zumino, Nucl. Phys. B244 (1984) 421.

[19] E.Guadagnini amd M.Mintchev, Pisa Univ. preprint, IFUP-TH 32/85, 1985.

SUPERSYMMETRIC LORENTZ CHERN-SIMONS INTERACTIONS FOR THE COMPACTIFIED SUPERSTRING(*)

S. Cecotti
Dipartimento di Fisica, Università di Pisa
and I.N.F.N., sez. di Pisa, Pisa.

ABSTRACT

We consider the N=1 4-dimensional Supergravity obtained from superstring compactification and we establish full supersymmetry of the Lorentz Chern-Simons terms induced by the string modification of the 10-dimensional theory. Supersymmetry requires an extra coupling of the chiral multiplet S (containing the axion) to a chiral multiplet which includes both the Hirzebruch and Euler index densities. We discuss the implications of the new terms for the topology of the field configurations in 4-dimensions and the related non-perturbative phenomena.

(*) The present talk is based on work done in collaboration with S. Ferrara, L. Girardello, and M. Porrati.

1. At the moment, Superstrings[1,2] are the only known candidates for a consistent theory of quantum gravity. They undergo spontaneous compactification[3] on a 6-dim. internal space K of the Calabi-Yau type[4], giving an effective flat 4-dim. space-time. In this case, we end up with a four dimensional Supergravity with one unbroken SUSY and $\frac{1}{2}|\chi|$ chiral generations of fermions.

Since Superstrings seem to be consistent quantum theories, potentially containing all interactions, the main point which remains to understand is if they describe the world we know. To extract experimentally detectable information out of the string, we need the corresponding effective 4-dim. Lagrangian for the light fields. Witten[5], has written down the first few terms of the 4 dim. effective Lagrangian obtained by dimensional reduction of the 10 dim. one, at least, for the simplest geometrical situation.

However, the correct effective Lagrangian cannot be completely computed by dimensional reduction, for the simple reason that we have not a full understanding of the 10 dim. case. For instance, we do not know the higher order terms required by the susy completion[(*)] of the Lorentz Chern-Simons (LCS) 3-form, which was added to the 3-index field strenght of the 2-form $B_{\mu\nu}$[1]

$$\hat{H} = dB + \frac{1}{30}\,\omega_{3YM} - \omega_{3L} \tag{1}$$

$$\omega_{3YM} = \mathrm{tr}\left(AF - \frac{1}{3}A^3\right) \tag{2.a}$$

(*) However, there are some partial results even for the 10 dim. case; see ref.(6).

$$\omega_{3L} = tr\left(\omega R - \frac{1}{3}\omega^3\right) \tag{2.b}$$

in order to cancel the anomalies (see ref.(1)). Without full supersymmetry, the (classical) spin-3/2 field equations are inconsistent, and this is not acceptable, even for an effective Lagrangian.

However, once we are in 4-dim., and one SUSY is unbroken, we have at our disposal the full, off-shell, tensor calculus for N=1 Supergravity, which enables us to construct N=1 supersymmetric invariant densities of any order in the gravitational curvatures and their derivatives. Then, we can construct the exact, supersymmetric completion of the Chern-Simons contribution to the effective Lagrangian in the framework of the N=1, 4 dim. tensor calculus. Of course, this gives us also important pieces of information for the original 10 dim. theory.

In this talk, we discuss the following points (see ref.(7) for full details):

i) What are the terms induced in 4 dim. by the Lorentz Chern-Simons (LCS) form added in the 10 dim. Lagrangian?

ii) What is their SUSY completion?

iii) It results that the Susy completion requires R^2 terms. Are there negative-metric states as it is usual with R^2-theories? What mechanism mayprevent the appearence of these ghost states?

iv) It seems that new classes of consistent 4-dim. supergravities, with higher derivatives, may exist. If this is true, we have to establish under what conditions they actually exist, and also construct them explicitly. The string effective Lagrangian is a first example of such a theory.

Contrary to the standard case, (see ref.(8)), it is unlikely that these theories are consistent for any choice of the Kähler-Hodge metric and/or superpotential. However, there is evidence that -if one restrict himself to the particular Kähler potentials predicted by superstring compactification-, the theory IS consistent. Of course, this may well be expected, since the theory we have started with, is supposed to have a consistent SUSY extension.

v) If in the effective Lagrangian, we neglect the LCS terms, the axion equations of motion would imply

$$\int_{M_4} F \wedge F = 0 \tag{3}$$

that is, no net instanton number. This truncation of the theory gives wrong semi-classical physics.

At the contrary, the complete effective model gives the following topological selection rule

$$\nu + 3\tau = 0 \tag{4}$$

where

$$\nu = \frac{1}{8\pi^2}\int \frac{1}{30}\ \mathrm{Tr}\ F\wedge F \tag{5}$$

is the negative of the Pontryagin number and

$$\tau = -\frac{1}{24\pi^2}\int \mathrm{Tr}\ R\wedge R \tag{6}$$

is the Hirzebruch signature.

Eq.(4) exactly corresponds to the 4-dim. projection of the original 10 dim. Bianchi Identity[9]

$$d\hat{H} = \frac{1}{30}\ F\wedge F - R\wedge R \tag{7}$$

that, since $\hat{H}$ is a globally defined 3-form, implies

$$\int_{M_4} \left(\frac{1}{30}\ \mathrm{Tr}\ F\wedge F - \mathrm{tr}\ R\wedge R\right) = 0 \tag{8}$$

If, as suggested in ref.(10), SUSY is broken by fermion condensates, the LCS modifications may be physically relevant, as they give clues to the non-perturbative physics of the string, and modifies the condensate effective potential.

2. Following Witten[5], we compactify the superstring theory on a six-torus, in an SU(3) invariant way. Then, we obtain the following

effective 4D Lagrangian

$$e^{-1}\mathcal{L} =$$

$$= -\frac{1}{2}R - 3\,\partial_\mu\sigma\,\partial^\mu\sigma - \frac{9}{16}\,\phi^{-1}\,\partial_\mu\phi\,\partial^\mu\phi$$

$$-\frac{3}{4}\,\phi^{-3/2}e^{6\sigma}\,\hat{H}_{\mu\nu\rho}\hat{H}^{\mu\nu\rho} -$$

$$-\frac{1}{4}\,\phi^{-3/4}\,e^{3\sigma}\,\frac{1}{30}\,\mathrm{Tr}\,F_{\mu\nu}F^{\mu\nu} + \ldots . \tag{9}$$

where we write only the terms which are relevant for our discussion of the LCS interactions.

In eq.(9), ϕ is the scalar field of the 10 dim. N=1 supergravity multiplet, σ is the breathing-mode of the internal Calabi-Yau manifold, i.e., the internal metric is chosen to be of the form

$$g_{IJ} = e^{\sigma}\delta_{IJ} \qquad I,J = 1,\ldots,6 . \tag{10}$$

and $\hat{H}_{\mu\nu\rho}$ is the improved field strenght defined in eq.(1).

With the duality transformation

$$e^{6\sigma}\,\phi^{-3/2}\,\hat{H}_{\mu\nu\rho} = \epsilon_{\mu\nu\rho\sigma}\,\partial^{\sigma}D \tag{11}$$

we trade the 2-form $B_{\mu\nu}$ for a pseudoscalar D; then, we define the complex field z as

$$z = e^{3\sigma} \phi^{-3/4} + i\, 3\sqrt{2}\, D \qquad (12)$$

Witten[5] has shown that this field z is the first component of a chiral multiplet

$$S = (z, \chi_L, h) \qquad (13)$$

for the effective N=1, 4D SUGRA. The Kinetic terms are then determined by a Kähler potential[5]

$$K = - \ln(S + \bar{S}) + \text{terms independent of } S \qquad (14)$$

which, as the entire theory, should be invariant under

$$S \rightarrow S + ic \qquad c \quad \text{real.} \qquad (15)$$

Working out the details of the duality transformation, eq.(11), and using the Bianchi identity, one sees that D has a coupling

$$\frac{3\sqrt{2}}{4} D \left[\frac{1}{30} Tr(F\tilde{F}) - tr(R\tilde{R}) \right] \qquad (16)$$

The first term in the bracket is already present in Lagrangian given by the Witten[5], and is rather standard in supergravity, whereas the second is new. This second term is the interaction induced in 4 dim. by the LCS term, in the $\hat{H}$ - field.

From eq.(11) we can rewrite all the spin-1 terms of the Lagrangian (9) as

$$- \frac{1}{4} \, Re(z) \, F_{\mu\nu} F^{\mu\nu} + \frac{1}{4} \, Im(z) \, F_{\mu\nu} \tilde{F}^{\mu\nu} \tag{17}$$

As is well known(8), eq.(17) is the bosonic part of the superfield expression

$$\left[f_{AB}(S) \, W^A W^B \right]_F + h.c. \tag{18}$$

where now $f_{AB} = \delta_{AB} S$, and W^A is the chiral multiplet containing the Yang-Mills field-strenght $F^A_{\mu\nu}$.

It is easy to see that all terms in eq.(9) have a SUSY completion in the framework of canonical supergravity(8), except for the new LCS interaction

$$e^{-1} \mathcal{L}_{LCS} = - \frac{1}{4} \, Im(z) \, R_{\mu\nu} \tilde{R}^{\mu\nu} \tag{19}$$

Our next problem will be to find a supersymmetric extension of the action density of eq.(19).

Since L_{LCS} is quadratic in the Riemann tensor, its supersymmetrization requires the knowledge of the specific properties of the N=1 gravitational multiplets. In the next point we give a short discussion of these multiplets.

3. <u>A quick review of N=1 Gravitational Multiplets</u>. - There are two, equivalent, methods to study the supergravitational multiplets: tensor calculus[11] and superspace geometry[12].

In the last approach[12], one solves the superspace Bianchi identities in terms of three basic multiplets:

Weyl Multiplet	$W_{\alpha\beta\gamma}$	Chiral
Scalar curvature multiplet	R	Chiral
Ricci (or Einstein) mult.	$E_{\alpha\dot{\alpha}}$	Real vector.

As the names suggest, $W_{\alpha\beta\gamma}$ contains the Weyl tensor, $C_{\rho\sigma\mu\nu}$, R the scalar curvature, and $E_{\alpha\dot{\alpha}}$ contains $R_{\mu\nu} - \frac{1}{6}\delta_{\mu\nu} R$.

These multiplets are also subjected to the following conditions

$$\bar{\nabla}_{\dot{\alpha}} R = \bar{\nabla}_{\dot{\alpha}} W_{\alpha\beta\gamma} = 0$$

$$\nabla^{\alpha} E_{\alpha\dot{\beta}} = \bar{\nabla}_{\dot{\beta}} \bar{R} \tag{20}$$

$$\nabla^{\alpha} W_{\alpha\beta\gamma} = \nabla_{\beta}{}^{\dot{\tau}} E_{\gamma\dot{\tau}} + \nabla_{\gamma}{}^{\dot{\tau}} E_{\beta\dot{\tau}} .$$

From these multiplets, it is possible to construct three chiral multiplets which are quadratic in the supercurvatures: $W_{\alpha\beta\gamma} W^{\alpha\beta\gamma}$, which is just the gravitational analogue of $W_{\alpha}W^{\alpha}$, $(\bar{\nabla}^2 - 8R)E^2$, $[(\bar{\nabla}^2 - 8R)\bar{R}]R$.

In the notation of the Poincarè tensor calculus[11], they are written as

$$\Sigma(W_L^2), \quad \Sigma(E^2), \quad \Sigma(R\,T(R)) \tag{21}$$

where T denotes the operation which acting on a chiral multiplet gives another chiral multiplet, known as the kinetic multiplet(11,13), if

$$S = (z, \chi_L, h) \tag{22}$$

then,

$$T(S) = \left(\bar{h} + \tfrac{1}{3}\bar{u}\bar{z},\ \psi_L,\ H\right) \tag{23}$$

$$\psi_L =$$

$$= \hat{\not{D}}\chi_R + \frac{i}{6}\not{A}\chi_R + \frac{1}{6}(\gamma\cdot R^P)\bar{z} \tag{24}$$

$$H = \hat{\Box} z - \frac{2}{3}u\left(\bar{h} + \frac{1}{3}\bar{u}\bar{z}\right). \tag{25}$$

and Σ is the operation which associate a chiral multiplet to the product of two multiplets (Σ is just the identity when acting on a product which is already a chiral multiplet, whereas -if the product is a vector multiplet V- we have(11,13)

$$V = (C, \zeta_L, H, B_m, \lambda_R, D) \tag{26}$$

$$\Sigma(V) =$$

$$= \left(-\bar{H},\ -i\lambda_L - i\hat{\not{D}}\zeta_R,\ D + \hat{\Box}C + i\hat{D}^m B_m\right) \tag{27}$$

In particular, one has[11,13],

$$Im\{[\Sigma(W_L^2)]_h\} = -Tr[R_{\mu\nu}\tilde{R}^{\mu\nu}] + \dots \tag{28}$$

Thus, if we insert the chiral multiplet $S\Sigma(W_L^2)$ in the formula for the action density[11,13],

$$\frac{1}{4}[S\Sigma(W_L^2)]_h + \dots + h.c. \tag{29}$$

we obtain a <u>locally</u> supersymmetric action containing the term we look for, $-\frac{1}{4}Im(z)R\tilde{R}$.

However, eq.(29) cannot be the correct answer! Indeed, eq.(29) also contains

$$Re(z)\, C_{\mu\nu\rho\sigma} C^{\mu\nu\rho\sigma}$$

which -as is well known- propagates negative metric states, and also the auxiliary fields u and A_μ.

Then, we look for a modification of eq.(29) which -while preserving the desidered couplings of Im z- does not propagate auxiliary fields. This condition is -by supersymmetry- equivalent to the requirement of no ghost states in the physical spectrum.

That these two conditions may be simultaneously satisfied is guaranteed by the fundamental theorem relating the three multiplets (21), i.e. the SUPER GAUSS-BONNET THEOREM.

The standard, N=0 Gauss-Bonnet theorem states that the particular combination

$$\sqrt{g}\left[(R_{abcd})^2 - 8\left(R^2_{ab} - \tfrac{1}{3}R^2\right)\right] = \qquad (30)$$
$$= \sqrt{g}\; {}^{*}R_{abcd}\, {}^{*}R^{abcd}$$

of the three possible scalar densities quadratic in the Riemann tensor is <u>locally</u> trivial, that is, has only topological contend. Indeed, eq.(30) is a total derivative, proportional to the Euler characteristic density.

The supersymmetric generalization of this theorem, discussed -at the linearized level- by Ferrara and Zumino[14] and then shown to hold to all orders by Townsend and van Niuvenhuizen[15], says that the supersymmetric action density

$$[\hat{\Sigma}(W^2_L)]_h + \bar{\psi}_R \cdot \gamma [\hat{\Sigma}(W^2_L)]_{\chi_L} + \qquad (31)$$
$$+ \bar{\psi}_{\mu R}\sigma^{\mu\nu}\psi_{\nu R}[\hat{\Sigma}(W^2_L)]_z$$

where $\hat{\Sigma}(W^2_L)$ is the special linear combination,

$$\hat{\Sigma}(W^2_L) = \Sigma(W^2_L) + \frac{1}{2}\Sigma(E^2) - 4\Sigma(RT(R)) \quad (32)$$

is a total superspace derivative, whose global contend is just the Gauss-Bonnet theorem, for the real part of eq.(32), and the Hirzebruch signatureindex theorem, for the imaginary part. That is

$$Re[\hat{\Sigma}(W^2_L)]_h =$$

$$= \epsilon_{abcd} R^{ab}_{\mu\nu} \tilde{R}^{\mu\nu cd} + \quad (33)$$

+ total derivative of
a gloabally defined quantity.

$$Im[\hat{\Sigma}(W^2_L)]_h = - \tilde{R}_{abcd} R^{abcd} + \quad (34)$$

+ total derivative of
a globally defined quantity.

4. The above arguments imply that the correct superspace Lagrangian term for the coupling of the axion-dilaton chiral superfield S, due to Lorentz and gauge Chern-Simons terms, is

$$\int d^2\Theta \, \mathcal{E} \left\{ S\left[\hat{\Sigma}(W^2_L) - \frac{1}{30}\Sigma(W^2_{YM}) \right] \right\} \quad (35)$$
$$+ h.c.$$

This terms does not introduce ghost states, since the quadratic piece of the R^2 terms add to a total derivative, because of the Gauss–Bonnet theorem, eqs.(33,34). The relation of the 4 dim. Gauss–Bonnet theorem to the absence of ghosts for the string theory was discussed by Zwiebach in ref.(16).

However, we must stress that the super Gauss-Bonnet theorem is only a necessary, but not a sufficient, condition for the supergravity theory with the new couplings (35), to be consistent. The generic such model

$$L = -3\int d^4\theta\, E\, e^{-\frac{1}{3}K} + \frac{1}{2}\int d^2\theta\, \mathcal{E}\, g$$
$$+ \int d^2\theta\, \mathcal{E}\, S\left[\, \hat{\Sigma}(W_L^2) - \frac{1}{30}\Sigma(W_{YM}^2)\right] \tag{36}$$
$$+ h.c.$$

will be -at most- consistent for some special form of the Kähler potential K and of the superpotential g. Consistency for generic K and g is rather unlikely; we need at least some sort of R-symmetry for which Im z plays the rôle of axion, as it is necessary in order the duality transformation to be implementable.

However, since that the above duality transformation and the no auxiliary field propagation condition, select <u>one and only one</u> possible Lagrangian for a given choice of K and g, and, because we know that the superstring is a consistent theory, we are guarateed that the theory specified by the Lagrangian (35) is consistent <u>at least</u> for the specific K and g predicted by string compactification. After all, in the superstring case, L_{LCS} is related to g by a

Kaluza-Klein symmetry. In the simplest string case[5]

$$K = -\ln(S+\bar{S}) - 3\ln(T+\bar{T} - 2\bar{C}^a C_a) \tag{36}$$

$$g = \lambda\, d^{abc} C_a C_b C_c \tag{37}$$

The first indications by explicit computation seems to agree with this general argument. Note that in x-space (contrary to superspace) the theory is non-polinomial in field derivatives and also non-local, but in some subtle way. However, we probably can make sense out of this, thanks to some "superstring miracle". Work on the subject is in progress.

The present results can be also used to guess the form of some of the unknown terms of the 10 dim. Lagrangian for the string, in the field limit. In particular, we must have a term of the form

$$-\frac{30}{4}\,\phi^{-3/4}\left[R^{ab}_{\mu\nu} R^{\mu\nu ab} + \ldots\right] \tag{38}$$

This term is actually needed to compactify the theory. Its presence is also proven in refs.(17,6).

5. As stated above, one of the most important consequence of the new LCS term in the effective Lagrangian is to enforce the corrected topological selection rule

$$\nu + 3\tau = 0 \tag{39}$$

in agreement with 10-dim. cohomological arguments. This can be seen from the equation of motion for the pseudoscalar field Im z. By the Gauss-Bonnet theorem, eq.(33,34) we have

$$D_\mu \frac{\partial \mathcal{L}}{\partial(\partial_\mu(\mathrm{Im}\, z))} =$$

$$= \frac{e^{-1}}{8}\left[\frac{1}{30}\,\mathrm{Tr}(F_{\mu\nu}\tilde{F}^{\mu\nu}) - \mathrm{tr}(R_{\mu\nu}\tilde{R}^{\mu\nu})\right] \tag{40}$$

$+$ total derivative of a globally defines quantity.

Integrating eq.(40) in d^4x and requiring the field configuration to be compactificable, we get

$$0 = \int_{M_4}\left[\frac{1}{30}\,\mathrm{Tr}(F\wedge F) - \mathrm{tr}(R\wedge R)\right] = \tag{41}$$

$$= \nu + 3\tau$$

Actually, one can show that this selection rule holds even off-shell, in the sense that all "field configurations" with give a net zero contribution to the Euclidean functional measure[6]. That this must be so, is already clear from the fact that the duality transformation, eq.(11), can be implemented, at the path integral level, just by a Gaussian integral.

Acknowledgments

All the results reported at this conference were obtained in collaboration with S. Ferrara, L. Girardello and M. Porrati, see ref.6.

I also thank A. Pasquinucci and C. Reina for discussions on specific problems related to LCS in 4 D string theory.

REFERENCES

(1) J. H. Schwarz, Phys. Rep. 89C (1982) 223.
M. B. Green, Sur. High Energy Phy. 3 (1984) 117.

(2) M. B. Green and J. H. Schwarz, Phys. Lett. 149B (1984) 117.

(3) P. Candelas, G. Horowitz, A. Strominger and E. Witten, Nucl. Phys. B258 (1985) 46.

(4) E. Calabi, in "Algebraic geometry and topology: a symposium in honor of S. Lefschetz" (Princeton University, (1957) 78).
S. T. Yau, Proc. Nat. Acad. Sci. 74 (1977) 1798.

(5) E. WItten, Phys. Lett. 155B (1985) 151.
M. Dine, R. Rohm, N. Seiberg and E. Witten, Phys. Lett. B156 (1985) 55.

(6) L. Romans and N. Warner, Caltech Preprint CALT-68-1291 (1985).
S.K. Han, J. K. Kim, Y. Tanii, preprint KAIST - 85/22, TIT-HET.

(7) S. Cecotti, S. Ferrara, L. Girardello and M. Porrati, Preprint CERN-TH 4253/85, Phys. Lett. to appear.

(8) E. Cremmer, S. Ferrara, L. Girardello and A. van Proyen, Phys. Lett. 116B (1982) 231, Nucl. Phys. B212 (1983).

(9) E. Witten, Phys. Lett. 149B (1984) 351.

(10) Second paper of ref.(5).

(11) S. Ferrara and P. van Nieuvenhuizen, Phys. Lett. 74B (1978) 333; 76B (1978) 404; 78B (1978) 573.
K.S. Stelle and P. C. West, Phys. Lett. 74B (1978) 330; 77B (1978) 376.

(12) See e. g. B. Zumino in "Recent Developments in Gravitation" Cargese 1978, ed. by M. Levy and S. Deser, p;405.
J. Wess and J. Bagger, in "Supersymmetry and aupergravity" Princeton series, 1983.
S. Gates J. R., M.T. Grisaru, M. Rocek and W. Siegel in "Superspace: one thousand and one lessons in Supersymmetry" (Frontiers in Physics, Benjamin, 1983).

(13) For recent reviews of N=1 tensor calculus, see, for instance:
A. van Proyen, lectures in the 1983 Karpacz school, ed. B. Milewski (World Scientific, 1983).
T. Kugo and S. Uehara, Nucl. Phys. B226 (1983) 49.

(14) S. Ferrara and B. Zumino, Nucl. Phys. B134 (1978) 301.

(15) P.K. Townsend and P. van Nieuvenhuizen, Phys. Rev. D19 (1979) 3592.

(16) B. Zwiebach, Berkeley preprint UCB-PTH-85/10.

(17) D. J. Gross, J. A. Harvey, E. Martinec, and R. Rhom, Princeton preprint (1985) "Heterotic string theory II: The interacting Heterotic String."
C. G. Callan, E. J. Martinec, M. J. Perry and D. Friedan, Princeton preprint (1985).

The vacuum states and their stability in N=1, D=10 anomaly-free Yang-Mills supergravity

J. Kowalski-Glikman[†],
NIKHEF-H / Amsterdam

ABSTRACT:

Using the positive energy expression we define the class of the background field configuration for N=1, D=10 anomaly free Yang-Mills supergravity and find conditions under which these backgrounds are stable.

---

† On leave of absence from the Institute of Theoretical Physics, University of Warsaw

The vacuum states and their stability in N=1, D=10 anomaly-free Yang-Mills supergravity

Recently the ten-dimensional N=1 supergravity coupled to $E_8 \times E_8$ or SO(32) Yang-Mills theory[1] became one of the most active fields in elementary particle physics. Indeed the discovery of anomaly cancellation in this theory[2] gives an attractive possibility that this model, that is a low-energy limit of superstrings theory, may provide a correct four-dimensional low-energy phenomenology after dimensional reduction.

In order to check this conjecture the dimensional reduction must be performed. However, the result of it depends on the choice of background field configuration. Usually it is assumed that such a so-called vacuum state, is a stable solution of field equations.

Recently Candelas et al.[3] proposed a vacuum state appropriate for dimensional reduction from ten to four dimensions. The authors of ref.[3] started from very natural assumptions and obtained a vacuum state of the form $M^4 \times K^6$ where M^4 is a four-dimensional Minkowski space and K^6 a so-called Calabi-Yau manifold. However this space is not a solution of the field equations (so compactification is not spontaneous) and nothing can be said about its stability.

In this paper we will analyze a problem of stability of ten-dimensional supergravity field configuration. We will use as a tool a positive-energy theorem[4], which connects the stability properties with the number of unbroken symmetries. It should be stressed however that our approach relies on the following assumptions about N=1, D=10 Yang-Mills supergravity.

[i] following Candelas et al.[3] we assume that the gravitino transformation law is

$$\delta\psi_\mu = \hat{\nabla}_\mu \varepsilon =$$

$$= \nabla_\mu(\omega(e))\varepsilon - {}^1/_{48}\, e^{-\frac{3}{4}\sigma}\left(\gamma_\mu{}^{\nu_1\nu_2\nu_3} - 9\delta_\mu{}^{\nu_1}\gamma^{\nu_2\nu_3}\right) H_{\nu_1\nu_2\nu_3} \qquad (1)$$

$$+ \text{ fermionic terms},$$

$$H = da - \omega_{3Y} + \omega_{3L} ,$$

$$H = H_{\nu_1\nu_2\nu_3}\, dx^{\nu_1} \wedge dx^{\nu_2} \wedge dx^{\nu_3},$$

$$\omega_{3Y} = \mathrm{tr}(A \wedge F - {}^1/_3\, A \wedge A \wedge A),$$

$$\omega_{3L} = \mathrm{tr}(\omega \wedge R - {}^1/_3\, \omega \wedge \omega \wedge \omega),$$

where A is a Yang-Mills gauge field, F its field strength, ω a connection and R its (Riemann) curvature.

ii] We assume also that the bosonic part of the action contains in addition to the terms described in [1] the following square curvature terms

$$\tfrac{1}{4}\sqrt{-g}\, e^{-\frac{3}{4}\sigma}\left(R_{\mu\nu\rho\sigma}R^{\mu\nu\rho\sigma} - 4R_{\mu\nu}R^{\mu\nu} + R^2\right) . \tag{2}$$

The presence of this term was recently obtained both from string[5] and 10-dimensional anomaly free supergravity[6] considerations. Results of the paper[6] indicate however that assumption [i] may not be true.

Now we are ready to analyze a positive-energy theorem. Consider a generalized Nester formula (since now ε is a commuting spinor)

$$\hat{E}^{\mu\nu} = 2\left(\overline{\hat{\nabla}_\rho \varepsilon}\gamma^{\mu\nu\rho}\varepsilon' - \bar{\varepsilon}\gamma^{\mu\nu\rho}\hat{\nabla}_\rho\varepsilon'\right). \tag{3}$$

Taking the covariant derivative of E and using Stokes theorem for a nine-dimensional space-like surface Σ we find

$$\begin{aligned}\int_{\partial\Sigma} \hat{E}^{\mu\nu}\, d\Sigma_{\mu\nu} &= \int_\Sigma (\text{Einstein})_\mu{}^\nu\, \bar{\varepsilon}\gamma^\mu\varepsilon'\, d\Sigma_\nu + 2\int_\Sigma (D_\mu \tilde{H}^{\mu\nu\rho})\,\bar{\varepsilon}\gamma_\nu\varepsilon'\, d\Sigma_\rho \\ &+ \tfrac{2}{3}\int_\Sigma (\partial_{\alpha_1}\tilde{H}_{\alpha_2\alpha_3\alpha_4})\,\bar{\varepsilon}\gamma^{\alpha_1\alpha_2\alpha_3\alpha_4\rho}\varepsilon'\, d\Sigma_\rho \\ &+ 4\int_\Sigma \overline{\hat{\nabla}_\mu\varepsilon}\,\gamma^{\mu\nu\rho}\hat{\nabla}_\nu\varepsilon\, d\Sigma_\rho + 4\int_\Sigma \overline{\delta\lambda}\,\gamma^\rho\delta\lambda'\, d\Sigma_\rho ,\end{aligned} \tag{4}$$

where

$$\begin{aligned}(\text{Einstein})_\mu{}^\nu &= 2(R_\mu{}^\nu - \tfrac{1}{2}\delta_\mu{}^\nu R) + \tfrac{9}{8}\left[\delta_\mu{}^\nu(\partial_\rho\sigma)^2 - 2(\partial_\mu\sigma)(\partial^\nu\sigma)\right] \\ &+ \tfrac{1}{3}\left[\delta_\mu{}^\nu\tilde{H}_{\alpha_1\alpha_2\alpha_3}\tilde{H}^{\alpha_1\alpha_2\alpha_3} - 6\tilde{H}_{\mu\alpha_1\alpha_2}\tilde{H}_\nu{}^{\alpha_1\alpha_2}\right],\end{aligned} \tag{5}$$

$$\tilde{H}_{\nu_1\nu_2\nu_3} = e^{-\frac{3}{4}\sigma} H_{\nu_1\nu_2\nu_3} \tag{6}$$

and

$$\delta\lambda = \tfrac{3}{8}\sqrt{2}\, i\gamma^\mu\partial_\mu\sigma\varepsilon + \tfrac{1}{12}\sqrt{\tfrac{1}{2}}\, i\gamma^{\nu_1\nu_2\nu_3}H_{\nu_1\nu_2\nu_3}\varepsilon . \tag{7}$$

The R.H.S. of (5) does not contain all terms that are present in the Einstein field equation of the theory. Thus we should add and subtract the following terms to the R.H.S. of (4)

$$\text{tr}\,\overline{\delta\chi}\gamma^\rho\delta\chi' - \tfrac{1}{2}\overline{\delta\rho}^{ac}\gamma^\rho\delta\rho'^{bd}\eta_{ab}\eta_{cd} + 3\overline{\delta\rho^a}\gamma^\rho\delta\rho'^b\eta_{ab} \tag{8}$$

where

$$\delta\chi = e^{-\frac{3}{8}\sigma} F_{\mu\nu}\gamma^{\mu\nu}\varepsilon, \; * \tag{9}$$

$$\delta\rho^{ab} = e^{-\frac{3}{8}\sigma} A_{\mu\nu}{}^{ab}\gamma^{\mu\nu}\varepsilon,$$

$$A_{\mu\nu}{}^{ab} = R_{\mu\nu}{}^{ab} - 2R_{[\mu}{}^{[a}e_{\nu]}{}^{b]}, \tag{10}$$

*) We will not write explicitly a gauge group index

$$\delta\rho^a = ie^{-3/8\sigma}\, R_\mu{}^a\gamma^\mu\varepsilon\,, \tag{11}$$

and η_{ab} is a flat Minkowski metric. Then we obtain

$$\begin{aligned}\partial\Sigma\int \hat{E}^{\mu\nu}d\Sigma_{\mu\nu} &= {}_\Sigma\int \widetilde{(\mathrm{Einstein})}_\mu{}^\nu\, \bar{\varepsilon}\gamma^\mu\varepsilon'\, d\Sigma_\nu \\ &+ 2\, {}_\Sigma\int (D_\mu \widetilde{H}^{\mu\nu\rho})\, \bar{\varepsilon}\gamma_\nu\varepsilon'\, d\Sigma_\rho \\ &+ {}^2/_3\, {}_\Sigma\int (\partial_{\alpha_1}\widetilde{H}_{\alpha_2\alpha_3\alpha_4})\, \bar{\varepsilon}\gamma^{\alpha_1\alpha_2\alpha_3\alpha_4\rho}\,\varepsilon' d\Sigma_\rho \\ &+ \tfrac{1}{2}\, {}_\Sigma\int e^{-\frac{3}{4}\sigma}\, \mathrm{tr}\,(F_{\alpha_1\alpha_2}F_{\alpha_3\alpha_4} - R_{\alpha_1\alpha_2}R_{\alpha_3\alpha_4})\, \bar{\varepsilon}\gamma^{\alpha_1\alpha_2\alpha_3\alpha_4\rho}\,\varepsilon' d\Sigma_\rho \\ &+ {}_\Sigma\int \big(4\overline{\hat{\nabla}_\mu\varepsilon}\gamma^{\mu\nu\rho}\hat{\nabla}\varepsilon' + 4\,\overline{\delta\lambda}\,\gamma^\rho\delta'\lambda + \tfrac{1}{8}\overline{\delta\chi}\,\gamma^\rho\delta'\chi \\ &- \tfrac{1}{2}\,\overline{\delta\rho^{ac}}\,\gamma^\rho\delta'\rho^{bd}\,\eta_{ab}\eta_{cd} + 3\overline{\delta\rho^a}\gamma^\rho\delta'\rho^b\eta_{ab}\big)\, d\Sigma_\rho \end{aligned} \tag{12}$$

where

$$\begin{aligned}\widetilde{(\mathrm{Einstein})}_\mu{}^\nu &= 2(R_\mu{}^\nu - \tfrac{1}{2}g_\mu{}^\nu R) + {}^9/_8\,[\delta_\mu{}^\nu(\partial_\rho\sigma)^2 - 2(\partial_\mu\sigma)(\partial^\nu\sigma)] \\ &+ {}^1/_3\,[\delta_\mu{}^\nu \widetilde{H}_{\alpha_1\alpha_2\alpha_3}\widetilde{H}^{\alpha_1\alpha_2\alpha_3} - 6\widetilde{H}_{\mu\alpha_1\alpha_2}\widetilde{H}^{\nu\alpha_1\alpha_2}] \\ &+ 2e^{-\frac{3}{4}\sigma}(F^{\nu\beta}F_{\mu\beta} - \tfrac{1}{4}\delta_\mu{}^\nu F_{\alpha\beta}F^{\alpha\beta}) \\ &- 2e^{-\frac{3}{4}\sigma}\big(R^{\nu\alpha\beta\gamma}R_{\mu\alpha\beta\gamma} - \tfrac{1}{4}\delta_\mu{}^\nu R_{\alpha\beta\gamma\delta}R^{\alpha\beta\gamma\delta} - 2R^{\nu\beta}R_{\mu\beta} - 2R^{\alpha\beta}R^\nu{}_{\alpha\mu\beta} + \\ &+ RR^\nu{}_\mu + R^{\alpha\beta}R_{\alpha\beta}\delta^\nu{}_\mu - \tfrac{1}{4}R^2\delta_\mu{}^\nu\big) = 0 \text{ (on-shell)}. \end{aligned} \tag{13}$$

We will define the background field configuration as such that

i) the R.H.S. of eq. (12) vanishes,
ii) equations of motion are satisfies.

Firstly we observe that the necessary and sufficient condition, for the integral on the R.H.S. of (12) to be surface Σ - independent, is that the covariant divergence of (13) vanishes. As a consequence of this fact we find that σ=constant (in order for H-field equation of motion $D_\mu(e^{-\frac{3}{4}\sigma}\widetilde{H}^{\mu\nu\rho})=0$ to be compatible with the result coming from covariant divergence calculations $D_\mu(\widetilde{H}^{\mu\nu\rho})=0$) and

$$R^{\alpha\beta}(D_\mu R_{\alpha\beta} + D_\alpha R_{\mu\beta}) = 0. \tag{14}$$

Next we assume that

$$\mathrm{tr}\big(F_{[\alpha_1\alpha_2}F_{\alpha_3\alpha_4]} - R_{[\alpha_1\alpha_2}R_{\alpha_3\alpha_4]}\big) = 0. \tag{15}$$

We can also assume that the background field configuration allows the following set of equations to possess a nontrivial solution

$$\hat{\nabla}_\mu \overset{\circ}{\varepsilon} = 0 \ , \tag{16}$$

$$H_{\mu\nu\rho}\gamma^{\mu\nu\rho}\overset{\circ}{\varepsilon} = 0, \tag{17}$$

$$F_{\mu\nu}\gamma^{\mu\nu}\overset{\circ}{\varepsilon} = 0, \tag{18}$$

$$A_{\mu\nu}{}^{ab}\gamma^{\mu\nu}\overset{\circ}{\varepsilon} = 0, \tag{19}$$

$$R_\mu{}^a\gamma^\mu\overset{\circ}{\varepsilon} = 0, \tag{20}$$

where (17) comes from $\delta\lambda=0$ and the fact that σ=constant. It is easy to see that if we assume the four-dimensional Lorentz invariance and the equality of Yang-Mills and Riemann connections, the resulting background state is of the form $M^4\times K^6$ where M^4 is a Minkowski space, K^6-Ricci flat, compact manifold with integrable almost complex structure and $H_{\mu\nu\rho}=0$ (because field equations were assumed to be satisfied).

Now we are ready to analyze the stability properties of our background. Assuming field equations of the theory and constraints (14), (15) are satisfied we have

$$\int_{\partial\Sigma} \hat{E}^{\mu\nu} d\Sigma_{\mu\nu} = \int_\Sigma T^\mu{}_\nu(F,R)\bar{\varepsilon}\gamma^\nu\varepsilon' d\Sigma_\mu + \int_\Sigma\left[4\overline{\hat{\nabla}_\mu\varepsilon}\gamma^{\mu\nu\rho}\hat{\nabla}_\nu\varepsilon' + 4\overline{\delta\lambda}\gamma^\rho\delta'\lambda\right] d\Sigma_\rho \ , \tag{21}$$

where

$$T_{\mu\nu}(F,R) = -2e^{-\frac{3}{4}\sigma}\Big[\mathrm{tr}\big(F_\mu{}^\rho F_{\nu\rho} - \tfrac{1}{4}g_{\mu\nu}F_{\rho\sigma}F^{\rho\sigma}\big) - R_{\mu\rho\sigma\delta}R_\nu{}^{\rho\sigma\delta} + \tfrac{1}{4}g_{\mu\nu}R_{\rho\sigma\delta\gamma}R^{\rho\sigma\delta\gamma} + 2R_\mu{}^\rho R_{\nu\rho} + 2R^{\rho\sigma}R_{\mu\rho\nu\sigma} - RR_{\mu\nu} - R^{\rho\sigma}R_{\rho\sigma}g_{\mu\nu} + \tfrac{1}{4}R^2 g_{\mu\nu}\Big]. \tag{22}$$

The last term in (21) is greater or equal to zero provided $\gamma^\mu\hat{\nabla}_\mu\varepsilon=0$ with asymptotic behaviour $\varepsilon \to \overset{\circ}{\varepsilon} + (1/r)$ $(r\to\infty)$ (r is a radial coordinate on the noncompact part of the space-time).

Now the positivity of the conserved charge $\int_{\partial\Sigma}\hat{E}^{\mu\nu}d\Sigma_{\mu\nu}$ for any fluctuation around the background relies on the fact that

$$T_{\mu\nu}(F,R)V^\mu W^\nu > 0 \tag{23}$$

for any time-like vectors V^μ and W^ν. If (23) is satisfied $\int_{\partial\Sigma}\hat{E}^{\mu\nu}d\Sigma_{\mu\nu}$ is a sum of two positive pieces. The problem whether the conditions (23) together with (14) and (15) are fulfilled for any fluctuation is very difficult and was not solved by the author. In addition it is by no means clear if the additional terms in the gravitino transformation laws (1) indicated by the results of paper [6] will automatically solve the stability problems (for example it turns out that even the Minkowski space is unstable in pure four-dimensional gravity if higher derivative terms are present[7]). Thus the problem of stability of the supersymmetric field configuration requires further analysis after the full N=1, D=10 anomaly free Yang-Mills supergravity will be known.

REFERENCES

[1] E. Bergshoeff, M. de Roo, B. de Wit, P. van Nieuwenhuizen, Nucl. Phys. B195 (1982) 97;
G.F. Chapline and N.S. Manton, Phys. Lett. 120B (1983) 105.

[2] M.B. Green and J.H. Schwarz, Phys. Lett. 149B (1984) 117.

[3] P. Candelas, G.T. Horowitz, A. Strominger and E. Witten, Nucl. Phys. B258 (1985) 46.

[4] E. Witten, Comm. Math. Phys. 80 (1984) 381;
J.M. Nester, Phys. Lett. 83A (1981) 259;
D.Z. Freedman, G.W. Gibbons, Nucl. Phys. B233 (1984) 24;
J. Kowalski-Glikman, Phys. Lett. B. (in press).

[5] B. Zwiebach, Phys. Lett. 156B (1985) 315.

[6] L.J. Romans and N.P. Warner, CALTECH preprint CALT-68-1291 (1985).

[7] G.T. Horowitz and R.M. Wald, Phys. Rev. D17 (1978) 414.

Part Six: Shorter Contributions

LARGE MASS EXPANSION IN THE HIGGS SYSTEM

Giampiero PASSARINO
Istituto di Fisica Teorica, Universita' di Torino, Italy
INFN, Sezione di Torino, Italy

Abstract

The Higgs system of the Standard Model for electroweak interactions is considered as an effective theory valid for energies low on the scale $\Lambda \approx 1TeV$. The limit where the bare mass parameter becomes large compared to the cutoff is analyzed by means of a continuum regulated strong-coupling expansion . In this context the shifted theory becomes unstable for all values of the field vacuum expectation value different from zero .

One of the fundamental ingredients of the Standard Model is the Higgs particle , but low energy weak interactions are screened from the dynamics of the Higgs field . The possibility that the Higgs sector is strongly interacting , suggested some time ago [1] , has received an increasing attention in recent years [2]. It has been shown by many authors that a manifestation of the Higgs system must show up around $1TeV$. Since self interactions in this sector are proportional to m_H^2 , heavy or strongly interacting are synonymous.

Explicit calculations suggest a breakdown of weak-coupling perturbation theory and an upper bound on m_H^2 can be fixed , above which weak interactions will become strong [3]. A guide for this breakdown is the violation of unitarity in tree level $W-W$ scattering which occurs for m_H above $1TeV$, or equivalently in the limit of large scalar self-coupling for fixed vacuum expectation value.

If the Higgs mass is very large compared to the energy in $W-W$ scattering, partial wave unitarity breaks down in Born approximation at $s \approx 1.8TeV$. Either a Higgs particle is found or vector boson scatttering at high energies will not be describable by perturbation theory.

On the other hand we can entertain the point of view , especially advocated by Veltman , that the Higgs system is just an effective way of describing the structure of particles and that ultimately we have to move beyond renormalization theory [4]. Following this approach we could regard the cutoff and the masses as independent and consider a strong coupling limit of the effective scalar sector of the theory .

In the following we analyze the Higgs system by assuming that the theory has a non trivial phisical content at $p^2 \ll \Lambda^2$. Therefore the regulated theory may be regarded as an effective field theory at energies which are low on the scale of Λ^2. As usual the vacuum expectation value is fixed by the weak scale $M_W^2 \ll \Lambda^2$. In this context we shall entertain the point of view that the parameter m , corresponding to the weak-coupling perturbative mass , is large on the scale of Λ. There are many papers dealing with the limit $m \to \infty$, mostly in reference to the non linear σ-model [2].

Instead of computing large corrections due to Higgs self-coupling we move to the problem of constructing a self consistent strong-coupling expansion for the Green's functions of the theory . Our expansion for large m will be a continuum regulated strong-coupling expansion a la Bender,Cooper,Kenway and Simmons [5] , based on the path-integral representation for the Green's functions with the kinetic energy as a perturbation. Treating the interaction terms locally we can evaluate path-integrals as the limit of a product of ordinary integrals on a hypercubical Euclidean lattice of spacing a. The formal continuum limit is obtained with the identification $a^{-4} = \delta_\Lambda(0)$ where the cutoff delta function is the Fourier transform of $\delta_\Lambda(p) = \theta(\Lambda - \mid p \mid)$. Feynman rules for the $1/m$ expansion will be first illustrated by means of a simple $O(1)$ model. The general formulation is described in details in ref.[5] . Consider the Lagrangian

$$\mathcal{L} = \partial_\mu \Phi \partial_\mu \Phi + \mu^2 \Phi^2 + \frac{1}{2}\lambda \Phi^4$$

Where Φ is a real field and v its vacuum expectation value

$$\Phi = \frac{1}{\sqrt{2}}(\phi + \sqrt{2}v)$$

Introducing new parameters $G = 1/4v^2$, $M^2 = \mu^2 + \lambda v^2$, $m^2 = \lambda/2G$ we have

$$\mathcal{L} = \frac{1}{2}\partial_\mu\phi\partial_\mu\phi + \frac{M^2}{\sqrt{2}G^{1/2}}\phi + \frac{1}{2}(m^2 + M^2)\phi^2 + \frac{1}{\sqrt{2}}G^{1/2}m^2\phi^3 + \frac{1}{4}Gm^2\phi^4$$

We investigate the regime $m \to \infty$ for G fixed ("weak scale") and with M^2 set by the condition $<\phi> = 0$. The formal generating functional for connected Green's functions will be $lnZ[J]$ where

$$Z[J] = N \int D\phi exp\Big\{ - \int d^4x \Big[\mathcal{L} - J\phi \Big] \Big\}$$

Introducing the operator $\Delta^{-1}(x,y) = -\partial^2\delta(x-y)$ we write

$$Z[J] = N\,D \int D\phi exp\Big\{-\int d^4x\Big[L_{int} - J\phi\Big]\Big\} \equiv D\,Z_0[J]$$

Where L_{int} is the interaction term of the Lagrangian and

$$D = exp\Big\{-\frac{1}{2}\int d^4x d^4y \frac{\delta}{\delta J(x)}\Delta^{-1}(x,y)\frac{\delta}{\delta J(y)}\Big\}$$

Going to a four dimensional Euclidean lattice of spacing a and lattice sites i we get

$$Z_0[J] = N'\prod_i \frac{F(J_i)}{F(0)} = N' exp\Big\{\sum_i ln\frac{F(J_i)}{F(0)}\Big\}$$

$$F(J) = \int_{-\infty}^{+\infty} dx exp\Big\{-a^4\Big[L_{int} - Jx\Big]\Big\}$$

Continuum regulation is obtained by replacing

$$a^{-4} \to \delta_\Lambda(0) = \frac{\Lambda^4}{32\pi^2} \qquad \Delta^{-1}(x,y) \to \Delta_\Lambda^{-1}(x,y) = -\partial^2\delta_\Lambda(x-y)$$

and finally by smearing the vertices via

$$J(x) \to J_\Lambda(x) = \int d^4y\,\delta_\Lambda(x-y)J(y)$$

Thus

$$Z_0[J] = N' exp\Big\{\frac{\Lambda^4}{32\pi^2}\int d^4x\, ln\frac{F[J]}{F[0]}\Big\}$$

The formal vertices of the theory are defined by the coefficients in the Taylor expansion of $lnZ_0[J]$

$$lnZ_0[J] = const + \int d^4x \sum_{n=1}^{\infty}\frac{V_n}{n!}J^n(x)$$

Given the vertices we construct strong-coupling Feynman diagrams in the following way . Each external line picks up a factor $\theta(\Lambda - |p|)$, while for each internal line there is a propagator $-\Delta^{-1}(p) = -p^2\theta(\Lambda - |p|)$. Finally for each loop we integrate $(2\pi)^{-4}\int d^4p$, with an overall momentum-conserving δ function at the vertices .

It is convenient to introduce one-particle irreducible (1PI) diagrams with respect to Δ^{-1} lines . Δ_n will denote the sum of all such diagrams with n external legs. In this way we can evaluate Δ_1 , the sum of 1PI tadpoles , and find a solution for M^2 of the equation $<\phi> = 0$. In lowest order in $1/m$ the solution is

$$M^2 = -\frac{3}{32\pi^2}G\Lambda^4 \ll m^2$$

Next to leading we find

$$M^2 = -\frac{3}{32\pi^2}G\Lambda^4 - \frac{57}{4096\pi^4}\frac{G^2\Lambda^8}{m^2} - \frac{1}{16\pi^2}\frac{G\Lambda^6}{m^2}$$

Therefore order by order we can adjust M^2 to make the $1/m$ perturbative vacuum stable. The two-point function for the ϕ-field becomes in terms of Δ_2

$$G_2 = \frac{\theta(\Lambda - |p|)}{\Delta^{-1}(p) + \Delta_2^{-1}}$$

which has a pole at

$$p^2 = -m_H^2 = -m^2\Big[1 + \frac{\Lambda^2}{m^2}(1 - \frac{69}{128}\frac{G\Lambda^2}{\pi^2}) + O(\frac{1}{m^4})\Big]$$

The method allows to compute the physical Higgs mass in the $m \to \infty$ regime. Higher order corrections are small and negligible . As a matter of fact for $m \gg \Lambda$ there is no pole at all , since all propagators contain a factor $\theta(\Lambda - \mid p \mid)$. This intriguing result would be particularly appealing if valid also for the more realistic $SU(2)$ doublet case . Indeed this would correspond to the Higgs system viewed as an effective low energy theory valid for $p^2 \ll \Lambda^2 \simeq 1TeV^2$ but with no physical scalar degrees of freedom ($m^2 \gg \Lambda^2$) . In the present scheme $m_H^2 = m^2$, the mass parameter , plus negligible corrections for $m^2 \to \infty$, on the contrary of what happens in weak-coupling expansion where the Higgs pole m_H migrates away from the perturbative value as the scalar self-coupling increases. The $O(1)$ model for $m^2 \to \infty$ is a computable empty model .

Next we turn to an $SU(2)$ Higgs-doublet Φ . The fields are H and $\vec{\omega}$ with a well known Lagrangian

$$\mathcal{L} = \frac{1}{2}\partial_\mu H \partial_\mu H + \frac{1}{2}m^2H^2 + \partial_\mu\omega^\dagger\partial_\mu\omega + \frac{M^2}{\sqrt{2}G^{1/2}}H + \frac{1}{2}M^2\phi^2 + \frac{1}{\sqrt{2}}G^{1/2}m^2H\phi^2 + \frac{1}{4}Gm^2(\phi^2)^2$$

where $\phi^2 = H^2 + 2\omega^\dagger\omega = H^2 + \omega_0^2 + 2\omega^+\omega^-$. If J is the source associated with a field ϕ we define D_ϕ as the functional differential operator which pulls out the corresponding kinematical term from the functional integral . The generating functional will be

$$Z[J,\vec{\Omega}] = N\, D_H D_{\vec{\omega}}\; Z_0[J,\vec{\Omega}]$$

$$Z_0[J,\vec{\Omega}] = N' exp\Big\{\frac{\Lambda^4}{32\pi^2}\int d^4x \ln\frac{F[J,\vec{\Omega}]}{F[0]}\Big\}$$

Again the function $F(J,\vec{\Omega})$ is written

$$F(J,\vec{\Omega}) = \int_{-\infty}^{+\infty} dH d\omega_0 \ldots d\omega_2 exp\Big\{-\frac{32\pi^2}{\Lambda^4}\Big[\mathcal{L}_{int} - JH - \vec{\Omega}\vec{\omega}\Big]\Big\}$$

Introducing polar coordinates for the ω fields the angular integration can be performed and the coefficients $f_{n,2l_0,2l_1,2l_2}$ of the Taylor expansion of F in powers of $J, \Omega_0^2, \Omega_1^2$ and Ω_2^2 are

$$f_{n,2l_0,\ldots,2l_2} = 2(\frac{32\pi^2}{\Lambda^4})^{n+2l}\frac{\prod_i \Gamma(l_i + \frac{1}{2})}{\Gamma(l + \frac{3}{2})} f_{n,2l}$$

Where $l = \sum_{i=0}^{2} l_i$. Also

$$f_{n,2l} = \int_{-\infty}^{+\infty} dH H^n \int_0^{+\infty} d\omega\,\omega^{2l+2} exp\Big\{-\frac{32\pi^2}{\Lambda^4}g(H,\omega^2)\Big\}$$

$$g(H,\omega^2) = \frac{1}{4}Gm^2\omega^4 + \frac{1}{2}Gm^2K(H)\omega^2 + \frac{1}{4}Gm^2K^2(H) - \frac{1}{4}\frac{M^4}{Gm^2}$$

being

$$K(H) = H^2 + \sqrt{2}G^{-1/2}H + \frac{M^2}{Gm^2}$$

The coefficients $f_{n,2l}$ can be expressed as an asymptotic expansion in $1/m$. For instance the $n = l = 0$ coefficient is

$$f_{00} = \frac{\pi}{4\sqrt{2}} \frac{\Lambda^3}{G^{3/2} m} \times (\textit{power series in } 1/m^2)$$

From the above equations we immediately notice that all the coefficients $f_{n,2l}$ and consequently all the vertices of the model are of the same order in $1/m$.

Moreover the adjustable parameter M^2 never appears in lowest order, which makes the $1/m$ expansion totally unreliable. To further discuss this point we start again from the definition of $f_{n,2l}$ and perform the ω —integration. We get

$$f_{n,2l} = \frac{1}{2}\Big(\frac{16\pi^2}{\Lambda^4} G m^2\Big)^{-l/2-3/4} \Gamma(l+3/2) \int_{-\infty}^{+\infty} dH H^n exp\Big\{-\frac{4\pi^2}{\Lambda^4} G m^2 K^2(H) + \frac{8\pi^2}{\Lambda^4}\frac{M^4}{G m^2}\Big\} \times D_{-l-3/2}\Big[\frac{4\pi}{\Lambda^2} G^{1/2} m K(H)\Big]$$

Where D denotes a parabolic cylinder function. A reliable perturbation theory can be constructed if the problem of computing $f_{n,2l}$ is reducible to the one of small field oscillations around a stable minimum.

The H-integral in $f_{n,2l}$ is computed for large m with Laplace's method. The terms in the exponential correspond to a double well and one of the minima is actually at $H = 0$. However the D-function destroys this behavior. Apart from constant terms the integrand can be written as

$$I_\nu(H) = exp\Big\{-\frac{\lambda^2}{4} K^2(H)\Big\} D_{-\nu}\Big[\lambda K(H)\Big]$$

$$\lambda = \frac{4\pi}{\Lambda^2} G^{1/2} m \to \infty \qquad \nu = l + 3/2$$

Using the well known properties of parabolic cylinder functions we derive

$$\frac{dI_\nu}{dH} = -\lambda(2H + \sqrt{2} G^{-1/2}) I_{\nu-1}(H)$$

For real ν, $D_{-\nu}(z)$ has $[1-\nu]$ real zeros, where $[1-\nu]$ denotes the greatest positive integer less than $1-\nu$ or zero if such positive integer does not exist. For $\nu = l + 3/2$ and $l = 0, 1 \ldots I_{\nu-1}$ has no real zeros and I_ν has only one minimum for $H = -\frac{1}{\sqrt{2}} G^{-1/2} = -\sqrt{2} v$.

We conclude from this analysis that regardless from M^2, $H = 0$ is never a minimum unless the vacuum expectation value v is zero. For large m the shifted theory seems to be inconsistent, at least from the point of view of a perturbative expansion and within the assumptions implicit in the present scheme.

We have considered the Higgs system as an effective low energy theory valid for $p^2 \ll \Lambda^2 \simeq 1 TeV^2$ Above the cutoff new physics must show up. In this scheme whenever the bare mass parameter m or equivalently the self-interaction parameter λ go to infinity then the strong-coupling perturbative vacuum of the shifted theory becomes unstable for all values of the vacuum expectation value v different from zero.

This result obtains from a continuum regulated strong-coupling expansion. It is clear that such a method implies a series of formal manipulations but however the outcome is simple and neat. The emerging result goes in the direction of an increasing skepticism about the Higgs system.

REFERENCES

[1] M.Veltman, Acta Physica Polonica B8(1977)475 ; Phys. Lett. 70B(1977)254.
B.Lee,C.Quigg and R.Thacker, Phys. Rev. Lett. 38(1977)883.

[2] T.Appelquist and R.Shankar, Nucl. Phys. B158(1979)317
T.Appelquist and C.Bernard, Phys. Rev. D22(1980)200
A.C.Longhitano, Phys. Rev. D22(1980)1166
T.Appelquist and C.Bernard, Phys. Rev. D23(1981)425
A.C.Longhitano, Nucl. Phys. B188(1981)118
R.Akhoury and Y.P.Yao, Phys. Rev. D25(1982)3361
J.J. van der Bij and M.Veltman, Nucl. Phys. B231(1984)205
J.J. van der Bij, Nucl. Phys. B248(1984)141
M.Einhorn, Nucl. Phys. B246(1984)75
M.Veltman, Phys. Lett. 139B(1984)307
R.Casalbuoni,D.Dominici and R.Gatto, Phys. Lett. 147B(1984)419
P.Q.Hung and H.B.Thacker, Phys. Rev. D31(1985)2866
R.Casalbuoni,S.De Curtis,D.Dominici and R.Gatto, Genevra Preprint UGVA-DPT 1985/02-456

[3] B.Lee,C.Quigg and R.Thacker, Phys. Rev. Lett. 38(1977)883.

[4] M.Veltman, Proc. Intern. High Energy Conf. (Brighton, July 1983)

[5] C.Bender,F.Cooper,R.Khenway and L.M.Simmons Jr., Phys. Rev. D24(1981)2693

Plane Waves in Supergravity Theories

A.A.Beler
Physics Department
Middle East Technical University
Ankara, Turkey

T.Dereli
TUBITAK Research Institute for Basic Sciences
P.O.Box 74, Gebze-Kocaeli, Turkey
and
Department of Science Education
Middle East Technical University
Ankara, Turkey

In this note, we present exact plane wave solutions of supergravity theories. These solutions generalise the parallel-plane wave solutions of Einstein's equations, however, they cannot be reduced to them by supersymmetry transformations. They are non-trivial in the above sense. The nice features of parallel-plane gravitational fields, such as the validity of superposition principle, are also enjoyed by these plane wave solutions of supergravity theories.

N=1 Simple supergravity:[1]

The simple supergravity field equations are written in terms of exterior differential forms as follows:

$$\frac{1}{2} R_{bc} * (e^a \wedge e^b \wedge e^c) = \frac{i}{2} \bar{\psi} \wedge \gamma_5 \gamma^a \, D\psi \tag{1}$$

$$\gamma \wedge D\psi = 0 \tag{2}$$

$$T^a = \frac{i}{4} \bar{\psi} \wedge \gamma^a \, \psi \, . \tag{3}$$

We use a natural system of units in which $\hbar = c = 8\pi G = 1$. $\psi = \psi_a e^a$ is an odd Grassmann real spinor valued 1-form where e^a label the orthonormal basis 1-forms. $D = d + 1/2 \; \omega^{ab} \; \sigma_{ab}$ is the exterior covariant derivative and $\gamma=\gamma_a e^a$. The connection 1-forms $\omega^a{}_b$ are determined from the Cartan structure equations

$$de^a+\omega^a{}_b\wedge e^b=T^a \quad , \tag{4}$$

and the curvature 2-forms $R^a{}_b$ are given by

$$R^a{}_b=d\omega^a{}_b+\omega^a{}_c\wedge\omega^c{}_b \ . \tag{5}$$

*denotes the Hodge map with $^*1=e^o\wedge e^1\wedge e^2\wedge e^3$.

We will employ in our calculations a complex null basis that enable us to exploit the analytic structure of supergravity equations better. The space-time metric is given in this basis by

$$g=2\ell\otimes n+2m\otimes\bar{m} \tag{6}$$

where the null basis 1-forms

$$\ell=\frac{1}{\sqrt{2}}\,(e^3+e^o),\ n=\frac{1}{\sqrt{2}}\,(e^3-e^o),\ m=\frac{1}{\sqrt{2}}\,(e^1+ie^2). \tag{7}$$

$\otimes$ denotes symmetric tensor product, and a bar over a symbol implies complex conjugation. We define complex connection 1-forms

$$\omega_+=\omega^1+i\omega^2,\quad \omega_-=\omega^1-i\omega^2,\quad \omega_o=\omega^3 \tag{8}$$

ω here

$$\omega^k=-\frac{1}{2}\,(i\omega^o{}_k+\frac{1}{2}\,\varepsilon_{kij}\omega^i{}_j)\ ,\ k=1,2,3. \tag{9}$$

The following expansion

$$\gamma=\gamma_v\ell+\gamma_u n+\gamma_+\bar{m}+\gamma_- m \tag{10}$$

defines the projection matrices

$$\gamma_u=\frac{1}{\sqrt{2}}\,(\gamma_3-\gamma_o)\ ,\ \gamma_v=\frac{1}{\sqrt{2}}\,(\gamma_3+\gamma_o)\ ,\ \gamma_\pm=\frac{1}{\sqrt{2}}(\gamma_1\pm i\gamma_2)\ . \tag{11}$$

To demonstrate our solutions, we first specify the space-time metric in a co-ordinate chart $(u,v,z,\bar{z})$ by setting

$$g=2H\,du^2+2dudv+2dzd\bar{z}\ . \tag{12}$$

$H(u,z,\bar{z})$ is real analytic function to be determined by the field equations. We identify the null basis 1-forms

$$\ell=du \ , \quad n=dv+H\,du, \ m=dz \ . \tag{13}$$

The above metric describes parallel-plane gravitational waves propagating in $\partial/\partial v$ direction with u=constant being the wave hypersurfaces. Next, we write an ansatz for the gravitino field

$$\psi = \xi\, du + \phi\, d\bar{z} + \phi^* dz. \tag{14}$$

$\xi(u,z,\bar{z})$ is a (odd Grassmann) real spinor and $\phi(u,\bar{z})$ is a (odd Grassmann) complex spinor. We have $\phi=\phi^1+i\phi^2$ and $\phi^*=\phi^1-i\phi^2$, ψ is required to be chiral $\gamma_u\psi=0$, transverse $d{*}\psi=0$, and traceless. $*\gamma_\wedge\psi=0$. Then the ansatz (14) describes progressive gravitino waves with definite helicity traveling in $\partial/\partial v$ direction.

The components of the torsion 2-forms computed from the ansatz (14) are

$$\frac{1}{\sqrt{2}}(T^3-T^o)=0, \ \frac{1}{\sqrt{2}}(T^3+T^o)=iK\, m_\wedge\bar{m}+N\, \ell_\wedge\bar{m}+\bar{N}\, \ell_\wedge m,$$
$$\frac{1}{\sqrt{2}}(T^1+iT^2)=M\, \ell_\wedge\bar{m} \tag{15}$$

where the (even Grassmann) functions $K=-1/\sqrt{2}\,\phi^+\phi$, $N=-i/\sqrt{2}\,\xi^+\phi$, $M=i/2\,\bar{\xi}\gamma_+\phi$. Putting the basis 1-forms (13) together with (15) into the structure equations (4), the following connection 1-forms are found:

$$\omega_o=-\frac{K}{4}\ell \ , \quad \omega_+=0 \ ,$$
$$\omega_-=i(H,_z+N)\ell + \frac{K}{2}\bar{m} + i\bar{M}m \ . \tag{16}$$

A comma after a symbol shows partial differentiation with respect to the coordinates that follow it. After having obtained a complete description of the space-time geometry, a straightforward calculation shows that the Rarita-Schwinger equation(2) reduces to

$$\gamma_\wedge D\psi=-(\gamma_+\xi,_z-\gamma_-\xi,_{\bar{z}})\ \ell_\wedge m_\wedge\bar{m} \ . \tag{17}$$

It is satisfied if we let $\gamma_+\xi_{,z}=0$. Similarly, the Einstein's equations (1) reduce to

$$2H_{,z\bar{z}} = \frac{i}{\sqrt{2}} (\phi^+\phi_{,u} - \phi^+_{,u}\phi) - 2(N_{,z} + \bar{N}_{,\bar{z}}) \tag{18}$$

$$K_{,z}=0 \ , \quad M_{,z}=0.$$

The last two relations are satisfied if we let $\phi_{,\bar{z}}=0$. The remaining Poissson's equation can be formally integrated:

$$H=H^{(o)} + \frac{i}{2\sqrt{2}} z\bar{z} \ (\phi^+\phi_{,u} - \phi^+_{,u}\phi) - \int^{\bar{z}} N \, d\bar{z} - \int^{z} \bar{N} \, dz. \tag{19}$$

$H^{(o)}$ is a solution to the homogeneous equation and specifies a vacuum Einstein space.

Suppose we started from a vacuum Einstein space described by a function $H^{(o)}$, and generated the gravitino field by the infinitesimal supersymmetry transformation

$$\delta e^a = \frac{i}{2} \bar{\varepsilon}\gamma^a\psi \quad . \ \delta\psi = D\varepsilon \tag{20}$$

with a real spinor parameter $\varepsilon(u,z,\bar{z})$ satisfying $\gamma_u\varepsilon=0$. Then, the generated gravitino ansatz will be such that

$$\xi_G = \varepsilon_{,u} \qquad , \ \phi_G = \varepsilon_{,\bar{z}} \ . \tag{21}$$

Therefore the trivial solutions can be characterised by the integral expression

$$\phi = \int^u \xi_{,z} \, du \ . \tag{22}$$

We note that both the AD-type solutions ($\phi \neq 0$, $\xi = 0$) and the U-type solutions ($\phi=0$, $\xi\neq0$) are non-trivial.

N=2 extended supergravity:[2]

N=2 extended supergravity theory is a locally supersymmetric generalisation of the coupled Einstein-Maxwell theory. N=2 supergravity field equations are written in terms of exterior forms as follows:

$$\frac{1}{2} R^{bc} \wedge {}^*(e_a \wedge e_b \wedge e_c) = \frac{i}{2} \bar{\psi}^k \wedge \gamma_5 \gamma_a D\psi^k + (F_{ac}F^c{}_b + \frac{1}{2}\eta_{ab}F_{cd}F^{cd})^* e^b$$

$$+ \frac{1}{\sqrt{2}}[\varepsilon^{kj}(\bar{\psi}^k{}_a \psi^j{}_c)F^c{}_b + \varepsilon^{kj}(\bar{\psi}^k{}_b \psi^j_c)F^c{}_a$$

$$+ \frac{1}{2}\eta_{ab}\varepsilon^{kj}(\bar{\psi}^k{}_c \psi^j{}_d)F^{cd}] \, {}^*e^b$$

$$- \frac{1}{2}[\varepsilon^{kj}(\bar{\psi}^k{}_a \psi^j{}_c)\varepsilon^{i\ell}(\bar{\psi}^{ic}\psi^\ell{}_b)$$

$$+ \frac{1}{4}\eta_{ab}\varepsilon^{kj}(\bar{\psi}^k{}_c \psi^j{}_d)\varepsilon^{i\ell}(\bar{\psi}^{ic}\psi^{\ell d})]^* e^b, \tag{23}$$

$$i\gamma_5\gamma \wedge D\psi^k - \frac{i}{2} T^a \wedge \gamma_5 \gamma_a \psi^k = -\frac{1}{\sqrt{2}} \varepsilon^{kj}\gamma_5\psi^j \wedge F + \frac{i}{\sqrt{2}} \varepsilon^{kj}\psi^j \wedge {}^*F$$

$$- \frac{1}{4} \varepsilon^{kj}\psi^j \wedge \varepsilon^{i\ell} {}^*(\bar{\psi}^i \wedge \psi^\ell)$$

$$+ \frac{1}{8} \varepsilon^{kj}\gamma_5\psi^i \wedge \varepsilon^{i\ell}(\bar{\psi}^i \wedge \psi^\ell)$$

$$- \frac{1}{8} \varepsilon^{kj}\psi^j \wedge \varepsilon^{i\ell}(\bar{\psi}^i \wedge \gamma_5 \gamma^\ell), \tag{24}$$

$$d^*F = -\frac{1}{2\sqrt{2}} \varepsilon^{kj} d^*(\bar{\psi}^k \wedge \psi^j) + \frac{1}{2\sqrt{2}} \varepsilon^{kj} d(\bar{\psi}^k \wedge \gamma_5 \psi^j), \quad dF=0 \tag{25}$$

$$T^a = \frac{i}{4} \bar{\psi}^k \wedge \gamma^a \psi^k . \tag{26}$$

Indices $i,j,k,...=1,2$ and refer to a global SO(2) symmetry. We take $\varepsilon^{12}=-\varepsilon^{21}=1$ and $\varepsilon^{11}=\varepsilon^{22}=0$. $F= 1/2\, F_{ab}e^a \wedge e^b$ is the Maxwell 2-form.

The plane wave solutions are described by the parallel-plane gravitational wave metric (12), together with a gravitino ansatz written below in a particular γ-matrix representation,

$$\psi^k = \begin{pmatrix} \mathrm{Re}\beta^k \\ \mathrm{Re}\beta^k \\ -\mathrm{Im}\beta^k \\ \mathrm{Im}\beta^k \end{pmatrix} du + \begin{pmatrix} \alpha^k \\ \alpha^k \\ i\alpha^k \\ -i\alpha^k \end{pmatrix} dz + \begin{pmatrix} \bar{\alpha}^k \\ \bar{\alpha}^k \\ -i\bar{\alpha}^k \\ i\bar{\alpha}^k \end{pmatrix} d\bar{z}, \quad k=1,2. \tag{27}$$

$\alpha^k(u,z)$ and $\beta^k(u,z,\bar{z})$ are (odd Grassmann) complex analytic functions.

The Maxwell 2-form

$$F = du \wedge \eta, \tag{28}$$

with $\eta(u,z,\bar{z})$ real analytic 1-form satisfying $d\eta=0$, characterises a circularly polarized electromagnetic wave traveling along $\partial/\partial v$. Then with the above assumptions N=2 extended supergravity equations (23)-(25) are satisfied provided

$$2H_{,z\bar{z}} = |\iota_{\partial/\partial z}\eta|^2 - i2\sqrt{2}(\beta^k{}_{,z}\bar{\alpha}^k - \alpha^k\bar{\beta}^k{}_{,\bar{z}}) - i\sqrt{2}(\alpha^k\bar{\alpha}^k{}_{,u} - \alpha^k{}_{,u}\bar{\alpha}^k)$$

$$\alpha^k{}_{,z} = 0$$

$$\beta^k{}_{,z} = \sqrt{2}\,\varepsilon^{kj}\alpha^j(\iota_{\frac{\partial}{\partial\bar{z}}}\eta)$$

$$d^*\eta = 0\,. \tag{29}$$

The above equations, if desired, may be integrated formally.

Trivial solutions are generated from a parallel-plane wave solution of the Einstein-Maxwell equations characterised by $H^{(o)}$ and η, by infinitesimal local supersymmetry transformations

$$\delta e^a = \frac{i}{2}\bar{\varepsilon}^k\gamma^a\psi^k\,, \quad \delta\psi^k = \hat{D}\varepsilon^k\,, \quad \delta A = i\sqrt{2}\,\varepsilon^{kj}(\bar{\varepsilon}^k\psi^j)\,, \tag{30}$$

with real spinor parameters $\varepsilon^k(u,z,\bar{z})$ such that $\gamma_u\varepsilon^k=0$. Then, the generated gravitino ansatz will be specified by $(\varepsilon=\varepsilon^1+i\varepsilon^2)$

$$\xi_G = \xi^1 + i\xi^2 = \varepsilon_{,u} - 4\sqrt{2}[(\iota_{\frac{\partial}{\partial z}}\eta)\gamma_+\varepsilon + (\iota_{\frac{\partial}{\partial z}}\eta)\gamma_-\varepsilon]$$

$$\phi_G = \varepsilon_{,\bar{z}}\,, \tag{31}$$

In this case the AD-type solutions are still non-trivial, but U-type solutions include some trivial solutions.

Eleven dimensional supergravity :[3]

The parallel-plane gravitational waves in a d-dimensional space-time are described by the metric

$$g = 2H(u,x^i)du^2 + 2du\,dv + \sum_{i=2}^{d} dx^i dx^i\,. \tag{32}$$

In what follows we restrict attention to U-type plane wave solutions of eleven dimensional supergravity equations. Accordingly, we write the gravitino ansatz

$$\psi = \xi(u,x^i)du \; . \tag{33}$$

ξ is an odd Grassmann 32 component real spinor satisfying $\Gamma_u\xi=0$. Then, all terms non-linear in the gravitino field cancel in the supergravity equations, and we are left with the bosonic part of the supergravity equations

$$R^{BC}\wedge {}^*(e_A\wedge e_B\wedge e_C) = \kappa(1_A F\wedge {}^*F - F\wedge 1_A {}^*F) \tag{34}$$

$$d^*F = \alpha F\wedge F \quad , \; dF=0$$

together with the background Rarita-Schwinger equation

$${}^*(\Gamma\wedge\Gamma\wedge\Gamma)\wedge D\psi=0 \; . \tag{35}$$

F is the Maxwell 4-form.
κ is the gravitational coupling constant in d=11 dimensions. α is a constant which is determined by local supersymmetry. We consider

$$F=du\wedge\eta \tag{36}$$

where $\eta(u,x^i)$ is a 3-form satisfying $d\eta=0$. Then the metric (32) together with the 4-form (36) satisfy the supergravity equations provided

$$d^*dH = \frac{\kappa}{840}\,\eta\wedge{}^*\eta \; ,$$

$$d^*\eta=0 \; .$$

References

[1] P.C.Aichelburg, T.Dereli, Phys.Rev.D18 (1978) 1754
L.Urrutia, Phys.Lett. 102B (1981) 393
A.Beler, T.Dereli, Class.Q Grav. 2 (1985) 147

[2] P.C.Aichelburg, T.Dereli, Phys.Lett. 80B (1979) 357
C.M.Hull, Phys.Rev. D30 (1984) 334
A.Beler, T.Dereli, Class.Q.Grav. 2 (1985)

[3] J.Kowalski-Glikman, Phys.Lett. 134B (1984) 194
C.M.Hull, Phys.Lett 139B (1984) 39
T.Dereli, M.Gürses "Generalised Kerr-Schild transform in eleven dimensional supergravity" preprint (1985)

MULTITEMPORAL CLASSICAL AND QUANTUM PARTICLE MECHANICS

Luca Lusanna

Sezione INFN di Firenze, Italy.

The relativistic and non-relativistic dynamics of two particles in direct interaction is formulated as a theory with first-class constraints and its canonical quantization without gauge-fixings is performed. A two-time description of relativistic bound-states with a well defined invariant scalar product is obtained.

It has been realized that the most fruitful approach to the dynamics of two relativistic particles in direct interaction[1] is by means of two first-class constraints[2] implying generalized mass-shell relations:

$$\chi_i = p_i^2 - m_i^2 - V(r_\perp^2, p^2, q^2, q\cdot r_\perp) \approx 0 \quad i=1,2 \qquad \{\chi_1, \chi_2\} = 0 \tag{1}$$

where $p^\mu = p_1^\mu + p_2^\mu$, $q^\mu = \frac{1}{2}(p_1^\mu - p_2^\mu)$, $r^\mu = x_1^\mu - x_2^\mu$, $r_\perp^\mu = (\eta^{\mu\nu} - \frac{p^\mu p^\nu}{p^2})\, r_\nu$, $\eta^{\mu\nu} = (+---)$.

This model has been extended to non-scalar interactions and to spin $\frac{1}{2}$ particles[3,4] with good phenomenological applications to the meson spectrum[5], by using a covariant generalization of static quark potentials.

Moreover G.Longhi and I discovered that the non-relativistic limit of eqs.(1) gives the following two first-class constraints (here and in the following discussion only the simplest case $V(r_\perp^2)$ will be considered):

$$\hat{\chi}_i = E_i - \frac{\vec{p}_i^{\,2}}{2m_i} - \frac{V(\vec{\rho}^{\,2})}{2m_i} \approx 0 \quad i=1,2 \quad \{\hat{\chi}_1, \hat{\chi}_2\} = 0 \quad \vec{\rho} = \vec{r} - (t_1 - t_2)\frac{\vec{P}}{m_1+m_2} \tag{2}$$

allowing an analogous reformulation of Newton mechanics.[1,6]

There is a twofold motivation for studying the models described by the constraints (1) and (2). The first one is to look for a classical basis for the Schwinger-Dyson equations and in particular for the relativistic bound-state problem. The first-class property $\{\chi_1, \chi_2\} = 0$ restricts the variables upon which the potential can depend to the ones shown in eqs.(1): as a result dynamical recoil effects and energy dependence, characteristic of field theory, emerge. For weak potentials Todorov's homogeneous quasipotential equations are recovered and for slow motions a canonical equivalence with the Darwin Hamiltonian, i.e. with the Fermi-Breit approximation to the Bethe-Salpeter equation, is present[3]. What has to be clarified is the role of the relative time and the two-time Newton mechanics of eqs.(2) is a simple laboratory where to investigate this problem, becouse in the non-relativistic case it is clear which is the one-time theory. This is connected to the second motivation for this study, that is the understanding of the classical and quantum aspects of the simplest non trivial models of gauge theories described by first-class constraints, which are the generators of canonical gauge transformations. In particular it was understood how to do the canonical quantization without gauge-fixings of these models when anomalies are not present.

Instead of eqs.(1) it is more useful to consider the following combinations of the first-class constraints:

$$\begin{cases} \chi_- = \frac{1}{2}(\chi_1 - \chi_2) = P\cdot q - \frac{1}{2}(m_1^2 - m_2^2) \approx 0 \\ \chi_+ = 2(\chi_1 + \chi_2) = P^2 + 4(q^2 - V(r_\perp^2)) - 2(m_1^2 + m_2^2) = \frac{1}{P^2}\left[(P^2 - M_+^2)(P^2 - M_-^2) + 4\chi_-(\chi_- + m_1^2 - m_2^2)\right] \approx 0 \end{cases} \tag{3}$$

with $M_\pm^2 = (M_1 \pm M_2)^2$, $M_i^2 = m_i^2 - (q_\perp^2 - V(r_\perp^2))$, i=1,2. χ_- determines the relative energy in the center-of-mass frame, while χ_+ determines the mass spectrum. The gauge variable conjugated to χ_- is some relative time: the associated gauge-fixing $\Phi_- \approx 0$ selects one kind of 'instantaneity' of the interaction, that is it says between which pairs of points the given potential V acts instantaneously. Instantaneity in the center-of-mass frame is given by $\Phi_- = p r \approx 0$ (as suggested by the Bethe-Salpeter equation of which this model is an approximation with direct interactions), implying that $r^\mu(\tau)$ is a space-like vector: τ is a scalar evolution parameter and in this case it is associated with a foliation of the Minkowski space by means of space-like surfaces. Instantaneity on the light-cone is given by $\Phi_- = r^2(\tau) \approx 0$. Here are hidden the various forms of the dynamics of Dirac[8] and their gauge equivalence. The gauge-fixing $\Phi_+ \approx 0$ identifies the parameter τ with some overall time variable relevant to the chosen form of the dynamics and so it breaks the reparametrization invariance of the theory.
In configuration space to each gauge-fixing Φ_- will correspond a different pair of world-lines: from each set of Cauchy data a world-sheet, spanned by all the gauge-equivalent pairs of world-lines, will emerge (see ref.[1,7,9] for the search of the associated singular Lagrangians and a comparison, for now only at the non-relativistic level, with the Fokker action). The same pattern is obtained in the non-relativistic case where the natural gauge-fixing is $\tilde{\Phi}_- = t_1 - t_2 \approx 0$, giving the Newton theory. As a conclusion in the two cases the physical world-lines are identified by $t_1 - t_2 \approx 0$ due to Newton mechanics and by $p\cdot r \approx 0$ due to the relativistic Bethe-Salpeter equation respectively. All the other gauge-equi-

valent world-lines correspond to describe the same dynamics by means of the introduction of some kind of 'delay' with respect to the chosen physical concept of instantaneity.

In ref.[1,6,7] the off-shell Hamilton-Dirac equations with two arbitrary gauge functions $\lambda_i(\tau)$ and the equivalent Droz-Vincent[10] formulation with two parameters τ_1, τ_2 are described. Then it is described the on-shell theory, that is the theory of the observables A, $\{A, \chi_i\}=0$, which are functions of six phase-space variables for each particle and of two time-parameters. The connection with predictive mechanics[10] is also given.

The canonical quantization without gauge-fixings[7] of the on-shell theory is possible becouse there are no ordering problems and the first-class property is preserved $[\hat{\chi}_1, \hat{\chi}_2]=0$. It amounts to replace the definition of the observables $\{A, \chi_i\}=0$ with the following pair of wave equations (integrable due to the first-class property):

$$\hat{\chi}_i \psi = 0 \tag{4}$$

Both in the non-relativistic and relativistic case the bilocal wave function ψ is a function of six position eigenvalues and of two time parameters. Each line in the two-time plane is in correspondence with a gauge-fixing Φ in a certain class and a scalar product exists which is constant in both times due to eqs.(4). In the non-relativistic case[1,6] it has been studied a variational principle for eqs.(4), the evolution operator and the probabilistic interpretation of $|\psi|^2$; the theory of Green's functions is now under investigation. In the relativistic case[11,12] eqs.(4) are integro-differential equations. They have been solved and solutions for the relativistic bound-states belonging to irreducible representations of the Poincarè group have been found. These representations are unitary, becouse the non trivial problem of the invariant scalar product[10,13] has been solved and also associated conserved currents have been found. A non-local canonical transformation, implementable at the quantum level, transforms eqs.(4) into differential equations: it allows a definition of the Cauchy problem and the Poincarè group becomes realized non-linearly.

Future developments will concentrate upon the following problems: i) to find a model with two first-class constraints describing a realistic light-cone dynamics of the Fokker-Feynman-Wheeler type; ii) to study more deeply the connections of eqs.(1) with the Bethe-Salpeter equation[14] (see ref.[1] for the non-relativistic case); iii) to understand how to implement 'separability' in the three body case; iv) to study the problem of the anomalies generated by the ordering problems, which destroy the quantum first-class property and therefore the integrability of eqs.(4): the string model seems to be the simplest example where to carry out this investigation; v) to apply this technique to the electromagnetic and Yang-Mills fields with the extra complication of the need of the Schrödinger representation of quantum field theory; vi) to look for a manifestly covariant Hamiltonian formulation of classical field theory based upon the use of first-class constraints.

REFERENCES.

1) For a review see: L.Lusanna, From relativistic mechanics toward Green's functions: multitemporal dynamics, in Proc. of the VII Workshop on High Energy Physics and Field Theory, Protvino, USSR, 1984; Non-relativistic and relativistic multitemporal dynamics, in AIP Conf.Proc. on Hadron Spectroscopy, 1985, ed.S.Oneda.
2) I.T.Todorov, Dynamics of relativistic point particles as a problem with constraints, Comm. of the JINR, E2-10125, Dubna, 1976.
A.Komar, Phys.Rev. D18 (1978) 1881.
3) H.W.Crater and P.Van Alstine, Ann.Phys.(N.Y.) 148 (1983) 57; Phys.Rev. D30 (1984) 2585.
4) H.Sazdjian, Relativistic wave equations for two interacting particles and zero mass bound states, Orsay preprint, IPNO/TH 84-46, 1984.
5) H.W.Crater and P.Van Alstine, Phys.Lett. 100B (1981) 166; Phys.Rev.Lett. 53 (1984) 1527.
H.W.Crater, Relativistic extensions of heavy quark static potentials through two-body Dirac equations, in AIP Conf.Proc. on Hadron Spectroscopy, 1985, ed.S.Oneda.
6) G.Longhi, Multitime approach to non-relativistic and relativistic quantum mechanics, Talk at the Encuentros Relativisticos Espanolos, 1984, St.Ander.
7) L.Lusanna, Multitemporal relativistic particle mechanics: a gauge theory without gauge-fixings, Talk given at the IV Marcel Grossmann Meeting, Roma, 1985.
8) P.A.M.Dirac, Rev.Mod.Phys. 21 (1949) 392.
9) L.Lusanna, Nuovo Cimento 65B (1981) 135.
10) P.Droz-Vincent, Rep.Math.Phys. 8 (1975) 79; Nuovo Cimento 58A (1980) 355; Phys.Rev. D29 (1984) 687.
11) G.Longhi, Non linear realizations of the Poincaré group and multitemporal quantum mechanics, Talk given at the Encuentros Relativisticos Espanolos, 1985, Maõ, Menorca.

12) G.Longhi and L.Lusanna, Bound-state solutions, invariant scalar products and conserved currents for a class of two-body relativistic systems, Firenze Univ. preprint, 1985.
13) V.A.Rizov, H.Sazdjian and I.T.Todorov, On the relativistic quantum mechanics of two interacting spinless particles, Orsay preprint IPNO/TH 84-39, 1984.
V.Iranzo, J.Llosa, F.Marqués and A.Molina, Ann.Inst.H.Poincaré 40 (1984) 1.
14) H.Sazdjian, The quantum mechanical transform of the Bethe-Salpeter equation, Orsay preprint IPNO/TH 84-80, 1984.

BLACK HOLES IN N=8 SUPERGRAVITY THEORY

R.Güven
TÜBİTAK Research Institute for Basic Sciences
P.O.Box 74, Gebze, Kocaeli, Turkey

ABSTRACT

Solutions of the bosonic field equations of the ungauged N=8 supergravity which describe black holes with no scalar hairs are considered. It is shown that, in contrast to the Einstein-Maxwell theory where uniqueness theorems exist, there are two distinct families of black holes in N=8 supergravity. There are also two distinct generalizations of Majumdar-Papapetrou solutions which describe the static equilibrium of multi black holes.

In Einstein-Maxwell theory one can prove uniqueness theorems about the possible equilibrium configurations of black holes. If a black hole settles down to a stationary state then, according to these theorems, its exterior geometry is described uniquely by a Kerr-Newman solution [1-4]. When one notes this remarkable simplicity of the classical black holes together with the impetus of the discovery of the Hawking effect [5] on quantum gravity, one naturally inquires into the black holes of the supergravity theories. It has been observed in the studies of gravitational solitons that black holes can be instrumental in our understanding of the non-perturbative aspects of the supergravity theories [6,7]. The generalizations of black holes within the supergravity theories have also been considered and these have lead us to black holes having scalar "hairs" in N=4 supergravity [8], a "superhair" in N=2 theory [9] and to locally compactifying solutions in N=1, d=11 supergravity [10].

In this talk we wish to examine how black holes are generalized in the framework of N=8 supergravity theory. For this purpose we shall restrict our attention to the bosonic sector of the ungauged N=8 theory in four dimensions [11] and neglect the scalar hairs by setting the scalar fields equal to zero. With these assumptions the field equations of N=8 supergravity reduce, in the symmetric SU(8) gauge, to a generalized Einstein-Maxwell system which possesses a rigid SO(8) symmetry as well as an SU(8) duality invariance. We shall exhibit two distinct families of exact black hole solutions of these equations. This will enable us to conclude that the black hole uniqueness theorems are transcended in N=8 supergravity even after neglecting the possible scalar and fermionic hairs. It will also be seen that there are at least two different generalizations of Majumdar-Papapetrou solutions which describe the static equilibrium of multi black holes in N=8 supergravity [12].

Consider the bosonic sector of the ungauged N=8 supergravity theory [11] which is described by the Lagrangian density

$$L = \frac{1}{4} VR - \frac{1}{8} V[F^{+}_{\mu\nu IJ}(2S^{IJ,KL}-\delta^{IJ}_{KL})F^{+\mu\nu}_{KL}+\text{h.c.}]$$

$$- \frac{1}{96} VA^{ijk\ell}_{\mu} A^{\mu}_{ijk\ell} \quad . \tag{1}$$

The field variables that enter into (1) are the vierbein V^a_μ, 28 Abelian spin-1 gauge fields A^{IJ}_μ and a set of 35 complex scalars. A complete account of (1) can be found in [13] whose conventions for the group indices and the permutation tensor δ^{IJ}_{KL} we follow [14]. We wish to consider here the field equations that follow from (1) only in the case where the scalar fields are zero. For this purpose it is convenient to work in the SU(8) symmetric gauge [11,13] where the scalar fields are represented as the 35 complex components $\phi_{ijk\ell}$ of a self-dual four-form in the internal space:

$$\phi_{ijk\ell} = \frac{1}{24}\eta\, \varepsilon_{ijk\ell mnpq}\, \bar{\phi}^{mnpq} , \quad \eta^2 = 1 . \tag{2}$$

Varying (1) with respect to V^a_μ, A^{ij}_μ and $\phi_{ijk\ell}$ and then setting in the resulting field equations

$$\phi_{ijk\ell} = 0 , \tag{3}$$

gives us a generalized Einstein-Maxwell system

$$G_{\mu\nu} = -2T_{\mu\nu} , \qquad \nabla^{\mu} F^{+}_{\mu\nu ij} = 0 , \tag{4}$$

which is subject to the algebraic condition

$$\delta^{ijk\ell}_{mnpq} F^{-mn} F^{-\mu\nu pq} + \frac{1}{24} \eta\varepsilon^{ijk\ell mnpq} F^{+}_{\mu\nu mn} F^{+\mu\nu}{}_{pq} = 0 . \tag{5}$$

Here $F^{+}_{\mu\nu\ ij}$ are the self-dual spin-one field strengths, $G_{\mu\nu}$ is the Einstein tensor and

$$T^{\nu}{}_{\mu} = \frac{1}{2} (F^{+}_{\mu\lambda ij} F^{-\lambda\nu ij} + F^{-ij}_{\mu\lambda} F^{+\lambda\nu}{}_{ij}) , \tag{6}$$

is the energy-momentum tensor of the spin-one fields. Having neglected the fermionic fields, ∇_{μ} denotes the covariant derivative with respect to the Christoffel connection. The algebraic condition (5) follows from tha scalar field equations when (3) is imposed.

An interesting feature of N=8 supergravity is that while (1) is invariant under the group SO(8)(global) x SU(8)(local), the field equations have a much richer, E_7(global) x SU(8)(local) symmetry [11]. In general the transformations which leave the field equations form-invariant can be written as

$$\mathcal{F}(x) \to E\mathcal{F}(x) ,$$

$$\mathcal{V}(x) \to U(x)\mathcal{V}(x)E^{-1} , \tag{7}$$

where $E \in E_7$, $U \in SU(8)$; $\mathcal{V}$ denotes the 56-bein which describes the scalar fields in on SU(8) covariant manner and $\mathcal{F}$ is a certain 56-dimensional column matrix with two-form entries constructed from $S^{IJ,KL}$ and $F^{+}_{\mu\nu IJ}$ [13]. In our problem it follows from (3) that

$$\mathcal{F} = \begin{bmatrix} F^{+}_{ij} \\ 0 \end{bmatrix} , \qquad \mathcal{V} = \begin{bmatrix} \delta^{k\ell}_{ij} & 0 \\ 0 & \delta^{mn}_{pq} \end{bmatrix} , \tag{8}$$

where $F^{+}_{ij} = \frac{1}{2} F^{+}_{\mu\nu\ ij} dx^{\mu} \wedge dx^{\nu}$, and in the search of black hole solutions of (4)-(6) we must take into account this large on-shell symmetry of N=8 supergravity. Clearly, all (F, V) which are reducible to (8) by (7) can be regarded as equivalent. Together with the tetrad such F and V will describe, at least classically, the same black hole in different gauges. Moreover, it should be noted that even within the symmetric gauge the duality invariance group of the solutions is E_7. Therefore, one is dealing with gauge equivalent sets of solutions even after picking the symmetric gauge. Within the symmetric gauge E_7 is realized in a non-linear fashion [11,13] and one can exhibit the covariance of the expressions only with respect to the rigid, diagonal SU(8) subgroup of $E_7 \times SU(8)$. Consequently, (3)-(6) preserve their form only under rigid SU(8) duality rotations. Because of this fact one can reduce even some solutions with non-zero, constant $\phi_{ijk\ell}$ to the form (3) and (8) by duality rotations. Conversely, if one wishes to utilize the duality symmetry within the symmetric gauge and without generating spurious $\phi_{ijk\ell}$, one should consider only the transformations

$$F^{+}_{ij} \rightarrow S_i^{\ k} S_j^{\ \ell} F^{+}_{k\ell} \tag{9}$$

where $S_i^{\ k}$ are spacetime independent SU(8) matrix elements.

The black hole solutions of (4)-(6) that we wish to report are displayed in a convenient manner by employing the Newman-Penrose formalism [15] where one takes, instead of the vierbein $V^a_{\ \mu}$, the null tetrad basis one-forms $(\ell,n,m,\bar{m})$ as the gravitational field variables. These basis one-forms can be related to $V^a = V^a_{\ \mu}dx^\mu$ by [16]

$$V^a \sigma_a = \sqrt{2}\begin{bmatrix} n & -\bar{m} \\ -m & \ell \end{bmatrix} \tag{10}$$

where the spatial components of σ_a are the Pauli spin matrices and σ_o is the two-dimensional identity maxtrix. Consider now, in the coordinate system (t,r,θ,ϕ), the null tetrad

$$\ell = dt - dr/\Delta\rho\bar{\rho} - a \sin^2 d\phi ,$$
$$n = \frac{1}{2}\Delta\rho\bar{\rho}(dt + dr/\Delta\rho\bar{\rho} - a \sin^2\theta\, d\phi) ,$$
$$m = (\bar{\rho}/\sqrt{2})[-ia \sin\theta\, dt + d\theta/\rho\bar{\rho} + i(r^2 + a^2)\sin\theta\, d\phi] \tag{11}$$

and the spin-one field strengths

$$F^{-jk} = \overline{F^{+}_{jk}} = -\frac{1}{2} z^{jk} \rho^2 (\ell \wedge n - m \wedge \bar{m}) , \qquad (12)$$

where $\Delta = r^2 - 2Mr + a^2 + \frac{1}{2} z^{jk} \bar{z}_{jk}$, $\rho = -(r - ia\cos\theta)^{-1}$ and M , a are real, z^{ik} are complex constants. The fields given in (11) and (12) are simple generalizations of the corresponding expressions of the Kerr-Newman solution in the Boyer-Lindquist coordinates and it is easy to see, with the aid of [17], that the field equations (4) are satisfied. The remaining task is to ensure that the condition (5), which is due to the supersymmetric coupling between the spin-zero and spin-one fields, is also fullfilled. When (12) is substituted (5) implies that either

$$z^{i[j} z^{k\ell]} = 0 \qquad (13)$$

and the parameter a is arbitrary, or a = 0 and

$$z^{i[j} z^{k\ell]} + \frac{1}{24} \eta \, \varepsilon^{ijk\ell mnpq} \bar{z}_{mn} \bar{z}_{pq} = 0 . \qquad (14)$$

We must therefore determine all solutions of (14) and distinguish the case where the stronger condition (13) also holds. This can be achieved by utilizing the duality symmetry (9) of the field equations. These SU(8) rotations of (12) just amount to the changes $Z \rightarrow SZS^T$ where Z is the 8x8 skew symmetric matrix whose elements are z^{jk} , $S \in SU(8)$ and T is the transposition. After bringing Z to its normal form [18] by a duality rotation and then substituting into (14) and (13) we find that (5) can be satisfied only in two distinct ways. These are [12]

Case I :

$$z^{12} = z , \quad \text{other } z^{ik} = 0 , \text{ a arbitrary,} \qquad (15)$$

where z is a complex parameter and

Case II:

$$z^{12} = z^{34} = z^{56} = z^{78} = b \exp[i(1+\eta)\pi/8] ,$$

$$\text{other} \quad z^{ik} = 0 \quad ; \quad a = 0 , \qquad (16)$$

where b is an arbitrary real constant.

We have thus obtained in the same SU(8) gauge two families of solutions to the bosonic field equations of N=8 supergravity.In this particular gauge the nature of the solutions and the distinction between the families is clear. The first family (Case I) is just the N=8 covariantized version of the usual Kerr-Newman solution. In other words Case I represents those solutions of the N=8 theory which are reducible to the Kerr-Newman family by E_7 x SU(8) transformations. (Notice that (15) is just the requirement for truncating the solution to the bosonic sector of N=2 supergravity.) This family is, of course, an expected one and survives all lower N truncations of the N=8 theory intact down to the Einstein-Maxwell limit. After demanding $M^2 \geqq a^2 + \frac{1}{2} z^{jk} \bar{z}_{jk}$ one can infer that in (11) and (12) M, a are, respectively, the mass and angular momentum parameters and z^{jk} are the complex sums of the central electric and magnetic charges of a black hole. Therefore, first family allows both static and rotating black holes. The second family (Case II) is, however, new and much more restrictive than the first one: it exists as a seperate family only in N=8 supergravity and contains only non-rotating holes. Although the spacetime geometry of these black holes are described by the Reissner-Nordström metrics and M, z^{jk} are again the mass and central charge parameters, the central charges z^{jk} are totally different from those of the first family. These central charges cannot be mapped to those of a static Case I black hole by duality rotations. (It can be checked that whereas (15) solves both (13) and (14), (16) does not satisfy the stronger condition (13).) Moreover, the truncation of the second family to any N < 7 supergravity is possible only at the expense of setting all central charges equal to zero and then the families coalesce. (N=7 supergravity is just a relabelling of the N=8 theory.) Because of this fact the two families transcend the black hole uniqueness theorems of [3] and [4] only in N=8 supergravity.

It is clear that each of these solutions will be represented in an arbitrary SU(8) gauge by performing an E_7 x SU(8) transformation (7) and the local SU(8) invariance of the theory will be more manifest in such a representation. Conversely, we may inquire whether all black hole solutions having no scalar hairs in the bosonic sector of N=8 supergravity can be reduced to the above two families. At present a partial affirmative answer to this question can be furnished. It can be shown that all such black hole solutions will be reducible to (11)-

(16) provided: (i) the Weyl tensor is of Petrov type D, (ii) F^{-ij} is of type D in the sense of [19] and (iii) the principal null vectors of F^{-ij} are aligned with those of the Weyl tensor.

Finally let us note that the Majumdar-Papapetrou solutions of the Einstein-Maxwell theory also have at least two different generalizations in N=8 supergravity. These solutions describe the static equilibrium of an arbitrary number of extreme Reissner-Nordström black holes [20] and can be interpreted as multisolitons [6, 7]. Consider a Majumdar-Papapetrou solution describing n extreme holes with masses M_B (B=1,2,...,n) and charges $e_B = M_B$:

$$ds^2 = W^2(\vec{x})dt^2 - W^{-2}(\vec{x})[(dx^1)^2 + (dx^2)^2 + (dx^3)^2] . \tag{17}$$

$$A \equiv A_\mu dx^\mu = W(\vec{x})dt, \tag{18}$$

where $\vec{x} = (x^1, x^2, x^3)$ and the function W is given by

$$W^{-1}(\vec{x}) = 1 + \sum_{B=1}^{n} \frac{M_B}{r_B} , \tag{19}$$

$$r_B = [(x^1 - x^1_B)^2 + (x^2 - x^2_B)^2 + (x^3 - x^3_B)^2]^{1/2} . \tag{20}$$

Let $F^-_{\mu\nu}$ be the corresponding anti-self-dual Maxwell tensor and take as the spin-one fields of N=8 supergravity

$$F^{-jk}_{\mu\nu} = \beta^{jk} F^-_{\mu\nu} , \tag{21}$$

where β^{jk} are complex SU(8) constants satisfying $\beta^{jk} \bar{\beta}_{jk} = 2$. Then one can easily see that (21) together with (17) constitute a solution to (4) by the virtue of the Einstein-Maxwell equations. Moreover, when (21) is substituted, (5) reduces once again to (14) but this time (14) becomes a condition to be satisfied by the central charges $z^{jk}_B = M_B \beta^{jk}$ of each black hole. The reduction to the form (15) or (16) is now applicable to β^{jk} and therefore, we may conclude that there are also two distinct families of multi black hole solutions in N=8 supergravity. The n=1 members of these families are the Case I and Case II extreme Reissner-Nordström black holes and their previous forms can be recovered by setting $M_1 = M$ and $r_1 = r-M$.

I thank International Centre for Theoretical Physics, Trieste and tha Institute for Scientific Interchange, Torino, for their support during this meeting.

References and Footnotes

[1] B.Carter, Commun.Math.Phys. 30 (1973) 261.

[2] D.C.Robinson, Phys.Rev.Lett. 34 (1975) 905.

[3] G.Bunting, Report to Australasian Math.Convention, Sydney 1981, (unpublished).

[4] P.O.Mazur, J.Phys. A15 (1982) 3173.

[5] S.W.Hawking, Commun.Math.Phys. 43 (1975) 199.

[6] P.Hajicek, Nucl.Phys. B185 (1981) 254.

[7] G.W.Gibbons, in:Heisenberg Symp., ed. P.Breitenlohner and H.P.Dürr, Lecture Notes in Physics, Vol.160(Springer, Berlin, 1982) p.145.

[8] G.W.Gibbons, Nucl.Phys. B207 (1982) 337.

[9] P.C.Aichelburg and R.Güven, Phys.Rev.Lett.51 (1983) 1613.

[10] P. van Baal, F.A.Bais and P. van Nieuwenhuizen,Nucl.Phys. B233 (1984) 477.

[11] E.Cremmer and B.Julia, Phys.Lett.80B (1978) 48; Nucl.Phys. B159 (1971) 141.

[12] R.Güven, Phys.Lett. 158B (1985) 468.

[13] B. de Wit and H.Nicolai, Nucl.Phys. B208 (1982) 323.

[14] Throughout the paper a bar over a quantity denotes the complex conjugation and $\wedge$ denotes the wedge product. Our other conventions are same as those of Ref.[12].

[15] E.Newman and R.Penrose, J.Math.Phys. 3 (1962) 566.

[16] R.Güven, in: Unified field theories of more than four dimensions, eds. V. de Sabbata and E.Schmutzer. (World Scientific, Singapore, 1983) p.395.

[17] S.K.Bose, J.Math.Phys. 16(1975) 773.

[18] C.N.Yang, Rev.Mod.Phys. 35(1962) 694; B. Zumino, J.Math.Phys. 3 (1962) 1055.

[19] J.Anandan and K.P.Tod,Phys.Rev. D18 (1978) 1144.

[20] J.B.Hartle and S.W.Hawking, Commun.Math. Phys. 26(1972) 87.

LIST OF PARTICIPANTS

E. Abdalla	University of Sao Paulo, Brasil
D. Amati	Division Théorique, CERN, Ginevra
M. Anselmino	Università di Torino
A. Ballestrero	INFN, sezione di Torino
R. Barbieri	Scuola Normale Superiore, Pisa
S. Bellucci	ICTP, Trieste
E. Bergshoeff	ICTP, Trieste
M. Caselle	INFN, sezione di Torino
L. Castellani	Istituto Nazionale di Fisica Nucleare, sez. Torino
S. Ceccotti	Università di Pisa
V. De Alfaro	Istituto di Fisica Teorica, Università di Torino
T. Derelli	ICTP, Trieste
B. De Wit	University of Utrecht, Nederland
M. Duff	Division Théorique, CERN, Ginevra
F. Englert	Université Libre de Bruxelles, Belgio
D. Z. Freedman	M. I. T., Cambridge, Massachusetts, U. S. A.
S. Fubini	Division Théorique, CERN, Ginevra
G. Gamberini	Università di Pisa
B. Gasperini	Università di Torino
R. Gatto	Inst. de Physique Théorique, Univ. de Geneve, Svizzera
L. Girardello	Istituto di Fisica Teorica, Università di Milano
F. Gliozzi	INFN, sezione di Torino
P. Goddard	DAMPT, Cambridge University N. K, U. S. A.
E. Guadagnini	Università di Pisa
R. Guven	ICTP, Trieste
G. Hull	DAMPT, Cambridge University, N. K., U. S. A.
L. Ibáñez	Division Théorique, CERN, Ginevra
K. Konishi	Università di Pisa
J. Kowalsky	University of Amsterdam, Nederland
J. Lukierski	University of Wrocław, Poland
K. T. Mahantappa	Boulder, University of Colorado, U. S. A.
E. Maina	Università di Torino
G. Marchesini	Università di Parma
G. Marino	Università di Pisa
P. Menotti	Università di Pisa
M. Minchen	Università di Pisa

H. Neto	ICTP, Trieste
F. Nicodemi	Università di Napoli
H. Nicolai	Division Theórique, CERN, Ginevra
N. K. Nielsen	University of Copenhagen, Denmark
C. Orzalesi	Università di Parma ed Ambasciata d' Italia U. S. A.
G. Paffuti	Università Pisa
A. Pasquinucci	Università di Milano
G. Passarino	Università di Torino
M. Pusterla	Università di Padova
L. Radicati	Scuola Normale Superiore di Pisa
L. Rytel	University of Wroclaw, Poland
S. Sciuto	Università di Torino
E. Sezgin	ICTP, Trieste
A. Shellekens	Division Theórique, CERN, Ginevra
G. Soliani	Università di Lecce
J. Sopcik	ICTP, Trieste
K. Stelle	Imperial College, Blackett Laboratory, London, U. K.
A. Taormina	Université de Mons, Belgique
K. Tamvakis	Division Théorique, CERN, Ginevra
M. Tonin	Università di Padova
P. K. Townsend	DAMPT, Cambridge University, N. Y., U. S. A.
A. Van Proeyen	University of Leuwen, Belgique
G. Veneziano	Division Théorique, CERN, Ginevra
N. Warner	Caltech, Pasadena, U. S. A.
P. West	King's College, London, U. K.